INTRODUCTION TO
CIRCUIT SYNTHESIS AND DESIGN

GABOR C. TEMES
Professor of Electrical Sciences and Engineering
University of California, Los Angeles

JACK W. LaPATRA
Associate Professor of Electrical Engineering
University of California, Davis

McGRAW-HILL BOOK COMPANY

New York   St. Louis   San Francisco   Auckland
Bogotá   Düsseldorf   Johannesburg   London   Madrid
Mexico   Montreal   New Delhi   Panama   Paris
São Paulo   Singapore   Sydney   Tokyo   Toronto

Library of Congress Cataloging in Publication Data

Temes, Gabor C   date
    Introduction to circuit synthesis and design.

    Includes index.
    1. Electric network synthesis. 2. Electric
circuits. I. LaPatra, Jack W., date   joint author.
II. Title.
TK454.2.T45   621.319'2   76–43370
ISBN 0–07–063489–0

INTRODUCTION TO CIRCUIT
SYNTHESIS AND DESIGN

  34567890 FGRFGR   83210

This book was set in Times New Roman.
The editors were Peter D. Nalle and Frances A. Neal;
the cover was designed by Nicholas Krenitsky;
the production supervisor was Robert C. Pedersen.
The drawings were done by Oxford Illustrators Limited.
Fairfield Graphics was printer and binder.

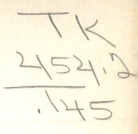

CONTENTS

PREFACE

Circuit design tasks performed by electrical engineers can be classified into two broad categories. One category involves empirical design, performed using tables, charts, rules of thumb, cut-and-try methods, or just sheer animal instinct, which may be checked a posteriori by bench tests or computer simulation. The other is mathematically based design, using an elaborate theoretical apparatus, special algorithms, and usually some large special-purpose computer programs to carry out the tedious calculations. The first category, while of great importance, is much too amorphous and varies too rapidly with changing technology to be taught in detail at engineering schools; it must be learned on the job. Mathematical design, on the other hand, is ideally suited for teaching. It uses elegant mathematics, it is systematic, and *it is useful* in practical applications. It shows students that their efforts devoted to mastering applied mathematics do have relevance in practical engineering problems, and it trains them to think through multistage logical arguments methodically. The subject of this book is this topic: mathematical circuit design, or *circuit synthesis*. We hope that we managed to convey our enthusiasm for the beauty and utility of the subject in the main text.

The primary purpose of the book is to serve as a new text for network synthesis courses taught at American universities today. We hope that the new topics introduced and the method of presentation will also make it an attractive reference for engineers who have already learned this material but would like to update their knowledge and for industrial circuit designers wishing to consolidate their theoretical foundations on the subject.

The structure of the book is the following: Chapter 1 explains the subject matter of the book and discusses some background material such as the classification of circuits, scaling, etc.; Chapter 2 describes some theoretical results related to the realizability of one-ports. Chapters 3 and 4 give synthesis techniques for passive lumped one-ports, and Chapters 5 and 6 for passive lumped two-ports. Chapter 7 deals with the design of lumped linear active two-ports and Chapter 8 with the synthesis of circuits containing also uniform lossless or RC transmission lines. Chapters 9 and 10 discuss tolerance and sensitivity analysis for resistive and dynamic circuits, respectively. Chapter 11 deals with the computer-aided optimization of circuits. Finally, Chapter 12 gives an overview of the most important topics in *approximation theory* for circuit synthesis.†

We considered, and rejected, adding a thirteenth chapter on digital filtering. We believe that that topic deserves a separate course and a separate text all its own.

This book grew out of lecture notes written for our classes at the University of California in Los Angeles and in Davis. The notes were written because we concluded that existing books on the subject were typically 10 years old‡ and omitted many important recent topics, such as active filters, computer-aided circuit design, etc., while discussing many results which had not stood the test of time. Also, their mathematical level tended to be too high, while they fell short on physical insight. We have attempted to construct an up-to-date, easily teachable collection of subjects. It was well received by our students; we hope that it will be liked by the larger community of university students and teachers at other schools.

The material contained in the book should be understandable to students of junior or senior standing who have taken previously at least an introductory-level circuit-analysis course based on a book such as "Basic Circuit Theory," by C. A. Desoer and E. S. Kuh (McGraw-Hill, New York, 1969) or "Circuit Theory with Computer Methods," by O. Wing (Holt, Rinehart and Winston, New York, 1972) or any of their numerous equivalents. The complete book can be taught comfortably in two semesters or three quarters, with four lecture hours per week. Alternatively, an abridged version which leaves out Chapters 8 to 11 can be presented in one semester or two quarters at a somewhat accelerated pace. Our experience was that the students appreciated the large number of numerical examples incorporated in the text and liked to have them worked out in class to acquire familiarity with the subject. This, however, can be done also in recitation sessions conducted by a teaching assistant.

We are indebted first and foremost to Professor H. J. Orchard for his wise and thoughtful advice involving many of the topics in this book. Useful reviews and criticisms have been provided also by Drs. R. Gadenz, D. Pederson, R. Rohrer, and G. Szentirmai. The many numerical calculations needed for the examples were carried out in part by our students K. M. Cho, R. Ebers, S.-C. Fan, R. Gregorian, and M. Rezai-Fakhr.

Gabor C. Temes
Jack W. LaPatra

† A more logical place for this material would have been following Chapter 7. However, it has become somewhat of a tradition to discuss approximation theory in the last chapter, and who are we to break a tradition?

‡ Although fine books have been published abroad recently, by K. Geher, W. Rupprecht, R. Unbehauen, and H. Watanabe, among others.

INTRODUCTION TO
CIRCUIT SYNTHESIS
AND DESIGN

WHAT IS NETWORK SYNTHESIS?

1-1 ANALYSIS VS. SYNTHESIS; NETWORK CLASSIFICATION

The reader's previous courses in the field of electric circuits have in all likelihood discussed basic circuit theory and analysis. However, the most important and exciting part of an engineer's work is the *creation* of circuits, i.e., the engineering *design*, rather than merely the analysis and study of systems designed by others. The purpose of this book is to introduce some fundamental concepts and methods which form the basis of scientific circuit design. The contrast between analysis and design is illustrated schematically in Fig. 1-1. As the figure shows, analysis typically involves the calculation of the response of a *known* circuit or system to a given excitation; design, on the other hand, consists of finding a *new* circuit which provides a desirable response to a given input excitation. While these two operations appear to be simply the inverses of each other, in fact there are three basic differences.

1. An analysis problem normally has a solution; by contrast, in design a solution may not exist at all. Consider, for example, the design problems illustrated in Figs. 1-2 and 1-3. In Fig. 1-2, the generator (containing an internal resistor R_1) is capable of supplying a maximum power of $E^2/4R_1 = 1$ W; yet the specified output is $V^2/R_2 = 2.5$ W. This is impossible if a passive two-port connects the generator to the load. As a second example, the circuit of Fig. 1-3 is required to have a response $v(t)$ which *precedes* (anticipates) the excitation $e(t)$. This feat is clearly beyond the capabilities of any physical device (except perhaps a crystal ball), since it requires a knowledge of future events. Hence, no physical circuit can satisfy the requirements of either Fig. 1-2 or 1-3.

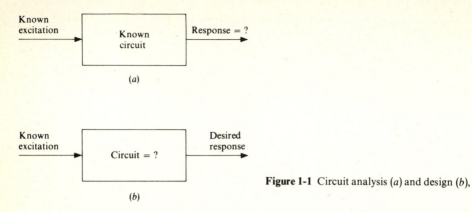

(a)

(b)

Figure 1-1 Circuit analysis (a) and design (b),

2. Analysis almost always leads to a *unique* solution, while if a design problem can be solved, it usually has *several* equivalent solutions. For example, the reader can readily verify that the four circuits shown in Fig. 1-4 all have the same $i(t)$ response for any excitation $e(t)$. *Hint:* Use Laplace transforms. Hence, if the specified property is a prescribed current response for a given voltage excitation, then clearly all four circuits are suitable if any one is. Thus, additional considerations (convenient element values, etc.) decide which of the networks should be used in a given application.
3. Linear-circuit analysis uses but a few basic methods, e.g., nodal and mesh analysis and state-variable analysis. Design, on the other hand, employs a wide variety of techniques, of which some eventually become obsolete while new ones are being invented even as this book is being written.

　　Many design techniques are based on empirical and/or cut-and-try methods; they will be ignored in this book. We shall concentrate instead on scientific, mathematically based design, which we shall call *synthesis*. Circuit synthesis is typically not a single operation but a sequence of several more or less distinct manipulations. These calculations depend on the kind of circuit we are trying to design. Hence, a classification of circuits is necessary. We shall assume that the reader has already been exposed to the categories into which circuits can be grouped. Therefore only a brief summary will be provided in what follows. References 1 and 2 are recommended for a detailed study of this topic.†

† References (usually denoted by numerical superscripts) are listed at the end of each chapter.

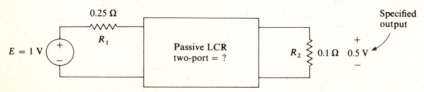

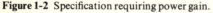

Figure 1-2 Specification requiring power gain.

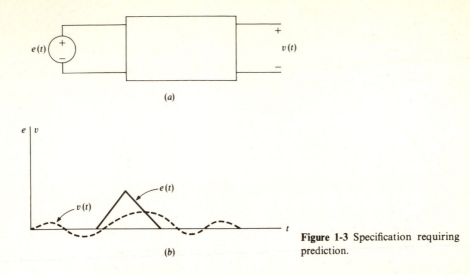

(a)

(b)

Figure 1-3 Specification requiring prediction.

A network will be called *passive* if it cannot generate or amplify energy, i.e., if the energy which it has supplied since the beginning of time cannot exceed the energy which was fed into it. Mathematically, using the notations of Fig. 1-5, a passive network must therefore satisfy

$$E(t) = \int_{-\infty}^{t} \sum_{i=1}^{n} [v_i(\tau) i_i(\tau)] \, d\tau \geq 0$$

Here, $E(t)$ is the difference between the energies entering and leaving the circuit, i.e., the *net* energy supplied to the network. If $E(t)$ equals the energy stored in the

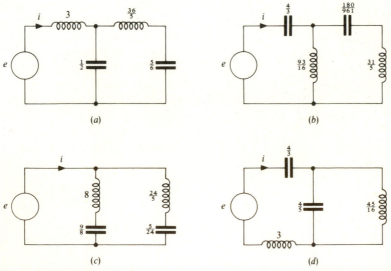

(a)

(b)

(c)

(d)

Figure 1-4 Equivalent circuits.

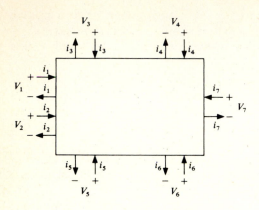

Figure 1-5 A circuit with $n = 7$ terminal pairs. The energy supplied to the circuit is

$$\int_{-\infty}^{t} \sum_{i=1}^{7} [v_i(\tau)i_i(\tau)] \, d\tau.$$

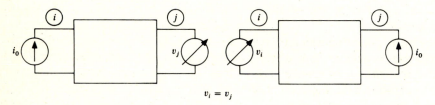

$$v_i = v_j$$

(a)

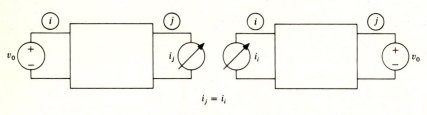

$$i_j = i_i$$

(b)

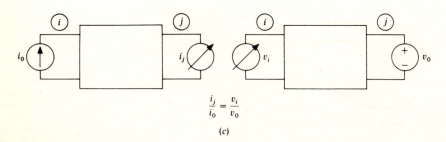

$$\frac{i_j}{i_0} = \frac{v_i}{v_0}$$

(c)

Figure 1-6 Reciprocity relations.

circuit, then the circuit is neither generating nor losing energy. Such a passive circuit is called a *lossless* network; otherwise the passive circuit is *lossy*. Conversely, if $E(t) < 0$ is possible, the circuit is *active*; such a circuit can supply excess energy to the outside world.

A network will be called *linear* if the amplitude of its response is always directly proportional to the amplitude of the excitation. For such a circuit, doubling the magnitude of the input signal, for example, will always double the magnitude of the output signal. If this proportionality does not always hold, the circuit is called *nonlinear*.

A circuit is *time-invariant* if the relation between its response and its excitation always remains the same, regardless of when the excitation is applied. Otherwise, the network is *time-varying*.

A circuit is *reciprocal* if all the relations illustrated in Fig. 1-6 hold for every possible choice of the branches *i* and *j*. What these relations indicate is that the ratio of the response to the excitation remains the same if the generator and meter change places as long as the impedances in the two branches are kept the same. A circuit which violates any of these conditions for any two of its branches is *nonreciprocal*.

A circuit can be considered *lumped* if the physical dimensions of all its components are negligible compared with the wavelength of the electromagnetic signal *inside the component*. In such a network, since Kirchhoff's voltage and current relations hold, the current within any branch is the same at all points of the branch between its terminating nodes (Fig. 1-7a). Also, the voltage along any

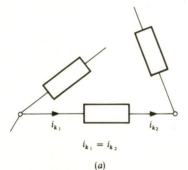

$$i_{k_1} = i_{k_2}$$

(a)

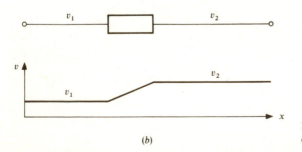

(b)

Figure 1-7 Properties of a lumped circuit.

wire (assumed to be perfect conductor) is constant; it can be considered to change only at the components (Fig. 1-7b).

If the physical dimensions of the elements are comparable to the signal wavelength, the spatial variations of voltages and currents along the wires and in the components must be taken into consideration. In such circuits, called *distributed networks*, Kirchhoff's laws are no longer valid, and the more general laws of Maxwell must be applied.

A circuit containing no energy-storage elements (L, C, M) is called *resistive* or *memoryless*; a circuit which does contain such elements is called *dynamic*. [An additional distinction must also be made between circuits carrying *continuous* signals and those processing *discrete* (sampled) signals; but since this book deals only with continuous-signal circuits, we omit any discussion of this topic.]

For example, a circuit containing only ideal R, L, C, and M elements is evidently passive, reciprocal, linear, dynamic, and time-invariant. If all resistors are removed, the circuit will also become lossless. If, on the other hand, the L, C, M elements are removed, the circuit becomes memoryless. If transistors are added to the network, then, in general, the properties of linearity, passivity, and reciprocity will no longer apply.

Whether or not the $RLCM$ circuit can be treated as a lumped one will depend on the ratio of the signal wavelength in each component (including wires) to the physical dimensions of the component. If this ratio is sufficiently large (say 100 or more), the circuit can be considered lumped.

In the early chapters of this book, the discussion will be restricted to continuous-signal, passive, linear, time-invariant, reciprocal, and lumped networks. We introduce active elements in Chap. 7 and distributed elements in Chap. 8. All circuits discussed will be dynamic.

1-2 SPECIFICATIONS; DESIGN EXAMPLE

The specifications for a circuit which carries only *steady-state sine-wave* signals usually prescribe the driving-point properties (input-output impedances, reflection factors, etc.) and/or the transfer characteristics (gain, loss, phase, delay, etc.) of the circuit. (The student is assumed to have some familiarity with these concepts; their exact definitions will be given later.) These properties are then prescribed as functions of frequency, i.e., in the *frequency domain*. On the other hand, circuits operating normally in the *transient* state are typically specified by their responses either to an impulse-function input (impulse response) or to a unit-step function input (step response) in the *time domain*. Such responses are illustrated in Figs. 1-8 and 1-9, along with some quantities (delay, rise time, overshoot, etc.) usually included in the specifications. Responses to other input signals, such as square waves and sawtooth waves, are also used in special applications, e.g., the transmission of data or of TV signals. Unfortunately, the definitions and the terminology used for describing time responses vary greatly in the literature.

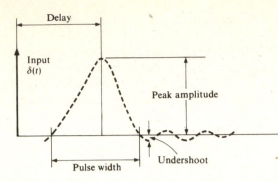

Figure 1-8 Impulse response.

The desirable *circuit diagram* is sometimes included in the specifications. More often, however, the designer establishes it a priori, on the basis of the specifications and usually some preliminary calculations, relying heavily on his past experience with similar problems. His decision will be influenced by economical considerations (he will attempt to find the cheapest and most efficient circuit to do the job), by the type of elements suitable for the given application, by the permissible size and weight, etc.

The choice of the circuit configuration is perhaps the most important and difficult part of the design process: it requires experience, as well as a thorough understanding of the application for which the circuit is intended, the relative advantages of various circuit types, the limitations of different circuit components, and even the manufacturing techniques used in the production of the final circuit. Although in this book an effort will be made to point out the practical advantages and limitations of the circuits discussed, at this step the practical experience of the designer is an essential and irreplaceable asset.

As an illustration of the design process, consider the example of a cable-equalizer design. (The student is urged to follow, at this stage, only the *logic* rather than the somewhat specialized *details* of this discussion. The details should become clearer as the reader progresses to later chapters.)

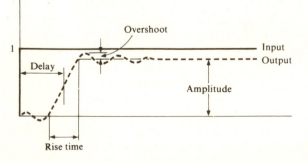

Figure 1-9 Step response.

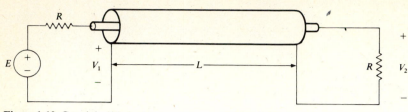

Figure 1-10 Coaxial-cable system.

Example 1-1 Consider the coaxial-cable system in Fig. 1-10. Measurements indicate that in the 1-kHz to 100-MHz range the impedance of the cable, to a good approximation, is a constant value $R = 50 \ \Omega$, while the loss α [defined here† as

$$\alpha = 20 \log \left| \frac{V_1}{V_2} \right| \tag{1-1}$$

and measured in decibels (dB)] is a function of frequency of the form

$$\alpha \approx kL\sqrt{\omega} \tag{1-2}$$

This loss is due to the skin effect in the cable.[3] In Eq. (1-2), L is the length of the cable and k is a constant. For the given cable, measurements give $\alpha_0 = 6$ dB at $\omega_0 = 200\pi$ Mrad/s (Fig. 1-11, continuous line) and hence α can be written in the form

$$\alpha = 6 \sqrt{\frac{\omega}{\omega_0}} \tag{1-3}$$

An important disadvantage of the loss response given by (1-2) or (1-3) is its rapid rise for quite low frequencies. As a consequence, the time responses are "slow" (Figs. 1-12 and 1-13, continuous-line curves); i.e., the pulse width and rise time are large.

To remedy this situation, a *loss equalizer* is needed which, when cascaded with the cable (Fig. 1-14), gives a flatter loss response, resulting in faster time responses. The anticipated equalized responses are shown in Figs. 1-11 to 1-13 as broken-line curves.

The design of this equalizer is described next. The specifications are the following:

† Throughout the book the customary abbreviations log for logarithm to the base 10 and ln for the natural logarithm (base e) are used.

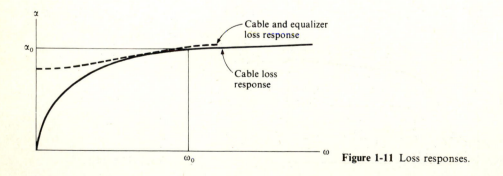

Figure 1-11 Loss responses.

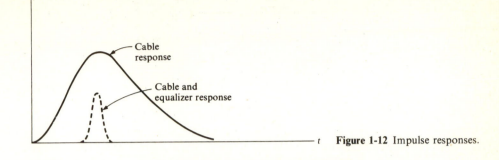

Figure 1-12 Impulse responses.

1. The equalized time responses should be fast but should contain no over- or undershoots; the decrease in signal amplitude (Figs. 1-12 and 1-13) should be as small as possible.
2. To avoid signal reflections, the input and output impedances at the equalizer ports should be R (Fig. 1-14).
3. For practical reasons, only passive (R, L, C) elements may be used, and no mutual inductances are allowed.
4. The frequency range of interest is 10^3 to 10^8 Hz. (Note that the upper limit of equalization is ω_0, the 6-dB radian frequency.)

From past experience it is known that the total equalized loss response

$$\alpha_t = K_1\omega^2 + K_2 \tag{1-4}$$

which corresponds to an impulse response†

$$h(t) = C_1 e^{-C_2 t^2} \tag{1-5}$$

and to a step response

$$g(t) = \int_0^t h(\tau)\, d\tau \tag{1-6}$$

† These time responses are valid only for zero phase shift. The function in (1-5) is often called a *gaussian function*.

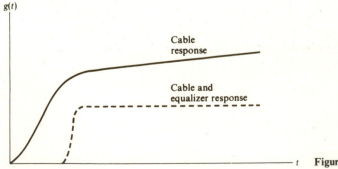

Figure 1-13 Step response.

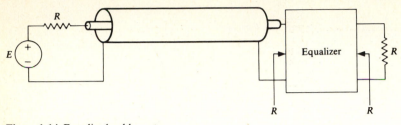

Figure 1-14 Equalized cable system.

is suitable. The corresponding curves are shown in broken line in Figs. 1-12 and 1-13. Also from experience we know that any of the circuits shown in Fig. 1-15 is potentially capable of realizing the required equalizer loss

$$\alpha_e = \alpha_t - \alpha = K_1 \omega^2 + K_2 - 6 \sqrt{\frac{\omega}{\omega_0}} \tag{1-7}$$

All six circuits are, in fact, equivalent in the sense that they all have the same input-output relation

$$\left| \frac{V_2}{V_1} \right| = \left| \frac{R}{R + Z} \right| \tag{1-8}$$

in the frequency domain. However, circuit (a) uses mutual inductance, while circuit (b) contains active components, violating specification 3; circuit (c) does not have either the required input or output impedance; and circuit (d) meets the impedance specifications only at its input port. This leaves only circuits (e) and (f) as candidates for realization. For reasons of economy, we choose circuit (f), a bridged-T network, for the equalizer. [Note, incidentally, that Fig. 1-15 does not list *all* possible equivalent circuits satisfying (1-8); several other practical circuits exist which can be used for this equalizer. This again illustrates the multiplicity of solutions for design problems.]

Returning to the required loss response $\alpha_e(\omega)$, we anticipate that the total equalized loss response $\alpha_t(\omega)$ is tangential to $\alpha(\omega)$ at ω_0 (Fig. 1-11) so that $\alpha_e = \alpha_t - \alpha$ has zero slope there. Then

$$\left. \frac{d\alpha_e}{d\omega} \right|_{\omega = \omega_0} = 2K_1 \omega_0 - 3\omega_0^{-1} = 0 \tag{1-9}$$

which gives $K_1 = 1.5/\omega_0^2$, and hence

$$\alpha_e = 1.5 \left(\frac{\omega}{\omega_0} \right)^2 - 6 \sqrt{\frac{\omega}{\omega_0}} + K_2 \tag{1-10}$$

Since we must have $\alpha_e(\omega) \geq 0$ (otherwise power gain is required from a passive circuit, an impossible demand), at $\omega = \omega_0$

$$\alpha_e(\omega_0) = 1.5 - 6 + K_2 \geq 0 \tag{1-11}$$

so that $K_2 \geq 4.5$ dB.

Next, the voltage ratio V_2/V_1 is selected to be the following function of $s = j(\omega/\omega_0)$:

$$\frac{V_2}{V_1} = \frac{a_3 s^3 + a_2 s^2 + a_1 s + a_0}{b_3 s^3 + b_2 s^2 + b_1 s + b_0} \tag{1-12}$$

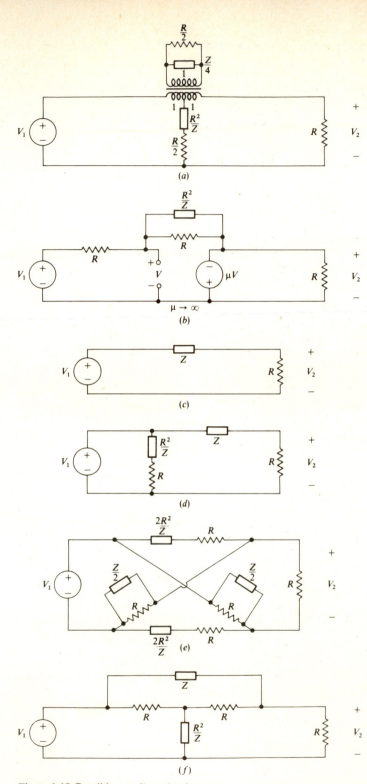

Figure 1-15 Possible equalizer circuits.

11

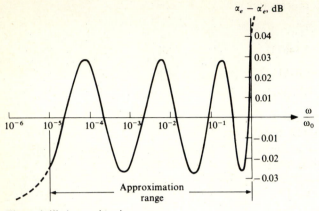

Figure 1-16 Approximation error.

Note that V_2/V_1 is a *rational* function of s; that is, it is a ratio of two polynomials in s. We know that such a rational function represents the general form of the network functions of a lumped, linear, time-invariant circuit. The *order* of V_2/V_1, defined as the highest exponent of s in the function, is here $n = 3$. This value can be found by trying and failing to solve the problem with $n = 1$ and 2. The larger n is, the more components needed in the equalizer. Clearly, the smallest suitable value should be chosen for n, which here is 3.

  The values of a_i and b_i can be found from a mathematical optimization procedure[4] which matches the desired loss response $\alpha_e(\omega)$ of the equalizer, given by (1-10), to the actual response

$$\alpha'_e = 20 \log \left| \frac{V_1}{V_2} \right| = -10 \log \left| \frac{a_3 s^3 + a_2 s^2 + a_1 s + a_0}{b_3 s^3 + b_2 s^2 + b_1 s + b_0} \right|^2_{s=j\omega/\omega_0} \tag{1-13}$$

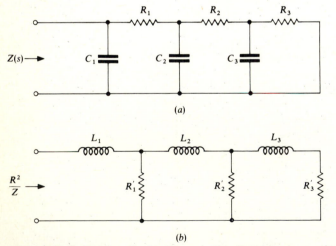

(a)

(b)

Figure 1-17 Preliminary circuit of loss equalizer impedances.

in an optimum manner. This process leads to the coefficient values

$$a_3 = 1 \qquad\qquad b_3 = 1$$

$$a_2 = 0.346732 \qquad b_2 = 0.470599$$

$$a_1 = 0.0161173 \qquad b_1 = 0.0256843$$

$$a_0 = 0.0000747510 \quad b_0 = 0.000128082$$

and to the value $K_2 = 4.8453$ dB. The approximation obtained is very good; the error $\alpha_e - \alpha'_e$ is plotted in Fig. 1-16. The maximum error is less than 0.03 dB; this is so small that it is less than the inevitable error due to inaccurate element values, imperfect tuning and testing, etc. Hence, the result is acceptable. Were this not the case, n would have to be increased to 4 and all calculations carried out again.

Now that we have thus found V_2/V_1 as a function of s, we can obtain $Z(s)$ from (1-8):

$$Z(s) = R(V_1/V_2 - 1) \qquad\qquad (1\text{-}14)$$

Next, the element values of $Z(s)$ and $R^2/Z(s)$ are found, using an exact, step-by-step synthesis method which will be discussed in Chap. 3. The diagrams of the resulting circuits are shown in Fig. 1-17. The element values of $Z(s)$ are $C_1 = 257$ pF, $C_2 = 3.72$ nF, $C_3 = 70.9$ nF, $R_1 = 22.9$ Ω, $R_2 = 8.59$ Ω, and $R_3 = 4.04$ Ω. The large spread in the capaci-

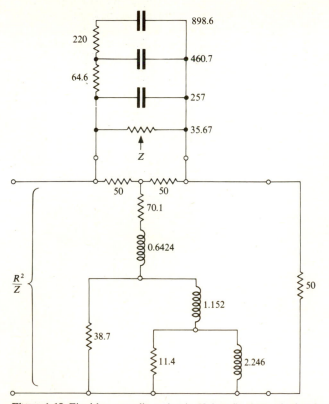

Figure 1-18 Final loss equalizer circuit. Values in ohms, picofarads, and microhenrys.

tance values makes the physical construction of this circuit inconvenient. (A similar spread exists in the inductance values of the R^2/Z circuit.) To avoid the practical problems which these element values would cause, a new design using a different circuit configuration is carried out. This results in the equalizer circuit of Fig. 1-18, which has more practical element values.

In a practical design situation, the circuit of Fig. 1-18 would next be analyzed (on a computer) to verify that no design errors have been made and that the circuit meets all specifications. Then the anticipated effects of parasitic elements (stray capacitances, etc.) can also be calculated. Thereafter, a laboratory prototype would be built and tested. If all these steps give satisfactory results, the design information and prototype are turned over to the manufacturing engineer for production.

Generalizing from the example, it is possible to identify the design steps which the engineer has to carry out in performing circuit synthesis. He has to:

1. Establish the specifications
2. Choose the circuit configuration
3. Calculate the circuit functions
4. Calculate the element values
5. If the element values are impossible or impractical, perform modifications or repeat some of the above steps
6. Analyze, build, and test the circuit

Note that (inevitably) there is some arbitrariness in this classification.

Step 3 is often called *approximation* and step 4 *realization*. As an exercise, the reader should identify the various design steps in the design of the loss equalizer.

1-3 OPTIMIZATION

Unfortunately, many important practical problems cannot be solved using the classical synthesis method described in general terms in the preceding section. This often happens because real-life circuit elements, such as solid-state devices, have many fixed parameters associated with them which are always present and which cannot be changed by the design process.

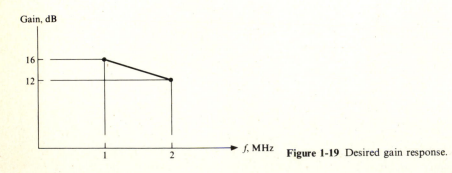

Figure 1-19 Desired gain response.

Example 1-2 Assume that an active amplifier-equalizer must be designed to provide the gain response shown in Fig. 1-19. (The gain is 20 log $|V_2/E|$, the negative of the loss α.) Gain, bandwidth, and price considerations dictate that we use as the active element an integrated-circuit operational amplifier with the following characteristics:

$$\text{Voltage gain} = 100 \text{ (or 40 dB)}$$

$$\text{Input impedance} = 10 \text{ k}\Omega$$

$$\text{Output impedance} = 100 \text{ }\Omega$$

The choice of the circuit is made easy by our familiarity with the two simple circuits shown in Fig. 1-20 and their responses. The gain response of circuit (a) has no slope; that of circuit (b) has too much for our present application. Hence, it is reasonable to expect that the circuit shown in Fig. 1-21a, which combines the features of the two simpler circuits, will be usable. The task is then to find the element values so as to give the best possible match between the specified response (Fig. 1-19) and the actual response of the circuit of Fig. 1-21a.

When the operational amplifier described above is chosen to realize the controlled source, the circuit of Fig. 1-21b is obtained. Here the value of $R_0 = 1000 \text{ }\Omega$ was chosen to be much smaller than the input impedance but much larger than the output impedance of the operational amplifier, to minimize their effects. The element values R and C are unknown; they must be found for optimum response.

Analysis of the circuit of Fig. 1-21b gives for the gain (in decibels)

$$F_1(\omega, R, C) = 10 \log \frac{(R_2 - RG)^2 + G^2/\omega^2 C^2}{\left[(G+1)R_0 + (R+R_2)\left(1 + \dfrac{R_0}{R_1}\right)\right]^2 + \dfrac{(1 + R_0/R_1)^2}{\omega^2 C^2}}$$

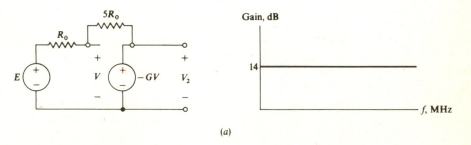

(a)

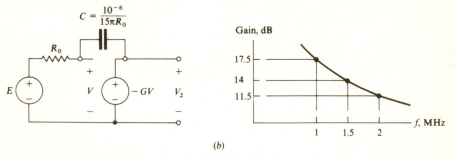

(b)

Figure 1-20 Circuits related to the active equalizer $(G \gg 1)$.

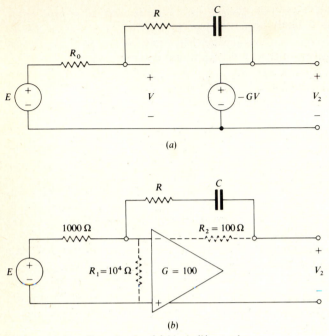

(a)

(b)

Figure 1-21 Equalizer circuits: (a) ideal; (b) actual.

Since F_1 is not linear in ω, while the specified function $F_2(\omega)$ shown in Fig. 1-19 is, $F_1(\omega)$ cannot exactly match $F_2(\omega)$ for any R and C. However, by finding appropriate values for R and C, the specified and actual responses can be made quite close. For example, if we choose $R = 2.884$ kΩ and $C = 27.27$ pF, the actual response becomes that shown in Fig. 1-22. The maximum error is about 0.3 dB, which may be small enough for many practical applications.

The constraints imposed by the prescribed constants of the operational amplifier preclude the use of a step-by-step design process. Instead, a different method, called *optimization*, can be used to find the best values for R and C. The values given above were obtained in this way.

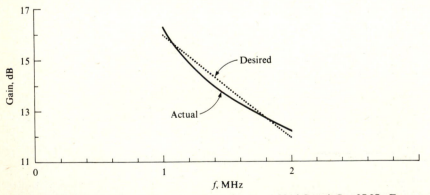

Figure 1-22 Desired and actual gain responses for $R = 2.884$ kΩ and $C = 27.27$ pF.

There are many similar situations which necessitate the use of optimization. For example, optimization is likely to be needed when:

1. The specifications are given in the form of a numerical table or a graph.
2. Nonideal elements (transistors, operational amplifiers, lossy reactances) must be used, and these nonideal effects cannot be neglected.
3. The configuration and/or some of the element values are constrained.

As an example of the last situation, if the circuit is to be realized in the form of a monolithic integrated circuit, the only permissible passive elements are resistors and capacitors. In addition, the values of these elements should normally be in the ranges $50 \ \Omega \leq R \leq 10^5 \ \Omega$ and $1 \ \text{pF} \leq C \leq 100 \ \text{pF}$ for convenient and reproducible fabrication.

Another situation where both the circuit configuration and the feasible element values are restricted involves coupled-resonator networks, such as microwave filters, crystal filters, mechanical filters, etc. Here the circuit configuration is prescribed, and the element values may be varied only within a fairly narrow range.

As will be seen later, the strict step-by-step methods of network synthesis cannot usually accommodate such restrictions and hence cannot be used under such circumstances. Therefore optimization must be used.

What is optimization? It is essentially an iterative process in which successive approximations to the desired response are evaluated and, if necessary, improved until a satisfactory response is obtained (or until the designer's time, computing funds, or patience runs out—whichever comes first). A flow chart of the process (Fig. 1-23) consists of the following steps:

1. An initial approximation (box 2) is obtained and compared with the specifications (box 1).

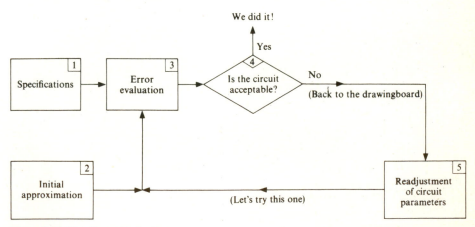

Figure 1-23 Iterative optimization.

2. The error of the approximation is evaluated (box 3), and a decision whether the error is tolerable is made (box 4). If it is, the process stops; if not, it continues.
3. The variable parameters (R and C in our previous example) are readjusted (box 5) in such a way that the remaining error is decreased. The readjustment is usually small and is based on fairly sophisticated mathematical theory. Thus a new circuit, representing a new approximation, is obtained.
4. We return to step 2.

Since the process must be repeated again and again (the number of iterations may be as high as 100), this is a laborious procedure. For circuits of comparable size, optimization may require 10 to 100 times the effort needed by synthesis. Hence, it should only be used when synthesis cannot solve the design problem.

Often a design requires a judicious mixing of optimization and synthesis techniques. In fact, the design of the cable loss equalizer of Figs. 1-16 to 1-18 represents such a case, since the calculation of the coefficients $a_0, \ldots, a_3$ and b_0, $\ldots, b_3$ of (1-12) to (1-13) was performed using optimization, while the rest of the design could be accomplished by synthesis. Carrying out the whole design by directly optimizing the element values would have been much more difficult and time-consuming.

It will be noted that optimization is similar to the empirical cut-and-try procedure sometimes used in a laboratory. In the laboratory procedure, variable components (resistors, capacitors, inductors) are "trimmed," one by one, by a screwdriver-wielding engineer or technician until the circuit performs in a satisfactory way. The differences between the two techniques are that:

1. Optimization usually utilizes sophisticated mathematical techniques and a computer rather than a screwdriver.
2. Optimization normally adjusts all variable parameters simultaneously.
3. As a result of 1 and 2, optimization can handle problems involving as many as 100 variable parameters. By contrast, empirical adjustment usually breaks down for more than five variables.

An introductory-level discussion of optimization will be given in Chap. 11.

1-4 SCALING, NORMALIZATION, AND DENORMALIZATION

The use of standard units (farads, hertz, henrys, ohms, seconds, radians per second, etc.) frequently leads to very large or very small numbers in circuit analysis and design; for example, 10^{-10} F, 10^9 rad/s. Such unwieldy numbers often cause human error in manual calculations, and when they are entered into computers for computer-aided analysis or design, they may result in over- or underflow. In order to avoid these complications, it is almost always preferable to scale all electrical quantities by some conveniently chosen factors. This process is called *normalization* or *scaling*. For example, if we decide to use 50 as the scale factor for all circuit impedances, then a 75-Ω resistance will have a scaled value of 1.5 Ω. In general if Z_0 is the impedance scale factor, then the normalized value z of an impedance Z is given by

$$z = \frac{Z}{Z_0} \tag{1-15}$$

and similarly the normalized values for frequency, radian frequency, capacitance, inductance, and time become

$$F = \frac{f}{f_0} \qquad \Omega = \frac{\omega}{\omega_0} \qquad c = \frac{C}{C_0} \qquad l = \frac{L}{L_0} \qquad T = \frac{t}{t_0} \tag{1-16}$$

respectively. Notice that unscaled quantities denoted by uppercase letters are denoted by lowercase symbols for the normalized circuit, and vice versa.

Since the physical relations which hold for the original circuit must remain valid for the electrical quantities of the scaled circuit, scale factors cannot all be arbitrary. For instance, the impedance of a capacitor in the unscaled circuit is given by

$$Z = \frac{1}{j\omega C} \tag{1-17}$$

In addition, the equation

$$z = \frac{1}{j\Omega c} \tag{1-18}$$

must hold for the normalized circuit. Substitution from (1-15) and (1-16) into (1-17) gives

$$zZ_0 = \frac{1}{j\Omega\omega_0 cC_0} \tag{1-19}$$

or

$$z = (\omega_0 C_0 Z_0)^{-1} \frac{1}{j\Omega c} \tag{1-20}$$

Hence (1-18) will hold only if

$$\omega_0 C_0 Z_0 = 1 \tag{1-21}$$

A similar argument based on the relation $Z = j\omega L$ reveals that the corresponding normalized equation $z = j\Omega l$ will be valid only if

$$\frac{\omega_0 L_0}{Z_0} = 1 \tag{1-22}$$

holds.

Consider now a sine-wave signal sin ωt. If the time axis of the original circuit is scaled by a factor t_0, say by $t_0 = 10^{-3}$, then the period t_p of the sine wave becomes t_p/t_0, that is, larger by the factor 1000. Consequently, the frequency changes (decreases by a factor of 1000) from $f = 1/t_p$ to $F = t_0/t_p$ and the radian frequency from $\omega = 2\pi/t_p$ to $\Omega = 2\pi t_0/t_p$. Comparison with (1-16) shows therefore that the relations

$$f_0 = \frac{1}{t_0} \tag{1-23}$$

and

$$\omega_0 = \frac{1}{t_0} \tag{1-24}$$

must hold. Thus, the frequency and the radian frequency are both scaled by $1/t_0$.

As a consequence of Eqs. (1-21) to (1-24), *of the six scale factors, Z_0, f_0, ω_0, C_0, L_0 and t_0 only two may be chosen independently*. In *circuit analysis*, they may be L_0 and C_0, for example. In *design*, the element values are not known in advance, but the impedance level and the operating frequency (or time) range usually are. Hence, Z_0 and f_0 (or t_0) can be so chosen that convenient values are obtained. The process will be illustrated with two simple examples.

Example 1-3 Carry out the frequency-domain analysis of the circuit of Fig. 1-24 using normalization. The element values are

$$C_2 = C_4 = 200 \text{ pF} \qquad L_3 = 3 \text{ } \mu\text{H} \qquad R_1 = R_4 = 75 \text{ } \Omega$$

It is clearly convenient to choose $C_0 = C_2$ and $Z_0 = R_1$. Then the other scale factors become, from (1-21) and (1-22),

$$\omega_0 = \frac{1}{C_0 Z_0} = \frac{1}{C_2 R_1} \approx 0.6667 \times 10^8$$

and

$$L_0 = \frac{Z_0}{\omega_0} = C_2 R_1^2 = 1.125 \times 10^{-6}$$

Hence, the normalized element values are

$$c_2 = c_4 = 1 \text{ F} \qquad l_3 = L_3/L_0 = \tfrac{8}{3} \text{ H} \qquad r_1 = r_4 = 1 \text{ } \Omega$$

The *symbolic* analysis of the scaled circuit proceeds as follows (Fig. 1-24):

$$I_3 = V_4\left(j\Omega c_4 + \frac{1}{r_4}\right) = V_4(1 + j\Omega)$$

$$V_2 = V_4 + I_3 \, j\Omega l_3 = V_4[1 + j\Omega l_3(1 + j\Omega)] = V_4(1 - \Omega^2 l_3 + j\Omega l_3)$$

$$I_2 = j\Omega c_2 V_2 = V_4[-\Omega^2 l_3 + j\Omega(1 - \Omega^2 l_3)]$$

$$I_1 = I_2 + I_3 = V_4[(1 - \Omega^2 l_3) + j\Omega(2 - \Omega^2 l_3)]$$

$$E = 1 = V_2 + I_1 r_1 = V_4[2(1 - \Omega^2 l_3) + j\Omega(2 + l_3 - \Omega^2 l_3)]$$

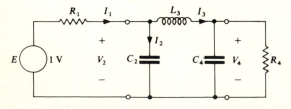

Figure 1-24 Circuit for analysis example.

From the last equation

$$V_4 = \frac{1}{2(1 - \Omega^2 l_3) + j\Omega(2 + l_3 - \Omega^2 l_3)}$$

and I_3, V_2, I_2, and I_1 can be found by back substitution.

It is evident that even this symbolic analysis was simplified by the absence of the (unit) values of c_2, c_4, r_1, and r_4. If actual numerical calculations are to be performed, the advantages become even greater, since all element values are either exactly equal to 1 or close to it. This causes all calculations to be well conditioned and prevents overflow.

The numerical *analysis* of the scaled circuit begins with the calculation of the normalized radian frequency Ω from the given unscaled frequency f. This can be done using the relation

$$\Omega = \frac{\omega}{\omega_0} = C_2 R_1 2\pi f = 3\pi 10^{-8} f$$

The values of Ω and $l_3 \approx 2.6667$ H are then entered into the formulas for V_4, I_3, ... to find their numerical values.

The reader should carry out the numerical analysis of both the scaled and unscaled circuits at $f = 10$ MHz to gain some appreciation of the advantages of normalized circuit analysis.

The next example will illustrate the use of normalization in *design*.

Example 1-4 The circuit shown in Fig. 1-25 must satisfy the following requirements:

$$|Z(j\omega)| \begin{cases} = 0 & \text{for } f = f_z = 572.5 \text{ kHz} \\ \to \infty & \text{for } f = f_p = 3f_z \\ = 93 \ \Omega & \text{for } f = f_i = 2f_z \end{cases}$$

The required behavior of $|Z|$ is illustrated in Fig. 1-26.

Here we have no advance knowledge of the element values; however, it is evident that the scale factors

$$\omega_0 = 2\pi f_z \approx 3.5971 \times 10^6 \qquad Z_0 = 93 \ \Omega$$

will serve the purpose. In terms of normalized quantities

$$z(j\Omega) = \frac{1 - \Omega^2 l(c + c_1)}{j\Omega c_1(1 - \Omega^2 lc)} \tag{1-25}$$

Hence the specifications translate into

$$1 - \Omega_z^2 l(c + c_1) = 0 \tag{1-26}$$

$$1 - \Omega_p^2 lc = 0 \tag{1-27}$$

$$|z(j\Omega_1)| = \left| \frac{1 - \Omega_1^2 l(c + c_1)}{\Omega_1 c_1(1 - \Omega_1^2 lc)} \right| = 1 \tag{1-28}$$

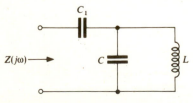

Figure 1-25 Circuit for design example.

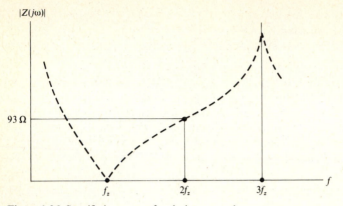

Figure 1-26 Specified response for design example.

where $\quad \Omega_z = \dfrac{\omega_z}{\omega_0} = 1$ rad/s $\qquad \Omega_p = \dfrac{2\pi f_p}{\omega_0} = 3$ rad/s $\qquad \Omega_1 = \dfrac{2\pi f_1}{\omega_0} = 2$ rad/s $\qquad$ (1-29)

Hence, from (1-26),

$$l(c + c_1) = 1 \tag{1-30}$$

from (1-27)

$$lc = \tfrac{1}{9} \tag{1-31}$$

and from (1-28)

$$\left| \frac{1 - 4l(c + c_1)}{2c_1(1 - 4lc)} \right| = 1 \tag{1-32}$$

Substitution of (1-30) and (1-31) into (1-32) gives directly

$$c_1 = 2.7 \text{ F}$$

Then, substituting back into (1-30) and (1-31) leads to

$$l = \tfrac{80}{243} \approx 0.3292 \text{ H} \qquad \text{and} \qquad c = \tfrac{27}{80} \approx 0.3375 \text{ F}$$

To find the physical (denormalized) values, we calculate from (1-21) and (1-22)

$$C_0 = \frac{1}{\omega_0 Z_0} \approx 2.989 \times 10^{-9} \qquad L_0 = \frac{Z_0}{\omega_0} \approx 2.5854 \times 10^{-5}$$

Hence, the actual values are

$$C_1 = c_1 C_0 \approx 8070 \text{ pF} \qquad C = cC_0 \approx 1009 \text{ pF} \qquad L = lL_0 \approx 8.512 \text{ } \mu\text{H}$$

Again, the reader can obtain an appreciation of the advantages of normalization by duplicating the above calculations *without* the benefit of scaling and then comparing the amount and nature of the unscaled computation with the normalized one.

For small problems, like the two examples discussed above, the use of normalization is advisable but not mandatory. However, for larger circuits, containing 10 or more elements, the computational saving gained by using scaling is so large that it is grossly wasteful not to apply it. Furthermore, the better numerical conditions which normalization achieves mean greater accuracy in the results. Some optimization processes will, in fact, converge *only* if proper scaling is used on all parameters.

An additional reason for the student to get well acquainted with the techniques of normalization (scaling) and denormalization, i.e., the calculation of the physical values from the normalized ones, is the existence of many useful tables of normalized circuits. These include filters, delay lines, pulse circuits, etc. A comprehensive and, at the time of its compilation, up-to-date listing of these tables is given in Ref. 5. To illustrate the use of such a table, consider the following example.

Example 1-5 Assume that we would like to design a delay circuit with a dc delay of 1 μs to work between two terminating resistors of 50 Ω. Suppose that the circuit shown in Fig. 1-27a with the delay vs. frequency response shown in Fig. 1-27b is to be changed into one satisfying our specifications. To accomplish this we must denormalize the time scale so that 1 s becomes 1 μs and the impedance scale so that 1 Ω becomes 50 Ω. Thus, choosing $t_0 = 10^{-6}$ and $Z_0 = 50$, from (1-24), (1-21), and (1-22), we obtain

$$\omega_0 = \frac{1}{t_0} = 10^6 \qquad C_0 = \frac{1}{\omega_0 Z_0} = 2 \times 10^{-8} \qquad L_0 = \frac{Z_0}{\omega_0} = 5 \times 10^{-5}$$

With these values the denormalized circuit and response become those shown in Fig. 1-28. This satisfies our specifications.

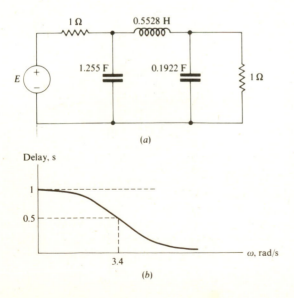

(a)

(b)

Figure 1-27 Normalized delay circuit and response. (*Data from L. Weinberg,* "*Network Analysis and Synthesis,*" *p.619, McGraw-Hill, New York, 1962; reprinted by Robert E. Krieger Publishing Company, Inc., Huntington, N.Y., 1975.*)

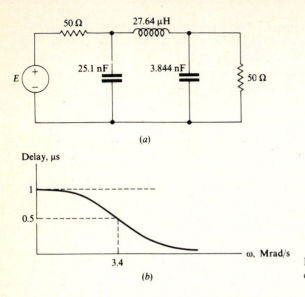

(a)

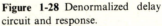

Figure 1-28 Denormalized delay circuit and response.

(b)

Example 1-6 A filter circuit is to be designed for operation between terminating resistors $R_1 = 150\ \Omega$ and $R_2 = 75\ \Omega$. Its loss response is to vary less than 0.5 dB in the frequency range $0 \le f \le 1.2$ MHz and to rise monotonically for higher frequencies.

Suppose that the circuit in Fig. 1-29a with the loss response of Fig. 1-29b is to be transformed into one which meets our requirements. We clearly have to use

$$\omega_0 = 2\pi 1.2 \times 10^6 \approx 7.54 \times 10^6 \qquad \text{and} \qquad Z_0 = 75$$

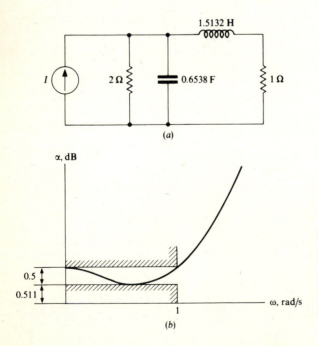

(a)

(b)

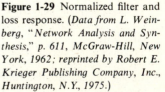

Figure 1-29 Normalized filter and loss response. (*Data from L. Wein-berg, "Network Analysis and Synthesis," p. 611, McGraw-Hill, New York, 1962; reprinted by Robert E. Krieger Publishing Company, Inc., Huntington, N.Y., 1975.*)

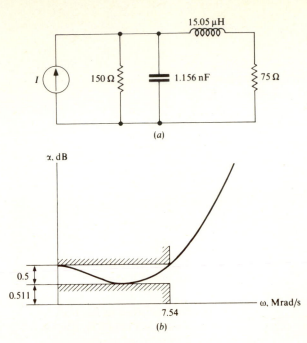

(a)

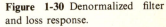

(b)

Figure 1-30 Denormalized filter and loss response.

Then the other scale factors must be

$$C_0 = \frac{1}{\omega_0 Z_0} \approx 1.7684 \times 10^{-9} \quad \text{and} \quad L_0 = \frac{Z_0}{\omega_0} \approx 9.947 \times 10^{-6}$$

With these values, the circuit and loss response of Fig. 1-30 are obtained. This circuit clearly has the required properties.

1-5 SUMMARY

In this chapter, the following subjects were discussed:

1. Synthesis was defined, and the main differences between analysis and design were explained (Sec. 1-1).
2. Circuits were classified into categories such as active and passive, lossless and lossy, linear and nonlinear, time-invariant and time-varying, reciprocal and nonreciprocal, lumped and distributed, resistive and dynamic, continuous and discrete (Sec. 1-1).
3. Circuit specifications were discussed, and the design process was illustrated by the description of the design of a loss equalizer for a cable system (Sec. 1-2).
4. The steps normally followed in circuit design were identified (Sec. 1-2).
5. Circuit optimization was defined and illustrated by the design of an active amplifier-equalizer (Sec. 1-3).
6. Situations requiring optimization were described, and the flow chart of the optimization process was discussed (Sec. 1-3).
7. Scaling, or normalization, was defined, and the relations between the scale factors were given (Sec. 1-4).

8. Examples were given for normalized-circuit analysis and design and for the use of tables containing normalized parameter values (Sec. 1-4).

All this material is general background for the design techniques discussed in later chapters. The next chapter will provide additional background theory needed in the design of one-port networks (impedances).

PROBLEMS

1-1 Verify that the four circuits shown in Fig. 1-4 have the same relation between $e(t)$ and $i(t)$. *Hint:* Use Laplace transforms to find the input impedances of the LC one-ports.

1-2 Express the constants C_1 and C_2 in Eq. (1-5) in terms of K_1 and K_2 used in Eq. (1-4).

1-3 Show that Eq. (1-8) holds for circuits b, c, d, and f shown in Fig. 1-15. If you are brilliant, show this also for circuits a and e.

1-4 Express $Z(s)$ in Fig. 1-17 in terms of $R_1, R_2, R_3, C_1, C_2, C_3$, and s. Show that the $Z(s)$ illustrated in Fig. 1-18 is of the same form.

1-5 Calculate the gain responses of the two circuits of Fig. 1-20. Show that the gain-vs.-frequency curves given are correct.

1-6 Construct the expressions for the gains of the circuits in Fig. 1-21a and b. What happens to the two expressions as $G \to \infty$? Evaluate the expressions for $R = 2884\ \Omega$, $C = 27.27$ pF, $f = 1.5$ MHz, and $G \to \infty$. Check your result against Fig. 1-22.

1-7 Prove Eq. (1-22).

1-8 Analyze the circuit of Fig. 1-24 for $C_2 = C_4 = 200$ pF, $L_3 = 3\ \mu$H, $R_1 = R_4 = 75\ \Omega$, and $f = 10$ MHz, with and without normalization. Which calculation is the more efficient and why?

1-9 Carry out the design of the circuit of Fig. 1-25 without normalization. The normalized design is discussed in Eqs. (1-25) to (1-32). Which process is more efficient? Why?

1-10 Redesign the circuit of Fig. 1-28 for a delay of 50 ns and terminating resistors of 2000 Ω.

1-11 Plot the loss response of the circuit of Fig. 1-30 if both L and C are multiplied by 1000.

1-12 Plot the impulse response of Eq. (1-5) for $C_1 = 1$ and $C_2 = 10^6$. Can any physical device have this impulse response? Why or why not? Which scale factor affects C_2?

1-13 Rescale the circuit of Fig. 1-28 so that the delay is 1 ns and the *total* capacitance in the circuit is 10 pF.

1-14 Rescale the elements of the circuit of Fig. 1-18 such that the 50-Ω resistors and the largest capacitor become unit-value elements.

REFERENCES

1. Desoer, C. A., and E. S. Kuh: "Basic Circuit Theory," McGraw-Hill, New York, 1969.
2. Wing, O.: "Circuit Theory with Computer Methods," Holt, New York, 1972.
3. Wigington, R. L., and N. S. Nahman: Transient Analysis of Coaxial Cables Considering the Skin Effect, *Proc. IRE*, February 1957, pp. 166–174.
4. Temes, G. C., and J. A. C. Bingham: Iterative Chebyshev Approximation Technique for Network Synthesis, *IEEE Trans. Circuit Theory*, March 1967, pp. 31–37.
5. Orchard, H. J., and G. C. Temes: Filter Design Using Transformed Variables, *IEEE Trans. Circuit Theory*, December 1968, pp. 385–408.
6. Weinberg, L.: "Network Analysis and Synthesis," McGraw-Hill, New York, 1962, reprint ed., R. E. Krieger Publishing Co., Inc., Huntington, N.Y., 1975.

POSITIVE REAL FUNCTIONS

2-1 THE COMPLEX FREQUENCY VARIABLE s

Most of the discussions in this book will utilize the complex variable s. This variable, as we shall see, can have two entirely different meanings, and it is very important to be aware which is being used. The dual role of s is seldom apparent to students, since time-domain and frequency-domain analyses are usually treated entirely separately. For this reason it is important to refresh our previously acquired familiarity with this important parameter. Consider the circuit of Fig. 2-1, where $v(t)$ is an arbitrary voltage, a function of the time t. This function can be of any form, e.g., that shown in Fig. 2-2. Summing the voltages of the individual elements and equating the sum to the generator voltage, we obtain

$$v = L \frac{di}{dt} + Ri + \frac{1}{C} \int_0^t i \, dt + v_c(0-) \tag{2-1}$$

where $v_c(0-)$ is the initial voltage across the capacitor C at time $t = 0-$. The solution of the differential equation (2-1) can be conveniently performed using Laplace transformation

$$V(s) = \left(sL + R + \frac{1}{sC}\right)I(s) - Li(0-) + \frac{v_c(0-)}{s} \tag{2-2}$$

Here, as is well known, the complex variable s is a purely abstract mathematical parameter, introduced for convenience and without any physical meaning. Also, $i(0-)$ is the value of the initial current through the inductor L.

Assuming

$$i(0-) = 0 \qquad v_c(0-) = 0 \tag{2-3}$$

27

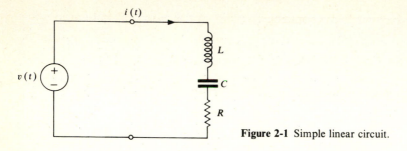

Figure 2-1 Simple linear circuit.

and solving for the unknown transformed current $I(s)$, we obtain

$$I(s) = \frac{V(s)}{sL + R + 1/sC} \tag{2-4}$$

The physical current $i(t)$ can now be obtained by carrying out the inverse Laplace transformation of $I(s)$.

Consider now the following, completely different problem. Let us assume that the circuit of Fig. 2-1 is excited by the sinusoidal voltage signal

$$v(t) = \hat{V} \cos (\omega t + \varphi) \tag{2-5}$$

Assume also that we are interested only in the steady-state behavior of the circuit. Then it is convenient to introduce complex-phasor notation and rewrite (2-5) in the form

$$v(t) = \text{Re} \ (Ve^{j\omega t}) = \tfrac{1}{2}(Ve^{j\omega t} + V^*e^{-j\omega t}) \tag{2-6}$$

where
$$V = \hat{V}e^{j\varphi} \tag{2-7}$$

is the complex phasor of the sinusoidal voltage and V^* its complex conjugate.†
Since the steady-state solution will also be sinusoidal (as is well known from introductory circuit analysis courses, see, for example, Ref. 1), we can write

$$i(t) = \text{Re} \ Ie^{j\omega t} = \tfrac{1}{2}(Ie^{j\omega t} + I^*e^{-j\omega t}) \tag{2-8}$$

where I is the unknown complex phasor of the sinusoidal current. Substituting into (2-1), and neglecting the terms corresponding to the initial conditions $i(0-)$

---

† The complex conjugate will be denoted by an asterisk (*) throughout the book. Re x denotes the real, Im x the imaginary part of x.

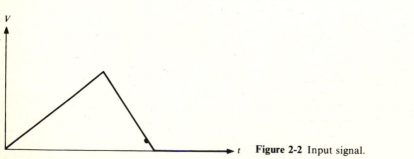

Figure 2-2 Input signal.

and $v_c(0-)$, which we know contribute only to the *transient* part of the solution, we obtain

$$\text{Re } Ve^{j\omega t} = \text{Re } j\omega LI e^{j\omega t} + \text{Re } RI e^{j\omega t} + \text{Re } \left(\frac{I}{j\omega C} e^{j\omega t} \right) \qquad (2\text{-}9)$$

After simplifications and cancellations[1] this equation will be of the form

$$V = j\omega LI + RI + \frac{I}{j\omega C} \qquad (2\text{-}10)$$

from which the unknown phasor I can be obtained:

$$I = \frac{V}{j\omega L + R + 1/j\omega C} \qquad (2\text{-}11)$$

We can now formally set $s \triangleq j\omega$, as an abbreviation.† Then (2-11) becomes identical to (2-4). It must be remembered, however, that the meaning of s in the two equations is completely different.

It appears now, that, at least for this example, the relation between the Laplace transforms of the general time functions $i(t)$ and $v(t)$ is formally the same as that connecting the complex phasors I and V for steady-state sine-wave signals. It will next be shown that this correspondence is valid in more general situations. Consider the circuit of Fig. 2-3. The nodal equations describing the circuit are

$$y_{11} v_1 + y_{12} v_2 + \cdots + y_{1k} v_k + \cdots + y_{1n} v_n = i_1$$

$$y_{21} v_1 + y_{22} v_2 + \cdots + y_{2k} v_k + \cdots + y_{2n} v_n = 0 \qquad (2\text{-}12)$$

$$\cdots\cdots\cdots\cdots\cdots\cdots\cdots\cdots\cdots\cdots\cdots\cdots\cdots\cdots\cdots$$

$$y_{n1} v_1 + y_{n2} v_2 + \cdots + y_{nk} v_k + \cdots + y_{nn} v_n = 0$$

where we assume that $i_1(t)$ is the only independent source in the network.

In (2-12), y_{11} denotes an *operator* such that

$$y_{11} v_1 \triangleq G_{11} v_1 + C_{11} \frac{dv_1}{dt} + \frac{1}{L_{11}} \int_0^t v_1(t)\, dt \qquad (2\text{-}13)$$

where G_{11} = sum of all conductances connected to node 1
$\quad\ C_{11}$ = sum of all capacitances connected to node 1
$\ 1/L_{11}$ = sum of all reciprocal inductances connected to node 1

† The symbol $\triangleq$ means "defined as."

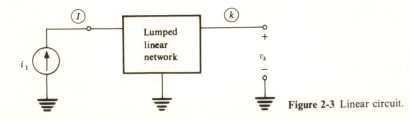

Figure 2-3 Linear circuit.

Similarly, y_{12} contains the conductances, capacitances, and reciprocal inductances connected between nodes 1 and 2, etc.† Here, we are interested in finding the output voltage $v_k(t)$. The solution of (2-12) can be carried out using Laplace transformation:

$$
\begin{aligned}
Y_{11}V_1(s) + Y_{12}V_2(s) + \cdots + Y_{1k}V_k(s) + \cdots + Y_{1n}V_n(s) &= I_1(s) \\
Y_{21}V_1(s) + Y_{22}V_2(s) + \cdots + Y_{2k}V_k(s) + \cdots + Y_{2n}V_n(s) &= 0 \\
&\ \ \ \cdots\cdots\cdots\cdots\cdots\cdots\cdots\cdots\cdots\cdots\cdots\cdots \\
Y_{n1}V_1(s) + Y_{n2}V_2(s) + \cdots + Y_{nk}V_k(s) + \cdots + Y_{nn}V_n(s) &= 0
\end{aligned}
\tag{2-14}
$$

where now, from (2-13),

$$
Y_{ij} = G_{ij} + sC_{ij} + \frac{1}{sL_{ij}} = \frac{s^2 L_{ij} C_{ij} + sL_{ij}G_{ij} + 1}{sL_{ij}}
\tag{2-15}
$$

The result, obtained using Cramer's rule, is

$$
V_k(s) = \frac{\Delta_{1k}(s)}{\Delta(s)} I_1(s)
\tag{2-16}
$$

Here, $\Delta(s)$ is the determinant formed from the coefficients $Y_{ij}(s)$. Also, $\Delta_{1k}(s)$ is the 1, k *cofactor* of $\Delta(s)$, obtained by erasing row 1 and column k of $\Delta(s)$ and multiplying the minor thus obtained by $(-1)^{1+k}$.

Next, we can define the transfer impedance $Z_T(s)$ as the output-voltage–input current ratio

$$
Z_T(s) \triangleq \frac{V_k(s)}{I_1(s)} = \frac{\Delta_{1k}(s)}{\Delta(s)}
\tag{2-17}
$$

in the Laplace-transform domain. Since all components Y_{ij} of Δ and Δ_{1k} are, by (2-15), *rational functions*‡ of s, and since the formation of determinant involves only addition, subtraction, and multiplication, $\Delta(s)$, $\Delta_{1k}(s)$, and $Z_T(s)$ are all rational functions of s. It should be reemphasized that all voltages and currents in the above derivations are *arbitrary* functions of time.

Next, assume that the input current is a *sine-wave* signal

$$
i_1(t) = \hat{I}_1 \cos(\omega t + \varphi) = \mathrm{Re}\, I_1 e^{j\omega t} = \frac{I_1 e^{j\omega t} + I_1^* e^{-j\omega t}}{2}
\tag{2-18}
$$

where $I_1 \triangleq \hat{I}_1 e^{j\varphi}$ is the complex phasor associated with $i_1(t)$. Then, in the *steady state*, the output voltage will also be sinusoidal

$$
v_k(t) = \mathrm{Re}\, V_k e^{j\omega t} = \frac{V_k e^{j\omega t} + V_k^* e^{-j\omega t}}{2}
\tag{2-19}
$$

where $V_k \triangleq \hat{V}_k e^{j\psi}$ is the complex phasor of $v_k(t)$.

---

† For $i \neq j$, y_{ij} has a negative sign; for $i = j$ it has a positive sign.
‡ A rational function is the ratio of two polynomials of the form

$$
\frac{p(s)}{q(s)} = \frac{p_0 + p_1 s + p_2 s^2 + \cdots + p_m s^m}{q_0 + q_1 s + q_2 s^2 + \cdots + q_n s^n}
$$

Next, the ratio of the phasors V_k and I_1 will be found. From (2-18), by Laplace transformation,

$$I_1(s) = \frac{I_1/2}{s - j\omega} + \frac{I_1^*/2}{s + j\omega} \tag{2-20}$$

Hence, from (2-16) [which is valid for any input $i_1(t)$ in the absence of initial stored energy in the circuit],

$$V_k(s) = \frac{I_1}{2} \frac{\Delta_{1k}(s)/\Delta(s)}{s - j\omega} + \frac{I_1^*}{2} \frac{\Delta_{1k}(s)/\Delta(s)}{s + j\omega} \tag{2-21}$$

Here, since I_1 and I_1^* are complex constants, $V_k(s)$ is a rational function of s. It therefore has the partial-fraction expansion

$$V_k(s) = \frac{I_1}{2} \left[\frac{\Delta_{1k}(s)}{\Delta(s)}\right]_{s=j\omega} \frac{1}{s - j\omega} + \frac{I_1^*}{2} \left[\frac{\Delta_{1k}(s)}{\Delta(s)}\right]_{s=-j\omega} \frac{1}{s + j\omega} + \sum_{i=1}^{n} \frac{k_i}{s - s_i} \tag{2-22}$$

where the s_i are the zeros of $\Delta(s)$, that is, the natural modes of the circuit. By inverse Laplace transformation,

$$v_k(t) = \frac{I_1}{2} \frac{\Delta_{1k}(j\omega)}{\Delta(j\omega)} e^{j\omega t} + \frac{I_1^*}{2} \frac{\Delta_{1k}(-j\omega)}{\Delta(-j\omega)} e^{-j\omega t} + \sum_{i=1}^{n} k_i e^{s_i t} \tag{2-23}$$

Comparison with (2-19) immediately shows that the first two terms represent the steady-state solution and that

$$V_k = I_1 \frac{\Delta_{1k}(j\omega)}{\Delta(j\omega)} \tag{2-24}$$

Defining the sine-wave transfer impedance $Z_T(j\omega)$ as the ratio of the complex phasors V_k and I_1, we get

$$Z_T(j\omega) \triangleq \frac{V_k(j\omega)}{I_1} = \frac{\Delta_{1k}(j\omega)}{\Delta(j\omega)} \tag{2-25}$$

Comparing (2-16) and (2-17) with (2-24) and (2-25) makes it apparent that *the relations between the Laplace transforms of the arbitrary time functions $i_1(t)$ and $v_k(t)$ are the same as those for the complex phasors of the corresponding functions under steady-state sine-wave conditions if s is set equal to $j\omega$*. Of course, the same conclusions remain valid if $k = 1$, so that we are dealing with the driving-point impedance

$$Z_D(j\omega) = \frac{V_1(j\omega)}{I_1} = \frac{\Delta_{11}(j\omega)}{\Delta(j\omega)} \tag{2-26}$$

or if the transfer function is, say, V_k/V_1 or I_k/I_1, etc.

The described relation is the result of the close mathematical correspondence between the Laplace and Fourier transformations. It enables us to carry out derivations which are valid for both transient and steady-state sine-wave signals. Whenever we invoke physical significance for the quantities V_i, I_k (such as calling $V_k I_k$ "power," etc.), however, we must restrict ourselves to $s = j\omega$ and hence to steady-state sine-wave conditions. This is because Laplace-transformed functions have no reality or physical interpretation, while the complex phasors can be simply related to physical quantities.

If, on the other hand, nonzero initial conditions are introduced, clearly time-domain or Laplace-transformed quantities are discussed.

2-2 TELLEGEN'S THEOREM

Next, an important law of circuit theory, *Tellegen's theorem*, will be introduced. This basic theorem will help us understand the fundamental properties of physically realizable impedance functions discussed later in this chapter. It will also be used much later in the book (in Chaps. 9 and 10) in connection with the sensitivity analysis of circuits.

Consider the circuit shown in Fig. 2-4a. With the notations of Fig. 2-4b, the Kirchhoff current laws (KCL) give

$$
\begin{aligned}
j_1 + j_2 \quad\quad\quad + j_5 \quad\quad &= 0 \\
- j_2 + j_3 + j_4 \quad\quad &= 0 \\
- j_4 - j_5 + j_6 &= 0
\end{aligned}
\tag{2-27}
$$

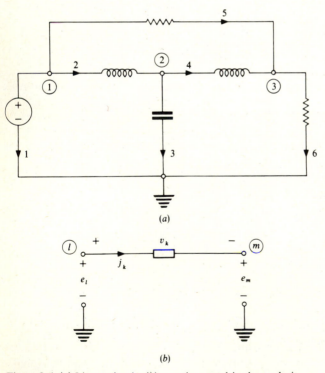

(a)

(b)

Figure 2-4 (a) Linear circuit; (b) notations used in the analysis

Note the use of *associated directions*[1] for v_k and j_k. Equations (2-27) can also be written in the form†

$$\mathbf{Aj} = 0 \tag{2-28}$$

where $\mathbf{A}$ is the *incidence matrix*

$$\mathbf{A} = \begin{bmatrix} 1 & 1 & 0 & 0 & 1 & 0 \\ 0 & -1 & 1 & 1 & 0 & 0 \\ 0 & 0 & 0 & -1 & -1 & 1 \end{bmatrix} \tag{2-29}$$

The elements of $\mathbf{A}$ are given by

$$a_{ij} = \begin{cases} +1 & \text{if branch } j \text{ leaves node } i \\ -1 & \text{if branch } j \text{ enters node } i \\ 0 & \text{if branch } j \text{ is not incident with node } i \end{cases} \tag{2-30}$$

Similarly, from the Kirchhoff voltage law (KVL)

$$
\begin{aligned}
v_1 &= e_1 \\
v_2 &= e_1 - e_2 \\
v_3 &= e_2 \\
v_4 &= e_2 - e_3 \\
v_5 &= e_1 - e_3 \\
v_6 &= e_3
\end{aligned}
\tag{2-31}
$$

or, using (2-29),

$$\mathbf{v} = \mathbf{A}^T \mathbf{e} \tag{2-32}$$

where $\mathbf{A}^T$ denotes the transpose of $\mathbf{A}$. Equations (2-28) and (2-32) are, of course, the general matrix forms of the KCL and KVL, respectively.[1]

Since all relations (2-27) to (2-32) involve only additions and subtractions, they remain valid if we perform any linear operations on the j_i, v_k, and e_l. For example, they hold also for the Laplace transforms $J_i(s)$, $V_k(s)$, and $E_l(s)$ or for the phasors J_i, V_k, and E_l, etc.

Let the branch power $v_k j_k$ be summed for all N branches of the circuit. Then, by (2-32) and (2-28),

$$\sum_{k=1}^{N} v_k j_k = \mathbf{v}^T \mathbf{j} = (\mathbf{A}^T \mathbf{e})^T \mathbf{j} = \mathbf{e}^T (\mathbf{A} \mathbf{j}) = 0 \tag{2-33}$$

Here the familiar rules $(\mathbf{Ae})^T = \mathbf{e}^T \mathbf{A}^T$, $(\mathbf{A}^T)^T = \mathbf{A}$, and $\mathbf{e}^T \mathbf{0} = 0$ of vector algebra have been used.

Equation (2-33) is equivalent, of course, to the conservation of power in the circuit, and thus its emergence from Kirchhoff's laws is not too surprising.

---

† Here, and in the rest of the book, x denotes a column vector and A denotes a matrix.

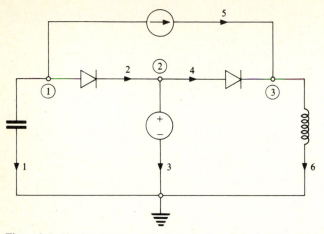

Figure 2-5 Circuit with the same configuration as that of Fig. 2-4 but with different elements.

Consider, however, the circuit of Fig. 2-5, which has the same topological configuration, same reference directions and numbering, and hence the same **A** as the circuit of Fig. 2-4. Hence, all the equations (2-27) to (2-32) remain valid for this circuit as well, with **A** remaining the same. Let the electric quantities of the circuit of Fig. 2-5 be $\mathbf{j}'$, $\mathbf{v}'$, and $\mathbf{e}'$. Then

$$\mathbf{A}\mathbf{j}' = \mathbf{0} \tag{2-34}$$

and

$$\mathbf{v}' = \mathbf{A}^T\mathbf{e}' \tag{2-35}$$

hold.

Next, let the physically meaningless quantity $\sum_{k=1}^{N} v_k j_k'$ be found:

$$\sum_{k=1}^{N} v_k j_k' = \mathbf{v}^T\mathbf{j}' = (\mathbf{A}^T\mathbf{e})^T\mathbf{j}' = \mathbf{e}^T(\mathbf{A}\mathbf{j}') = 0 \tag{2-36}$$

where (2-32) and (2-34) were used. While the left-hand side of (2-36) does have the dimension of watts, it does not correspond to physical power since v_k and j_k' exist in two different circuits.

An entirely analogous derivation gives

$$\sum_{k=1}^{N} v_k' j_k = \mathbf{v}'^T\mathbf{j} = 0 \tag{2-37}$$

Equations (2-36) and (2-37) are general forms and (2-33) is a special form of *Tellegen's theorem*. The general forms have great significance (as will be shown in Chaps. 9 and 10) in the calculation of circuit sensitivities.

Consider now a linear passive RLCM one-port (Fig. 2-6). By (2-32), using Laplace transformation,

$$\mathbf{V}(s) = \mathbf{A}^T\mathbf{E}(s) \tag{2-38}$$

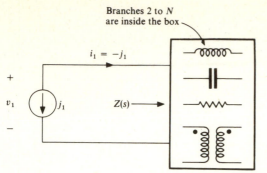

Figure 2-6 *RLCM* one-port.

and from (2-28), using Laplace transformation and taking the complex conjugate,

$$\mathbf{AJ^*}(s) = \mathbf{0} \tag{2-39}$$

Hence

$$\mathbf{V}^T\mathbf{J^*} = \mathbf{E}^T\mathbf{AJ^*} = \mathbf{0} \tag{2-40}$$

or

$$\sum_{k=1}^{N} V_k(s)J_k^*(s) = 0 \tag{2-41}$$

From Fig. 2-6 and Eq. (2-41), using $i_1 = -j_1$, we get

$$- V_1(s)J_1^*(s) = V_1(s)I_1^*(s) = \sum_{k=2}^{N} V_k J_k^* \tag{2-42}$$

Note that branches 2 to N are *inside* the one-port.

Defining the *impedance* of the one-port as the ratio of $V_1(s)$ and $I_1(s)$,

$$Z(s) \triangleq \frac{V_1(s)}{I_1(s)} = \frac{V_1(s)I_1^*(s)}{I_1(s)I_1^*(s)} = \frac{V_1(s)I_1^*(s)}{|I_1(s)|^2}$$

leads to

$$Z(s) = \frac{1}{|I_1(s)|^2} \sum_{k=2}^{N} V_k(s)J_k^*(s) \tag{2-43}$$

It is easy to show, using an entirely analogous derivation, that the dual relation

$$Y(s) \triangleq \frac{I_1(s)}{V_1(s)} = \frac{1}{|V_1(s)|^2} \sum_{\substack{\text{all internal} \\ \text{branches}}} V_k^*(s)J_k(s) \tag{2-44}$$

holds for the input *admittance* $Y(s)$ of the one-port.

Equations (2-43) and (2-44) are fundamental to the analysis and design of one-ports, as will be shown in the next section.

2-3 THE POSITIVE REAL PROPERTY

In the early chapters of this book, we shall be dealing with the design of one-ports (impedances) composed exclusively of resistors, inductors, capacitors, and coupled inductors. A basic question which must be answered before design can start is that of *realizability*; i.e., given a $Z(s)$ or $Y(s)$ function, is it possible to find any realization for it if only R, L, C, and M elements may be used?

The answer (given by Otto Brune in a classical paper in 1931) is as follows: *An impedance function Z(s) is realizable using lumped RLCM elements (where all R, L, M, and C are positive) if and only if it is a positive real rational function in s*, i.e., if the following conditions hold:

(a) $Z(s)$ is a real rational function of s:

$$Z(s) = \frac{a_0 + a_1 s + a_2 s^2 + \cdots + a_n s^n}{b_0 + b_1 s + b_2 s^2 + \cdots + b_m s^m} \tag{2-45}$$

where all coefficients a_i, b_j are real, and hence $Z(s)$ is real if s is real.

(b) If s (which is generally complex) has a nonnegative real part, then so does $Z(s)$; that is, using Re s to denote the real part of s,

$$\text{Re } Z(s) \geq 0 \quad \text{if} \quad \text{Re } s \geq 0 \tag{2-46}$$

Thus, any point in the closed right half plane† (RHP) of the s plane maps in the closed RHP of the Z plane. This is illustrated for some points in Fig. 2-7. Note that points which are on the positive *real s* axis (positive σ axis) must map to a point on the positive *real Z* axis (positive R axis), by condition a.

PROOF The proof of Brune's theorem follows. To prove *condition a*, we note that for the special choice $k = 1$, as a comparison of the circuits of Figs. 2-3

---

† The term *closed* for a region means that its boundary (in this case the $j\omega$ axis) is included.

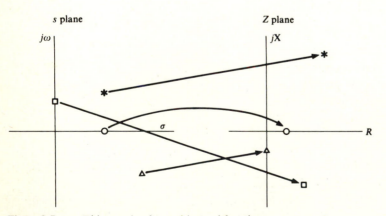

Figure 2-7 s-to-$Z(s)$ mapping for positive real functions.

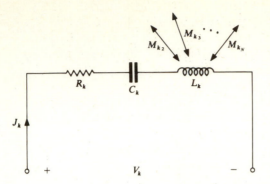

Figure 2-8 General branch of an *RLCM* one-port.

and 2-6 shows, Eqs. (2-12) to (2-17) give the input (driving-point) impedance

$$\frac{V_1(s)}{I_1(s)} = Z(s) = \frac{\Delta_{11}(s)}{\Delta(s)} \tag{2-47}$$

Here the elements Y_{ij} of the determinants Δ_{11} and Δ are given by Eq. (2-15) and hence are real rational functions of s. Since Δ_{11}, Δ, and Z are all constructed from the Y_{ij} by multiplications, additions, and subtractions, the real rational property holds also for these functions of s. This proves condition *a* for $Z(s)$. Since $Y(s) = 1/Z(s)$, it, too, satisfies condition *a*.

Condition b follows from Eq. (2-43). Since the most general branch of the *RLCM* circuit is that shown in Fig. 2-8, we have

$$V_k = \left(R_k + sL_k + \frac{1}{sC_k} \right) J_k + \sum_{\substack{l=2 \\ l \neq k}}^{N} sM_{kl} J_l \tag{2-48}$$

Substituting into (2-43) gives

$$Z(s) = \frac{1}{|I_1(s)|^2} \left[\sum_{k=2}^{N} R_k |J_k|^2 + \frac{1}{s} \sum_{k=2}^{N} \frac{1}{C_k} |J_k|^2 \right.$$
$$\left. + s \sum_{k=2}^{N} J_k^* \left(L_k J_k + \sum_{\substack{l=2 \\ l \neq k}}^{N} M_{kl} J_l \right) \right] \tag{2-49}$$

Since the summations entering (2-49) (frequently but misleadingly called *energy functions*) play an important role in circuit synthesis, they will be denoted by special symbols. Since all $R_k \geq 0$,

$$F_0(s) \triangleq \sum_{k=2}^{N} R_k |J_k(s)|^2 \geq 0 \tag{2-50}$$

for all values of s. Similarly, if there are no mutual couplings, i.e., if all $M_{kl} = 0$,

$$T_0(s) \triangleq \sum_{k=2}^{N} L_k |J_k(s)|^2 \geq 0 \tag{2-51}$$

and
$$V_0(s) \triangleq \sum_{k=2}^{N} \frac{1}{C_k} |J_k(s)|^2 \geq 0 \tag{2-52}$$

These quantities are closely related to the power and stored energies of the circuit under steady-state sine-wave conditions. The *power* dissipated as heat in the resistors, averaged over one period of the sine wave, is

$$P_{av} = \frac{1}{2} \sum_{k=2}^{N} R_k |J_k|^2 = \tfrac{1}{2} F_0 \tag{2-53}$$

The average stored *magnetic energy* in all inductors, in the absence of mutual couplings, is

$$\mathscr{E}_M = \frac{1}{4} \sum_{k=2}^{N} L_k |J_k|^2 = \tfrac{1}{4} T_0 \tag{2-54}$$

while the average *electric energy* stored in the capacitors is

$$\mathscr{E}_E = \frac{1}{4\omega^2} \sum_{k=2}^{N} \frac{1}{C_k} |J_k|^2 = \frac{1}{4\omega^2} V_0 \tag{2-55}$$

Since P_{av}, $\mathscr{E}_M$, and $\mathscr{E}_E$ are, by definition, nonnegative, these relations also show the nonnegative real character of F_0, T_0, and V_0 already proved by Eqs. (2-50) to (2-52).

Next, we consider the case when mutual couplings *are* present. For steady-state sine-wave signals, the average *magnetic energy* stored in coupled inductors is

$$\mathscr{E}_M = \frac{1}{4} \sum_{k=2}^{N} L_k J_k J_k^* + \frac{1}{4} \sum_{\substack{k=2 \\ k \neq l}}^{N} \sum_{l=2}^{N} M_{kl} J_k^* J_l \tag{2-56}$$

and hence, defining

$$M_0 \triangleq \sum_{k=2}^{N} L_k J_k J_k^* + \sum_{\substack{k=2 \\ k \neq l}}^{N} \sum_{l=2}^{N} M_{kl} J_k^* J_l \tag{2-57}$$

we find

$$M_0 = 4\mathscr{E}_M \tag{2-58}$$

Therefore, M_0 is also real and nonnegative. [The proof of formulas (2-53) to (2-56) giving P_{av}, $\mathscr{E}_M$, and $\mathscr{E}_E$ is implied in Prob. 2-2.]

In conclusion, $Z(s)$ can be written

$$Z(s) = \frac{1}{|I_1(s)|^2} \left[F_0(s) + \frac{1}{s} V_0(s) + s M_0(s) \right] \tag{2-59}$$

where $F_0(s)$, $V_0(s)$, and $M_0(s)$ are real and nonnegative quantities for any value of s, real or complex. If no coupled inductors are present, M_0 is replaced by T_0, which is also real nonnegative.

An entirely dual derivation for $Y(s)$, based on Eq. (2-44), gives

$$Y(s) = \frac{1}{|V_1(s)|^2} \left[F_0(s) + \frac{1}{s^*} V_0(s) + s^* M_0(s) \right] \tag{2-60}$$

where F_0, V_0, and M_0 are defined as before. If no mutual coupling is present, T_0 again replaces M_0.

Next, choose $s = \sigma + j\omega$ and let $\sigma \geq 0$. Then, by (2-59),.

$$\text{Re } Z(s) = \frac{1}{|I_1|^2} \left(F_0 + \frac{\sigma}{\sigma^2 + \omega^2} V_0 + \sigma M_0 \right) \tag{2-61}$$

Since F_0, V_0, M_0, and σ are all nonnegative, $\text{Re } Z(\sigma + j\omega) \geq 0$ for $\sigma \geq 0$. The same conclusion is valid, by (2-60), for $Y(s)$. Hence, both $Z(s)$ and $Y(s)$ satisfy condition b. Since it was shown earlier that they also satisfy condition a, it is now clear that the positive real rational conditions are *necessary* for any realizable $Z(s)$ or $Y(s)$ function. It will be shown later that these conditions are also *sufficient* for realizability, i.e., that it can be guaranteed that a realization with real positive R, L, C, and M elements exists for any $Z(s)$ or $Y(s)$ meeting conditions a and b.

For brevity, positive real is usually abbreviated PR. We could use PRR as an abbreviation for "positive real rational"; however, it is clear that as long as we restrict the discussion to lumped linear elements (active or passive), the rational character of all network functions will be maintained.

2-4 EQUIVALENT PR CONDITIONS

The PR conditions a and b given in Sec. 2-3 are admirably elegant and concise, but from a practical viewpoint, condition b is almost useless since for a given $Z(s)$ function it is very hard to ascertain whether or not it is satisfied. For this reason, we shall now replace conditions a and b by the following set of equivalent conditions:

(a') Identical to condition a.
(b') For all real ω,

$$\text{Re } Z(j\omega) \geq 0 \tag{2-62}$$

(c') All poles of $Z(s)$ are in the *closed* LHP of the s plane,† i.e., inside the LHP or on the $j\omega$ axis. All $j\omega$-axis poles must be simple, with positive real residues. Since $s = 0$ and $s \to \infty$ lie on the $j\omega$ axis, this holds also for poles at the origin and at infinity.

To show the equivalence of conditions a and b, on the one hand, and conditions a', b', and c', on the other, we must prove that either set of conditions can be derived from the other. To derive a', b', and c' from a and b, we note that a' and b'

---

† LHP denotes "left half plane," RHP "right half plane" throughout the book.

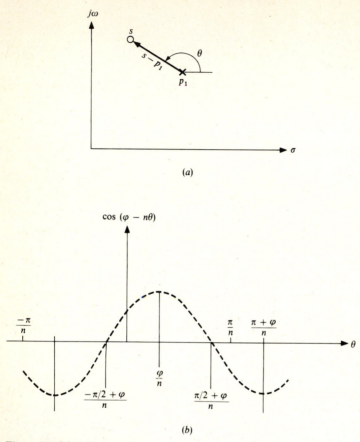

Figure 2-9 (a) RHP pole of $Z(s)$; (b) the $\cos(\varphi - n\theta)$ curve.

follow directly from a and b, respectively. To prove c', assume first that it does *not* hold. Then, an nth-order RHP pole of $Z(s)$ may exist at some p_1 (Fig. 2-9a). Evaluating $Z(s)$ at a point s close to p_1 in the RHP using a Laurent-series expansion[2] gives

$$Z(s) = \frac{k_{-n}}{(s - p_1)^n} + \frac{k_{-n+1}}{(s - p_1)^{n-1}} + \cdots + \frac{k_{-1}}{s - p_1}$$

$$+ k_0 + k_1(s - p_1) + k_2(s - p_1)^2 + \cdots \quad (2\text{-}63)$$

As $s \to p_1$, the first term will dominate and $Z(s) \to k_{-n}/(s - p_1)^n$. Hence, introducing polar coordinates

$$k_{-n} = k e^{j\varphi} \qquad k \triangleq |k_{-n}| > 0 \qquad s - p_1 = r e^{j\theta} \qquad r \triangleq |s - p_1| > 0 \quad (2\text{-}64)$$

we get

$$Z(s) \approx \frac{k}{r^n} e^{j(\varphi - n\theta)} \qquad (2\text{-}65)$$

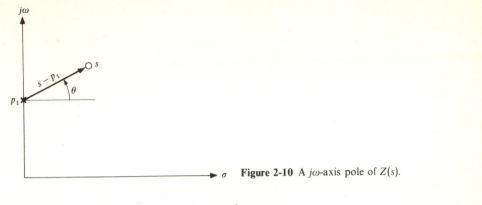

Figure 2-10 A $j\omega$-axis pole of $Z(s)$.

and hence

$$\text{Re } Z(s) \approx \frac{k}{r^n} \cos{(\varphi - n\theta)} \tag{2-66}$$

Since s is in the RHP, by condition b the inequality $\cos{(\varphi - n\theta)} \geq 0$ must hold for $-\pi \leq \theta \leq \pi$. However, as Fig. 2-9$b$ illustrates, this is impossible for any $n \geq 1$, since the period of this cosine function is $2\pi/n$. Thus, a $Z(s)$ with a RHP pole cannot be PR.

Assuming a $j\omega$-axis pole (Fig. 2-10), a similar argument can be used. At a RHP s value close to p_1, by (2-66) again $\cos{(\varphi - n\theta)} \geq 0$ is required if $Z(s)$ is to be PR. However, now s will only be in the RHP if $-\pi/2 \leq \theta \leq \pi/2$, as Fig. 2-10 illustrates. Since by Fig. 2-9b, $\cos{(\varphi - n\theta)}$ is nonnegative for $(-\pi/2 + \varphi)/n \leq \theta \leq (\pi/2 + \varphi)/n$, the PR condition *can* be met if and only if $\varphi = 0$ and $n = 1$. The $n = 1$ condition by (2-63) implies a simple (first-order) pole; $\varphi = 0$, by (2-64), means that the residue k_{-1} is real and positive, as condition c' specifies. The proof clearly holds also if the pole lies at $\omega = 0$.

If $Z(s)$ has an nth-order pole at $s \to \infty$, then for a sufficiently large s value $Z(s) \approx k_n s^n$. When $s = re^{j\theta}$ and $k_n = ke^{j\varphi}$ are used, $\text{Re } Z(s) = kr^n \cos{(n\theta + \varphi)}$. Since $\text{Re } s \geq 0$ corresponds to $-\pi \leq \theta \leq \pi$, by an argument closely following that given for finite $j\omega$-axis poles $n = 1$ and $\varphi = 0$ can be deduced.

This completes the derivation of condition c' from conditions a and b.

Next, we shall show that conditions a and b can, in turn, be derived from a', b', and c'. Since a is the same as a', we only have to worry about b. To prove b from a', b', and c', we use the maximum-modulus theorem from the theory of complex variables.[2]

> **Theorem** If a function $F(s)$ is analytical in a region R, then its modulus $|F(s)|$ will reach its maximum for R on the boundary.

Let $F(s)$ be chosen as $\exp{[-Z(s)]}$. Then $F(s)$ will be analytical in any region where $Z(s)$ has no poles, and $|F(s)| = \exp{[-\text{Re } Z(s)]}$ will reach its regional maximum on the boundary. Hence, if $Z(s)$ has no poles in a region R, the minimum of its real part occurs on the boundary of the region. Choose the RHP as R; also, let the boundary, illustrated in Fig. 2-11, coincide with the $j\omega$ axis every-

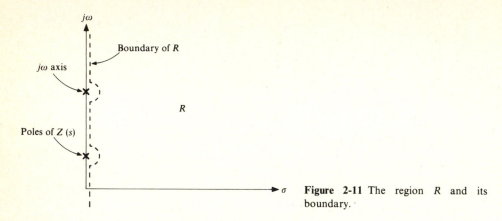

Figure 2-11 The region R and its boundary.

where except at the (simple) $j\omega$-axis poles of $Z(s)$, where it detours in the RHP. Then, by c', R is free from poles of $Z(s)$. Also by c', Re $Z(s) \geq 0$ on the semicircles drawn around the $j\omega$-axis poles; this can be seen by reversing the steps proving c': specifically, since we know that the $j\omega$-axis pole is simple and the residue positive real, by (2-66) near the pole

$$\text{Re } Z(s) \approx \frac{k}{r} \cos \theta \qquad (2\text{-}67)$$

and hence on the semicircle where $-\pi/2 \leq \theta \leq \pi/2$ holds, Re $Z(s) \geq 0$. On the rest of the boundary, i.e., on the $j\omega$ axis, the Re $Z(j\omega) \geq 0$ condition is guaranteed by b'.

Thus, Re $Z(s) \geq 0$ holds everywhere on the boundary of R; therefore, by the maximum-modulus theorem, it holds in the whole RHP, as specified by condition b. This completes the derivation of a and b from a', b', and c'. Since we have already derived a', b', and c' from a and b, the equivalence of these conditions is now established.

Note that the *necessity* of condition b' also follows from the fact that the average power entering the one-port in the steady-state sine-wave condition is

$$P_{\text{av}} = \tfrac{1}{2}|I_1|^2 \text{ Re } Z(j\omega) \qquad (2\text{-}68)$$

and for a passive circuit is nonnegative. The necessity of condition c' can also be shown on physical grounds: if the input current is chosen as $j_1(t) = \delta(t)$ (the impulse function), the voltage becomes

$$v_1(t) = \sum_i k_i e^{p_i t} + \sum_j (k_{j0} + k_{j1} t + \cdots + k_{jn_j} t^{n_j}) e^{p_j t} \qquad (2\text{-}69)$$

where the first sum includes the single poles and the second the multiple poles of $Z(s)$. If any pole has a positive real part, the corresponding term in (2-69) will diverge for large t (Fig. 2-12a); similarly, if a multiple (say double) pole has zero real part, the corresponding expression in the second summation will be divergent (Fig. 2-12b). Both voltage responses are clearly impossible for a passive $RLCM$ circuit.

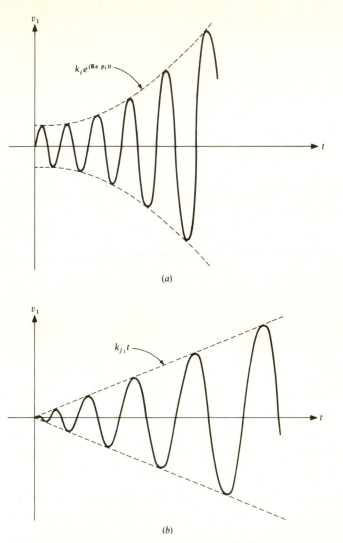

v_1

$k_i e^{(\mathrm{Re}\ p_i)t}$

t

(a)

v_1

$k_{j_1} t$

t

(b)

Figure 2-12 Divergent voltage responses: (a) $k_i e^{p_i t}$, $\mathrm{Re}\ p_i > 0$; (b) $k_{j1} t e^{p_j t}$, $\mathrm{Re}\ p_j = 0$.

Note that both $Z(s)$ and $Y(s)$ are PR if $Z(s)$ is; hence, applying condition c' to $Y(s)$ shows that the *zeros* of $Z(s)$ must also be in the closed LHP and that if they are on the $j\omega$ axis, they must be simple. This, however, follows from the other PR conditions on $Z(s)$ and need not be included in the checking of $Z(s)$ for realizability.

We repeat that so far only the *necessity* of conditions a and b (or a', b', and c') has been proved. The *sufficiency* of these conditions will follow in Chap. 4, where we shall show how to realize any PR function $Z(s)$ or $Y(s)$ in the form of a passive *RLCM* one-port.

2-5 HURWITZ POLYNOMIALS; SIMPLE TESTS FOR PR FUNCTIONS

We have seen that condition c' restricts the poles of $Z(s)$ to the closed LHP of the s plane. Hence, when $Z(s)$ is written as the ratio of two polynomials $N(s)$ and $D(s)$

$$Z(s) = \frac{N(s)}{D(s)} \tag{2-70}$$

$D(s)$ must have its zeros restricted to the closed LHP. We now introduce some definitions:

1. A polynomial with its zeros restricted to the *closed* LHP, i.e., the LHP including the $j\omega$ axis, will be called a *Hurwitz polynomial*.
2. A polynomial with its zeros restricted to the *inside* of the LHP (excluding the $j\omega$ axis) will be called a *strictly Hurwitz polynomial*.
3. A polynomial *not* satisfying condition 1 will be called *non-Hurwitz*.

An illustration of some possible zero locations for these types is given in Fig. 2-13, which also indicates all complex and imaginary zeros occurring in complex conjugate pairs. This is necessary if the coefficients of the polynomial are to be real, as conditions a and a' require for both $N(s)$ and $D(s)$. For example, a complex zero $z_1 = \sigma_1 + j\omega_1$ corresponds to a factor $(s - z_1) = (s - \sigma_1 - j\omega_1)$ in the polynomial. If $z_1^* = \sigma_1 - j\omega_1$ is also a zero, the two factors can be combined to give

$$(s - z_1)(s - z_1^*) = (s - \sigma_1)^2 + \omega_1^2 \tag{2-71}$$

which has only real coefficients. Otherwise, the $s - \sigma_1 - j\omega_1$ factor will introduce complex coefficients into the polynomial. Real zeros (including those at $s = 0$) may occur singly.

For a strictly Hurwitz polynomial with real coefficients the only possible zeros occur either on the negative σ axis or in conjugate pairs inside the LHP. The real zeros introduce factors $s - \sigma_i$ where $\sigma_i < 0$; complex ones, by (2-71), factors $s^2 - 2\sigma_i s + \sigma_i^2 + \omega_i^2$, where again $\sigma_i < 0$ and, of course, $\sigma_i^2 + \omega_i^2 > 0$. Since both types of factors have only positive coefficients and no missing terms, i.e., the coefficients of s^0, s^1, ... are all greater than zero, the strictly Hurwitz polynomial which is the product of these factors has this same property. Thus, $s^3 + 3s^2 + 2s + 1$ may be a strictly Hurwitz polynomial; $s^3 + 3s + 1$ or $s^3 - 3s^2 + 2s + 1$ cannot possibly be one.

However, positive coefficients represent only a necessary and *not* a sufficient condition of the strictly Hurwitz character;† e.g., the zeros of $P(s) = s^5 + s^4 + 6s^3 + 6s^2 + 25s + 25$, which has only positive coefficients, are shown in Fig. 2-14. Clearly, $P(s)$ is non-Hurwitz.

---

† The only exceptions are first- and second-degree polynomials. For these, as can easily be shown, positive coefficients guarantee the strictly Hurwitz property.

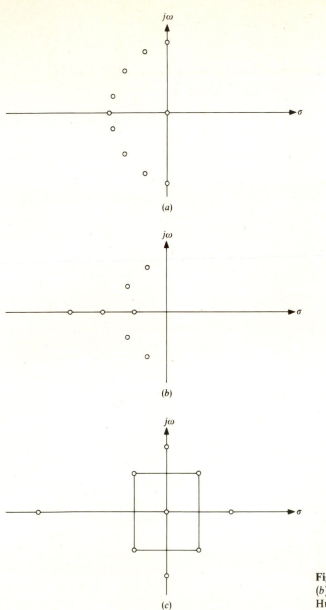

Figure 2-13 Zeros of (a) Hurwitz; (b) strictly Hurwitz; (c) non-Hurwitz polynomials.

The $j\omega$-axis zeros permissible in a Hurwitz (but not in a strictly Hurwitz) polynomial introduce factors of the form $(s - j\omega_i)(s + j\omega_i) = s^2 + \omega_i^2$. A zero at the origin corresponds to a factor s. If the Hurwitz polynomial is a product of one or more $s^2 + \omega_i^2$ factors and a strictly Hurwitz polynomial, it will have the same appearance as a strictly Hurwitz one. For example,

$$s^3 + s^2 + 4s + 4 = (s^2 + 4)(s + 1)$$

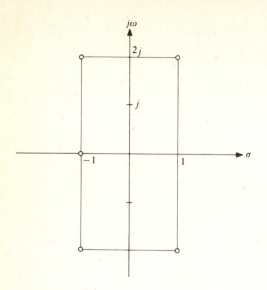

Figure 2-14 Zeros of the polynomial $s^5 + s^4 + 6s^3 + 6s^2 + 25s + 25$.

is Hurwitz. If a zero occurs at the origin, the constant term will be missing, as in $s^4 + s^3 + 4s^2 + 4s = s(s^2 + 4)(s + 1)$. If the Hurwitz polynomial contains *only* $j\omega$-axis zeros, it will have the form $s^k(s^2 + \omega_1^2)(s^2 + \omega_2^2) \cdots$, k an integer. In such a polynomial, the lowest-order term will be $a_k s^k$; also every second term $a_{k+1} s^{k+1}$, $a_{k+3} s^{k+3}$, ... will be missing. The nonzero coefficients of the polynomial, however, will still be positive; thus $s^4 + 2s^2 + 1$ or $s^5 + 3s^3$ may be a Hurwitz polynomial, but $s^5 - 4s^3 + s$ cannot be one.

In conclusion, an nth-degree strictly Hurwitz polynomial must be in the form

$$P(s) = a_n s^n + a_{n-1} s^{n-1} + \cdots + a_1 s + a_0 \tag{2-72}$$

where all $a_i > 0$. A Hurwitz polynomial may have $a_0 = a_1 = \cdots a_{k-1} = 0$ due to an s^k factor; also, every second term may be absent if only $j\omega$-axis zeros occur. Thus, $a_n s^n + a_{n-2} s^{n-2} + \cdots + a_{k+2} s^{k+2} + a_k s^k$ may be a Hurwitz polynomial if a_n, $a_{n-2}, \ldots, a_{k+2}, a_k$ are all positive.

All the above criteria are necessary but not sufficient. They may be used to discover by inspection some non-Hurwitz polynomials but not to ascertain the Hurwitz or strictly Hurwitz property in all cases.

There are two basic methods for establishing the Hurwitz character of a polynomial. The first one uses the obvious technique of factoring the polynomial numerically, a task which requires either infinite patience or the use of a computer for polynomials of degree 5 or higher. Then by inspection each factor can be classified as Hurwitz, strictly Hurwitz, or non-Hurwitz. If all factors are Hurwitz (strictly Hurwitz), the polynomial is also Hurwitz (strictly Hurwitz). Otherwise, it is not. The second method, which is more elegant and less laborious, relies on some properties of *LC* impedances, to be discussed in the next chapter. Hence, a description of this method will be postponed until Chap. 4.

Utilizing our discussion of Hurwitz polynomials, we can rephrase condition c' on PR functions as follows:

(c') The denominator $D(s)$ of $Z(s)$ must be a Hurwitz or strictly Hurwitz polynomial. If $D(s)$ is only Hurwitz, its $j\omega$-axis zeros must be simple and the residues of $Z(s)$ at these poles must be positive real. Poles of $Z(s)$ at $s \to \infty$ [manifested by a higher degree of $N(s)$ than that of $D(s)$] must also be simple and have a positive residue.

We are now in a position to carry out the PR test on some simple functions.

Example 2-1 Is $Z(s) = (2s + 3)/(s + 1)$ PR?

By inspection, $Z(s)$ meets conditions a' and c' (it has a strictly Hurwitz denominator). To test for condition b', we write

$$\operatorname{Re} Z(j\omega) = \operatorname{Re} \frac{2j\omega + 3}{j\omega + 1} = \operatorname{Re} \left(\frac{2j\omega + 3}{j\omega + 1} \right) \left(\frac{-j\omega + 1}{-j\omega + 1} \right)$$

$$\operatorname{Re} Z(j\omega) = \operatorname{Re} \frac{(3 + 2\omega^2) + j(2\omega - 3\omega)}{\omega^2 + 1} = \frac{2\omega^2 + 3}{\omega^2 + 1}$$

Clearly, for all values of ω, $\operatorname{Re} Z(j\omega) \geq 0$ and hence b' is also satisfied. Thus, $Z(s)$ is **PR**.

Example 2-2 Is $Z(s) = (s^2 + 2s + 25)/(s + 4)$ PR?

Again, conditions a' and c' are obviously satisfied. To ascertain whether condition b' is met, we write

$$\operatorname{Re} Z(j\omega) = \operatorname{Re} \left(\frac{-\omega^2 + 2j\omega + 25}{j\omega + 4} \right) \left(\frac{-j\omega + 4}{-j\omega + 4} \right)$$

$$\operatorname{Re} Z(j\omega) = \operatorname{Re} \frac{(-2\omega^2 + 100) + j(\omega^3 - 17\omega)}{\omega^2 + 16} = \frac{-2\omega^2 + 100}{\omega^2 + 16}$$

Here, for values of ω^2 greater than 50, $\operatorname{Re} Z(j\omega) < 0$. Hence, Condition b' is violated, and $Z(s)$ is not PR.

Noting that even powers of s become real while odd powers become imaginary when s is replaced by $j\omega$, we can simplify the real-part test somewhat. Let the subscript e denote the even part and the subscript o the odd part of a function. Then we can write

$$Z(s) = \frac{N(s)}{D(s)} = \frac{N_e(s) + N_o(s)}{D_e(s) + D_o(s)}$$

$$= \frac{N_e(s) + N_o(s)}{D_e(s) + D_o(s)} \frac{D_e(s) - D_o(s)}{D_e(s) - D_o(s)}$$

$$= \frac{(N_e D_e - N_o D_o) + (N_o D_e - N_e D_o)}{D_e^2 - D_o^2} \qquad (2\text{-}73)$$

Now when s is replaced by $j\omega$, even functions will become real and odd ones imaginary. Hence

$$\text{Re } Z(j\omega) = \frac{N_e(j\omega)D_e(j\omega) - N_o(j\omega)D_o(j\omega)}{[D_e(j\omega)]^2 - [D_o(j\omega)]^2} \tag{2-74}$$

Since $[D_e(j\omega)]^2 \geq 0$ and $[D_o(j\omega)]^2 \leq 0$, the denominator of $\text{Re } Z(j\omega)$ cannot become negative. Therefore, it is enough to test

$$P(\omega^2) \triangleq N_e(j\omega)D_e(j\omega) - N_o(j\omega)D_o(j\omega) \tag{2-75}$$

for $P(\omega^2) \geq 0$, for all real ω.

Example 2-2 (continued) By inspection, $N_e = s^2 + 25$, $N_o = 2s$, $D_e = 4$, $D_o = s$; hence $P(\omega^2) = (-\omega^2 + 25)(4) - (2j\omega)(j\omega) = -2\omega^2 + 100$, and the same conclusion is reached as before.

Example 2-3 Is $Z(s) = (s^3 + 2s)/(s^2 + 1)$ PR?
Condition a' is met. Since $N_e = D_o = 0$, $P(\omega^2) \equiv 0$. Hence, $\text{Re } Z(j\omega) \equiv 0$, and condition b' is met in a trivial fashion.

Next, condition c' is considered. Here, $D(s)$ is clearly Hurwitz; all its zeros are on the $j\omega$ axis, at $j\omega = \pm j$. Furthermore, there is a pole at $s \to \infty$, since the degree of $N(s)$ is 1 higher than that of $D(s)$. All three $j\omega$-axis poles are simple; hence $Z(s)$ is indeed PR if its residues at all poles are real and positive. Breaking $Z(s)$ into partial fractions, we get

$$Z(s) = k_1 s + \frac{k_2}{s - j} + \frac{k_3}{s + j} = \frac{s^3 + 2s}{s^2 + 1}$$

where

$$k_1 = \left[\frac{Z(s)}{s}\right]_{s \to \infty} = \left(\frac{s^2 + 2}{s^2 + 1}\right)_{s \to \infty} = 1$$

$$k_2 = [(s - j)Z(s)]_{s \to j} = \left(\frac{s^3 + 2s}{s + j}\right)_{s = j} = \frac{1}{2}$$

$$k_3 = [(s + j)Z(s)]_{s \to -j} = \left(\frac{s^3 + 2s}{s - j}\right)_{s = -j} = \frac{1}{2}$$

All three residues are real and positive; hence, $Z(s)$ meets condition c' and is therefore PR.

It will be noticed that $k_2 = k_3$. This is not a coincidence. In general, for partial-fraction terms corresponding to conjugate $j\omega$-axis poles we have

$$\frac{k_1}{s - j\omega_1} + \frac{k_2}{s + j\omega_1} = \frac{(k_1 + k_2)s + j\omega_1(k_1 - k_2)}{s^2 + \omega_1^2} \tag{2-76}$$

Hence, unless $k_2 = k_1$, the function will violate conditions a and a'. For $k_2 = k_1$,

$$\frac{k_1}{s - j\omega_1} + \frac{k_1}{s + j\omega_1} = \frac{2k_1 s}{s^2 + \omega_1^2} \tag{2-77}$$

To avoid complex arithmetic, partial-fraction terms corresponding to $j\omega$-axis poles should be removed using (2-77). For the previous example, this results in

$$Z(s) = k_1 s + \frac{2k_2 s}{s^2 + 1} = \frac{s^3 + 2s}{s^2 + 1}$$

and

$$k_2 = \left[\frac{s^2 + 1}{2s} Z(s)\right]_{s^2 \to -1} = \left(\frac{s^2 + 2}{2}\right)_{s^2 = -1} = \frac{1}{2}$$

as before.

Some more advanced methods will be given for PR testing in Chap. 4, where the sufficiency of the PR conditions will also be discussed.

2-6 SUMMARY

The topics discussed in this chapter were the following:

1. The dual meaning of the complex frequency variable s, as an abstract mathematical parameter introduced by the Laplace transformation of arbitrary signals *or* as an abbreviation for $j\omega$ in the phasor representation of sinusoid signals (Sec. 2-1).
2. Tellegen's theorem, relating the branch voltages and currents either of a single circuit, or of two different circuits with the same topology (Sec. 2-2).
3. The positive real (PR) property, which all passive $RLCM$ impedance must possess, was described, and its necessity (but not its sufficiency) was proved; in the process, the energy functions F_0, T_0, V_0, and M_0 were defined (Sec. 2-3).
4. Equivalent conditions, less compact but more easily applicable, were given for the PR property; their equivalence to the original conditions was proved by deriving them from the original conditions and vice versa (Sec. 2-4).
5. Hurwitz polynomials, strictly Hurwitz polynomials, and non-Hurwitz polynomials were defined according to the locations of their zeros; the testing of impedance functions for the PR property was illustrated by simple examples (Sec. 2-5).

The theorems and PR conditions learned in this chapter will enable us to derive the properties of various impedance functions and then to design a circuit from any realizable function. This will be discussed in Chaps. 3 and 4.

PROBLEMS

2-1 Derive Eq. (2-44). *Hint:* Follow steps similar to those leading to Eq. (2-43).

2-2 Prove relations (2-53) to (2-56) giving P_{av}, $\mathscr{E}_M$, and $\mathscr{E}_E$. *Hint:* The following equations are needed:

Power in a branch:

$$P_k(t) = v_k(t)i_k(t)$$

Energy stored:

$$\mathscr{E}_k = \int_0^t P_k(\tau)\, d\tau$$

Branch relations:

R:

$$v_k = Ri_k$$

C:
$$v_k = \frac{1}{C} \int_0^t i_k(\tau) \, d\tau$$

L, M:
$$v_k = L_k \frac{di_k}{dt} + \sum_{l \neq k} M_{kl} \frac{di_l}{dt}$$

Phasor relation: $\quad i_k = I_k \cos(\omega t + \varphi_k) = \text{Re } I_k e^{j\varphi_k} e^{j\omega t} = \text{Re } J_k e^{j\omega t}$

2-3 Prove Eq. (2-60). *Hint:* Follow the steps yielding Eq. (2-59) but start from (2-44) rather than (2-43).

2-4 Let $Z(s)$ be a PR function. Prove, using only $Y(s) = 1/Z(s)$ and conditions a and b, that $Y(s)$ is also PR.

2-5 Let $Z(s)$ and $F(s)$ be PR functions. Prove, using conditions a and b, that $Z[F(s)]$ is also PR.

2-6 Use the result proved in Prob. 2-5 to show that a pole of $Z(s)$ at infinity must be simple, with positive real residue. *Hint:* Use $F(s) = 1/s$.

2-7 Show that the zeros shown in Fig. 2-14 belong to $s^5 + s^4 + 6s^3 + 6s^2 + 25s + 25$.

2-8 Decide by inspection which of the following polynomials *cannot* be strictly Hurwitz polynomials and which ones cannot even be Hurwitz polynomials:

(a) $s^4 + s^2 + s + 2$ (b) $s^3 + 3s^2 + 5s + 1$
(c) $s^7 + 2s^6 + 2s^5 + 2s^4 + 4s^3 + 8s^2 + 8s + 4$
(d) $s^4 + 3s^3 + 3s^2 + 2s + 1$ (e) $s^5 + 4s^3 + 2s$
(f) $s^6 - 3s^4 + 2s^2 + 1$ (g) $s^8 + 2s^6 + 4s^4$

2-9 Show that the sum of two PR functions must be PR.

2-10 Is the product or ratio of two PR functions necessarily PR? *Hint:* Try it on some low-order PR functions.

2-11 Which of the following $Z(s)$ or $Y(s)$ functions are PR? Why or why not?

(a) $\dfrac{s^2 + 2s + 24}{s^2 + 5s + 16}$ (b) $\dfrac{s + 3}{s^2 + 2s + 1}$

(c) $\dfrac{s^3 + 2}{s^2 + 2s + 1}$ (d) $\dfrac{s^3 + 2s^2 + 3s + 1}{s + 1}$

2-12 For what values of the coefficient a will the function $Z(s) = (s^2 + as)/(s^2 + 6s + 8)$ be PR?

2-13 An admittance $Y(s)$ has zeros at $-1 \pm j2$, poles at -1 and -3, and $Y(0) = 1$. Is $Y(s)$ PR?

2-14 Is

$$Z(s) = 1 - \frac{1}{s+1} + \frac{s^2 + 1}{s(s^2 + 2)}$$

PR?

2-15 Show that in steady-state sine-wave condition and for a 1-A input current the input impedance of Fig. 2-6 is $Z(j\omega) = 2P_{av} + 4j\omega(\mathscr{E}_M - \mathscr{E}_E)$, where P_{av}, $\mathscr{E}_M$, and $\mathscr{E}_E$ are defined in Eqs. –2-53) to (2-55). Verify this relation for the circuit of Fig. 2-1. When is $Z(j\omega)$ real?

2-16 Show that linear and quadratic polynomials with all positive coefficients are strictly Hurwitz. Show, by creating a counterexample, that this property does not carry over to cubic polynomials.

REFERENCES

1. Desoer, C. A., and E. S. Kuh: "Basic Circuit Theory," McGraw-Hill, New York, 1969.
2. Kreyszig, E.: "Advanced Engineering Mathematics," Wiley, New York, 1962.
3. Balabanian, N.: "Network Synthesis," Prentice-Hall, Englewood Cliffs, N.J., 1958.

THREE

SYNTHESIS OF ONE-PORTS WITH
TWO KINDS OF ELEMENTS

3-1 THE PROPERTIES OF *LC* IMMITTANCES†

In this chapter we shall restrict our attention to the design of one-ports which contain only two types of elements: L and C, *or R* and C, *or R* and L. Since simple and infallible design techniques are available for such circuits, they serve to illustrate the power of mathematically based design techniques; they will also be used to introduce such concepts as pole removal and ladder expansion, which will be found useful in more complicated synthesis problems. Finally, they demonstrate the variety of synthesis methods available for solving the same problem, which is characteristic of circuit design.

First, we shall discuss *LC one-ports*. In such circuits, all $R_k = 0$, and hence also $F_0(s) \equiv 0$. Therefore, from (2-59) and (2-60),

$$Z(s) = \frac{1}{|I_1|^2}\left[\frac{V_0(s)}{s} + sM_0(s)\right] \tag{3-1}$$

and

$$Y(s) = \frac{1}{|V_1|^2}\left[\frac{V_0(s)}{s^*} + s^*M_0(s)\right] \tag{3-2}$$

Here, $|I_1|^2$, $|V_1|^2$, V_0, and M_0 are all functions of s; however, they are all real and nonnegative for any value of s, real or complex.

---

† Immittance is a word coined by Bode[1] to describe either an *impedance* or an ad*mittance*.

Next, we explore the properties of the zeros of $Z(s)$. At a zero s_z, assuming†
finite $|I_1|$,

$$\frac{V_0(s_z)}{s_z} + s_z M_0(s_z) = 0 \qquad s_z^2 = -\frac{V_0(s_z)}{M_0(s_z)} \tag{3-3}$$

Since $V_0 \geq 0$, $M_0 \geq 0$ hold for $s = s_z$ (as well as for any other s value), we conclude
that s_z^2 is a nonpositive real number and therefore s_z is imaginary: $s_z = j\omega_z$. By the
argument used in Sec. 2-5, such a zero must be accompanied by its complex
conjugate $s_z^* = -j\omega_z$ to give real coefficients in the numerator $N(s)$ of $Z(s) =$
$N(s)/D(s)$. The only exceptions are zeros at $\omega = 0$ and $\omega \to \infty$, which can occur
singly. Furthermore, since these zeros of $Z(s)$ represent $j\omega$-axis poles of the PR
function $Y(s)$, all zeros of $Z(s)$ must be simple.

An exactly similar argument based on (3-2) shows that the same properties
hold also for the zeros of $Y(s)$, that is, the poles of $Z(s)$. Hence, both $N(s)$ and $D(s)$
must be of the general form

$$P(s) = s(s - j\omega_1)(s + j\omega_1)(s - j\omega_2)(s + j\omega_2) \cdots$$
$$= s(s^2 + \omega_1^2)(s^2 + \omega_2^2) \cdots \tag{3-4}$$

where the first factor s, representing a zero at $\omega = 0$, may or may not be present.
Hence these polynomials contain either only odd powers of s (if the factor s is
included) or only even powers (if it is not). But since odd powers of s become
imaginary when $s = j\omega$, while even powers become real, $Z(j\omega)$ would become real
if both $N(s)$ and $D(s)$ contained only even powers of s or if both contained only odd
powers. However, a glance at (3-1) shows that $Z(j\omega)$ must be pure imaginary.
Hence, $N(s)$ and $D(s)$ cannot both be even or both odd. One must be even and the
other odd; therefore, the two possibilities are

$$Z(s) = \frac{N(s)}{D(s)} = \frac{s(s^2 + \omega_{z1}^2)(s^2 + \omega_{z2}^2) \cdots}{(s^2 + \omega_{p1}^2)(s^2 + \omega_{p2}^2) \cdots} \quad \text{or} \quad \frac{(s^2 + \omega_{z1}^2)(s^2 + \omega_{z2}^2) \cdots}{s(s^2 + \omega_{p1}^2)(s^2 + \omega_{p2}^2) \cdots} \tag{3-5}$$

Thus, $Z(s)$ is an odd function of s, that is, $Z(-s) = -Z(s)$.

Consider now the partial-fraction expansion of $Z(s)$. By (3-5), in general,

$$Z(s) = k_n s^n + k_{n-1} s^{n-1} + \cdots k_2 s^2 + k_1 s + k_0 + \frac{k_{-1}}{s} + \frac{k_{p1}}{s - j\omega_{p1}} + \frac{k'_{p1}}{s + j\omega_{p1}}$$
$$+ \frac{k_{p2}}{s - j\omega_{p2}} + \frac{k'_{p2}}{s + j\omega_{p2}} + \cdots \tag{3-6}$$

holds. But $Z(s)$, a PR function, cannot have multiple poles for $s \to \infty$; hence,
$k_n = k_{n-1} = k_{n-2} = \cdots = k_3 = k_2 = 0$. Also, k_0 cannot be present, since it would
destroy the purely odd character of $Z(s)$. Finally, the sum of the terms corresponding to conjugate poles

$$\frac{k_{pi}}{s - j\omega_{pi}} + \frac{k'_{pi}}{s + j\omega_{pi}} = \frac{(k_{pi} + k'_{pi})s + j\omega_{pi}(k_{pi} - k'_{pi})}{s^2 + \omega_{pi}^2} \tag{3-7}$$

† This can be assured by driving the one-port with a finite-valued current source, as in Fig. 2-6.

must be odd in s and must have real coefficients. Both conditions make it necessary that $k'_{pi} = k_{pi}$ for all i.

In conclusion, $Z(s)$ must be in the form

$$Z(s) = k_1 s + \frac{k_{-1}}{s} + \frac{2k_{p_1} s}{s^2 + \omega_{p_1}^2} + \frac{2k_{p_2} s}{s^2 + \omega_{p_2}^2} + \cdots \tag{3-8}$$

or, revising our notation for convenience,

$$Z(s) = K_\infty s + \frac{K_0}{s} + \frac{K_1 s}{s^2 + \omega_1^2} + \frac{K_2 s}{s^2 + \omega_2^2} + \cdots + \frac{K_n s}{s^2 + \omega_n^2} \tag{3-9}$$

where $K_\infty \triangleq k_1$, $K_0 \triangleq k_{-1}$, and $K_i \triangleq 2k_{pi}$, $i = 1, 2, \ldots, n$; also $\omega_i \triangleq \omega_{pi}$. Since k_1, k_{-1}, and the k_{pi} are all residues of $Z(s)$ at its $j\omega$-axis poles, they are all nonnegative real numbers; hence so are K_∞, K_0, and the K_i.

Next, let $s = j\omega$. Then, by (3-9)

$$Z(j\omega) = j \left[K_\infty \omega - \frac{K_0}{\omega} + \frac{K_1 \omega}{\omega_1^2 - \omega^2} + \frac{K_2 \omega}{\omega_2 - \omega^2} + \cdots \right] \tag{3-10}$$

Here the function in the brackets

$$X(\omega) \triangleq K_\infty \omega - \frac{K_0}{\omega} + \frac{K_1 \omega}{\omega_1^2 - \omega^2} + \frac{K_2 \omega}{\omega_2 - \omega^2} + \cdots \tag{3-11}$$

is the *reactance* of the LC one-port.

It will next be shown that $X(\omega)$ is a monotone increasing function of ω. In fact,

$$\frac{dX(\omega)}{d\omega} = K_\infty + \frac{K_0}{\omega^2} + \frac{K_1(\omega_1^2 + \omega^2)}{(\omega_1^2 - \omega^2)^2} + \frac{K_2(\omega_2^2 + \omega^2)}{(\omega_2^2 - \omega^2)^2} + \cdots \tag{3-12}$$

Every term is positive for any finite value of ω; hence

$$\frac{dX(\omega)}{d\omega} > 0 \qquad \omega < \infty$$

$$\tag{3-13}$$

$$\frac{dX(\omega)}{d\omega} \to K_\infty \geq 0 \qquad \omega \to \infty$$

It was shown earlier that all zeros and poles of $Z(s)$, and hence of $X(\omega)$, are on the $j\omega$ axis. Assuming two adjacent zeros without a pole between them leads to one of the situations depicted in Fig. 3-1a; clearly, condition (3-13) is violated. Similarly, two poles without an intervening zero results in one of the curves shown in Fig. 3-1b; these too are impossible. The only possible situation is therefore the one illustrated in Fig. 3-1c, where *the zeros and poles of $X(\omega)$ are interlaced*.

In conclusion, it is evident that:

1. $Z(s)$ is the ratio of an even $N(s)$ and an odd $D(s)$ polynomial, or vice versa.
2. The degrees of $N(s)$ and $D(s)$ differ by exactly 1, since $Z(s)$ or $Y(s)$ can have only a simple pole for $s \to \infty$.
3. At $s = 0$ there is either a zero (if $K_0 = 0$) or a pole (if $K_0 > 0$).

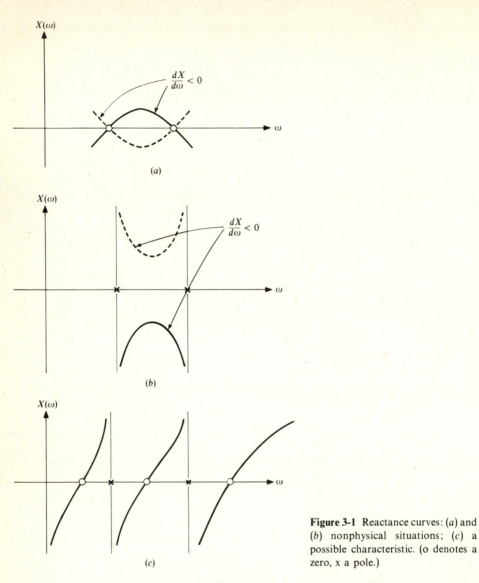

Figure 3-1 Reactance curves: (a) and (b) nonphysical situations; (c) a possible characteristic. (o denotes a zero, x a pole.)

4. At $s \to \infty$ there is a zero (if $K_\infty = 0$) or a pole (if $K_\infty > 0$).
5. $Z(s)$ has only simple poles and zeros; all are located interlaced on the $j\omega$ axis.
6. The residues at all poles are real and positive.

The four possible behavior patterns at zero and infinite frequencies are illustrated in Fig. 3-2 for the four simple impedance functions

$$Z_1(s) = \frac{s}{s^2 + 1} \qquad Z_2(s) = \frac{s(s^2 + 2)}{s^2 + 1}$$

$$Z_3(s) = \frac{s^2 + 1}{s(s^2 + 2)} \qquad Z_4(s) = \frac{s^2 + 1}{s} \qquad (3\text{-}14)$$

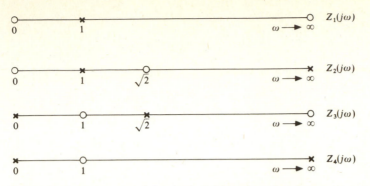

Figure 3-2 Pole-zero patterns for simple LC impedance functions.

Only the zeros and poles are shown; the reactance curves can be readily visualized with Eq. (3-13) in mind.

If one starts out from (3-2), an entirely similar derivation results in finding identical properties for the LC admittance $Y(s)$. Defining $B(\omega)$, the susceptance, via $Y(j\omega) = jB(\omega)$, we find that $B(\omega)$ can also be written in the form given for $X(\omega)$ in (3-11). The constants entering the expressions will, of course, be different for the $B(\omega)$ and $X(\omega)$ functions of the circuit, but the forms of the two functions and hence the behavior of the frequency characteristics will be similar.

3-2 FOSTER REALIZATIONS FOR LC CIRCUITS

The partial-fraction expansion (3-9) immediately suggests a realization for $Z(s)$. Since all constants K_∞, K_0, and K_i are positive, the terms in the expansion can be directly identified with simple one- or two-element impedances, as illustrated in Fig. 3-3. Since the sum of these terms forms $Z(s)$, these impedances must be connected in series to obtain the overall $Z(s)$ impedance; hence, the realization is that shown in Fig. 3-4.

This circuit is called the *first Foster realization* or *Foster 1 realization*, after R. M. Foster, the engineer who introduced it in 1924.

Since, as discussed earlier, the expansion for $Y(s)$ is of the same general form† as for $Z(s)$, that is,

$$Y(s) = K'_\infty s + \frac{K'_0}{s} + \frac{K'_1 s}{s^2 + \omega'^2_1} + \frac{K'_2 s}{s^2 + \omega'^2_2} + \cdots \qquad (3\text{-}15)$$

where all K' are nonnegative and the ω'_i are the poles of $Y(s)$ and hence the zeros of $Z(s)$. Identifying each term with a simple admittance (Fig. 3-5) gives the overall realization of Fig. 3-6. This circuit is called the *second Foster realization* or *Foster 2 realization* of the one-port.

The design process will be illustrated by a simple design problem.

---

† Of course, if $Z(s)$ has a pole for $s \to \infty$, then $Y(s) = 1/Z(s)$ has a zero there, and vice versa. Hence, if $K_\infty > 0$, then $K'_\infty = 0$. Similarly, either $K_0 > 0$ or $K'_0 > 0$, but not both.

$$L_\infty = K_\infty$$

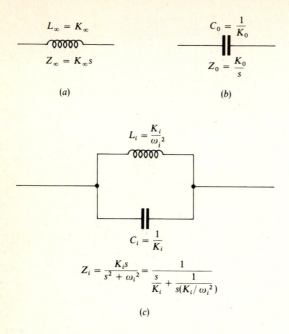

$$Z_\infty = K_\infty s$$

(a)

$$C_0 = \frac{1}{K_0}$$

$$Z_0 = \frac{K_0}{s}$$

(b)

$$L_i = \frac{K_i}{\omega_i^2}$$

$$C_i = \frac{1}{K_i}$$

$$Z_i = \frac{K_i s}{s^2 + \omega_i^2} = \frac{1}{\dfrac{s}{K_i} + \dfrac{1}{s(K_i/\omega_i^2)}}$$

(c)

Figure 3-3 Impedances corresponding to the individual terms in the partial-fraction expansion of an LC impedance.

Example 3-1 Find the two Foster realizations of

$$Z(s) = \frac{2s^3 + 8s}{s^2 + 1}$$

For the *Foster 1 realization* we expand $Z(s)$ into partial fractions. Since the poles of $Z(s)$ are clearly at $s \to \infty$ and $s^2 = -1$,

$$Z(s) = K_\infty s + \frac{K_1 s}{s^2 + 1}$$

Here $\qquad K_\infty = \left[\frac{Z(s)}{s} \right]_{s \to \infty} = 2 \qquad$ and $\qquad K_1 = \left[\frac{s^2 + 1}{s} Z(s) \right]_{s^2 = -1} = 6$

Hence $\qquad Z(s) = 2s + \dfrac{6s}{s^2 + 1} = 2s + \dfrac{1}{s/6 + 1/6s}$

and the circuit is that shown in Fig. 3-7a.

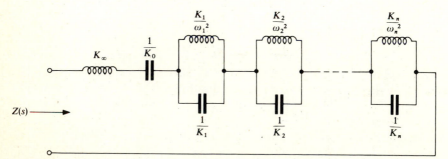

Figure 3-4 Foster 1 realization of an LC impedance.

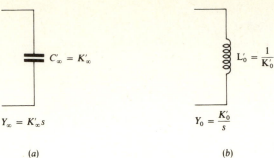

(a)

(b)

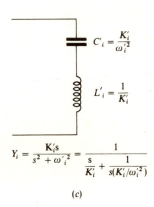

$$Y_i = \frac{K'_i s}{s^2 + {\omega'_i}^2} = \frac{1}{\dfrac{s}{K'_i} + \dfrac{1}{s(K'_i/{\omega'_i}^2)}}$$

(c)

Figure 3-5 Admittances corresponding to the individual terms in the partial-fraction expansion of an LC admittance.

For the *Foster 2 realization*, we expand

$$Y(s) = \frac{1}{Z(s)} = \frac{s^2 + 1}{2s^3 + 8s}$$

The poles are at $s = 0$ and $s^2 = -4$. Hence

$$Y(s) = \frac{K'_0}{s} + \frac{K'_1 s}{s^2 + 4}$$

where

$$K'_0 = [sY(s)]_{s=0} = \tfrac{1}{8}$$

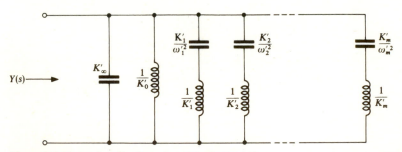

Figure 3-6 Foster 2 realization of an LC admittance.

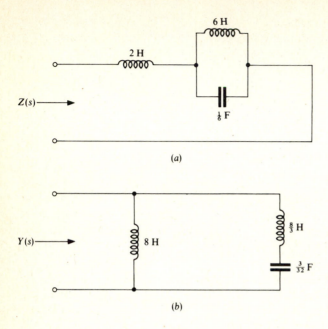

(a)

(b)

Figure 3-7 (a) Foster 1 and (b) Foster 2 realizations of $Z(s) = (2s^3 + 8s)/(s^2 + 1)$.

and

$$K_1' = \left[\frac{s^2 + 4}{s} Y(s)\right]_{s^2 = -4} = \frac{3}{8}$$

Hence

$$Y(s) = \frac{1}{8s} + \frac{\frac{3}{8}s}{s^2 + 4} = \frac{1}{8s} + \frac{1}{\frac{8}{3}s + 1/\frac{3}{32}s}$$

and the circuit is thus that shown in Fig. 3-7b.

A simple and useful test of our results can be performed by comparing the asymptotic behavior of the synthesized circuits with the asymptotic values of $Z(s)$ for $s = 0$ and $s \to \infty$. From $Z(s) = (2s^3 + 8s)/(s^2 + 1)$, we have $Z(s) \to 8s$ for $s \to 0$ and $Z(s) \to 2s^3/s^2 = 2s$ for $s \to \infty$. From Fig. 3-7a, as $s \to 0$, the admittance of the capacitor becomes negligible compared with that of the 6-H inductor; hence $Z(s) \to 2s + 6s = 8s$. For $s \to \infty$, on the other hand, the capacitor shorts out the 6-H inductor, and hence $Z(s) \to 2s$. Similarly, for $s \to 0$ the $\frac{3}{32}$-F capacitor opens the right-hand branch of the circuit of Fig. 3-7b, and hence the impedance becomes $8s$; at $s \to \infty$ the capacitor behaves like a short circuit and hence $Z(s) \to \{[(8)(\frac{8}{3})]/(8 + \frac{8}{3})\}s = 2s$. All values agree, and hence the check indicates no error.

With a little practice, the student will be able to perform this test by inspection. It is thus useful for the quick checking of even moderately large circuits.

Next, we shall show that the Foster realizations give the most economical circuits possible for a given $Z(s)$; such circuits are often called *canonical*. We recall that in the Laplace-transform interpretation of s each factor s^k is introduced by replacing a kth derivative $(d/dt)^k$ in the time-domain circuit equations. Thus if the highest power of s in $Z(s)$ is s^N, this indicates that after eliminating all integrals in the original integrodifferential equations an Nth-order differential equation results. N is often called the *degree* of the immittance.

Since each L and C performs *one* differentiation $[v_L(t) = L\, di_L(t)/dt,\ i_c(t) = C\, dv_c(t)/dt]$, the circuit must contain at least N reactive (L and C) elements. In conclusion, an Nth-degree LC immittance requires N elements.

Combining the terms in (3-9), we obtain

$$Z(s) = \frac{K_\infty s^{2n+2} + \cdots + K_0 \prod_{i=1}^{n} \omega_i^2}{s(s^2 + \omega_1^2)(s^2 + \omega_2^2) + \cdots + (s^2 + \omega_n^2)}$$

$$= \frac{a_N s^N + a_{N-2} s^{N-2} + \cdots + a_0}{b_{N-1} s^{N-1} + b_{N-3} s^{N-3} + \cdots + b_1 s} \tag{3-16}$$

where we assumed for convenience that K_∞ and K_0 are both nonzero. Then, $N = 2n + 2$. A look at Fig. 3-4 reveals that the number of elements there is the same as N, that is, the smallest possible. [If K_∞ and/or K_0 is zero, the degree of $Z(s)$ and the number of elements decrease by equal amounts.] This proves the canonical nature of the Foster 1 realization.

Next we shall give an alternative proof of this canonical property, which will also offer some insight into the calculation process. Suppose that $Z(s)$ is given and we start the design by the realization of $K_\infty s$. This requires one element (Fig. 3-8a). At the same time, the order of the *remainder impedance* $Z'(s) \triangleq$

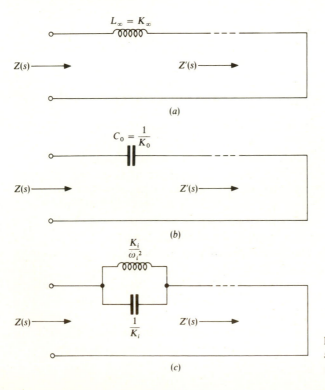

Figure **3-8** Pole removal at (a) $s \to \infty$; (b) $s = 0$; and (c) $s = \pm j\omega_i$.

$Z(s) - K_\infty s$ is reduced by 1. This can easily be seen. From (3-16), $K_\infty = [Z(s)/s]_{s\to\infty} = a_N/b_{N-1}$, and hence

$$Z'(s) = Z(s) - \frac{a_N}{b_{N-1}} s$$

$$= \frac{(a_{N-2} - b_{N-3} a_N/b_{N-1})s^{N-2} + \cdots}{b_{N-1} s^{N-1} + b_{N-3} s^{N-3} + \cdots} \tag{3-17}$$

Hence, the order of $Z'(s)$ is $N - 1$, that is, the realization of $K_\infty s$ reduced the degree by 1.

An analog calculation shows that as the term K_0/s is realized, in the remainder impedance $Z'(s)$ the degrees of the numerator and denominator are both reduced by 1. Of course, the realization of such a term costs again one element (Fig. 3-8b). The proof is left as an exercise.

Finally, let

$$Z(s) = \frac{N(s)}{D(s)} = \frac{N(s)}{(s^2 + \omega_i^2)D_1(s)} \tag{3-18}$$

Then, realization of a term $K_i s/(s^2 + \omega_i^2)$ in the partial-fraction expansion of $Z(s)$ involves the calculation of the residue

$$K_i = \left[\frac{s^2 + \omega_i^2}{s} Z(s)\right]_{s^2 = -\omega_i^2} = \left[\frac{N(s)}{sD_1(s)}\right]_{s^2 = -\omega_i^2} \tag{3-19}$$

and of the remainder impedance

$$Z'(s) = Z(s) - \frac{K_i s}{s^2 + \omega_i^2} = \frac{N(s)}{(s^2 + \omega_i^2)D_1(s)} - \frac{K_i s}{s^2 + \omega_i^2}$$

$$= \frac{N(s) - K_i s D_1(s)}{(s^2 + \omega_i^2)D_1(s)} \tag{3-20}$$

For $s^2 = -\omega_i^2$, the numerator by (3-19) becomes equal to zero. This shows that both the new numerator $N(s) - K_i s D_1(s)$ and the denominator have a factor $s^2 + \omega_i^2$. Cancellation of this factor reduces the degrees of both polynomials by 2, and hence also the order of $Z(s)$ by 2. Thus, the realization of the term $K_i s/(s^2 + \omega_i^2)$ which is at the cost of *two* elements (Fig. 3-8c), reduces the order by 2.

It is now clear that the Foster 1 realization of an Nth-order impedance requires N elements, which is the minimum possible number. This again proves the canonical nature of the Foster 1 expansion. For the Foster 2 expansion an exactly dual proof gives the same result.

It is readily seen that in the remainder impedance $Z'(s)$ of (3-17) the pole which $Z(s)$ had at $s \to \infty$ is no longer present. Similarly, $Z'(s) = Z(s) - K_0/s$ no longer has the pole at $s = 0$; and the $Z'(s)$ of (3-20) no longer has the poles which $Z(s)$ had at $s = j\omega_i$ and $-j\omega_i$ after the common factor $s^2 + \omega_i^2$ is canceled. For this reason, the described operations are often called *full pole removals* at $s \to \infty$, $s = 0$, and $s = \pm j\omega_i$, respectively.

Example 3-2 Carry out a full pole removal of the poles at $s = \pm j3$ from the function

$$Z(s) = \frac{s^3 + 2s}{(s^2 + 9)(s^2 + 1)}$$

The residue is

$$K_1 = \left[\frac{s^2 + 9}{s} Z(s) \right]_{s^2 = -9} = \left(\frac{s^2 + 2}{s^2 + 1} \right)_{s^2 = -9} = \frac{7}{8}$$

and hence the remainder is given by

$$Z'(s) = Z(s) - \frac{K_1 s}{s^2 + 9} = \frac{(s^3 + 2s) - \frac{7}{8}s(s^2 + 1)}{(s^2 + 9)(s^2 + 1)}$$

$$= \frac{\frac{1}{8}s^3 + \frac{9}{8}s}{(s^2 + 9)(s^2 + 1)} = \frac{(s/8)(s^2 + 9)}{(s^2 + 1)(s^2 + 9)} = \frac{\frac{1}{8}s}{s^2 + 1}$$

where the cancellation of the common factor $s^2 + 9$ can be observed.

Clearly, the Foster expansion can be regarded as a sequential removal of all poles of the function $Z(s)$ or $Y(s)$.

Finally, we point out that both Foster expansions give element values of the forms K_i, $1/K_i$, or K_i/ω_i^2. Since all $K_i \geq 0$, these element values are always given by nonnegative real numbers. Thus, *both Foster expansions are always physically realizable from any odd PR function $Z(s)$ or $Y(s)$*. A circuit form which has the property of always giving physical-element values from a realizable circuit function is also often called canonical; to avoid confusion, we shall call such circuits *feasible realizations*.

It should be noted that feasibility is a theoretical concept. The realization of, say, a capacitor of 10 F, or 10^{-15} F, while theoretically feasible, is quite impractical. On the other hand, a -100-pF capacitor can be built if active elements (excluded in this chapter) can be employed.

3-3 LADDER REALIZATIONS AND CAUER FORMS FOR *LC* IMMITTANCES

As shown above, each pole removal, i.e., each realization of a term in the partial-fraction expansion, reduces the degree of the function $Z(s)$. If the removed pole occurs at $s = 0$ or $s \to \infty$, the reduction equals 1; if it is at some $s = \pm j\omega_i$, $0 < \omega_i < \infty$, the reduction is by 2. Naturally, as can be seen from (3-8), the remainder impedance $Z'(s)$ still has the basic realizability property of $Z(s)$: it can still be written in the partial-fraction form (3-8) with all positive residues, only one of the terms will be missing.

We proceed with the realization of $Z'(s)$ by continuing to remove its remaining poles: this will give the Foster 1 realization. We may decide instead, however, to find $Y'(s) \triangleq 1/Z'(s)$ and perform a pole removal on that. This amounts to the realization of a term in the partial-fraction expansion of $Y'(s)$, that is, to a step in a Foster 2 type realization. This leaves a remainder $Y''(s)$, which is still realizable. If

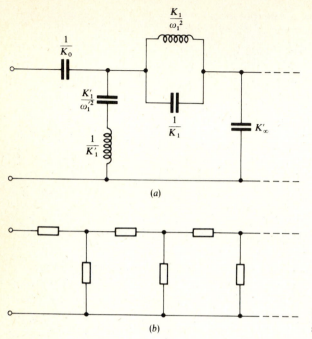

(a)

(b)

Figure 3-9 Ladder circuit: (a) special and (b) general form.

now we again interrupt the Foster 2 process, calculate $Z''(s) \triangleq 1/Y''(s)$, and perform a Foster 1 step on it, etc., the circuit of Fig. 3-9a results. Clearly, the pole removals occurred as $s = 0$, $s = \pm j\omega_1'$, $s = \pm j\omega_1$, $s \to \infty$, etc. The circuit thus obtained is a special case of the general configuration of Fig. 3-9b, called (for obvious reasons) a *ladder circuit*. The ladder may start and end with either a series or a parallel branch. It provides an alternative realization for an *LC* one-port. It can be derived, as shown above, by a sequence of alternating pole removals from $Z(s)$ and $Y(s)$, and by its construction it is a canonical and always realizable (feasible) circuit.

Two special forms of the ladder, called *Cauer realizations*, are of special interest. The *first Cauer realization* (or Cauer 1 realization) is obtained by removing poles exclusively at $s \to \infty$ in the course of designing the ladder. Thus, assuming that $Z(s)$ has a pole at infinity, the sequence is

$$Z_1(s) \triangleq Z(s) = K_\infty s + Z_2(s) = K_\infty s + \frac{1}{Y_2(s)} \tag{3-21}$$

Here, due to the pole removal, $Z_2(s)$ has a zero at $s \to \infty$, and hence $Y_2(s)$ has a pole there. Therefore we can repeat the process

$$Y_2(s) = K'_{\infty 2} s + Y_3(s) = K'_{\infty 2} s + \frac{1}{Z_3(s)} \tag{3-22}$$

$$Z_3(s) = K_{\infty 3} s + Z_4(s) = K_{\infty 3} s + \frac{1}{Y_4(s)} \tag{3-23}$$

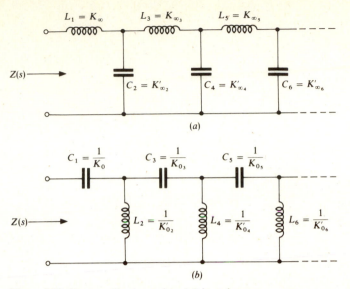

Figure 3-10 Cauer realizations of an LC impedance.

The circuit thus obtained is shown in Fig. 3-10a. By construction, all $K_{\infty i} > 0$ and $K'_{\infty i} > 0$ if $Z(s)$ is a PR odd function of s. Hence, the circuit is *feasible*. Since each pole removal requires one element and reduces the degree of $Z(s)$ by 1, it is also *canonical*.

If $Z(s) \to 0$ for $s \to \infty$, the process starts with the step

$$Y_1(s) \triangleq Y(s) = K'_{\infty 1} s + Y_2(s) = K'_{\infty 1} s + \frac{1}{Z_2(s)} \tag{3-24}$$

The circuit thus obtained is similar to that shown in Fig. 3-10a, except that the first element is a shunt capacitor rather than a series inductor.

When Eqs. (3-21) to (3-23) are combined, it is clear that

$$Z(s) = K_\infty s + \cfrac{1}{K'_{\infty 2} s + \cfrac{1}{K_{\infty 3} s + \cfrac{1}{K'_{\infty 4} s + \cdots}}} \tag{3-25}$$

The form is called the *continued-fraction expansion* of $Z(s)$ around infinity. If $Z(s) \to 0$ for $s \to \infty$, the first term $K_\infty s$ will be absent.

Example 3-3 Find the Cauer 1 realization of

$$Z(s) = \frac{2s^3 + 8s}{s^2 + 1}$$

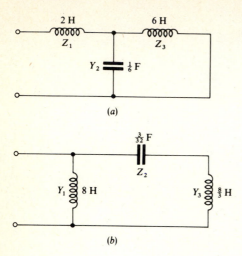

(a)

(b)

Figure 3-11 (a) Cauer 1 and (b) Cauer 2 expansions of $Z(s) = (2s^3 + 8s)/(s^2 + 1)$.

In the process, (3-21) to (3-23) are followed. The residues K_∞, $K'_{\infty 2}$, ... are simply the ratios of the highest-order polynomial coefficients

$$Z(s) = 2s + Z_2(s) \qquad Z_2(s) = \frac{2s^3 + 8s}{s^2 + 1} - 2s = \frac{6s}{s^2 + 1}$$

$$Y_2(s) = \frac{s^2 + 1}{6s} = \frac{s}{6} + Y_3(s) \qquad Y_3(s) = \frac{s^2 + 1}{6s} - \frac{s}{6} = \frac{1}{6s}$$

Hence
$$Z(s) = 2s + \frac{1}{s/6 + 1/6s}$$

The circuit is therefore that given in Fig. 3-11a.†

Example 3-4 Find the Cauer 1 realization of

$$Z(s) = \frac{s^3 + 4s}{2s^4 + 20s^2 + 18}$$

Proceeding as in Example 3-3 but (since $Z(s) \to 0$ as $s \to \infty$) starting from $Y(s)$ gives

$$Y(s) = 2s + Y_2(s) \qquad Y_2(s) = \frac{2s^4 + 20s^2 + 18}{s^3 + 4s} - 2s = \frac{12s^2 + 18}{s^3 + 4s}$$

$$Z_2(s) = \frac{s}{12} + Z_3(s) \qquad Z_3(s) = \frac{s^3 + 4s}{12s^2 + 18} - \frac{s}{12} = \frac{2.5s}{12s^2 + 18} \qquad Y_3(s) = 4.8s + Y_4(s)$$

$$Y_4(s) = \frac{12s^2 + 18}{2.5s} - 4.8s = \frac{18}{2.5s} = \frac{1}{\frac{5}{36}s}$$

† For degrees $N \leq 3$, the Foster and Cauer expansions lead to the same circuits. Hence, this circuit is identical to that of Fig. 3-7a.

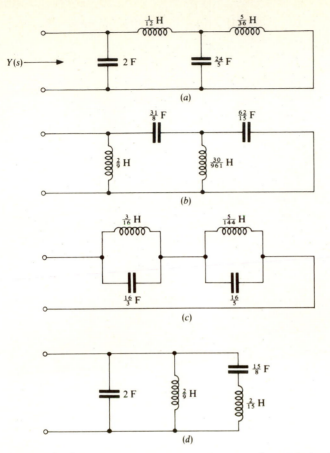

Figure 3-12 Cauer and Foster expansions of $Z(s) = (s^3 + 4s)/(2s^4 + 20s^2 + 18)$.

Hence

$$Z(s) = \cfrac{1}{2s + \cfrac{1}{\cfrac{s}{12} + \cfrac{1}{\cfrac{24}{5}s + \cfrac{1}{\cfrac{5}{36}s}}}}$$

and the circuit is as shown in Fig. 3-12a.

The *second Cauer (Cauer 2) realization* of an immittance function consists of removing poles at $s = 0$ alternatively from the $Z(s)$ and $Y(s)$ functions. The result is the continued-fraction expansion around zero

$$Z(s) = \cfrac{K_0}{s} + \cfrac{1}{\cfrac{K'_{0_2}}{s} + \cfrac{1}{\cfrac{K_{0_3}}{s} + \cfrac{1}{\cfrac{K'_{0_4}}{s} + \cdots}}} \qquad (3\text{-}26)$$

This results in the circuit of Fig. 3-10b. If $Z(0) = 0$, the first element is a shunt inductor rather than a series capacitor.

Example 3-5 Find the Cauer 2 expansion of

$$Z(s) = \frac{2s^3 + 8s}{s^2 + 1}$$

Since $Z(0) = 0$, the pole-removal process must start with $Y(s)$. Here the residues are obtained as ratios of the *lowest-order* coefficients

$$Y(s) = \frac{s^2 + 1}{2s^3 + 8s} = \frac{1}{8s} + Y_2(s)$$

$$Y_2(s) = \frac{s^2 + 1}{2s^3 + 8s} - \frac{1}{8s} = \frac{\frac{3}{4}s^2}{2s^3 + 8s} = \frac{\frac{3}{4}s}{2s^2 + 8}$$

$$Z_2(s) = \frac{2s^2 + 8}{\frac{3}{4}s} = \frac{32}{3s} + Z_3(s)$$

$$Z_3(s) = \frac{2s^2 + 8}{\frac{3}{4}s} - \frac{32}{3s} = \frac{2s^2}{\frac{3}{4}s} = \frac{8s}{3}$$

Hence

$$Z(s) = \cfrac{1}{\cfrac{1}{8s} + \cfrac{1}{\cfrac{1}{\frac{3}{32}s} + \cfrac{1}{\frac{8}{3}s}}}$$

and the circuit is as shown in Fig. 3-11b. For this low degree, it coincides with the Foster 2 realization of the circuit (Fig. 3-7b). For $N > 3$, the circuits obtained by Cauer and Foster realizations are different.

For larger circuits with realistic (noninteger) coefficients, the partial- and continued-fraction expansions required by the Foster and Cauer realizations are tedious and prone to both human and truncation errors. These calculations should therefore be carried out on a digital computer. If hand calculation is necessary, an efficient organization of the calculations can save some of the writing effort. One possible scheme is illustrated in the following example.

Example 3-6 The Cauer expansions of the impedance

$$Z(s) = \frac{2s^3 + 8s}{s^2 + 1}$$

have been obtained earlier. These calculations are now repeated using a more efficient scheme, based on long division, which it is hoped, is self-explanatory.

Cauer 1 expansion

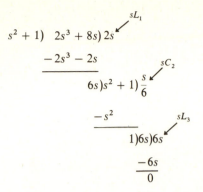

Here, L_1, C_2, L_3 represent the elements in branches 1, 2, and 3, respectively (Fig. 3-11a).

Cauer 2 expansion

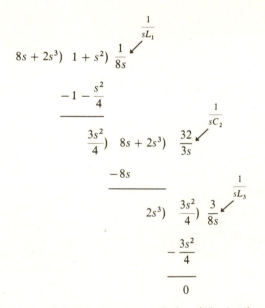

The circuits obtained are of course the same as before (Fig. 3-11).

Note that within all polynomials the powers of s are arranged in *decreasing* order for the Cauer 1 expansion and in *increasing* order for the Cauer 2 expansion. The polynomials themselves are arranged according to the scheme:

Divisor) dividend) ratio of leading terms = immittance

Example 3-7 Find the Cauer realizations of

$$Z(s) = \frac{s^4 + 10s^2 + 9}{s^3 + 2s}$$

Cauer 1 expansion

$$s^3 + 2s)\ \ s^4 + 10s^2 + 9)s \quad \overset{sL_1}{\swarrow}$$

$$\underline{-s^4 - 2s^2}$$

$$8s^2 + 9)\ \ s^3 + 2s)\ \frac{s}{8} \quad \overset{sC_2}{\swarrow}$$

$$\underline{-s^3 - \frac{9s}{8}}$$

$$\frac{7s}{8})\ \ 8s^2 + 9)\ \tfrac{64}{7}s \quad \overset{sL_3}{\swarrow}$$

$$\underline{-8s^2}$$

$$9)\ \frac{7s}{8})\ \tfrac{7}{72}s \quad \overset{sC_4}{\swarrow}$$

$$\underline{-\frac{7s}{8}}$$

$$0$$

Cauer 2 expansion

$$2s + s^3)\ 9 + 10s^2 + s^4)\frac{9}{2s} \quad \overset{\frac{1}{sC_1}}{\swarrow}$$

$$\underline{-9 - \tfrac{9}{2}s^2}$$

$$\tfrac{11}{2}s^2 + s^4)\ 2s + s^3)\frac{4}{11s} \quad \overset{\frac{1}{sL_2}}{\swarrow}$$

$$\underline{-2s - \tfrac{4}{11}s^3}$$

$$\tfrac{7}{11}s^3)\tfrac{11}{2}s^2 + s^4)\frac{121}{14s} \quad \overset{\frac{1}{sC_3}}{\swarrow}$$

$$\underline{-\tfrac{11}{2}s^2}$$

$$s^4)\ \tfrac{7}{11}s^3)\frac{7}{11s} \quad \overset{\frac{1}{sL_4}}{\swarrow}$$

$$\underline{-\tfrac{7}{11}s^3}$$

$$0$$

The circuits are shown in Fig. 3-13.

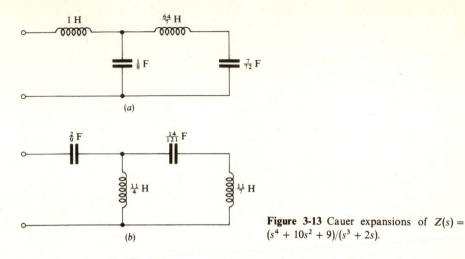

Figure 3-13 Cauer expansions of $Z(s) = (s^4 + 10s^2 + 9)/(s^3 + 2s)$.

The $s = 0$ and $s \to \infty$ asymptotic behavior can (and should) be checked for the Cauer circuits as well as for the Foster ones. For $s \to 0$, $Z(s) \to 9/2s$; for $s \to \infty$, $Z(s) \to s$. From Fig. 3-13a, for $s = 0$ the circuit degenerates into two parallel-connected capacitors: $Z(s) \approx 1/(\frac{1}{8} + \frac{7}{72})s = 1/\frac{2}{9}s$. For $s \to \infty$, the capacitors behave like short circuits, and hence $Z(s) \to s$. Thus the Cauer 1 expansion satisfies this simple test.

In the Cauer 2 circuit, for $s \to 0$ the inductors become short circuits, and hence $Z(s) \to 1/sC_1 = 1/\frac{2}{9}s$. For $s \to \infty$, the capacitors behave like short circuits and hence

$$Z(s) \to \frac{(\frac{11}{4})(\frac{11}{7})}{\frac{11}{4} + \frac{11}{7}} s = s$$

Thus, the Cauer 2 circuit passes the test as well.

3-4 THE PROPERTIES OF RC IMMITTANCES

We now consider one-ports built exclusively from *resistors and capacitors*. Since there are no inductors in the circuit, the energy functions M_0 and T_0, defined in Sec. 2-3, are both zero and hence, by Eqs. (2-59) and (2-60),

$$Z(s) = \frac{1}{|I_1(s)|^2} \left[F_0(s) + \frac{1}{s} V_0(s) \right] \tag{3-27}$$

and

$$Y(s) = \frac{1}{|V_1(s)|^2} \left[F_0(s) + \frac{1}{s*} V_0(s) \right] \tag{3-28}$$

As in Sec. 3-1 for LC immittances, we shall establish the properties of RC immittances from these relations. At a zero s_z of $Z(s)$, for $|I_1| < \infty$,

$$F_0(s_z) + \frac{V_0(s_z)}{s_z} = 0 \qquad s_z = -\frac{V_0(s_z)}{F_0(s_z)} \tag{3-29}$$

Since F_0 and V_0 are real nonnegative numbers for any s, s_z *is a real nonpositive quantity*. A similar argument can be based on (3-28) to show that the zeros of $Y(s)$, that is, the poles of $Z(s)$, also have real nonpositive values. Hence, *all zeros and poles of $Z(s)$ lie on the negative σ axis.*

At some $s = \sigma + j\omega$, the real part R and the imaginary part X of $Z(s)$ are found from (3-27):

$$Z(\sigma + j\omega) = R + jX = \frac{F_0 + V_0/(\sigma + j\omega)}{|I_1|^2} \tag{3-30}$$

and hence
$$R = \frac{F_0 + \sigma V_0/(\sigma^2 + \omega^2)}{|I_1|^2} \qquad X = -\frac{\omega V_0/|I_1|^2}{\sigma^2 + \omega^2} \tag{3-31}$$

Next, Eq. (3-31) will be used to derive additional conditions on the poles of $Z(s)$. From (3-31), the sign of $X(\sigma + j\omega)$ is the opposite of that of ω. Consider now the behavior of $Z(s)$ for some s near one of its nth-order negative σ-axis poles (Fig. 3-14). In the Laurent-series expansion, for $s \rightarrow p_1$,

$$Z(s) \rightarrow \frac{k_{-n}}{(s - p_1)^n} \tag{3-32}$$

[compare Eq. (2-63) and the derivation which followed it], and hence in polar coordinates

$$k_{-n} = ke^{j\varphi} \qquad k \triangleq |k_{-n}| > 0 \qquad s - p_1 = re^{j\theta} \qquad r \triangleq |s - p_1| > 0 \tag{3-33}$$

we have

$$Z(s) \approx \frac{k}{r^n} e^{j(\varphi - n\theta)} \tag{3-34}$$

and
$$X(s) = \frac{k}{r^n} \sin (\varphi - n\theta) \tag{3-35}$$

From Fig. 3-14, $\omega = r \sin \theta$. The condition that $X(s)$ and ω have opposite signs for all θ values can be met only if $n = 1$ and $\varphi = 0$. Then $X(s) = -kr^{-n-1}\omega = -(k/r^2)\omega$. Here $n = 1$ means that *the pole must be simple*; $\varphi = 0$, by (3-33), implies *a real positive residue.*

The argument clearly holds also for a pole at $s = 0$.

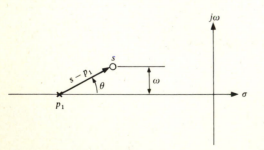

Figure 3-14 The behavior of an RC impedance near one of its poles.

Next, it will be shown that $Z(s)$ cannot have a pole for $s \to \infty$. If such pole existed, it would be simple and the residue positive, since $Z(s)$ is PR. Then for a sufficiently large s

$$Z(s) \approx K_\infty s \tag{3-36}$$

and hence, with $s = \sigma + j\omega$,

$$X = K_\infty \omega \qquad K_\infty > 0 \tag{3-37}$$

which contradicts (3-31).

Hence, *all poles of $Z(s)$ lie on the finite negative σ axis. All poles are simple and all residues real and positive.*

A dual derivation can be performed for the finite poles of $Y(s)$ from (3-28). Since, however, $s^* = \sigma - j\omega$ rather than $s = \sigma + j\omega$ enters that equation, the susceptance $B(s)$ turns out to have the *same* sign as ω (rather than the opposite sign, as was the case for X). Hence, *while the finite poles of $Y(s)$ also turn out to be simple and the residues real, the sign of these residues is negative.*

$Y(s)$ can also have a pole at $s \to \infty$; since it is PR, the pole must be simple and the residue there positive. However, $Y(s)$ *cannot have a pole at $s = 0$*; otherwise its asymptotic form

$$Y(s) \approx \frac{K_0}{s} \qquad K_0 > 0 \tag{3-38}$$

would lead to $B(\sigma + j\omega) \approx -K_0 \omega/(\sigma^2 + \omega^2)$, which contradicts the identical signs of ω and $B(s)$ stated above.

From the described properties of $Z(s)$ it is easy to see that its partial-fraction expansion must be in the form

$$Z(s) = K_\infty + \frac{K_0}{s} + \frac{K_1}{s + \sigma_1} + \frac{K_2}{s + \sigma_2} + \cdots + \frac{K_n}{s + \sigma_n} \tag{3-39}$$

where $K_0, K_\infty \geq 0$ and $K_i, \sigma_i > 0$, $i = 1, 2, \ldots, n$. Also, from the properties of $Y(s)$ we have

$$Y(s) = K'_\infty s + K'_0 + \frac{K'_1}{s + \sigma'_1} + \frac{K'_2}{s + \sigma'_2} + \cdots + \frac{K'_{n'}}{s + \sigma'_{n'}} \tag{3-40}$$

where $K'_\infty \geq 0$, $K'_i < 0$, $\sigma'_i > 0$, $i = 1, 2, \ldots, n'$. For $s = 0$, all capacitors in the RC one-port become open-circuited, and the circuit becomes a one-port composed of positive resistors only. Hence, from (3-40),

$$Y(0) = K'_0 + \frac{K'_1}{\sigma'_1} + \frac{K'_2}{\sigma'_2} + \cdots + \frac{K'_{n'}}{\sigma'_{n'}} \geq 0 \tag{3-41}$$

where all $K'_i/\sigma'_i < 0$, and therefore

$$K'_0 \geq -\sum_{i=1}^{n'} \frac{K'_i}{\sigma'_i} = \sum_{i=1}^{n'} \left| \frac{K'_i}{\sigma'_i} \right| > 0 \tag{3-42}$$

Because of the mixed signs of its residues, the expansion of $Y(s)$ is unsuitable as a basis for synthesis. Let us instead expand $Y(s)/s$. Since

$$\frac{Y(s)}{s} = K'_\infty + \frac{K'_0}{s} + \sum_{i=1}^{n'} \frac{K'_i}{s(s + \sigma'_i)} \tag{3-43}$$

where

$$\frac{K'_i}{s(s + \sigma'_i)} = \frac{K'_i/\sigma'_i}{s} + \frac{-K'_i/\sigma'_i}{s + \sigma'_i} \tag{3-44}$$

introducing the new constants

$$k'_0 \triangleq K'_0 + \sum_{i=1}^{n'} \frac{K'_i}{\sigma'_i} \tag{3-45}$$

$$k'_\infty \triangleq K'_\infty \tag{3-46}$$

$$k'_i \triangleq -\frac{K'_i}{\sigma'_i} \qquad i = 1, 2, \ldots, n' \tag{3-47}$$

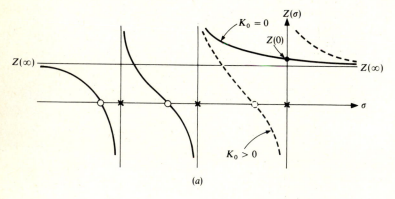

(a)

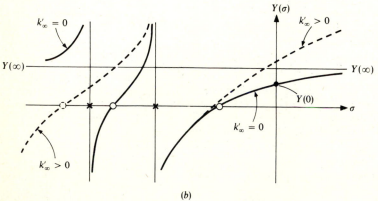

(b)

Figure 3-15 The behavior of $Z(s)$ and $Y(s)$ along the σ axis.

we obtain the expansions

$$\frac{Y(s)}{s} = k'_\infty + \frac{k'_0}{s} + \sum_{i=1}^{n'} \frac{k'_i}{s + \sigma'_i}$$

$$Y(s) = k'_\infty s + k'_0 + \sum_{i=1}^{n'} \frac{k'_i s}{s + \sigma'_i} \tag{3-48}$$

Here, all constants are nonnegative, k'_0 due to (3-42) and the k'_i due to the condition $K'_i/\sigma'_i < 0$.

Along the real (σ) axis, from (3-39),

$$\frac{dZ(\sigma)}{d\sigma} = -\frac{K_0}{\sigma^2} - \sum_{i=1}^{n} \frac{K_i}{(\sigma + \sigma_i)^2} < 0 \tag{3-49}$$

and in the absence of a pole at zero, i.e., for $K_0 = 0$,

$$Z(0) = K_\infty + \sum_{i=1}^{n} \frac{K_i}{\sigma_i} > Z(\infty) = K_\infty \tag{3-50}$$

Hence, the behavior of $Z(\sigma)$ is as shown in Fig. 3-15a. On the other hand, from (3-48),

$$\frac{dY(\sigma)}{d\sigma} = k'_\infty + \sum_{i=1}^{n'} \frac{k'_i \sigma'_i}{(\sigma + \sigma'_i)^2} > 0 \tag{3-51}$$

Table 3-1

RC impedances	RC admittances
1. All poles and zeros are simple and lie on the $-\sigma$ (real negative) axis of the s plane	1. All poles and zeros are simple and lie on the $-\sigma$ (real negative) axis of the s plane
2. Poles and zeros alternate on the $-\sigma$ axis	2. Poles and zeros alternate on the $-\sigma$ axis
3. The first critical frequency, i.e. the smallest in absolute value, on the $-\sigma$ axis is a pole; it may be at $s = 0$	3. The first critical frequency on the $-\sigma$ axis is a zero; it may be at $s = 0$
4. The last critical frequency is a zero, which may be at infinity	4. The last critical frequency is a pole, which may be at infinity
5. The residues at the poles of $Z(s)$ are real and positive	5. The residues at the poles of $Y(s)$ are real and negative; the residues of $Y(s)/s$ are real and positive

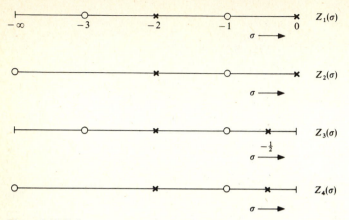

Figure 3-16 Pole-zero patterns for RC impedances.

and if there is no pole at $s \to \infty$,

$$Y(\infty) = k'_0 + \sum_{i=1}^{n'} k'_i > Y(0) = k'_0 \tag{3-52}$$

The behavior of $Y(\sigma)$ is indicated in Fig. 3-15b.

It is clear from the responses that the zeros and poles are interlaced along the negative σ axis.

In conclusion, the properties of RC immittances can be summarized as shown in Table 3-1.

The four possible behavior patterns at zero and infinite values of s, illustrated in Fig. 3-16, correspond to the functions

$$Z_1(s) = \frac{(s+1)(s+3)}{s(s+2)} = \frac{s^2 + 4s + 3}{s^2 + 2s}$$

$$Z_2(s) = \frac{s+1}{s(s+2)} = \frac{s+1}{s^2 + 2s}$$

$$Z_3(s) = \frac{(s+1)(s+3)}{(s+0.5)(s+2)} = \frac{s^2 + 4s + 3}{s^2 + 2.5s + 1}$$

$$Z_4(s) = \frac{s+1}{(s+0.5)(s+2)} = \frac{s+1}{s^2 + 2.5s + 1}$$

3-5 THE SYNTHESIS OF RC IMMITTANCES

The design theory of RC immittances parallels that of LC immittances. The Foster 1 expansion follows directly from (3-39) if we identify each term with a simple impedance (Fig. 3-17). The resulting circuit is shown in Fig. 3-18. The Foster 2

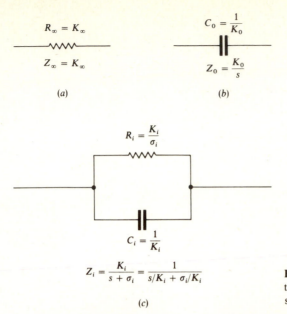

Figure 3-17 Impedances corresponding to the terms in the partial-fraction expansion of an RC impedance.

expansion is immediately obtained from (3-48), using the identifications shown in Fig. 3-19. The circuit itself is shown in Fig. 3-20. The calculations will be illustrated with a simple example.

Example 3-8 Find the Foster realizations of

$$Z(s) = \frac{2(s + 1)(s + 3)}{s(s + 2)} = 2\frac{s^2 + 4s + 3}{s^2 + 2s}$$

The poles of $Z(s)$ are at $s = 0$ and $s = -2$; hence

$$Z(s) = K_\infty + \frac{K_0}{s} + \frac{K_1}{s + 2}$$

where

$$K_\infty = [Z(s)]_{s \to \infty} = 2 \qquad K_0 = [sZ(s)]_{s \to 0} = 3 \qquad K_1 = [(s + 2)Z(s)]_{s \to -2} = 1$$

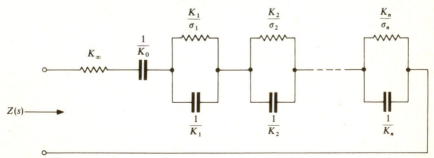

Figure 3-18 Foster 1 realization of an RC impedance.

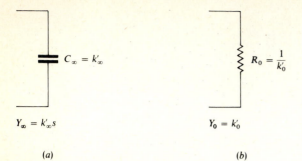

$Y_\infty = k'_\infty s$

(a)

$Y_0 = k'_0$

(b)

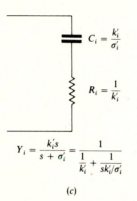

$$Y_i = \frac{k'_i s}{s + \sigma'_i} = \frac{1}{\dfrac{1}{k'_i} + \dfrac{1}{s k'_i / \sigma'_i}}$$

(c)

Figure 3-19 Admittances corresponding to the terms in the partial-fraction expansion of an *RC* admittance.

Hence

$$Z(s) = 2 + \frac{3}{s} + \frac{1}{s + 2}$$

and the circuit is that shown in Fig. 3-21a. An analog expansion of $Y(s)$ gives

$$Y(s) = \frac{s(s + 2)}{2(s + 1)(s + 3)} = \frac{1}{2} + \frac{-\frac{1}{4}}{s + 1} + \frac{-\frac{3}{4}}{s + 3}$$

which is useless for realization. Instead we should expand $Y(s)/s$ as in Eq. (3-48):

$$\frac{Y(s)}{s} = \frac{s + 2}{2(s + 1)(s + 3)} = \frac{\frac{1}{4}}{s + 1} + \frac{\frac{1}{4}}{s + 3}$$

Figure 3-20 Foster 2 expansion of an *RC* admittance.

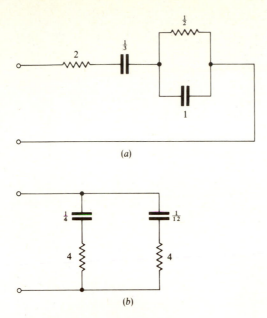

(a)

(b)

Figure 3-21 Foster realizations of $Z(s) = 2(s^2 + 4s + 3)/(s^2 + 2s)$.

Hence

$$Y(s) = \frac{\frac{1}{4}s}{s + 1} + \frac{\frac{1}{4}s}{s + 3}$$

which gives the realization of Fig. 3-21b.

From the impedance function, we see that $Z(s) \to 3/s$ for $s \to 0$ and $Z(s) \to 2$ for $s \to \infty$. In the Foster 1 realization the $C_0 = \frac{1}{3}$ capacitor dominates for $s \to 0$ and the $R_\infty = 2\text{-}\Omega$ resistor for $s \to \infty$. This circuit therefore passes the test at the extreme frequencies. The Foster 2 circuit degenerates into two parallel capacitors as $s \to 0$; hence $Z(s) \to 1/(\frac{1}{4} + \frac{1}{12})s = 1/\frac{1}{3}s$. For $s \to \infty$, the capacitors become short circuits, and $Z(s) = [(4) \times (4)]/(4 + 4) = 2$. Thus, this circuit also checks out.

Since K_0, K_∞, K_i, and σ_i are all nonnegative, it is clear that the Foster 1 realization of $Z(s)$ is *feasible*. To explore whether or not it is also *canonical*, we note that $Z(s)$, as given in (3-39), is of degree $n + 1$ if $K_0 > 0$ and of degree n if $K_0 = 0$. This agrees with the number of capacitors in the two Foster circuits (Figs. 3-18 and 3-20). By an argument similar to that given to show the canonical property of the Foster realization of LC immittances, it is easy to see that this is also the minimum possible number of capacitors. To show that the total number of elements (and thus also the number of resistors) is also the smallest possible, we note that from (3-39)

$$Z(s) = \frac{a_{n+1}s^{n+1} + a_n s^n + \cdots + a_1 s + a_0}{s(s + \sigma_1)(s + \sigma_2) \cdots (s + \sigma_n)} = \frac{a_{n+1}s^{n+1} + \cdots + a_1 s + a_0}{s^{n+1} + b_n s^n + \cdots + b_1 s} \quad (3\text{-}53)$$

This function contains $2n + 2$ prescribed constants ($n + 2$ numerator and n denominator coefficients). The same is true for the form of $Z(s)$ given in (3-39). There the residues K_0, K_∞, and the K_i represent $n + 2$ prescribed constants, and the poles σ_i represent n additional ones. Clearly, we need at least $2n + 2$ "design-able" element values to match these $2n + 2$ constants. This is exactly the total number of elements in the Foster realizations. This total number is thus the minimum possible. Therefore, both Foster realizations are feasible and canonical.

The Cauer expansions also exist for RC immittances. As before, the *Cauer 1 expansion* is based on the behavior of $Z(s)$ and $Y(s)$ for $s \to \infty$. From (3-39), $Z(s) \to K_\infty \geq 0$ for $s \to \infty$. Hence, when K_∞ is realized as a series resistor R_1 (Fig. 3-22a), the remainder impedance is

$$Z_2(s) \triangleq Z(s) - K_\infty \tag{3-54}$$

Clearly, the remainder, which is given by

$$Z_2(s) = \frac{K_0}{s} + \sum_{i=1}^{n} \frac{K_i}{s + \sigma_i} \tag{3-55}$$

is a realizable RC impedance with the property that $Z_2(s) \to 0$ for $s \to \infty$. Hence by Eq. (3-48), $Y_2(s) \triangleq 1/Z_2(s)$ must be of the form

$$Y_2(s) = k'_{\infty 2}s + k'_{02} + \sum_{i=1}^{n'} \frac{k'_{i2} s}{s + \sigma'_{i2}} \tag{3-56}$$

where $k'_{\infty 2} > 0$. When $k'_{\infty 2} s$ is realized as a shunt capacitor C_2 (Fig. 3-22), the remainder $Y_3(s) \triangleq Y_2(s) - k'_{\infty 2} s$ is now constant for $s \to \infty$. Hence, $Z_3(s) \triangleq 1/Y_3(s)$ is again in the form

$$Z_3(s) = K_{\infty 3} + \frac{K_{03}}{s} + \sum_{i=1}^{n3} \frac{K_{i3}}{s + \sigma_{i3}} \tag{3-57}$$

Here $K_{\infty 3} > 0$, and again it can be realized as a series resistor R_3. The remainder $Z_4(s)$ now satisfies $Z_4(s) \to 0$ for $s \to \infty$. Hence, $Y_4(s)$ has a pole at infinite frequency, which can be removed by realizing a shunt capacitor $C_4 = k'_{\infty 4}$. Proceeding this way, we obtain the circuit of Fig. 3-22a.

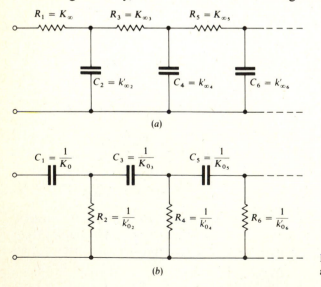

(a)

(b)

Figure 3-22 Cauer realizations of an RC impedance.

Since each operation reduces the number of constants in $Z(s)$ or $Y(s)$ by one and costs one element, this realization is also canonical. Realization of the resistors R_i does not reduce the degree of $Z_i(s)$; realization of each C_k, however, reduces the degree of $Y_k(s)$ by 1. Thus, *this circuit is also canonical*. By its construction, *it is also feasible*.

From Fig. 3-22a, this circuit represents the continued-fraction expansion

$$Z(s) = K_\infty + \cfrac{1}{k'_{\infty 2} s + \cfrac{1}{K_{\infty 3} + \cfrac{1}{k'_{\infty 4} s + \cdots}}} \qquad (3\text{-}58)$$

If $Z(\infty) = 0$, $K_\infty = 0$ and hence the first element is a shunt capacitor.

Example 3-9 Find the Cauer 1 realization of

$$Z(s) = \frac{2s^2 + 8s + 6}{s^2 + 2s}$$

Here $\quad R_1 = K_{\infty 1} = [Z(s)]_{s \to \infty} = 2 \; \Omega \qquad Z_2(s) = Z(s) - R_1 = \dfrac{4s + 6}{s^2 + 2s}$

$$Y_2(s) = \frac{s^2 + 2s}{4s + 6} = sC_2 + Y_3(s) \qquad C_2 = \left[\frac{Y_2(s)}{s}\right]_{s \to \infty} = \tfrac{1}{4} \text{F}$$

$$Y_3(s) = Y_2(s) - sC_2 = \frac{\tfrac{1}{2}s}{4s + 6} \qquad Z_3(s) = \frac{4s + 6}{\tfrac{1}{2}s} = R_3 + Z_4(s)$$

$$R_3 = [Z_3(s)]_{s \to \infty} = 8 \; \Omega \qquad Z_4(s) = Z_3(s) - R_3 = \frac{6}{\tfrac{1}{2}s}$$

$$Y_4(s) = \frac{s}{12} = sC_4 \qquad C_4 = \tfrac{1}{12} \text{F}$$

Hence, the circuit is that shown in Fig. 3-23a. As in the case of LC realization, we can perform the continued-fraction expansion using long division

$$s^2 + 2s \overline{)2s^2 + 8s + 6} \quad 2 \nearrow^{R_1}$$
$$\underline{-2s^2 - 4s}$$

$$4s + 6 \overline{)s^2 + 2s} \quad \frac{s}{4} \nearrow^{sC_2}$$
$$\underline{-s^2 - \tfrac{3}{2}s}$$

$$\tfrac{1}{2}s \overline{)4s + 6} \quad 8 \nearrow^{R_3}$$
$$\underline{-4s}$$

$$6 \overline{)\tfrac{s}{2}} \quad \frac{s}{12} \nearrow^{sC_4}$$
$$\underline{-\tfrac{s}{2}}$$
$$0$$

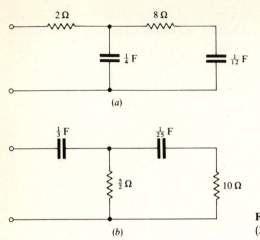

Figure 3-23 Cauer realizations of $Z(s) = (2s^2 + 8s + 6)/(s^2 + 2s)$.

It is clear from Fig. 3-22a or 3-23a that the last branch of the Cauer 1 circuit will contain a resistor if $Z(0) < \infty$ and a capacitor if $Z(0) \to \infty$.

The *Cauer 2 expansion* is based on the behavior of the immittances for $s = 0$. From (3-39),

$$Z_1(s) \triangleq Z(s) = \frac{1}{C_1 s} + Z_2(s) \tag{3-59}$$

where $C_1 = 1/K_0$ and $Z_2(0) < \infty$. Thus $Y_2(s) \triangleq 1/Z_2(s)$ has a finite positive value for $s = 0$. By (3-48), therefore,

$$Y_2(s) = \frac{1}{R_2} + Y_3(s) \tag{3-60}$$

where $R_2 = 1/k'_{0_2}$ and $Y_3(s)$ has the property that $Y_3(0) = 0$. Hence, $Z_3(s) \triangleq 1/Y_3(s)$ has a pole at $s = 0$, and it can be written in the form

$$Z_3(s) = \frac{1}{C_3 s} + Z_4(s) \tag{3-61}$$

where $C_3 = 1/K_{0_3}$ and $Z_4(0) < \infty$. Proceeding this way, we obtain the continued-fraction expansion

$$Z(s) = \frac{K_0}{s} + \cfrac{1}{k'_{0_2} + \cfrac{1}{\cfrac{K_{0_3}}{s} + \cfrac{1}{k'_{0_4} + \cdots}}} \tag{3-62}$$

This corresponds to the circuit of Fig. 3-22b. If $Z(0) < \infty$, C_1 is absent and the first element is a shunt resistor $1/k'_0$.

Example 3-10 Find the Cauer 2 realization of

$$Z(s) = \frac{2s^2 + 8s + 6}{s^2 + 2s}$$

Following the process described above gives

$$Z(s) = \frac{1}{sC_1} + Z_2(s) \qquad C_1 = \left[\frac{1}{sZ(s)}\right]_{s \to 0} = \tfrac{1}{3} F$$

$$Z_2(s) = Z(s) - \frac{3}{s} = \frac{2s^2 + 5s}{s^2 + 2s} = \frac{2s + 5}{s + 2} \qquad R_2 = [Z_2(s)]_{s=0} = \tfrac{5}{2}\,\Omega$$

$$Y_2(s) \triangleq \frac{1}{Z_2(s)} = \frac{1}{R_2} + Y_3(s) \qquad Y_3(s) = Y_2(s) - \frac{1}{R_2} = \frac{s+2}{2s+5} - \frac{2}{5} = \frac{s/5}{2s+5}$$

$$Z_3(s) = \frac{10s + 25}{s} = \frac{25}{s} + 10 \qquad C_3 = \tfrac{1}{25} F \qquad R_4 = 10\,\Omega$$

The circuit is shown in Fig. 3-23b.

The above calculations can be carried out more compactly by using the continued-fraction scheme

$$
\begin{array}{r}
\cfrac{1}{C_1 s} \nearrow \\
2s + s^2 \Big)\ 6 + 8s + 2s^2 \Big)\ \dfrac{3}{s} \\
-6 - 3s \\
\hline
\end{array}
$$

$$
\begin{array}{r}
\dfrac{1}{R_2} \nearrow \\
5s + 2s^2 \Big)\ 2s + s^2 \Big)\ \tfrac{2}{5} \\
-2s - \tfrac{4}{5}s^2 \\
\hline
\end{array}
$$

$$
\begin{array}{r}
\dfrac{1}{C_3 s} \nearrow \\
\dfrac{s^2}{5} \Big)\ 5s + 2s^2 \Big)\ \dfrac{25}{s} \\
-5s \\
\hline
\end{array}
$$

$$
\begin{array}{r}
\dfrac{1}{R_4} \nearrow \\
2s^2 \Big)\ \dfrac{s^2}{5} \Big)\ \tfrac{1}{10} \\
-\dfrac{s^2}{5} \\
\hline
0
\end{array}
$$

The reader is urged to carry out the checks for $s = 0$ and $s \to \infty$ for both Cauer realizations.

A comparison of the Cauer realization algorithms for LC and RC immittances reveals that while for LC circuits every step in the calculation is a full pole removal, for RC circuits we alternate between removing a *pole* (by realizing a capacitor) and removing a *constant* (by realizing a resistor). This provides a wonderful opportunity for mistakes, since we have to know when to remove what.

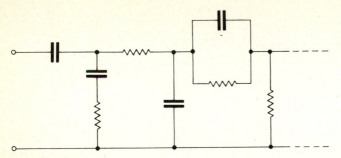

Figure 3-24 Mixed ladder realization of an *RC* impedance.

Removal of the wrong element leaves a non-PR remainder and sooner or later results in a negative or nonphysical element. The best way to avoid mixups is to start by drawing the appropriate circuit diagram from Fig. 3-22*a* or *b* and to follow all calculations, branch by branch, on the diagram.

Example 3-11 The impedance

$$Z_2(s) = \frac{2s + 5}{s + 2}$$

was one of the remainders in the previous example, where we removed $Y_2(0) = \frac{2}{3}$ from its reciprocal. If instead we unthinkingly remove $Z_2(0) = 5/2$ from $Z_2(s)$, that is, remove the resistor in series, rather than in parallel, we obtain

$$Z_3(s) = Z_2(s) - \frac{5}{2} = \frac{-s/2}{s + 2}$$

which is non-PR.

As for the *LC* immittances, it is also possible to obtain *mixed RC ladder realizations* by removing constants and poles at finite σ values as well as at $s = 0$ and $s \to \infty$. This may result, for example, in a circuit like that shown in Fig. 3-24. All such circuits are feasible and canonical.

3-6 *RL* IMMITTANCES

The remaining class of two-element one-ports is the one which contains only resistors and inductors. Since there are no capacitors in the circuit, $V_0(s) \equiv 0$ and hence, by Eqs. (2-59) and (2-60),

$$Z(s) = \frac{1}{|I_1(s)|^2} [F_0(s) + sM_0(s)] \qquad (3\text{-}63)$$

and

$$Y(s) = \frac{1}{|V_1(s)|^2} [F_0(s) + s^*M_0(s)] \qquad (3\text{-}64)$$

where the energy functions $F_0(s)$ and $M_0(s)$ are both real and nonnegative. The properties of RL immittances can be derived from (3-63) and (3-64) in a way similar to that in which the properties of RC immittances were found from (3-27) and (3-28). It is simpler, however, to introduce the new function $M'_0(s) \triangleq |s^2| M_0(s) = ss^* M_0(s)$. Substituting into (3-63) and (3-64) gives

$$Z(s) = \frac{1}{|I_1(s)|^2} \left[F_0(s) + \frac{1}{s^*} M'_0(s) \right]$$

(3-65)

and

$$Y(s) = \frac{1}{|V_1(s)|^2} \left[F_0(s) + \frac{1}{s} M'_0(s) \right]$$

(3-66)

A comparison with (3-27) and (3-28) shows now that the *RL impedance* has the same form and thus the same properties as an *RC admittance* and that the *RL admittance* has the same form and properties as the *RC impedance*. Accordingly, the RL immittances have the partial-fraction expansions

$$Z(s) = K_\infty s + K_0 + \sum_{i=1}^{n} \frac{K_i s}{s + \sigma_i}$$

(3-67)

and

$$Y(s) = K'_\infty + \frac{K'_0}{s} + \sum_{i=1}^{n'} \frac{K'_i}{s + \sigma'_i}$$

(3-68)

where all constants are nonnegative.

The properties of RL immittances are summarized in Table 3-2.

Table 3-2

RL impedances	RL admittances
1. All poles and zeros are simple and lie on the $-\sigma$ (real negative) axis of the s plane	1. All poles and zeros are simple and lie on the $-\sigma$ (real negative) axis of the s plane
2. Poles and zeros alternate on the $-\sigma$ axis	2. Poles and zeros alternate on the $-\sigma$ axis
3. The first critical frequency on the $-\sigma$ axis is a zero; it may be at $s = 0$	3. The first critical frequency on the $-\sigma$ axis is a pole; it may be at $s = 0$
4. The last critical frequency is a pole, which may be at infinity	4. The last critical frequency is a zero, which may be at infinity
5. The residues at the poles of $Z(s)$ are real and negative; the residues of $Z(s)/s$ are real and positive	5. The residues at the poles of $Y(s)$ are real and positive

The realization of RL immittances closely parallels that of RC immittances. From (3-67) and (3-68) the Foster realizations shown in Fig. 3-25 are obtained. The Cauer forms are shown in Fig. 3-26. The synthesis will be illustrated by a simple example.

Example 3-12 Find the Cauer 2 realization of

$$Z(s) = \frac{(s + 1)(s + 3)}{(s + 2)(s + 6)} = \frac{s^2 + 4s + 3}{s^2 + 8s + 12}$$

From Fig. 3-26 and Eq. (3-67),

$$Z(s) = R_1 + Z_2(s) \qquad R_1 = [Z(s)]_{s=0} = \tfrac{1}{4}\,\Omega$$

$$Z_2(s) = Z(s) - R_1 = \frac{\tfrac{3}{4}s^2 + 2s}{s^2 + 8s + 12}$$

$$Y_2(s) = \frac{s^2 + 8s + 12}{\tfrac{3}{4}s^2 + 2s} = \frac{1}{sL_2} + Y_3(s)$$

$$L_2 = \left[\frac{1}{sY_2(s)}\right]_{s \to 0} = \tfrac{1}{6}\,\text{H}$$

$$Y_3(s) = Y_2(s) - \frac{6}{s} = \frac{s^2 + \tfrac{7}{2}s}{\tfrac{3}{4}s^2 + 2s} = \frac{s + \tfrac{7}{2}}{\tfrac{3}{4}s + 2}$$

$$Z_3(s) = \frac{\tfrac{3}{4}s + 2}{s + \tfrac{7}{2}} = R_3 + Z_4(s)$$

$$R_3 = [Z_3(s)]_{s=0} = \tfrac{4}{7}\,\Omega$$

$$Z_4(s) = Z_3(s) - \frac{4}{7} = \frac{\tfrac{5}{28}s}{s + \tfrac{7}{2}}$$

$$Y_4(s) = \frac{28}{5} + \frac{98}{5s} = \frac{1}{L_4 s} + \frac{1}{R_5}$$

$$L_4 = \tfrac{5}{98}\,\text{H} \qquad R_5 = \tfrac{5}{28}\,\Omega$$

The calculations can be performed more compactly using the long-division scheme. The continued-fraction form of $Z(s)$ is

$$Z(s) = \frac{1}{4} + \cfrac{1}{\dfrac{6}{s} + \cfrac{1}{\dfrac{4}{7} + \cfrac{1}{\dfrac{98}{5s} + \dfrac{5}{28}}}}$$

The circuit is shown in Fig. 3-27. Performing our usual test at $s = 0$, we find $Z(0) = \tfrac{1}{4}\,\Omega$ from both $Z(s)$ and the circuit. For $s \to \infty$, $Z(s) \to 1$; from the circuit, $Z(s) \to \tfrac{1}{4} + \tfrac{4}{7} + \tfrac{5}{28} = 1$. Hence, the circuit passes these simple checks.

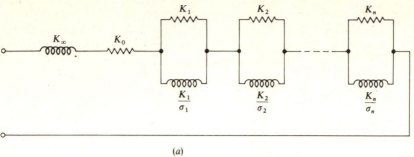

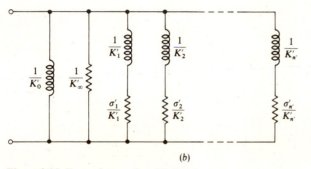

(a)

(b)

Figure 3-25 Foster forms of an *RL* immittance.

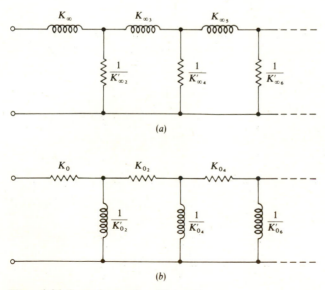

(a)

(b)

Figure 3-26 Cauer forms of an *RL* immittance.

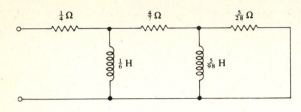

Figure 3-27 Cauer 2 form of $Z(s) = (s^2 + 4s + 3)/(s^2 + 8s + 12)$.

3-7 SUMMARY

In this chapter, the following material was discussed:

1. The properties of realizable LC impedance functions were established and tabulated (Sec. 3-1).
2. On the basis of these properties, two different realizations based on the partial-fraction expansion of either $Z(s)$ or $Y(s)$ were derived; these are the Foster 1 and Foster 2 realizations. The design was illustrated by numerous examples (Sec. 3-2).
3. Ladder realization of LC impedances was introduced, and two special ladders, synthesized by repeated pole removals at zero or infinite frequency, discussed. These are the Cauer realizations; the design process was again illustrated by several numerical examples (Sec. 3-3).
4. The properties of realizable RC immittance functions were derived and tabulated (Sec. 3-4).
5. The design of RC immittances using the Foster expansions, the Cauer expansions, and mixed ladder realization was described and illustrated with numerous examples (Sec. 3-5).
6. The properties of RL immittances were derived, and the analogy between RL and RC immittances established. The realization process was illustrated by an example (Sec. 3-5).

   This chapter illustrates synthesis techniques at the peak of their power. As the reader has probably noted, the described synthesis techniques enable us to design *any* realizable RC or RL or LC impedance function with a minimum number of physical elements. Furthermore, there is a variety of design procedures available, resulting in a variety of circuits—all economical and realizable. Finally, none of the circuits derived requires the use of coupled inductors, only uncoupled L, C, and R elements. This happy situation is going to change in the design of general passive impedance functions, to be discussed in the next chapter.

PROBLEMS

3-1 Show from Eq. (3-2) that all zeros of $Y(s)$ of an LC one-port are located on the $j\omega$ axis and are simple.

3-2 Show that the reactance function satisfies $dX(\omega)/d\omega \geq |X(\omega)/\omega|$. Find a graphical interpretation for this inequality. *Hint:* Use Eq. (3-1).

3-3 Show that a pole removal at $s = 0$ (Fig. 3-8b) reduces the order of the impedance by 1.

3-4 Show that the Foster 2 expansion of an LC admittance is canonical.

3-5 When is the last branch of a Cauer 1 expansion of an LC impedance a series element? When is it a shunt element?

3-6 Find all Foster and Cauer expansions of $Z(s) = (3s^2 + 3)/(s^3 + 4s)$.

3-7 Carry out the Cauer 2 expansion of $Z(s) = (s^3 + 4s)/(2s^4 + 20s^2 + 18)$. (The resulting circuit is shown in Fig. 3-12b.)

3-8 Carry out the Foster expansions of the $Z(s)$ of Prob. 3-7. (These circuits are shown in Fig. 3-12c and d.)

3-9 Find the Foster realizations of $Z(s) = (s^4 + 10s^2 + 9)/(s^3 + 2s)$.

3-10 Using full pole removals at all finite as well as infinite poles, how many different ladder realizations of the $Z(s)$ of Prob. 3-9 can be obtained? Can you generalize your result? Are all circuits feasible and canonical?

3-11 Which of the following functions are LC immittances? Why?

(a) $\dfrac{s(s^2 + 1)(s^2 + 16)}{(s^2 + 9)(s^2 + 25)}$ (b) $\dfrac{s^2 + s}{s^4 + 5s^2 + 6}$

(c) $\dfrac{(s^2 + 1)(s^2 + 9)}{s(s^2 + 4)}$ (d) $\dfrac{s^4 + 3s^2 + 2}{s^3 + 3s}$

(e) $\dfrac{s^4 + 5s^2 + 4}{s^3 + 2s}$ (f) $\dfrac{s(s^2 + 2)}{(s^2 + 1)(s^2 + 4)}$

3-12 An LC impedance is zero at 2 kHz and infinite at 1 and 5 kHz. It has the value $|Z| = 100 \ \Omega$ at 3 kHz. Find the Foster 1 realization. *Hint:* Use scaling.

3-13 What can you state about an LC impedance which contains four capacitors and three inductors?

3-14 For what ranges of the constants a, b are the functions

$$Z_1(s) = \frac{s^2 + a}{2s^3 + 8s} \qquad Z_2(s) = \frac{s^4 + bs^2 + 9}{2s^3 + 8s}$$

realizable LC impedances?

3-15 A common error has been made in the Cauer 2 development of $Z(s) = (s^2 + 1)/(2s^3 + 8s)$ given below:

$$2s^3 + 8s)s^2 + 1)\frac{1}{2s} \swarrow^{Y_1}$$
$$\frac{-s^2 - 4}{}$$
$$- 3)2s^3 + 8s) - \tfrac{2}{3}s^3 \swarrow^{Z_2}$$

Z_2 is clearly nonphysical. Find out what went wrong and work out the correct expansion.

3-16 Show that for $s = j\omega$ the real part $R(\omega)$ of an RC impedance is monotone decreasing, that is, $dR(\omega)/d\omega < 0$.

3-17 When is the last branch in a Cauer 2 form for an RC impedance (Fig. 3-22b) a series capacitor? When is it a shunt resistor?

3-18 Find all Foster and Cauer realizations of $Z(s) = (s^2 + 6s + 8)/(s^2 + 4s + 3)$.

3-19 Show that removing $R = Z(0)$ in series from a PR RC impedance leaves a non-PR remainder. *Hint:* Use Eq. (3-50).

3-20 Sort the following impedance functions into five categories:

1. LC impedances
2. RC impedances
3. RL impedances
4. $RLMC$ impedances
5. Non-PR functions

Explain your decisions.

(a) $\dfrac{s^2 + 4s + 3}{2s^2 + 4}$ (b) $\dfrac{s^4 + 5s^2 + 10}{s^3 + 4s}$

(c) $\dfrac{s^2 + s + 10}{s^2 + 4s + 2}$ (d) $\dfrac{3s^2 + 3}{s^3 + 4s}$

(e) $\dfrac{s^2 + 6s + 8}{s^2 + 4s + 3}$ (f) $\dfrac{s^2 + 1}{s^2 + 3s + 4}$

(g) $\dfrac{s^2 + 2s}{s^2 + 4s + 3}$ (h) $\dfrac{s^2 + 2s + 1}{s^2 + 4s + 2}$

REFERENCES

1. Bode, H. W.: "Network Analysis and Feedback Amplifier Design," Van Nostrand, New York, 1945.
2. Rupprecht, W.: "Netzwerksynthese," Springer, Berlin, 1972.
3. Balabanian, N.: "Network Synthesis," Prentice-Hall, Englewood Cliffs, N.J., 1958.
4. Calahan, D. A.: "Computer-aided Network Design," rev. ed., McGraw-Hill, New York, 1972.

FOUR

SYNTHESIS OF *RCLM* ONE-PORTS

4-1 PR FUNCTIONS REVISITED

In this chapter, we climax our study of one-port synthesis techniques by investigating design methods for *RCLM* immittances. As we found in Secs. 2-3 and 2-4, a function $Z(s)$ is realizable as the impedance of a one-port built from positive R, C, L, and M elements only if it is a rational PR function, i.e., if it satisfies Brune's conditions:

(*a*) $Z(s)$ is a real rational function of *s*.
(*b*) Re $Z(s) \geq 0$ for Re $s \geq 0$.

or the equivalent conditions

(*a'*) $Z(s)$ is a real rational function of *s*.
(*b'*) Re $Z(j\omega) \geq 0$ for all real ω.
(*c'*) All poles of $Z(s)$ are in the closed LHP; all $j\omega$-axis poles are simple, with positive real residues.

We have described in Sec. 2-5 some simple techniques for testing a $Z(s)$ function to see whether or not it satisfies these conditions. In this section, first we introduce a third set of equivalent conditions, more convenient for testing; then we introduce some new test algorithms, more sophisticated and less laborious than those given in Sec. 2-5.

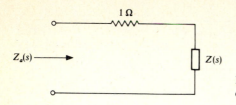

Figure 4-1 The augmented impedance $Z_a(s)$ associated with $Z(s)$.

The new set of equivalent PR conditions involves the *augmented impedance*

$$Z_a \triangleq Z(s) + 1 \qquad (4\text{-}1)$$

which has the interpretation illustrated in Fig. 4-1. Now $Z(s)$ is a PR function if and only if its augmented impedance satisfies the following three conditions:

(a'') $Z_a(s)$ is a real rational function of s.
(b'') Re $Z_a(j\omega) \geq 1$.
(c'') All *zeros* of $Z_a(s)$ are inside the *open* LHP.†

We next show that conditions a'', b'', and c'' are completely equivalent to a', b', and c'; that is, either set of conditions can be derived from the other. The equivalence of a'' and b'', on the one hand, and a' and b', on the other, is evident in view of Eq. (4-1). To derive c'' from a', b', and c', we note that at a zero s_z of $Z_a(s)$ we have

$$Z(s_z) = Z_a(s_z) - 1 = -1 \qquad (4\text{-}2)$$

but if $Z(s)$ is PR, then Re $Z(s_z) < 0$ implies Re $s_z < 0$ and $s_z < \infty$. This completes the derivation of a'', b'', and c'' from a', b', and c'.

To show that a', b', and c' follow from a'', b'', and c'', all we need to prove is that if $Z_a(s)$ satisfies a'', b'', and c'', then $Z(s) = Z_a(s) - 1$ satisfies c' (it obviously satisfies a' and b'). Consider

$$Y_a(s) \triangleq \frac{1}{Z_a(s)} \qquad (4\text{-}3)$$

Since $Z_a(s)$ meets a'', $Y_a(s)$ itself is a real rational function. Since furthermore $Z_a(s)$ meets b'', Re $Y_a(j\omega)$ satisfies

$$\text{Re } Y_a(j\omega) = \frac{\text{Re } Z_a(j\omega)}{[\text{Re } Z_a(j\omega)]^2 + [\text{Im } Z_a(j\omega)]^2} \geq 0 \qquad (4\text{-}4)$$

Finally, since $Z_a(s)$ meets c'', the poles of $Y_a(s)$ are inside the open LHP. Hence, $Y_a(s)$ meets a', b', and c' and is therefore PR. Hence, so is $Z_a(s)$ (compare

† That is, all zeros s_z satisfy Re $s_z < 0$ and $s_z < \infty$.

Prob. 2-4). It follows that the poles and residues of $Z_a(s)$ satisfy c'. Since, however, the poles and residues of $Z(s) = Z_a(s) - 1$ are the same as those of $Z_a(s)$, $Z(s)$ also meets c'. Thus, if $Z_a(s)$ meets a'', b'', and c'', then $Z(s)$ meets a', b', and c', as stated.

We have at this stage shown that a'', b'', and c'' are entirely equivalent to the PR conditions introduced in Chap. 2. Next, we shall describe methods to test for these new conditions.

A test for *condition a''* merely involves an inspection of $Z(s)$. If it is a rational function of s with real constant coefficients, then a'' is satisfied; otherwise it is not.

Condition b'' is equivalent to condition b'. Hence, the procedure described in Sec. 2-5, in connection with Eqs. (2-73) to (2-75) is applicable. Specifically, the polynomial in ω^2

$$P(\omega^2) = N_e(j\omega)D_e(j\omega) - N_o(j\omega)D_o(j\omega) \tag{4-5}$$

must be nonnegative for all $\omega^2 \geq 0$. With the notation $x \triangleq \omega^2$, the condition is

$$P(x) = a_n x^n + a_{n-1} x^{n-1} + \cdots + a_1 x + a_0 \geq 0 \qquad \text{for all } x \geq 0 \tag{4-6}$$

Since $P(x) \to a_0$ for $x \to 0$, we must have $a_0 \geq 0$ for (4-6) to hold; since also $P(x) \to a_n x^n$ for $x \to \infty$, we must have $a_n > 0$. Hence, a polynomial with $a_n < 0$ or $a_0 < 0$ fails the test. Also, if *all* a_k are nonnegative, then certainly $P(x) \geq 0$ for $x \geq 0$ and no further checking is needed.

If $a_0 > 0$ and $a_n > 0$ but $a_i < 0$ for some i $(1 \leq i \leq n-1)$, a possible behavior of $P(x)$ is illustrated in Fig. 4-2. Clearly, $P(x) \geq 0$ for all $x \geq 0$ requires that $P(x)$ have no simple zeros (or, in general, odd-order zeros) on the positive x axis except at $x = 0$. Double (or, in general, even-order) x-axis zeros are permissible; so are complex zeros.

Our test is thus reduced to checking the polynomial $P(x)$ for simple or odd-order zeros on the positive x axis. This can, of course, be performed by factoring $P(x)$ to find all its zeros and then checking the multiplicity of all positive real zeros. This usually requires a computer for $n \geq 5$. If we are restricted to hand

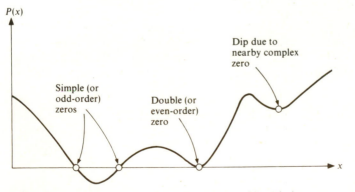

Figure 4-2 A possible response for the numerator of Re $Z(j\omega)$.

calculation, the *Sturm test*[1] may be used instead. The steps in the Sturm test are the following.

Step 1 Develop a sequence of polynomials P_0, P_1, P_2, ... (called *Sturm functions*) as follows. Define

$$P_0(x) \triangleq P(x) \quad \text{and} \quad P_1(x) \triangleq \frac{dP(x)}{dx} = P'_0(x)$$

To find P_2, P_3, ... divide $P_0(x)$ by $P_1(x)$ using long division; stop when the remainder is 1 degree lower than $P_1(x)$. Call the remainder (which is a polynomial of degree $n - 2$) $-P_2(x)$. Repeat by dividing $P_2(x)$ into $P_1(x)$ and defining the remainder as $-P_3(x)$, etc. Hence, the algorithm is described by

$$P_0(x) = q_1 P_1(x) - P_2(x)$$
$$P_1(x) = q_2 P_2(x) - P_3(x)$$
$$\cdots\cdots\cdots\cdots\cdots\cdots\cdots\cdots \tag{4-7}$$
$$P_{k-3}(x) = q_{k-2} P_{k-2}(x) - P_{k-1}(x)$$
$$P_{k-2}(x) = q_{k-1} P_{k-1}(x) - P_k(x)$$

Here, the general form of the quotient q_i is $q_i(x) = b_i x + c_i$ although $c_i = 0$ may occur. The process terminates when the remainder $P_k(x)$ becomes a constant or zero. The former situation ($P_k = \text{const} \neq 0$) occurs for $k = n$ and indicates that all zeros of $P(x)$ are simple; the latter ($P_k = 0$) can occur for $k \leq n$ and shows, as will be seen later, that $P(x)$ has one or more multiple zeros.

Step 2 Assuming that $P_k = \text{const}$ terminates step 1, the number of simple real zeros which $P(x)$ has in the region $a \leq x \leq b$ is given by *Sturm's theorem*.

Sturm's theorem This number is equal to the number of sign changes in the sequence $P_0(a)$, $P_1(a)$, $P_2(a)$, ..., $P_k(a)$ minus the number of sign changes in the sequence $P_0(b)$, $P_1(b)$, $P_2(b)$, ..., $P_k(b)$.

Since in our application we are interested in the presence or absence of any simple zero *anywhere* on the positive x axis, we shall use $a = 0$, $b = \infty$ and hence merely observe the signs of the lowest- and highest-order coefficients of the $P_i(x)$.

Example 4-1 Test whether $P(x) = x^4 - 15x^2 + 10x + 24$ satisfies $P(x) \geq 0$ for $x \geq 0$. Since $a_n = a_4 = 1$ and $a_0 = 24$, it is possible that $P(x)$ meets (4-6). However, since $a_2 < 0$, this ain't necessarily so. Hence, the Sturm test is needed to establish condition b''.
   From (4-6)

$$P_0(x) = x^4 - 15x^2 + 10x + 24$$
$$P_1(x) = 4x^3 - 30x + 10$$

From (4-7), by long division

$$
4x^3 - 30x + 10 \quad) \quad x^4 - 15x^2 + 10x + 24 \quad \dfrac{x}{4} + 0 \overset{q_1(x)}{\nwarrow}
$$

$$
\begin{array}{l}
\underline{-x^4 + 7.5x^2 - 2.5x} \qquad \overset{-P_2(x)}{\nwarrow} \\
\quad - 7.5x^2 + 7.5x + 24
\end{array}
$$

$$
\begin{array}{l}
P_2(x) \\
\downarrow \\
7.5x^2 - 7.5x - 24 \quad)\quad 4x^3 \qquad - 30x + 10) \; 0.5333x + 0.5333 \overset{q_2(x)}{\nwarrow} \\
\qquad \underline{-4x^3 + 4x^2 + 12.8x} \\
\qquad\qquad 4x^2 - 17.2x + 10 \\
\qquad\qquad \underline{-4x^2 + \quad 4x + 12.8} \qquad \overset{-P_3(x)}{\nwarrow} \\
\qquad\qquad\qquad -13.2x + 22.8
\end{array}
$$

$$
\begin{array}{l}
P_3(x) \qquad\qquad\qquad\qquad\qquad\qquad q_3(x) \\
\downarrow \qquad\qquad\qquad\qquad\qquad\qquad\quad \downarrow \\
13.2x - 22.8 \quad)\; 7.5x^2 - \quad 7.5x - 24) \; 0.56818x + 0.41322 \\
\qquad \underline{-7.5x^2 + 12.954x} \\
\qquad\qquad 5.4545x - 24 \\
\qquad\qquad \underline{-5.4545x + \quad 9.4215} \\
\qquad\qquad\qquad -14.578 \; \longleftarrow \; -P_4(x)
\end{array}
$$

Since P_4 is a constant and $k = n = 4$, the polynomial $P_0(x) = P(x)$ has only simple zeros. The sequence of lowest-order coefficients is $+24, +10, -24, -22.8, +14.578$, which changes signs *two* times; the sequence of highest-order coefficients $+1, +4, +7.5, +13.2, +14.578$ *zero* times. By Sturm's theorem, $P(x)$ has hence two real positive zeros. Indeed, $P(x) = (x + 1)(x - 2)(x - 3)(x + 4)$, with zeros at $x = 2$ and $x = 3$.

Therefore $P(x) < 0$ for $2 < x < 3$ and thus fails the test.

Returning to the case when $P_k(x) \equiv 0$† for some $k \le n$, the last equation of (4-7) indicates that in this case $P_{k-2} = q_{k-1} P_{k-1}$; that is, P_{k-1} is a factor of P_{k-2}. But then the last equation but one shows that P_{k-1} is a factor also of P_{k-3}. Working our way back to the first equation in (4-7) this way, we find that P_{k-1} is a common factor of $P_0 \triangleq P(x)$ and $P_1 \triangleq P'(x)$. Hence, it corresponds to a factor $(x - x_1)^m$ of $P_0(x)$, that is, to a multiple zero. From the formulas

$$
P(x) = (x - x_1)^m Q(x)
$$
$$
P'(x) = (x - x_1)^{m-1}[mQ(x) + (x - x_1)Q'(x)] \tag{4-8}
$$

we see that $P_{k-1}(x) = (\text{const})(x - x_1)^{m-1}$ so that the multiplicity of the zero x_1 in $P(x)$ is 1 higher than in $P_{k-1}(x)$.

† The sign $\equiv$ is read "is identically equal to."

[In general, $P(x)$ may have several multiple zeros, causing a common factor $P_{k-1}(x) = (\text{const})(x - x_1)^{m_1 - 1}(x - x_2)^{m_2 - 1} \cdots .]$

If m is odd and $x_1 > 0$, Fig. 4-2 shows that $P(x) < 0$ for some $x > 0$. Then $P(x)$ fails the test. Otherwise, for m even and/or $x_1 < 0$, $(x - x_1)^m > 0$ for all $x > 0$, and the remaining factors of $P(x)$ determine its sign along the positive axis. The number of zeros which $P(x)$ has for $0 \leq x < \infty$ can still be established from the signs of the lowest- and highest-order coefficients of the $P_i(x)$, as shown before; however, *each multiple zero is counted only once*.

Example 4-2 Test $P(x) = x^4 - 7x^3 + 17x^2 - 17x + 6$ for its sign along the positive x axis.

Inspection reveals nothing: $a_4 > 0$ and $a_0 > 0$; hence $P(x)$ *may* be nonnegative for all $x \geq 0$; but $a_3 < 0$ and $a_1 < 0$, hence it *need* not. Therefore, the Sturm test will be used. Now

$$P_0 \equiv P(x) = x^4 - 7x^3 + 17x^2 - 17x + 6$$

$$P_1 \equiv P'(x) = 4x^3 - 21x^2 + 34x - 17$$

and by long division

$q_1(x)$

$$4x^3 - 21x^2 + 34x - 17)x^4 - \quad 7x^3 \quad + 17x^2 \quad - 17x + 6) \frac{x}{4} - 0.4375$$

$$-x^4 + 5.25x^3 \quad - 8.5x^2 \quad + 4.25x$$

$$\overline{\qquad\qquad -1.75x^3 \quad + 8.5x^2 \quad - 12.75x + 6}$$

$$+1.75x^3 - 9.1875x^2 + 14.875x - 7.4375$$

$$-P_2(x) \longrightarrow \qquad \overline{\quad -0.6875x^2 \quad + 2.125x - 1.4375}$$

$P_2(x)$ $q_2(x)$

$$0.6875x^2 - 2.125x + 1.4375) \quad 4x^3 - 21x^2 + 34x - 17) \quad 5.8182x - 12.562$$

$$-4x^3 + 12.364x^2 - 8.364x$$

$$\overline{\qquad\qquad -8.6364x^2 + 25.6364x - 17}$$

$$+8.6364x^2 - 26.6942x + 18.0578$$

$$-P_3(x) \longrightarrow \qquad \overline{\qquad\qquad - 1.0578x + 1.0578}$$

$P_3(x)$ $q_3(x)$

$$1.0578x - 1.0578) \quad 0.6875x^2 - 2.125x + 1.4375) \quad 0.6499x - 1.3589$$

$$0.6875x^2 + 0.6875x$$

$$\overline{\qquad\qquad -1.4375x + 1.4375}$$

$$+1.4375x - 1.4375$$

$$\overline{\qquad\qquad P_4(x) \rightarrow 0}$$

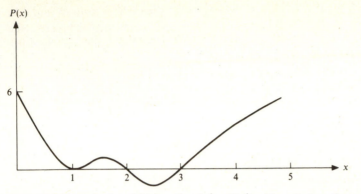

Figure 4-3 The behavior of $P(x) = x^4 - 7x^3 + 17x^2 - 17x + 6$ along the positive x axis.

Since $P_4(x) \equiv 0$, $P_3(x) = 1.0578(x - 1)$ must be a factor of all the $P_i(x)$, $i = 0, 1, 2, 3$. Consequently, as (4-8) shows, $(x - 1)^2$ is a factor of $P_0(x)$. Since here $m = 2$, this factor is nonnegative. Hence, here $P(x) \geq 0$ for $x \geq 0$ may be possible. Therefore we examine the signs of the lowest- and highest-order coefficients of the $P_i(x)$ to determine the location of the zeros of the other factors of $P(x)$. The lowest-order coefficients are $+6$, -17, $+1.4375$, -1.0578, which change signs three times; the highest-order ones are $+1$, $+4$, $+0.6875$, $+1.0578$, which do not change signs at all. Therefore, $P(x)$ has three zeros for $0 \leq x < \infty$, where the double zero at $x = 1$ is counted once. We must therefore have two simple positive x-axis zeros. Indeed,

$$P(x) = (x - 1)^2(x - 2)(x - 3)$$

and hence $P(x) < 0$ for $2 < x < 3$ (Fig. 4-3). It thus fails the test.

We are not including a proof of Sturm's theorem; the interested reader should consult Ref. 3 or any standard text on the numerical analysis of polynomials. A clear treatment is also given in Ref. 5.

Returning now to our original topic, the testing of $Z(s)$ for conditions a'', b'', and c'', we see that the Sturm test is efficient for checking whether $Z(s)$ satisfies b'', the real-part condition. Turning to *condition c''*, recall that

$$Z_a(s) = Z(s) + 1 = \frac{N(s)}{D(s)} + 1 = \frac{N(s) + D(s)}{D(s)} \tag{4-9}$$

must have all its zeros inside the open LHP. Hence, $N(s) + D(s)$ must be *strictly Hurwitz*. Such a polynomial must not have any terms missing, and all its coefficients must be positive. This, however, is not sufficient, as the example of Fig. 2-14 illustrated. As was the case with condition b'', the test for condition c'' can be settled by brute force: we can find (preferably using a computer) all zeros of $Q(s) \triangleq N(s) + D(s)$ and inspect the real parts of the zeros. If all real parts are negative, condition c'' is satisfied; if any one is zero or positive, it is not.

A different and more sophisticated approach can be based on the following important theorem.

Theorem If a polynomial $Q(s)$ is strictly Hurwitz, the ratio of its even and odd parts $Q_e(s)$ and $Q_o(s)$ is a realizable reactance function,† and, vice versa, the sum of the numerator and denominator of a realizable reactance function forms a strictly Hurwitz polynomial.

PROOF To prove this theorem, we note that the impedance function

$$Z_a(s) \triangleq 1 + \frac{Q_e(s)}{Q_o(s)} = \frac{Q(s)}{Q_o(s)} \tag{4-10}$$

obviously meets conditions a'', b'', and c''. Hence, $Q_e(s)/Q_o(s)$ is PR. But $Q_e(s)/Q_o(s)$ is purely odd and hence imaginary for $s = j\omega$; therefore it is a reactance function.‡

The second part of the theorem also follows from (4-10). If $Q_e(s)/Q_o(s)$ is PR, then so is $Z_a(s)$. Therefore $Q(s)$ is *at least* Hurwitz if not strictly Hurwitz. But for $s = j\omega$, $Q_e(s)/Q_o(s)$ is pure imaginary, and hence $Z_a(s) = 1 + Q_e(s)/Q_o(s)$ cannot become zero. Hence, $Q(s)$ cannot have any $j\omega$-axis zeros and must therefore be *strictly* Hurwitz.

The theorem clearly indicates a convenient test for condition c'': design a feasible LC reactance circuit from $Q_e(s)/Q_o(s)$. If all element values are positive, $Q(s)$ is strictly Hurwitz; if they are not, it is not.

Example 4-3 Is $Q(s) = s^4 + s^3 + 5s^2 + 3s + 2$ strictly Hurwitz?

Continued-fraction expansion around infinity gives the Cauer 1 form of the hypothetical reactance

$$\frac{Q_e(s)}{Q_o(s)} = \frac{s^4 + 5s^2 + 2}{s^3 + 3s} = s + \cfrac{1}{\cfrac{s}{2} + \cfrac{1}{s + 1/s}}$$

(The reader should work out the details as an exercise.) Since all "element values" $1, \frac{1}{2}, 1, 1$ are positive, the "reactance" is realizable, and hence, by the theorem, $Q(s)$ is strictly Hurwitz.

Note that the Cauer development is usually preferable to the Foster form, since the latter would require factoring $Q_o(s)$ or $Q_e(s)$, while the former requires only the long division of polynomials. For low-order polynomials ($n < 5$), either development is convenient.

An anomaly which may occur in this strictly Hurwitz test will be illustrated by an example.

---

† As before, "realizable" is short for "realizable with real positive element values."
‡ So, of course, is $Q_o(s)/Q_e(s)$.

Example 4-4 Test $Q(s) = s^4 + 4s^3 + 5s^2 + 8s + 6$ for the strictly Hurwitz property.

Developing the reactance in the Cauer 1 form gives

$$4s^3 + 8s \overline{\smash{\big)}\ s^4 + 5s^2 + 6} \quad \frac{s}{4}$$
$$\underline{-s^4 - 2s^2}$$

$$3s^2 + 6 \overline{\smash{\big)}\ 4s^3 + 8s} \quad \frac{4s}{3}$$
$$\underline{-4s^3 - 8s}$$

$$0$$

Hence

$$\frac{Q_e(s)}{Q_o(s)} = \frac{s^4 + 5s^2 + 6}{4s^3 + 8s} = \frac{s}{4} + \frac{1}{\frac{4}{3}s}$$

The continued-fraction development terminates prematurely, since two terms, rather than the usual one term, cancel in the last step. As a result, the fourth-degree reactance function appears to be realizable using only two elements. This is an impossibility; hence, $Q_e(s)/Q_o(s)$ must be of second degree to start with. Therefore, there must be a common quadratic factor in $Q_e(s)$ and $Q_o(s)$. It is easy to verify that this factor is $s^2 + 2$. Canceling it leaves

$$\frac{Q_e(s)}{Q_o(s)} = \frac{s^2 + 3}{4s}$$

which is indeed a second-degree function.

Since $s^2 + 2$ is a factor in both $Q_e(s)$ and $Q_o(s)$, it is also a factor of $Q(s)$. As $s^2 + 2 = (s - j2)(s + j2)$, $Q(s)$ has $j\omega$-axis zeros and cannot be strictly Hurwitz.

Since the coefficients in the reduced $Q_e(s)/Q_o(s)$ are positive, all factors of $Q(s)$ other than $s^2 + 2$ must be strictly Hurwitz.

Indeed, factoring shows that

$$Q(s) = (s^2 + 2)(s + 1)(s + 3)$$

is Hurwitz but not strictly Hurwitz, even though the "element values" $\frac{1}{4}$ and $\frac{4}{3}$ were positive.

It is easy to see that a common factor of $Q_e(s)$ and $Q_o(s)$ must be a pure even function of s; otherwise it would introduce odd terms into $Q_e(s)$ and even terms into $Q_o(s)$. Such a pure even factor in $Q(s)$ can cause $Q_e(s)/Q_o(s)$ to be a realizable reactance even though $Q(s)$ is not strictly Hurwitz. Fortunately the presence of such factor is always made obvious by the premature termination of the Cauer or Foster development and hence by a smaller number of "circuit elements" than the degree of $Q(s)$. When this occurs, we can immediately conclude that $Q(s)$ is not strictly Hurwitz.

At this stage, it should be obvious why c'' is a more convenient criterion to check than c'. In testing for c', we have to establish the Hurwitz (rather than strictly Hurwitz) character of $D(s)$, the denominator of $Z(s)$. Pure even factors detected in test c' do not disqualify $D(s)$, since they may correspond to $j\omega$-axis rather than RHP zeros. Hence, any such even factor must be isolated, its zeros located, and (if they are all on the $j\omega$ axis) the residues of $Z(s) = N(s)/D(s)$ calculated and examined at these poles. All this takes a lot more effort than the simple strictly Hurwitz test which we need for ascertaining property c''.

We can now perform the PR test efficiently by checking conditions a'', b'', and c'' with our newly found algorithms.

Example 4-5 Is

$$Z(s) = \frac{2s^3 + 3s^2 + 2s + 3}{s^3 + 3s^2 + 4s + 1}$$

a PR function?

Condition a'' is obviously met. To see whether b'' holds, we test

$$P(\omega^2) = N_e(j\omega)D_e(j\omega) - N_o(j\omega)D_o(j\omega)$$
$$= (-3\omega^2 + 3)(-3\omega^2 + 1) - (-2j\omega^3 + 2j\omega)(-j\omega^3 + 4j\omega)$$
$$= 2\omega^6 - \omega^4 - 4\omega^2 + 3$$

or, using $x \triangleq \omega^2$,

$$P(x) = 2x^3 - x^2 - 4x + 3$$

Since $P(x)$ has positive highest- and lowest-order coefficients, it may meet test b'', but since two of the coefficients are negative, we need the Sturm test to determine whether $P(x) \geq 0$ for all $x \geq 0$. We have

$$P_0(x) = 2x^3 - x^2 - 4x + 3$$
$$P_1(x) = 6x^2 - 2x - 4$$

and, as before, P_2, P_3, ... are found from

$$6x^2 - 2x - 4 \quad) \quad 2x^3 - x^2 - 4x + 3 \quad) \quad \frac{x}{3} - \frac{1}{18} \nwarrow q_1(x)$$

$$\underline{-2x^3 + \tfrac{2}{3}x^2 + \tfrac{4}{3}x}$$

$$-\tfrac{1}{3}x^2 - \tfrac{8}{3}x + 3$$

$$\underline{+\tfrac{1}{3}x^2 - \tfrac{1}{9}x - \tfrac{2}{9}} \qquad -P_2(x)$$

$$-\tfrac{25}{9}x + \tfrac{25}{9}$$

$$P_2(x)$$
$$\downarrow$$

$$\tfrac{25}{9}x - \tfrac{25}{9} \quad) \quad 6x^2 - 2x - 4 \quad) \quad \tfrac{54}{25}x + \tfrac{36}{25} \swarrow q_2(x)$$

$$\underline{-6x^2 + 6x}$$

$$4x - 4$$

$$\underline{-4x + 4}$$

$$0 \quad \leftarrow -P_3(x)$$

$P_3(x) \equiv 0$, showing that $P_2(x) = \tfrac{25}{9}(x - 1)$ is a factor of P_1 and P_0. Therefore, as discussed before, P_0 contains an $(x - 1)^2$ factor. The lowest-order coefficients of P_0, P_1,

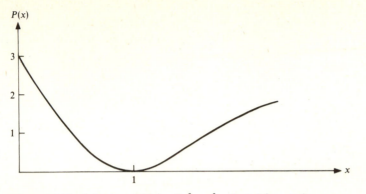

Figure 4-4 The behavior of $P(x) = 2x^3 - x^2 - 4x + 3$ for $x \geq 0$.

and P_2 are 3, -4, $-\frac{25}{9}$; the highest-order ones 2, 6, $\frac{25}{9}$. The difference in the numbers of sign changes is 1. Hence, the total number of zeros which $P(x)$ has along the positive x axis is one. This number includes the double zero at $x = 1$ counted as one; therefore there can be no other zeros for $x > 1$. Indeed, factoring shows that $P(x) = (x - 1)^2(2x + 3)$.

The behavior of $P(x)$ along the positive x axis is illustrated in Fig. 4-4. Clearly, condition b'' holds.

To test whether or not $Z(s)$ meets also condition c'', we must examine the numerator of the augmented impedance

$$Z_a = Z(s) + 1 = \frac{N(s)}{D(s)} + 1 = \frac{N(s) + D(s)}{D(s)}$$

for the strictly Hurwitz property. Writing

$$Q(s) \triangleq N(s) + D(s) = 3s^3 + 6s^2 + 6s + 4$$

we can analyze the ratio

$$\frac{Q_o(s)}{Q_e(s)} = \frac{3s^3 + 6s}{6s^2 + 4}$$

to see whether it is a realizable reactance function.†

By continued-fraction expansion

$$6s^2 + 4 \quad) \quad 3s^3 + 6s \quad) \quad 0.5s$$
$$\underline{-3s^3 - 2s}$$
$$4s \quad) \quad 6s^2 + 4 \quad) \quad 1.5s$$
$$\underline{-6s^2}$$
$$4 \quad) \quad 4s \quad) \quad s$$
$$\underline{-4s}$$
$$0$$

† For Cauer 1 expansion, it is convenient to place the higher-degree polynomial (here Q_o) in the numerator.

The coefficients in the continued-fraction expansion

$$\frac{3s^3 + 6s}{6s^2 + 4} = 0.5s + \frac{1}{1.5s + 1/s}$$

are 0.5, 1.5, and 1; they are all positive. This shows that $Q(s)$ is strictly Hurwitz and hence condition c'' holds.

This completes the PR test. Since all three conditions a'', b'', and c'' hold, $Z(s)$ is a PR function.

4-2 MINIMUM IMPEDANCES

The two techniques used in the synthesis of two-element one-ports were pole removal and constant removal. The former operation realized a series reactance branch when used on a $Z(s)$ function and a shunt susceptance branch when performed on a $Y(s)$ function. The latter realized a series resistor from a $Z(s)$ function and a shunt conductance from a $Y(s)$ function. In some instances, these methods can also be used in the design of $RCLM$ impedances.

Example 4-6 Consider again the impedance tested for the PR property above

$$Z(s) = \frac{2s^3 + 3s^2 + 2s + 3}{s^3 + 3s^2 + 4s + 1}$$

First, we look for $j\omega$-axis poles to remove. Since the degrees of the numerator and the denominator are the same, there is no pole or zero for $s \to \infty$. For $s \to 0$, $Z(0) = 3$; hence there is no pole or zero at $s = 0$ either. To find a $j\omega$-axis pole or zero at some finite nonzero frequency ω_i, we must test $D(s)$ and $N(s)$ for $j\omega$-axis zeros. Of course, this can be done by factoring $D(s)$ and $N(s)$, but recalling that such a zero corresponds to a pure even factor $s^2 + \omega_i^2$ and that such a factor becomes evident during continued-fraction expansion, we can use a less tedious process. Expanding $D_o(s)/D_e(s)$ using the Cauer 1 procedure gives

$$
\begin{array}{r}
3s^2 + 1 \;) \; s^3 + 4s \;) \quad \dfrac{s}{3} \\[1mm]
-s^3 - \dfrac{s}{3} \\[1mm]
\hline
\tfrac{11}{3}s \;) \; 3s^2 + 1 \;) \; \tfrac{9}{11}s \\[1mm]
-3s^2 \\[1mm]
\hline
1 \;) \; \tfrac{11}{3}s \;) \; \tfrac{11}{3}s \\[1mm]
-\tfrac{11}{3}s \\[1mm]
\hline
0
\end{array}
$$

All coefficients are positive, and there is no cancellation; hence $D(s)$ is strictly Hurwitz, and $Z(s)$ has no $j\omega$-axis poles.

Testing $N(s)$ the same way gives

$$3s^2 + 3 \;) \quad 2s^3 + 2s \quad) \quad \tfrac{2}{3}s$$
$$\underline{-2s^3 - 2s}$$
$$0$$

Here there is a premature termination. Hence N_e and N_o have a common even factor. By inspection, this factor is found to be $s^2 + 1$; hence $N(s)$ has zeros at $\pm j1$. Therefore, $Y(s) = 1/Z(s)$ has poles there. These poles can be removed by partial-fraction expansion:

$$Y(s) = \frac{K_1 s}{s^2 + 1} + Y_1(s)$$

where the residue is

$$K_1 = \left[\frac{s^2 + 1}{s} Y(s) \right]_{s=j1} = \left(\frac{s^2 + 1}{s} \frac{s^3 + 3s^2 + 4s + 1}{2s^3 + 3s^2 + 2s + 3} \right)_{s=j}$$

$$K_1 = \left[\frac{s^3 + 3s^2 + 4s + 1}{s(2s + 3)} \right]_{s=j} = \frac{-j - 3 + 4j + 1}{j(2j + 3)} = 1$$

Hence, the realization can begin with a shunt LC branch (Fig. 4-5a). The remainder admittance is

$$Y_1(s) = Y(s) - \frac{K_1 s}{s^2 + 1} = \frac{s^3 + 3s^2 + 4s + 1}{(s^2 + 1)(2s + 3)} - \frac{s}{s^2 + 1}$$

$$Y_1(s) = \frac{(s^2 + 1)(s + 1)}{(s^2 + 1)(2s + 3)} = \frac{s + 1}{2s + 3}$$

$Y_1(s)$ can be recognized as an RC admittance and could be treated as such. Instead, however, we proceed in a more systematic fashion. We inspect $Y_1(s)$ for $j\omega$-axis zeros (its $j\omega$-axis poles have been removed): it has none. Next we examine the behavior of its real part along the $j\omega$ axis:

$$\text{Re } Y_1(j\omega) = \text{Re } \frac{j\omega + 1}{j2\omega + 3} = \frac{2\omega^2 + 3}{4\omega^2 + 9} = \frac{1}{3} + \frac{2}{3} \frac{\omega^2}{4\omega^2 + 9}$$

Clearly, the minimum value of Re $Y_1(j\omega)$ is $\tfrac{1}{3}$, reached at $\omega = 0$. This value is next removed as a shunt conductance (Fig. 4-5b). The remainder is

$$Y_2(s) = Y_1(s) - \tfrac{1}{3} = \frac{s}{6s + 9}$$

Now $Z_2(s) = 1/Y_2(s)$ has a pole at $s = 0$, which can be removed:

$$Z_2(s) = \frac{6s + 9}{s} = \frac{9}{s} + Z_3(s) \qquad Z_3(s) = 6$$

Hence, the circuit is as shown in Fig. 4-5c.

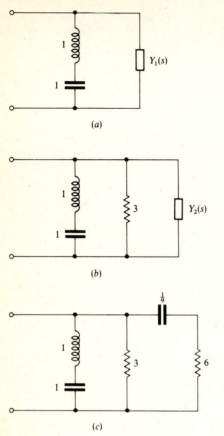

(a)

(b)

(c)

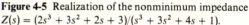

Figure 4-5 Realization of the nonminimum impedance $Z(s) = (2s^3 + 3s^2 + 2s + 3)/(s^3 + 3s^2 + 4s + 1)$.

As the example illustrates, the following steps can be attempted in the design of an *RCLM* impedance:

1. Test $Z(s)$ for $j\omega$-axis poles. If there are any, remove them, thereby realizing a series reactance branch of the circuit.
2. Test the remainder $Z_1(s)$ for $j\omega$-axis zeros. If there are any, remove the corresponding poles from $Y_1(s) = 1/Z_1(s)$. Return to step 1 and repeat steps 1 and 2 until the remainder has no more $j\omega$-axis poles or zeros. Then proceed to step 3.
3. Find the minimum value which the real part of the remainder $Z(s)$ [or $Y(s)$] assumes along the $j\omega$ axis. Remove this value as a series resistance (or shunt conductance). Return to step 1 to test the new remainder for $j\omega$-axis zeros and poles.

It will now be shown that for a PR $Z(s)$ these operations always give positive element values and leave a PR remainder. Specifically, a PR function $Z(s)$ has a real positive residue at any $j\omega$-axis pole (condition c'). Hence the corresponding partial-fraction term K_0/s or $K_\infty s$ or $K_1 s/(s^2 + \omega_1^2)$ can be realized as a branch containing a positive C or a positive L or an LC tuned circuit composed of a positive L and C. Thus the removal of such a pole gives positive elements in the realized branch. The impedance left after the pole removal has the same real part along the $j\omega$ axis as the original impedance since $K_0/j\omega$ or $K_\infty j\omega$ or $K_1 j\omega/(-\omega^2 + \omega_1^2)$ are all pure imaginary. Hence if condition b', that is, Re $Z(j\omega) \geq 0$, holds for the original impedance, it holds also for the remainder. Also, the removal of a $j\omega$-axis pole affects neither the location nor the residue of the other poles of $Z(s)$; hence if $Z(s)$ meets condition c', the remainder will too. Therefore, if $Z(s)$ is PR, so is the remainder.

In conclusion, step 1 gives positive element values in the realized series reactance branch and leaves a PR remainder if $Z(s)$ is PR. A dual argument proves the same result for step 2.

Next, step 3 reduces the real part of $Z(s)$ [or $Y(s)$] by its (positive) $j\omega$-axis minimum value. This leaves condition b' valid for a PR function $Z(s)$ [or $Y(s)$] and has no effect on its poles and residues. Hence, the resistor realized in the step will have a positive element value, and the remainder will be PR if $Z(s)$ or $Y(s)$ was PR.

Note that step 1 or 2 reduces the degree by 1 if the removed pole is at $\omega = 0$ or ∞ and by 2 if it is at some finite nonzero frequency ω_1. The removal of a constant (R or G) leaves the degree unchanged but sets the stage for a subsequent pole removal. The purpose of all three steps is of course to whittle $Z(s)$ away by gradually reducing its degree to zero.

A constant removal always makes a pole removal possible if the minimum value of Re $Z(j\omega)$ [or Re $Y(j\omega)$] occurs for $\omega = 0$ or ∞. For example, let the minimum of Re $Z(j\omega)$ be located at $\omega = 0$ (Fig. 4-6). Then $Z(j\omega) - R_1$ is pure imaginary at $\omega = 0$. However, since even powers of s are real while odd powers are imaginary for $s = j\omega$, it follows that Im $Z(j\omega)$ is an odd rational function in ω. Such a function is in one of the forms

$$\text{Im } Z(j\omega) = \omega \frac{P_1(\omega^2)}{P_2(\omega^2)} \quad \text{or} \quad \text{Im } Z(j\omega) = \frac{P_1(\omega^2)}{\omega P_2(\omega^2)} \tag{4-11}$$

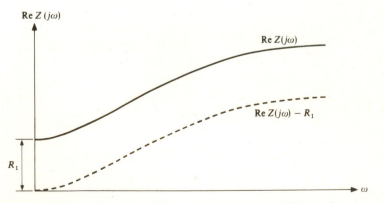

Figure 4-6 Real-part function with its minimum R_0 assumed for $\omega = 0$.

where $P_1(\omega^2)$ and $P_2(\omega^2)$ are even polynomials in ω. If the first form holds, both the real and imaginary parts of the remainder $Z_1(s) \triangleq Z(s) - R_1$ will be zero at $\omega = 0$. Hence, $Y_1(s) = 1/Z_1(s)$ will have a removable pole at $s = 0$. If the second form holds,† then $Z_1(s)$ has a pole at $s = 0$ which can be removed.

Similar arguments hold for the case when the minimum of the real part occurs for $\omega \to \infty$ and, of course, when $Y(s)$ rather than $Z(s)$ is being reduced.

Hence, if either of the following conditions holds:

1. $Z(s)$ has a $j\omega$-axis pole or zero.
2. Re $Z(j\omega)$ or Re $Y(j\omega)$ has its minimum at $\omega = 0$ or ∞.

the degree of $Z(s)$ can be reduced by an immediate pole removal or by a constant removal followed by a pole removal.

Now we introduce a number of new definitions. An impedance function $Z(s)$ which has no $j\omega$-axis poles and from which therefore no series reactance can be extracted using pole removal will be called a *minimum-reactance* function. Dually, an admittance function without any $j\omega$-axis poles will be called a *minimum-susceptance* function. A function $Z(s)$ [or $Y(s)$] whose real part is zero at any point along the $j\omega$ axis will be called a *minimum-real-part* function; from such function no series resistance (or shunt conductance) can be extracted without destroying its PR character.

Finally, a PR immittance function which is simultaneously minimum reactance, minimum susceptance, and minimum real part will be called simply a *minimum function*. Clearly, a minimum function has neither zeros nor poles on the $j\omega$ axis; its real part reaches a minimum of value 0 at some frequency ω_1 along the $j\omega$ axis. In addition, as our discussions accompanying Eq. (4-11) showed, the condition $0 < \omega_1 < \infty$ must hold for the location of the real-part minimum. Figure 4-7 illustrates a possible behavior for the real part of a minimum function.

† This cannot happen in the sequence of steps described earlier since the pole of $Z(s)$ at $s = 0$ would have been removed in step 1.

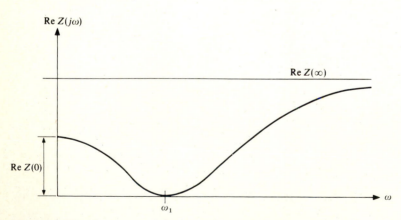

Figure 4-7 The real part of a minimum function $Z(s)$ along the $j\omega$ axis.

By definition, no pole removal or constant removal can be carried out on a minimum function, and hence such an immittance function can no longer be reduced in degree by any of the operations familiar to us at this stage.

Example 4-7 Test whether the impedance

$$Z(s) = \frac{5s^2 + 18s + 8}{s^2 + s + 10}$$

is a minimum function.

By inspection, $Z(s)$ is seen to have no $j\omega$-axis zeros or poles. Its real part along the $j\omega$ axis is

$$\text{Re } Z(j\omega) = \frac{5\omega^4 - 40\omega^2 + 80}{(10 - \omega^2)^2 + \omega^2} = 5\frac{(\omega^2 - 4)^2}{\omega^4 - 19\omega^2 + 100}$$

Clearly, Re $Z(j\omega)$ has a double zero at $\omega_1 = 2$; it is positive everywhere else. Its behavior is therefore as shown in Fig. 4-7, with $\omega_1 = 2$, Re $Z(0) = 0.8$, and Re $Z(\infty) = 5$.

Hence, $Z(s)$ is a minimum function.

For more complicated functions we need of course a continued-fraction expansion to reveal any $j\omega$-axis zeros and poles of $Z(s)$ and the Sturm test to find any double zeros of Re $Z(j\omega)$.

4-3 BRUNE'S IMPEDANCE SYNTHESIS METHOD

Consider an impedance function $Z(s)$ which is minimum reactance and minimum susceptance and whose real part has a $j\omega$-axis minimum at a finite nonzero frequency ω_1 (Fig. 4-8). Removal of the constant R_1 results in the realization of a series resistor (Fig. 4-9a) and a PR minimum-function remainder $Z_1(s) = Z(s) - R_1$ such that Re $Z_1(j\omega_1) = 0$ (Fig. 4-8). Consequently, $Z_1(j\omega_1)$ is pure imaginary and can be written as

$$Z_1(j\omega_1) = jX_1 \tag{4-12}$$

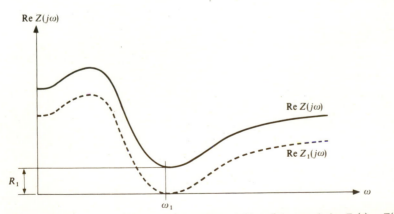

Figure 4-8 Real-part responses of an impedance $Z(s)$ and the remainder $Z_1(s) = Z(s) - R_1$.

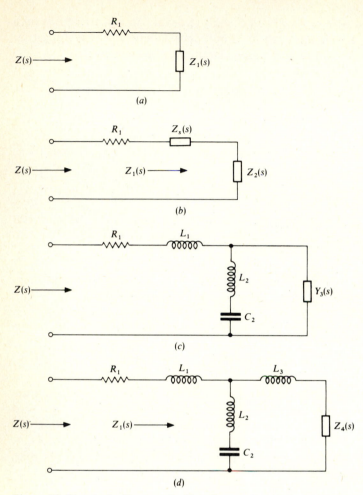

Figure 4-9 Steps in the Brune cycle, $X_1 < 0$.

where X_1 is real. Since $Z_1(s)$ is a minimum function, it is no longer subject to either pole removal or constant removal, as discussed in the preceding section. However, an imaginative extension of the pole-removal process, due to Otto Brune, manages to get around the defenses of the minimum function and to realize a physical impedance from it. The details of the derivation differ slightly according to the sign of X_1 in Eq. (4-12). Hence, the three cases $X_1 < 0$, $X_1 = 0$, and $X_1 > 0$ will be discussed separately.

Case 1: $X_1 < 0$

We first try to extract a series element $Z_s(s)$ such that the remainder $Z_2(s)$ is zero at $j\omega_1$ (Fig. 4-9b). Therefore we require

$$Z_s(j\omega_1) = Z_1(j\omega_1) - Z_2(j\omega_1) = jX_1 \qquad (4\text{-}13)$$

Since the unknown element $Z_s(s)$ thus has a negative reactance X_1 at ω_1, the "obvious" choice is a capacitor of value C_1 such that $1/\omega_1 C_1 = |X_1|$. However, the remainder $Z_2(s) = Z_1(s) - 1/sC_1$ will then have a pole at $s = 0$ with a residue $-1/C_1$ (recall that Z_1, a minimum function, has no pole at $s = 0$ or, for that matter, anywhere else on the $j\omega$ axis). Hence, $Z_2(s)$ would not be PR, and we would not know how to handle the rest of the design.

Instead of going the obvious way, it turns out to be wiser to bite the bullet immediately, and to extract a *negative inductor* of value $L_1 < 0$. Then (4-13) gives

$$j\omega_1 L_1 = jX_1 \qquad L_1 = \frac{X_1}{\omega_1} < 0 \tag{4-14}$$

The negative inductor will, of course, have to be eliminated later on if the circuit is to be realized with passive elements.

Now the remainder is

$$Z_2(s) = Z_1(s) - sL_1 = Z_1(s) + s|L_1| \tag{4-15}$$

which is a sum of two PR functions and thus PR itself. Since it also has a zero at $s_1 = j\omega_1$ and a pole at $s \to \infty$, we can now perform two pole removals. First, $Y_2(s) = 1/Z_2(s)$ has poles at $\pm j\omega_1$, which can be removed the usual way:

$$Y_2(s) = \frac{K_1 s}{s^2 + \omega_1^2} + Y_3(s) \tag{4-16}$$

where

$$K_1 = \left[\frac{s^2 + \omega_1^2}{s} Y_2(s) \right]_{s = j\omega_1} \tag{4-17}$$

is a positive real residue. Writing

$$\frac{K_1 s}{s^2 + \omega_1^2} = \frac{1}{\dfrac{s}{K_1} + \dfrac{1}{sK_1/\omega_1^2}} = \frac{1}{sL_2 + \dfrac{1}{sC_2}} \tag{4-18}$$

we get $L_2 = 1/K_1$ and $C_2 = K_1/\omega_1^2$ (Fig. 4-9c). Note that $L_2 > 0$ and $C_2 > 0$ since $K_1 > 0$. Furthermore, as discussed in the preceding section, the remainder

$$Y_3(s) = Y_2(s) - \frac{s/L_2}{s^2 + \omega_1^2} \tag{4-19}$$

is PR since it results from a step 2 operation on the PR function $Y_2(s)$.

Next, we perform another pole removal. As mentioned earlier, $Z_2(s)$ has a pole at $s \to \infty$, and so $Y_2(s)$ has a zero there. It then follows from (4-19) that $Y_3(s) \to 0$ as $s \to \infty$. Hence, $Z_3(s) \triangleq 1/Y_3(s)$ has a pole at infinite s. Since $Z_3(s)$ is a PR function, this pole can be removed by extracting a positive inductance L_3 (Fig. 4-9d). The remainder

$$Z_4(s) = Z_3(s) - sL_3 \tag{4-20}$$

will again be PR.

From Fig. 4-9d, by inspection

$$Z_1(s) = sL_1 + \cfrac{1}{\cfrac{1}{sL_2 + 1/sC_2} + \cfrac{1}{sL_3 + Z_4(s)}} \tag{4-21}$$

Now let $s \to \infty$. Since then $1/sC_2 \to 0$ and $Z_4(s)$ remains finite, we have

$$Z_1(s) \to sL_1 + \cfrac{1}{\cfrac{1}{sL_2} + \cfrac{1}{sL_3}} = s\left(L_1 + \frac{L_2 L_3}{L_2 + L_3}\right) \tag{4-22}$$

Hence, it appears that $Z_1(s)$ has a pole at infinite frequency with a residue

$$K_\infty = L_1 + \frac{L_2 L_3}{L_2 + L_3} \tag{4-23}$$

Recall, however, that $Z_1(s)$ was a minimum function: it *cannot* have a pole at $s \to \infty$. Hence, we must have $K_\infty = 0$, or

$$L_1 L_2 + L_1 L_3 + L_2 L_3 = 0 \qquad \frac{1}{L_1} + \frac{1}{L_2} + \frac{1}{L_3} = 0 \tag{4-24}$$

The value of L_3 can be obtained from (4-24):

$$L_3 = \frac{-L_1 L_2}{L_1 + L_2} = \frac{|L_1| L_2}{-|L_1| + L_2} \tag{4-25}$$

Next, to absorb the negative inductor L_1, we utilize the potential equivalence of the two two-ports shown in Fig. 4-10a and b. The voltage-current relations for the coupled inductor are (Ref. 4, chap. 8):

$$V_1 = sL_p I_1 + sM I_2 \qquad V_2 = sM I_1 + sL_s I_2 \tag{4-26}$$

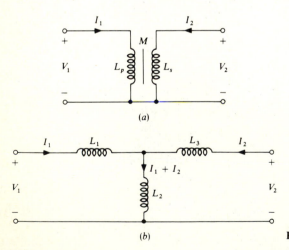

(a)

(b)

Figure 4-10 Equivalent two-ports.

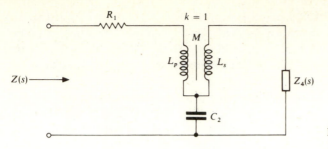

Figure 4-11 Brune section.

while for the T circuit of Fig. 4-10b we can find by inspection the equations

$$V_1 = sL_1 I_1 + sL_2(I_1 + I_2) \qquad V_2 = sL_3 I_2 + sL_2(I_1 + I_2) \tag{4-27}$$

or $\qquad V_1 = s(L_1 + L_2)I_1 + sL_2 I_2 \qquad V_2 = sL_2 I_1 + s(L_2 + L_3)I_2 \tag{4-28}$

For the two-ports to be equivalent, the relations (4-26) and (4-28) must be the same. Hence, we must have

$$L_p = L_1 + L_2 \qquad M = L_2 \qquad L_s = L_2 + L_3 \tag{4-29}$$

Therefore, if (4-29) holds, the inductive T composed of L_1, L_2, and L_3 in Fig. 4-9d can be replaced by the coupled-inductor circuit of Fig. 4-10a. The result is the circuit shown in Fig. 4-11. The first part of the circuit, which consists of R_1, C_2, and the coupled inductor and which we realized above, is called a *Brune section*. The realization of a Brune section is called a *Brune cycle*.

If the Brune section of Fig. 4-11 is to be realizable with a physical transformer, the self-inductances L_p and L_s must be positive:

$$L_p = L_1 + L_2 = -|L_1| + L_2 > 0 \qquad L_s = L_2 + L_3 > 0 \tag{4-30}$$

Also, the *coefficient of coupling*

$$k \triangleq \frac{|M|}{\sqrt{L_p L_s}} \tag{4-31}$$

must satisfy $0 \le k \le 1$.

To show that (4-30) holds, we use (4-24):

$$L_p = L_2 - |L_1| = L_2 - \frac{L_2 L_3}{L_2 + L_3} = \frac{L_2^2}{L_2 + L_3} > 0 \tag{4-32}$$

Of course, $L_s = L_2 + L_3$ is also positive, since L_2 and L_3 are both positive.

To show that the value of k is between 0 and 1, (4-29) and (4-24) are used:

$$k = \frac{|M|}{\sqrt{L_p L_s}} = \frac{L_2}{\sqrt{(L_1 + L_2)(L_2 + L_3)}}$$
$$k = \frac{L_2}{\sqrt{L_1 L_2 + L_1 L_3 + L_2 L_3 + L_2^2}} = \frac{L_2}{L_2} = 1 \tag{4-33}$$

Hence the inductors are *closely coupled;* i.e., all magnetic flux is coupled to all turns of both windings. Transformers with closely coupled windings can be realized using pot cores or toroid cores, and they can be built reliably for a wide frequency range (about 100 Hz to 1 MHz). Hence, the Brune section can actually be used for practical applications in that range.

Of course, for the Brune cycle to be useful, it must decrease the degree of $Z(s)$. Let $Z(s)$ be a ratio of two nth-degree polynomials.† Abbreviating this statement in the form $Z(s) \doteq (n)/(n)$, we obtain for the degrees of the other immittances the following:

$$Z_1(s) = Z(s) - R_1 \doteq \frac{(n)}{(n)}$$

$$Z_2(s) = Z_1(s) + s|L_1| \doteq \frac{(n+1)}{(n)}$$

$$Y_3(s) = Y_2(s) - \frac{s/L_2}{s^2 + \omega_1^2} \doteq \frac{(n-2)}{(n-1)}$$ (4-34)

$$Z_4(s) = Z_3(s) - sL_3 \doteq \frac{(n-2)}{(n-2)}$$

Here we used our knowledge that a pole removal at $\pm j\omega_1$ decreases both numerator and denominator degrees by 2, while a pole removal at infinite

† Since $Z(s)$, a minimum function, has neither a pole nor a zero at $s \to \infty$, the degrees of its numerator and denominator must be the same.

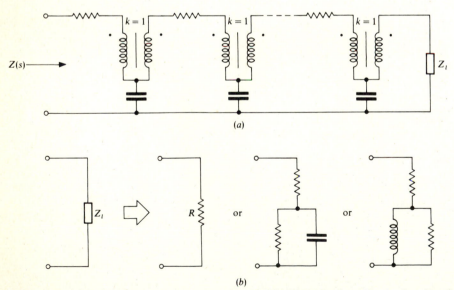

Figure 4-12 (a) Brune realization of an impedance $Z(s)$; (b) realizations for the last impedance Z_l.

frequency reduces only the numerator degree by 1 (recall the discussions in Sec. 3-2).

As (4-34) shows, the remainder $Z_4(s)$ of the Brune cycle (Fig. 4-11) has a degree which is lower by 2 than that of $Z(s)$. Hence, repeating the Brune cycle enough times means that the degree of $Z(s)$ can be reduced to 0 (if n is even) or to 1 (if n is odd). The last impedance Z_l is then either simply a resistor (if n is even) or a three-element RC or RL impedance (if n is odd). The circuit realizing $Z(s)$ is illustrated in Fig. 4-12a;† the possible realizations of Z_l are shown in Fig. 4-12b.

Example 4-8 Synthesize

$$Z(s) = \frac{8s^2 + s + 4}{24s^3 + 11s^2 + 20s + 1}$$

Clearly $Z(s)$ is not a minimum impedance: it has a zero for $s \to \infty$. Removing the corresponding pole from $Y(s) = 1/Z(s)$ gives

$$Y(s) = sC_1 + Y_1(s) \qquad C_1 = \left[\frac{Y(s)}{s} \right]_{s \to \infty} = \frac{24}{8} = 3 \text{ F}$$

(Fig. 4-13a). This leaves a remainder

$$Y_1(s) = \frac{8s^2 + 8s + 1}{8s^2 + s + 4}$$

All the zeros and poles of $Z_1(s) = 1/Y_1(s)$ are inside the LHP. Its real part for $s = j\omega$ is

$$\text{Re } Z_1(j\omega) = \frac{(-8\omega^2 + 1)(-8\omega^2 + 4) + 8\omega^2}{(-8\omega^2 + 1)^2 + 64\omega^2}$$

or

$$\text{Re } Z_1(j\omega) = \frac{64\omega^4 - 32\omega^2 + 4}{(-8\omega^2 + 1)^2 + 64\omega^2} = \frac{(8\omega^2 - 2)^2}{(-8\omega^2 + 1)^2 + 64\omega^2}$$

† Note that at this stage we have proved only that the circuit of Fig. 4-12a is applicable if $X_1 < 0$ in all Brune cycles. It will be shown soon that the Brune section of Fig. 4-11 is valid also for $X_1 \geq 0$ and hence Fig. 4-12a is generally valid.

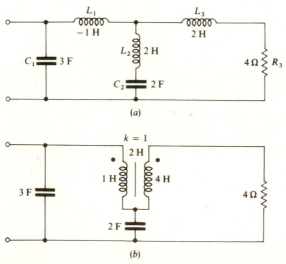

(a)

(b)

Figure 4-13 Brune synthesis of the impedance $Z(s) = (8s^2 + s + 4)/(24s^3 + 11s^2 + 20s + 1)$.

Clearly, Re $Z_1(j\omega)$ has double zeros at $\omega_1 = 0.5$ and at $-\omega_1$. It is therefore a minimum impedance. Also,

$$Z_1(j\omega_1) = -j0.5$$

Hence, $X_1 = -0.5 < 0$ here, and thus the Brune process described before is applicable. From (4-14)

$$L_1 = \frac{X_1}{\omega_1} = -\frac{0.5}{0.5} = -1 \text{ H}$$

(Fig. 4-13a). From (4-15)

$$Z_2(s) = Z_1(s) + s1 = \frac{8s^3 + 16s^2 + 2s + 4}{8s^2 + 8s + 1}$$

By its construction, $Z_2(s)$ must have zeros at $\pm j\omega_1$. Hence, using (4-16),

$$Y_2(s) = \frac{8s^2 + 8s + 1}{8(s^2 + 0.25)(s + 2)} = \frac{K_1 s}{s^2 + 0.25} + Y_3(s)$$

where

$$K_1 = \left[\frac{s^2 + 0.25}{s} Y_2(s) \right]_{s = j0.5} = \tfrac{1}{2}$$

From (4-18), therefore (Fig. 4-13a),

$$L_2 = \frac{1}{K_1} = 2 \text{ H} \qquad C_2 = K_1 \omega_1^{-2} = 2 \text{ F}$$

The remainder is

$$Y_3(s) = Y_2(s) - \frac{K_1 s}{s^2 + 0.25} = \frac{0.5}{s + 2}$$

where an $s^2 + 0.25$ factor was canceled in numerator and denominator. Now $Z_3(s)$ has a pole at $s \to \infty$ with a residue equal to 2:

$$Z_3(s) = 2s + 4$$

Removing this pole gives (Fig. 4-13a)

$$L_3 = 2 \text{ H}$$

and leaves a remainder

$$R_3 = 4 \ \Omega$$

The circuit of Fig. 4-13a can next be transformed into the final circuit utilizing the equivalence of Fig. 4-10. The parameters of the coupled coil replacing L_1, L_2, and L_3 are, by (4-29),

$$L_p = L_1 + L_2 = -1 + 2 = 1 \text{ H}$$

$$M = L_2 = 2 \text{ H} \qquad L_s = L_2 + L_3 = 4 \text{ H}$$

The resulting circuit is shown in Fig. 4-13b.

All the above derivations referred to the case when X_1 was negative. Next, the other two possibilities will be examined.

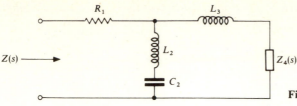

Figure 4-14 Brune section for $X_1 = 0$.

Case 2: $X_1 = 0$†

For this case, the remainder $Z_1(s) = Z(s) - R_1$ is zero at $s = j\omega_1$. Hence, $L_1 = 0$, $Y_2(s) = Y_1(s)$, and when Eqs. (4-16) to (4-20) are used, the circuit of Fig. 4-14 is obtained. Since this circuit contains only positive elements, there is no need to introduce coupled inductors.

Case 3: $X_1 > 0$

The process followed for this case is the exact dual of that used for $X_1 < 0$. After extracting R_1 (Fig. 4-15a), the value of the admittance $Y_1(s) = 1/Z_1(s)$ at $s = j\omega_1$ is

$$Y_1(j\omega_1) = \frac{1}{jX_1} = -j\frac{1}{X_1} \qquad (4-35)$$

† Strictly speaking, this cannot occur if $Z(s)$ is a minimum impedance.

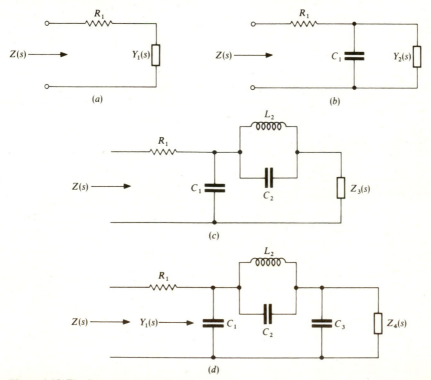

Figure 4-15 The Brune cycle for $X_1 > 0$.

i.e., it is negative imaginary. Extracting from $Y_1(s)$ a negative capacitance C_1 such that

$$j\omega_1 C_1 = Y_1(j\omega_1) = \frac{1}{jX_1} \qquad C_1 = -\frac{1}{\omega_1 X_1} < 0 \qquad (4\text{-}36)$$

leaves a PR remainder (Fig. 4-15b)

$$Y_2(s) = Y_1(s) - sC_1 \qquad (4\text{-}37)$$

which is zero at $s = \pm j\omega_1$. Hence, $Z_2(s) = 1/Y_2(s)$ has poles there, which can be removed the usual way

$$Z_2(s) = \frac{K_1 s}{s^2 + \omega_1^2} + Z_3(s) \qquad (4\text{-}38)$$

where $K_1 > 0$. Since the first term can be written

$$\frac{K_1 s}{s^2 + \omega_1^2} = \frac{1}{\dfrac{s}{K_1} + \dfrac{1}{sK_1/\omega_1^2}} = \frac{1}{sC_2 + \dfrac{1}{sL_2}} \qquad (4\text{-}39)$$

this corresponds to a parallel-tuned circuit (Fig. 4-15c) with positive L_2 and C_2. Finally, since $Y_2(s)$ has a pole at $s \to \infty$, $Z_2(s)$ and the PR remainder

$$Z_3(s) = Z_2(s) - \frac{K_1 s}{s^2 + \omega_1^2} \qquad (4\text{-}40)$$

have zeros there. Thus, $Y_3(s)$ has a pole at infinite s which can be removed by extracting a shunt capacitor C_3 (Fig. 4-15d). All elements in the circuit are positive, with the exception of C_1, and the remainder $Z_4(s)$ is PR.

Since $Y_1(s)$ has no pole at $s \to \infty$, the residue of its apparent pole which can readily be found from Fig. 4-15d to be

$$K_\infty = C_1 + \frac{C_2 C_3}{C_2 + C_3} \qquad (4\text{-}41)$$

must equal zero. This gives

$$C_1 C_2 + C_1 C_3 + C_2 C_3 = 0 \qquad \frac{1}{C_1} + \frac{1}{C_2} + \frac{1}{C_3} = 0 \qquad (4\text{-}42)$$

which is the dual relation of (4-24).

Finally, the negative capacitor C_1 must be absorbed to obtain a realizable circuit. This is accomplished by making use of the potential equivalence of the two circuits in Fig. 4-16. From Fig. 4-16a, by inspection,

$$I = I_1 - sC_1 V_1 = sC_3 V_2 - I_2 = (V_1 - V_2)\left(sC_2 + \frac{1}{sL_2}\right) \qquad (4\text{-}43)$$

From the last two equalities,

$$sC_1 V_1 + sC_3 V_2 = I_1 + I_2$$

$$-\left(sC_2 + \frac{1}{sL_2}\right)V_1 + \left(sC_2 + \frac{1}{sL_2} + sC_3\right)V_2 = I_2 \qquad (4\text{-}44)$$

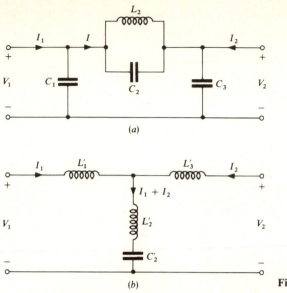

(a)

(b)

Figure 4-16 Equivalent two-ports.

We shall use Cramer's rule to solve (4-44) for V_1 and V_2. The determinant of the left-hand-side coefficients is

$$D = \begin{vmatrix} sC_1 & sC_3 \\ -sC_2 - \dfrac{1}{sL_2} & sC_2 + \dfrac{1}{sL_2} + sC_3 \end{vmatrix}$$

$$= s^2 C_1 C_2 + s^2 C_1 C_3 + \frac{C_1}{L_2} + s^2 C_2 C_3 + \frac{C_3}{L_2}$$

$$= \frac{C_1 + C_3}{L_2} \tag{4-45}$$

where (4-42) was used.

Hence, by Cramer's rule

$$V_1 = \frac{1}{D} \begin{vmatrix} I_1 + I_2 & sC_3 \\ I_2 & sC_2 + \dfrac{1}{sL_2} + sC_3 \end{vmatrix}$$

$$= I_1 \left(s \frac{C_2 + C_3}{C_1 + C_3} L_2 + \frac{1}{s} \frac{1}{C_1 + C_3} \right) + I_2 \left(s \frac{C_2}{C_1 + C_3} L_2 + \frac{1}{s} \frac{1}{C_1 + C_3} \right) \tag{4-46}$$

$$V_2 = \frac{1}{D} \begin{vmatrix} sC_1 & I_1 + I_2 \\ -sC_2 - \dfrac{1}{sL_2} & I_2 \end{vmatrix}$$

$$= I_1 \left(s \frac{C_2}{C_1 + C_3} L_2 + \frac{1}{s} \frac{1}{C_1 + C_3} \right) + I_2 \left(s \frac{C_1 + C_2}{C_1 + C_3} L_2 + \frac{1}{s} \frac{1}{C_1 + C_3} \right)$$

On the other hand, the voltage-current relations of the circuit of Fig. 4-16b are, by inspection,

$$V_1 = sL_1'I_1 + \left(sL_2' + \frac{1}{sC_2'}\right)(I_1 + I_2)$$

$$= \left[s(L_1' + L_2') + \frac{1}{sC_2'}\right]I_1 + \left(sL_2' + \frac{1}{sC_2'}\right)I_2 \qquad (4\text{-}47)$$

$$V_2 = sL_3'I_2 + \left(sL_2' + \frac{1}{sC_2'}\right)(I_1 + I_2)$$

$$= \left(sL_2' + \frac{1}{sC_2'}\right)I_1 + \left[s(L_2' + L_3') + \frac{1}{sC_2'}\right]I_2$$

The two circuits will be equivalent if we can arrange to make their voltage-current relations the same.

Equating the corresponding coefficients of I_1 and I_2 in (4-46) and (4-47) gives directly

$$L_1' + L_2' = \frac{C_2 + C_3}{C_1 + C_3}L_2 \qquad C_2' = C_1 + C_3$$

$$L_2' = \frac{C_2}{C_1 + C_3}L_2 \qquad L_2' + L_3' = \frac{C_1 + C_2}{C_1 + C_3}L_2 \qquad (4\text{-}48)$$

From (4-48), using (4-42), we have

$$C_2' = C_1 + C_3 = \frac{C_3^2}{C_2 + C_3} > 0$$

$$L_2' = \frac{C_2 L_2}{C_1 + C_3} = \frac{C_2(C_2 + C_3)}{C_3^2}L_2 > 0$$

$$L_1' = \frac{C_2 + C_3}{C_1 + C_3}L_2 - L_2' = \frac{C_2 + C_3}{C_3}L_2 > 0 \qquad (4\text{-}49)$$

$$L_3' = \frac{C_1 + C_2}{C_1 + C_3}L_2 - L_2' = -\frac{C_2}{C_3}L_2 < 0$$

Hence, in the circuit of Fig. 4-16b all elements except L_3' are positive. Furthermore, as in (4-24), we have

$$\frac{1}{L_1'} + \frac{1}{L_2'} + \frac{1}{L_3'} = \frac{C_3}{L_2}\left[\frac{1}{C_2 + C_3} + \frac{C_3}{C_2(C_2 + C_3)} - \frac{1}{C_2}\right] = 0 \qquad (4\text{-}50)$$

From the equivalence shown in Fig. 4-10 we can now conclude, as we did for the $X_1 < 0$ case, that the circuit of Fig. 4-16b (and hence of Fig. 4-16a) is equivalent to the coupled-inductor section shown in Fig. 4-11. Using (4-30) and (4-49), we can obtain the final element values

$$L_p = L_1' + L_2' = \left(\frac{C_2 + C_3}{C_3}\right)^2 L_2 > 0$$

$$L_s = L_2' + L_3' = \left(\frac{C_2}{C_3}\right)^2 L_2 > 0 \tag{4-51}$$

$$M = L_2' = \frac{C_2(C_2 + C_3)}{C_3^2} L_2 > 0$$

Also, from (4-31) and (4-50), we find

$$k = \frac{M}{\sqrt{L_p L_s}} = \frac{L_2'}{\sqrt{(L_1' + L_2')(L_2' + L_3')}} = 1 \tag{4-52}$$

Hence, the transformer is again closely coupled.

We conclude that the circuit of Fig. 4-11 is applicable to both the $X_1 < 0$ and $X_1 > 0$ cases. In fact, it can also be used for $X_1 = 0$. (However, since we usually try to avoid coupled inductors, the circuit of Fig. 4-14 is preferred for that case.) Hence, the representation of Fig. 4-12a is generally valid. Furthermore, Eqs. (4-18), (4-30), (4-49), and (4-51) reveal that *in all sections all element values are positive in the circuit, if Z(s) is PR.*

Therefore, the PR conditions are seen to be not only *necessary* (as proved in Chap. 2) but also *sufficient* for the realization of an impedance $Z(s)$ using only positive *RCLM* elements.

It should finally be noted that the lengthy derivation given for the $X_1 > 0$ case was necessary solely to prove the realizability of the Brune section of Fig. 4-11 for this situation. In actual numerical design we have two options. We can follow the derivations of Eqs. (4-35) to (4-40) and then use (4-49), (4-51), and (4-52) to get the final element values C_2', L_p, L_s, and M. Or we can immediately assume the existence of the circuit of Fig. 4-16b and steer the design process accordingly. The latter procedure is somewhat simpler and is the one normally used.

Example 4-9 Synthesize

$$Z(s) = \frac{6s^2 + 19s + 18}{s^2 + s + 10}$$

Since the zeros and poles of $Z(s)$ are inside the open LHP, we have to find and examine the behavior of its real part along the $j\omega$ axis. Now

$$\text{Re } Z(j\omega) = \frac{(-6\omega^2 + 18)(-\omega^2 + 10) + 19\omega^2}{(-\omega^2 + 10)^2 + \omega^2}$$

$$= \frac{6\omega^4 - 59\omega^2 + 180}{\omega^4 - 19\omega^2 + 100}$$

Next, we replace ω^2 by x and find the minimum of Re $Z(x)$ the familiar way, by equating $d[\text{Re } Z(x)]/dx$ to zero. The numerator of

$$\frac{d}{dx}[\text{Re } Z(x)] = \frac{d}{dx}\frac{6x^2 - 59x + 180}{x^2 - 19x + 100}$$

is $-55x^2 + 840x - 2480$, which is zero at $x_1 = 4$ and at $x_2 = 620/55 \approx 11.273$. Since Re $Z(x_1) = 1$ and Re $Z(x_2) \approx 21.51$ (Fig. 4-17), we conclude that the minimum lies at $\omega_1^2 = 4$. We have to resort to Brune synthesis, since $Z(s)$ has no poles or zeros on the $j\omega$ axis and the minimum of its $j\omega$-axis real part lies at a finite nonzero ω_1. Therefore extracting

$$R_1 = \text{Re } Z(j\omega_1) = 1 \ \Omega$$

(Fig. 4-18a) leaves a remainder

$$Z_1(s) = Z(s) - R_1 = \frac{5s^2 + 18s + 8}{s^2 + s + 10}$$

Since $Z_1(j\omega_1) = Z_1(j2) = j6$, here $X_1 = 6 > 0$. Assuming directly the existence of the Brune section of Fig. 4-16b, we extract (Fig. 4-18a)

$$L_1' = \frac{X_1}{\omega_1} = 3 \ \text{H}$$

The remainder

$$Z_2(s) = Z_1(s) - sL_1 = \frac{-3s^3 + 2s^2 - 12s + 8}{s^2 + s + 10}$$

is now clearly *not* PR, but since we know that the final circuit *will* be realizable,† we proceed nevertheless. $Z_2(s)$ must have a zero at $s = \pm j2$; in fact

$$Z_2(s) = \frac{(s^2 + 4)(-3s + 2)}{s^2 + s + 10}$$

Hence we can remove the corresponding pole from $Y_2(s)$

$$Y_2(s) = \frac{K_1 s}{s^2 + 4} + Y_3(s)$$

which gives $K_1 = 0.5$ and leaves a remainder

$$Y_3(s) = \frac{2.5}{-3s + 2}$$

Now $L_2' = 1/K_1 = 2 \ \text{H}$ and $C_2' = K_1/\omega_1^2 = 0.125 \ \text{F}$ (Fig. 4-18a). Finally,

$$Z_3(s) = -1.2s + 0.8$$

giving $L_3' = -1.2 \ \text{H}$ and $R_3 = 0.8 \ \Omega$. As a check, we note that

$$\frac{1}{L_1'} + \frac{1}{L_2'} + \frac{1}{L_3'} = \frac{1}{3} + \frac{1}{2} - \frac{1}{1.2} = 0$$

To obtain the final circuit (Fig. 4-18b) we find $L_p = L_1' + L_2' = 5 \ \text{H}$ and $L_s = L_2' + L_3' = 0.8 \ \text{H}$.

† This is why we needed the tedious derivation for the $X_1 > 0$ case.

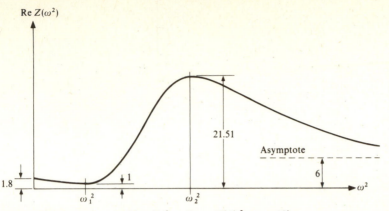

Figure 4-17 Real part of $Z(s) = (6s^2 + 19s + 18)/(s^2 + s + 10)$.

Counting each coupled inductor (transformer) as a single reactance, it is readily seen that the Brune realization (Fig. 4-12a) of an nth-degree impedance requires n reactances. In this sense, the Brune realization is *canonical*. As we pointed out earlier, it is also a *feasible* realization.

Other feasible realizations exist for *RCLM* one-ports. One which is of some theoretical interest is the *Bott-Duffin circuit*. It completely avoids the mutually coupled inductors inherent in the Brune realizations and still guarantees positive element values. A stiff price is paid, however, in terms of the large number of circuit elements required. As an illustration, the impedance realized in the last example (Fig. 4-18) has the Bott-Duffin realization shown in Fig. 4-19. The number of elements is about three times higher than for the Brune circuit. The disparity becomes rapidly larger for higher degrees, since the number of elements

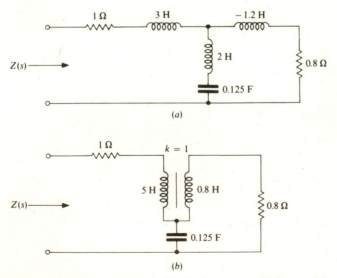

Figure 4-18 Brune synthesis of $Z(s) = (6s^2 + 19s + 18)/(s^2 + s + 10)$.

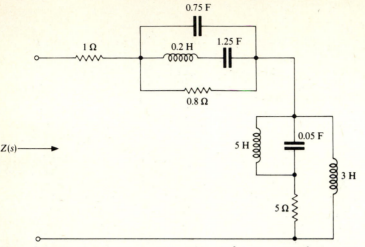

Figure 4-19 The Bott-Duffin circuit for $Z(s) = (6s^2 + 19s + 18)/(s^2 + s + 10)$.

in the Bott-Duffin circuit grows exponentially with n while it grows only linearly for the Brune realization. For this reason, the Bott-Duffin circuit is of limited usefulness and will not be discussed here. The reader interested in the topic and in other impedance-synthesis procedures should consult Refs. 1, 2, and 5 to 7. Extensions of the Brune process can also be found in Ref. 7.

4-4 SUMMARY

In this chapter the following subjects were discussed:

1. A new set of PR conditions, based on the augmented impedance Z_a was derived, and its equivalence to the sets of conditions given in Chap. 2 was proved (Sec. 4-1).
2. The Sturm test for polynomial zeros on the positive real axis was described (but not proved) and illustrated by working out examples involving simple and multiple zeros (Sec. 4-1).
3. An important theorem linking strictly Hurwitz polynomials and realizable reactance functions was proved and then used to reduce Hurwitz testing to Cauer expansion. The process was illustrated by examples (Sec. 4-1).
4. The above-mentioned results were used to assemble an efficient process of PR testing (Sec. 4-1).
5. It was shown how pole removals and constant removals can be used to reduce some impedance functions. Minimum-reactance, minimum-susceptance, minimum-real-part, and minimum functions were defined (Sec. 4-2).
6. Brune's impedance synthesis process was described and illustrated with examples; the Bott-Duffin realization technique was mentioned but not discussed.

   The design techniques discussed in this chapter are quite powerful, in that they allow the realization of *any* PR function with physical elements. They thus prove the sufficiency of the PR conditions whose *necessity* was established earlier, in Chap. 2.
   The circuits realized using the Brune synthesis need, in general, closely coupled inductors. This is somewhat of a nuisance in practical situations.
   This chapter concludes our discussions of lumped passive one-ports. The next two chapters are devoted to passive lumped two-port synthesis.

PROBLEMS

4-1 Perform the Sturm test on $P(x) = x^5 - 5x^4 + x^3 - 7x^2 + 2x + 1$ to determine whether $P(x) \geq 0$ holds for all $x \geq 0$.

4-2 Determine by inspection which of the following polynomials:

  1. Obviously satisfies $P(x) \geq 0$ for all $x \geq 0$.
  2. Cannot possibly satisfy the above condition.
  3. May or may not satisfy the condition.

  (a) $P(x) = x^7 + 5x^6 - x^5 + 3x^4 - x^3 + 11x^2 + x - 7$
  (b) $P(x) = 5x^5 + 7x^4 + 2x^3 + x^2 + 12x + 1$
  (c) $P(x) = x^4 - 7x^3 + x^2 - 9x + 15$
  (d) $P(x) = -x^4 + 3x^3 + 2x^2 + x + 1$
  (e) $P(x) = x^5 - 7x^4 + 9x^3 + x^2 + 17x$

4-3 Find and test the real part of

$$Z(s) = \frac{s^4 + 3s^3 + 5s^2 + 6s + 2}{s^4 + 6s^3 + 9s^2 + 11s + 6}$$

for condition b''.

4-4 Is $Q(s) = s^5 + s^4 + 6s^3 + 6s^2 + 25s + 25$ strictly Hurwitz, Hurwitz, or non-Hurwitz?

4-5 Is $Q(s) = s^4 + s^3 + 2s^2 + 3s + 1$ strictly Hurwitz, Hurwitz, or non-Hurwitz?

4-6 Show that if $Q(s)$ is either a pure even *or* a pure odd function of s, then it is Hurwitz if and only if

$$Z(s) \triangleq \frac{dQ(s)/ds}{Q(s)}$$

is a realizable reactance.

4-7 Apply the Hurwitz test described in Prob. 4-6 to $Q(s) = s^6 + 6s^4 + 11s^2 + 6$. Is $Q(s)$ Hurwitz?

4-8 Are the following polynomials strictly Hurwitz, Hurwitz, or non-Hurwitz?

  (a) $s^4 + 4s^3 + 4s^2 + 4s + 3$
  (b) $s^5 + 3s^4 + 3s^3 + 9s^2 + 2s + 6$
  (c) $s^5 + 6s^4 + 12s^3 + 12s^2 + 11s + 6$
  (d) $s^7 + 2s^6 + 5s^5 + 10s^4 + 7s^3 + 14s^2 + 3s + 4$

4-9 Which of the following functions are realizable impedances?

  (a) $\dfrac{s + 4}{(s + 1)^2(s + 2)^2}$ (b) $\dfrac{s^2 + 3s + 1}{s^2 + s}$

  (c) $\dfrac{s + 4}{s^3 + 2s^2 + 9}$ (d) $\dfrac{s^2 + 4s + 3}{s^2 + 8s + 12}$

  (e) $\dfrac{s^2 + 3}{s^2 + 1}$ (f) $\dfrac{s^4 + 3s^3 + 3s^2 + 2s + 1}{s^2 + 2s + 10}$

  (g) $\dfrac{s^2 + 2s + 25}{s + 4}$ (h) $\dfrac{s^3 + 2s + 3}{s^2 + 2s + 5}$

4-10 (a) How many Brune cycles are needed to realize an nth-degree minimum impedance $Z(s)$?

  (b) Show that for an odd-degree impedance $Z(s)$, the last impedance Z_l of the Brune realization (Fig. 4-12a) is in the form

$$Z_l(s) = \frac{a_1 s + a_0}{b_1 s + b_0}$$

  (c) Show that $Z_l(s)$ has the *RC* or *RL* realization illustrated in Fig. 4-12b. Which applies when?

4-11 Which of the following impedances are minimum functions? Find the location of the minimum of Re $Z(j\omega)$ for each function.

(a) $Z(s) = \dfrac{3s^2 + 2s + 1}{s^2 + 2s + 3}$ (b) $Z(s) = \dfrac{5s^2 + 4s + 1}{s^2 + 4s + 5}$

(c) $Z(s) = \dfrac{2s^2 + s + 3}{3s^2 + s + 2}$

4-12 Synthesize the impedances given in Prob. 4-11.

4-13 Find the Brune realizations for

(a) $Y(s) = \dfrac{s^2 + s + 1}{4s^2 + s + 1}$ (b) $Z(s) = \dfrac{3s^2 + 2s + 1}{s^2 + 2s + 3}$

4-14 Assume that the function in part (a) of Prob. 4-13 is an impedance and the one in part (b) an admittance. Repeat the realizations.

4-15 Synthesize

$$Z(s) = \frac{6s^2 + 19s + 18}{5s^2 + 18s + 8}$$

4-16 Let $Z_1 = \dfrac{s^2 + s + 1}{s + 1}$.

(a) Synthesize $Z_1(s)$.

(b) Find an impedance $Z_2(s)$ such that the parallel combination of Z_1 and Z_2 is a constant impedance R_0.

(c) Find the largest value of R_0 for a realizable $Z_2(s)$.

(d) Using the maximum value of R_0, synthesize $Z_2(s)$.

4-17 Synthesize

$$Z(s) = \frac{15s^2 + 30s + 60}{s^2 + s + 1}$$

4-18 Synthesize

$$Z(s) = \frac{3}{5} + \frac{\frac{6}{5}}{s + 1} - \frac{\frac{12}{5}}{s + 3}$$

REFERENCES

1. Balabanian, N.: "Network Synthesis," Prentice Hall, Englewood Cliffs, N.J., 1958.
2. Rupprecht, W.: "Netzwerksynthese," Springer, Berlin, 1972.
3. Ralston, A.: "A First Course in Numerical Analysis," McGraw-Hill, New York, 1965.
4. Desoer, C. A., and E. S. Kuh: "Basic Circuit Theory," McGraw-Hill, New York, 1969.
5. Guillemin, E. A.: "Synthesis of Passive Networks," Wiley, New York, 1957.
6. Van Valkenburg, M. E.: "Introduction to Modern Network Synthesis," Wiley, New York, 1960.
7. Unbehauen, R.: "Synthese elektrischer Netzwerke," Oldenbourg, Munich, 1972.

CHAPTER FIVE

LOSSLESS TWO-PORTS

5-1 TWO-PORT NETWORKS

Two-ports are the most commonly used circuits in electronics and communications. A two-port has an input port (terminal pair) and an output port. It is usually driven at the input port by a generator with internal impedance Z_G and terminated at its output port by a load impedance Z_L (Fig. 5-1). In the figure, the generator is represented by its Thevenin equivalent; the Norton equivalent, consisting of a parallel combination of a current source (with source current $I = E/Z_G$) and the impedance Z_G, could equally well have been used.

In the connection illustrated in Fig. 5-1, the two-port is used to modify the generator signal E in some useful way. For example, E may be small, and the two-port may amplify it to obtain $V_2 = GE$, where $|G| \gg 1$. Or E may contain a mixture of useful signal and noise, and the two-port may be used to transmit the signal and suppress the noise. Or, again, the two-port may be used merely to provide a delayed facsimile of the input signal. In the first example, the two-port operates as an *amplifier*, in the second, as a *filter*, and in the third, as a *delay line*. Many other applications, e.g., the loss equalizer discussed in Sec. 1-2, are also possible.

The four-terminal circuit shown between the two terminal pairs in Fig. 5-1 will be called a two-port if, and only if, the four currents at its terminals satisfy the *two-port constraints*

$$I_1' = I_1 \qquad I_2' = I_2 \tag{5-1}$$

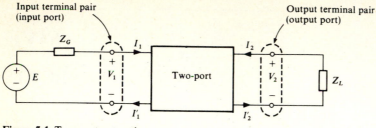

Figure 5-1 Two-port network.

When the four-terminal circuit is connected to an arbitrary network (Fig. 5-2a), the two-port constraints do not necessarily hold. Consider, for example, Fig. 5-2b. The terminal currents of the four-terminal network can be found by inspection: clearly, $I_1 = -I_2 = 1$ A, while $I_1' = -I_2' = 0$. Thus, (5-1) is violated and the four-terminal network does not function as a two-port.

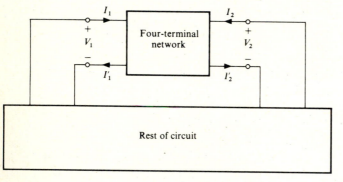

(a)

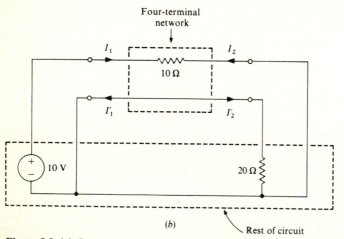

(b)

Figure 5-2 (a) General four-terminal network. (b) Four-terminal network for which the two-port constraints are violated.

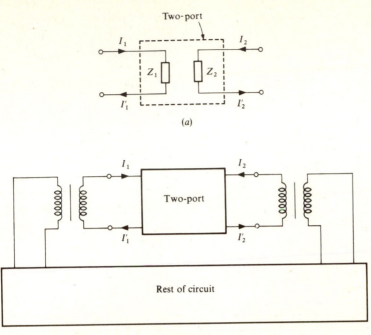

Figure 5-3 Circuits whose structure assures the satisfaction of the two-port conditions. (*a*) The internal structure which forces (5-1) to hold. (*b*) The transformers at the two ports.

Except for trivial four-terminal networks whose structure guarantees that (5-1) holds (like that shown in Fig. 5-3*a*), it is the rest of the circuit which determines whether or not the two-port conditions are satisfied. For the usual situation depicted in Fig. 5-1, Eq. (5-1) is certainly valid. For more general arrangements, the conditions can still be met by including transformers in the circuit, as illustrated in Fig. 5-3*b*.

If the two-port conditions are satisfied, then by the substitution theorem (Ref. 1, p. 654) the rest of the circuit can be replaced by two equivalent current sources without disturbing the electrical conditions inside the two-port. Thus the general circuit shown in Fig. 5-2*a* can be replaced by that indicated in Fig. 5-4*a*, and all internal branch voltages and currents of the two-port remain unchanged. Furthermore, by the superposition theorem (Ref. 1, pp. 658–659), if the two-port contains only linear-circuit elements and no independent sources, the effect of the two current sources I_1 and I_2 can be found by analyzing the two simpler circuits shown in Fig. 5-4*b* and *c*. For the circuit of Fig. 5-4*b*, by linearity, the port voltages† V_1^1 and V_2^1 are proportional to I_1; that is,

$$V_1^1 = z_{11} I_1 \qquad V_2^1 = z_{21} I_1 \tag{5-2}$$

---

† Here, the superscripts 1 and 2 refer to the excited ports and are not exponents.

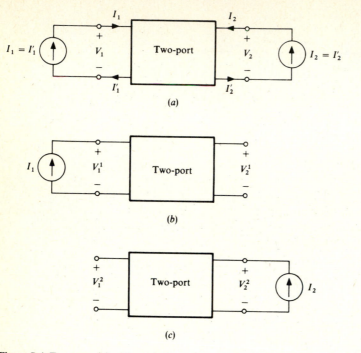

(a)

(b)

(c)

Figure 5-4 Two-port (a) with equivalent current sources, (b) excited at its input port only, (c) excited at its output port only.

and from Fig. 5-4c similarly

$$V_1^2 = z_{12} I_2 \qquad V_2^2 = z_{22} I_2 \tag{5-3}$$

Superposition gives therefore for the circuit of Fig. 5-4a

$$V_1 = V_1^1 + V_1^2 = z_{11} I_1 + z_{12} I_2 \qquad V_2 = V_2^1 + V_2^2 = z_{21} I_1 + z_{22} I_2 \tag{5-4}$$

Here, the parameters $z_{11}, z_{12}, z_{21},$ and z_{22} clearly depend only on the internal branch structure of the two-port. If the two-port is reciprocal, so that the relation shown in Fig. 1-6a holds, then for $I_1 = I_2$ also $V_2^1 = z_{21} I_1$ equals $V_1^2 = z_{12} I_2 = z_{12} I_1$. Consequently, for a reciprocal two-port

$$z_{21} = z_{12} \tag{5-5}$$

The quantities z_{ij} (which are functions of s) are called the *open-circuit impedance parameters* of the two-port; the matrix

$$\mathbf{Z}(s) \triangleq \begin{bmatrix} z_{11} & z_{12} \\ z_{21} & z_{22} \end{bmatrix} \tag{5-6}$$

is called the *open-circuit impedance matrix*. As we shall see, $\mathbf{Z}$ plays much the same role for the two-port as the impedance Z plays for a one-port.

When we define the port-voltage vector

$$\mathbf{V} \triangleq \begin{bmatrix} V_1 \\ V_2 \end{bmatrix} \tag{5-7}$$

as well as the port-current vector

$$\mathbf{I} \triangleq \begin{bmatrix} I_1 \\ I_2 \end{bmatrix} \tag{5-8}$$

Eq. (5-4) can be rewritten in the simple form

$$\mathbf{V} = \mathbf{ZI} \tag{5-9}$$

The substitution theorem also enables us to replace all parts of the circuit external to the two-port by two voltage sources (Fig. 5-5). Then a derivation entirely dual to that leading to (5-9) gives the relation

$$\mathbf{I} = \mathbf{YV} \tag{5-10}$$

where

$$\mathbf{Y} = \begin{bmatrix} y_{11} & y_{12} \\ y_{21} & y_{22} \end{bmatrix} \tag{5-11}$$

is the *short-circuit admittance matrix*. The *short-circuit admittance parameters* y_{ij} are defined by the relations

$$I_1^1 = y_{11} V_1 \qquad I_2^1 = y_{21} V_1 \tag{5-12}$$

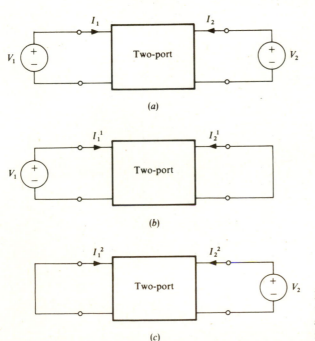

(a)

(b)

(c)

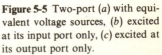

Figure 5-5 Two-port (a) with equivalent voltage sources, (b) excited at its input port only, (c) excited at its output port only.

(Fig. 5-5b) and

$$I_1^2 = y_{12} V_2 \qquad I_2^2 = y_{22} V_2 \tag{5-13}$$

(Fig. 5-5c). Due to the opposing reference directions chosen for I_1 and I_2 (as in Fig. 5-5a), y_{12} and y_{21} often turn out to be negative.

Example 5-1 For the simple circuit of Fig. 5-6a, from (5-12) and Fig. 5-6b,

$$y_{11} \triangleq \frac{I_1^1}{V_1} = \frac{1}{R_1 + R_2} \qquad y_{21} = \frac{I_2^1}{V_1} = -\frac{1}{R_1 + R_2}$$

and similarly, from (5-13) and Fig. 5-6c,

$$y_{12} = \frac{I_1^2}{V_2} = -\frac{1}{R_1 + R_2} \qquad y_{22} = \frac{I_2^2}{V_2} = \frac{1}{R_1 + R_2}$$

For this same two-port, there are no port currents if either port is open-circuited. Hence, Eqs. (5-2) and (5-3) lead to relations of the form

$$V_1^1 = (z_{11})(0) \qquad V_2^1 = (z_{21})(0)$$
$$V_1^2 = (z_{12})(0) \qquad V_2^2 = (z_{22})(0)$$

We conclude that the open-circuit impedance parameters do not exist for this simple circuit.

Unless the determinant

$$\Delta_Y \triangleq \det \mathbf{Y} = y_{11} y_{22} - y_{12} y_{21} \tag{5-14}$$

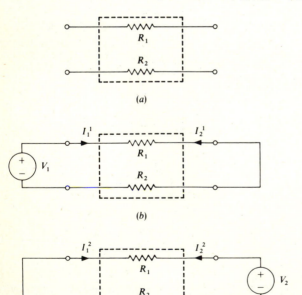

(a)

(b)

(c)

Figure 5-6 Simple resistive two-port and the calculation of its admittance parameters.

happens to be zero, (5-10) can be solved for **V**:

$$\mathbf{V} = \mathbf{Y}^{-1}\mathbf{I} \tag{5-15}$$

or, using the formula for inverting a 2×2 matrix,

$$\begin{bmatrix} V_1 \\ V_2 \end{bmatrix} = \begin{bmatrix} \dfrac{y_{22}}{\Delta_Y} & \dfrac{-y_{12}}{\Delta_Y} \\ \dfrac{-y_{21}}{\Delta_Y} & \dfrac{y_{11}}{\Delta_Y} \end{bmatrix} \begin{bmatrix} I_1 \\ I_2 \end{bmatrix} \tag{5-16}$$

Comparing (5-16) with (5-9) gives the relations

$$z_{11} = \frac{y_{22}}{\Delta_Y} \qquad z_{12} = \frac{-y_{12}}{\Delta_Y}$$

$$z_{21} = \frac{-y_{21}}{\Delta_Y} \qquad z_{22} = \frac{y_{11}}{\Delta_Y} \tag{5-17}$$

If $\Delta_Y = 0$, the impedance parameters do not exist.

Example 5-2 For the two-port of Fig. 5-6a,

$$\Delta_Y = \frac{1}{R_1 + R_2} \frac{1}{R_1 + R_2} - \left(-\frac{1}{R_1 + R_2} \right)^2 \qquad \Delta_Y = 0$$

Hence, the z_{ij} parameters do not exist, as we concluded earlier.

Example 5-3 Find **Y** and **Z** for the circuit of Fig. 5-7a.
From Fig. 5-7b (where we have chosen for simplicity $V_1 = 1$ V), by inspection

$$y_{11} \triangleq \frac{I_1^1}{V_1} = I_1^1 = sC_1 + \frac{1}{R} \qquad y_{21} \triangleq \frac{I_2^1}{V_1} = I_2^1 = -\frac{1}{R}$$

and from Fig. 5-7c, with $V_2 = 1$ V,

$$y_{12} \triangleq \frac{I_1^2}{V_2} = I_1^2 = -\frac{1}{R} \qquad y_{22} \triangleq \frac{I_2^2}{V_2} = I_2^2 = sC_2 + \frac{1}{R}$$

Hence

$$\mathbf{Y} = \begin{bmatrix} sC_1 + \dfrac{1}{R} & -\dfrac{1}{R} \\ -\dfrac{1}{R} & sC_2 + \dfrac{1}{R} \end{bmatrix}$$

Therefore

$$\Delta_Y = \left(sC_1 + \frac{1}{R} \right)\left(sC_2 + \frac{1}{R} \right) - \frac{1}{R^2} = s^2 C_1 C_2 + \frac{s(C_1 + C_2)}{R}$$

and, by (5-17),

$$\mathbf{Z} = \begin{bmatrix} \dfrac{sC_2 + 1/R}{\Delta_Y} & \dfrac{1/R}{\Delta_Y} \\ \dfrac{1/R}{\Delta_Y} & \dfrac{sC_1 + 1/R}{\Delta_Y} \end{bmatrix}$$

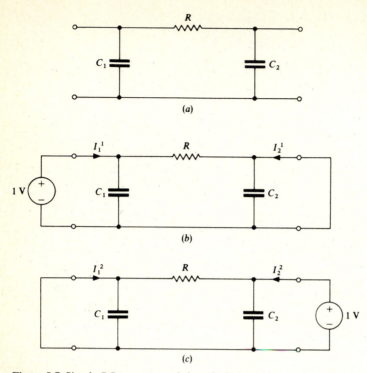

Figure 5-7 Simple RC two-port and the calculation of its admittance parameters.

Of course, **Z** can also be obtained directly, using the method illustrated in Fig. 5-4. For our example, this gives

$$z_{11} = \frac{V_1^1}{I_1} = \cfrac{1}{sC_1 + \cfrac{1}{R + 1/sC_2}}$$

$$z_{11} = \frac{sRC_2 + 1}{s^2 RC_1 C_2 + s(C_1 + C_2)} = \frac{sC_2 + 1/R}{s^2 C_1 C_2 + s(C_1 + C_2)/R}$$

which agrees with the result obtained from **Y**. The direct calculation of z_{12} and z_{22} is left as an exercise.

Since, by (5-10) and (5-15),

$$\mathbf{Z} = \mathbf{Y}^{-1} \qquad \mathbf{Y} = \mathbf{Z}^{-1} \tag{5-18}$$

the y_{ij} can be obtained from

$$y_{11} = \frac{z_{22}}{\Delta_z} \qquad y_{12} = -\frac{z_{12}}{\Delta_z} \qquad y_{21} = -\frac{z_{21}}{\Delta_z} \qquad y_{22} = \frac{z_{11}}{\Delta_z} \tag{5-19}$$

where

$$\Delta_z \triangleq \det \mathbf{Z} = z_{11} z_{22} - z_{12} z_{21} \tag{5-20}$$

Example 5-4 Find **Z** and **Y** for the two-port shown in Fig. 5-8.

With an input current $I_1 = 1$ A, from (5-2) and Fig. 5-4b,

$$z_{11} = V_1^1 = \frac{sL_1 R_1}{sL_1 + R_1} + \frac{R_2/sC}{R_2 + 1/sC} = \frac{s}{s+1} + \frac{1/s}{1+1/s} = 1$$

and

$$z_{21} = V_2^1 = \frac{R_2/sC}{R_2 + 1/sC} = \frac{1/s}{1+1/s} = \frac{1}{s+1}$$

Similarly, from (5-3) and Fig. 5-4c,

$$z_{12} = V_1^2 = \frac{R_2/sC}{R_2 + 1/sC} = \frac{1}{s+1}$$

and

$$z_{22} = V_2^2 = \frac{sL_2 R_3}{sL_2 + R_3} + \frac{R_2/sC}{R_2 + 1/sC} = 1$$

Hence,

$$\Delta_z = 1 - \frac{1}{(s+1)^2} = \frac{s^2 + 2s}{s^2 + 2s + 1}$$

and by (5-19)

$$y_{11} = \frac{z_{22}}{\Delta_z} = \frac{s^2 + 2s + 1}{s^2 + 2s}$$

$$y_{12} = y_{21} = -\frac{z_{12}}{\Delta_z} = -\frac{s+1}{s^2 + 2s}$$

$$y_{22} = \frac{z_{11}}{\Delta_z} = \frac{s^2 + 2s + 1}{s^2 + 2s}$$

Again, y_{11} will be checked using the scheme of Fig. 5-5b. For $V_1 = 1$ V,

$$y_{11} = I_1^1 = \frac{1}{\dfrac{sL_1 R_1}{sL_1 + R_1} + \dfrac{1}{1/R_2 + sC + 1/R_3 + 1/sL_2}}$$

$$= \frac{1}{\dfrac{s}{s+1} + \dfrac{1}{2 + s + 1/s}} = \frac{s^2 + 2s + 1}{s^2 + 2s}$$

as before.

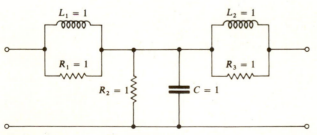

Figure 5-8 *RLC* two-port example.

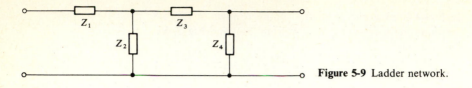

Figure 5-9 Ladder network.

Notice that for simple circuits where the first and last branches are connected in series with the ports, it is simpler to find the z_{ij} than the y_{ij}; for simple two-ports with branches connected across the ports, the y_{ij} are easier to calculate.

Example 5-5 Find the z_{ij} for the two-port shown in Fig. 5-9.
From Fig. 5-4, we obtain

$$z_{11} = Z_1 + \frac{Z_2(Z_3 + Z_4)}{Z_2 + Z_3 + Z_4}$$

$$z_{12} = z_{21} = \frac{Z_2 Z_4}{Z_2 + Z_3 + Z_4}$$

$$z_{22} = \frac{Z_4(Z_2 + Z_3)}{Z_2 + Z_3 + Z_4}$$

The reader should fill in the details.

Hitherto, all two-port examples contained reciprocal circuits. Hence, by (5-5), $z_{12} = z_{21}$ held, and so did

$$y_{12} = y_{21} \tag{5-21}$$

Equation (5-21) can be obtained either from Eqs. (5-5) and (5-19) or from Fig. 1-6b. For circuits containing active elements, (5-5) and (5-21) need not hold.

Example 5-6 For the circuit of Fig. 5-10a, **Y** can readily be found using the scheme of Fig. 5-5. With the output port short-circuited and $V_1 = 1$ V (Fig. 5-10b),

$$y_{11} = I_1^1 = \frac{1}{R_1} + \frac{1}{R_2} \qquad y_{21} = I_2^1 = -\frac{G}{R_3} - \frac{1}{R_2}$$

whereas if the input port is short-circuited and $V_2 = 1$ V (Fig. 5-10c),

$$y_{12} = I_1^2 = -\frac{1}{R_2} \qquad \text{and} \qquad y_{22} = \frac{1}{R_3} + \frac{1}{R_2}$$

Here, $y_{12} \neq y_{21}$, except if $G = 0$ or $R_3 \rightarrow \infty$, that is, if the active element (which here is a voltage-controlled voltage source) is removed from the circuit.

So far, we have used **Z** and/or **Y** to characterize the two-port. A third important two-port parameter set is given by the *chain parameters*, also called *cascade parameters* or *transmission parameters*. These can be obtained, for example, by solving (5-4) for V_1 and I_1. Rearranging (5-4) gives

$$V_1 - z_{11}I_1 = z_{12}I_2 \qquad -z_{21}I_1 = z_{22}I_2 - V_2 \tag{5-22}$$

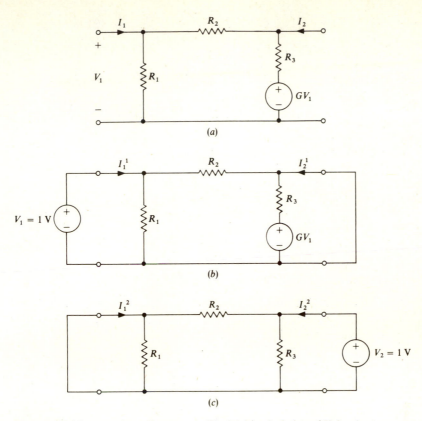

Figure 5-10 (a) Nonreciprocal two-port; (b) and (c) calculation of **Y** for the two-port.

leading directly to

$$I_1 = \frac{1}{z_{21}} V_2 - \frac{z_{22}}{z_{21}} I_2$$

$$V_1 = z_{11} I_1 + z_{12} I_2 = \frac{z_{11}}{z_{21}} V_2 - \left(\frac{z_{11} z_{22}}{z_{21}} - z_{12} \right) I_2 \qquad (5\text{-}23)$$

Equation (5-23) will be written in the form

$$V_1 = A V_2 - B I_2 \qquad I_1 = C V_2 - D I_2 \qquad (5\text{-}24)$$

A, B, C, and D are called the *chain parameters;* their matrix

$$\mathbf{T} = \begin{vmatrix} A & B \\ C & D \end{vmatrix} \qquad (5\text{-}25)$$

is the *chain matrix*. From (5-23) and (5-24), using (5-17) and (5-19), we find

$$A = \frac{z_{11}}{z_{21}} = -\frac{y_{22}}{y_{21}} \qquad B = \frac{\Delta_z}{z_{21}} = -\frac{1}{y_{21}}$$

$$C = \frac{1}{z_{21}} = -\frac{\Delta_Y}{y_{21}} \qquad D = \frac{z_{22}}{z_{21}} = -\frac{y_{11}}{y_{21}} \qquad (5\text{-}26)$$

Hence, **T** can be found from either **Z** or **Y**. From (5-24), also

$$A = \left(\frac{V_1}{V_2}\right)_{I_2=0} \qquad B = \left(\frac{V_1}{-I_2}\right)_{V_2=0} \qquad C = \left(\frac{I_1}{V_2}\right)_{I_2=0} \qquad D = \left(\frac{I_1}{-I_2}\right)_{V_2=0}$$

$$(5\text{-}27)$$

Hence, **T** can be found from the schemes shown in Fig. 5-11.

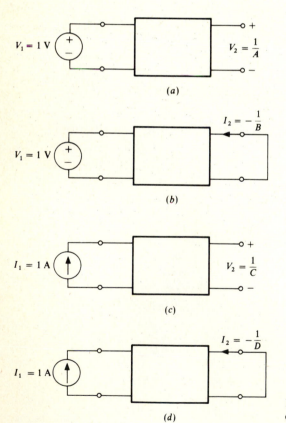

(a)

(b)

(c)

(d)

Figure 5-11 Schemes for calculating the chain parameters of a two-port.

Example 5-7 For the two-port of Fig. 5-7a, using Fig. 5-11, we find

$$A = \left(\frac{1}{V_2}\right)_{\substack{V_1=1 \text{ V} \\ I_2=0}} = \left(\frac{1/sC_2}{R + 1/sC_2}\right)^{-1} = sRC_2 + 1$$

$$B = \left(-\frac{1}{I_2}\right)_{\substack{V_1=1 \text{ V} \\ V_2=0}} = R$$

$$C = \left(\frac{1}{V_2}\right)_{\substack{I_1=1 \text{ A} \\ I_2=0}} = \left(\frac{1}{V_1 \dfrac{1/sC_2}{R + 1/sC_2}}\right)_{\substack{I_1=1 \text{ A} \\ I_2=0}}$$

$$= \frac{1}{\dfrac{(1/sC_1)(R + 1/sC_2)}{(1/sC_1 + R + 1/sC_2)} \dfrac{1/sC_2}{R + 1/sC_2}} = s^2 RC_1 C_2 + s(C_1 + C_2)$$

and
$$D = \left(-\frac{1}{I_2}\right)_{\substack{I_1=1 \text{ A} \\ V_2=0}} = \left(\frac{1/R}{sC_1 + 1/R}\right)^{-1} = sRC_1 + 1$$

When (5-26) is used in conjunction with the values found earlier for the **Z** and **Y** of this circuit, the above results can be confirmed.

From (5-26), we can also readily express **Z** and **Y** in terms of **T**. The result is

$$\mathbf{Z} = \begin{bmatrix} \dfrac{A}{C} & \dfrac{\Delta_T}{C} \\[2ex] \dfrac{1}{C} & \dfrac{D}{C} \end{bmatrix} \tag{5-28}$$

and
$$\mathbf{Y} = \begin{bmatrix} \dfrac{D}{B} & \dfrac{-\Delta_T}{B} \\[2ex] \dfrac{-1}{B} & \dfrac{A}{B} \end{bmatrix} \tag{5-29}$$

where Δ_T is the determinant of the chain matrix **T**.

From (5-28) and (5-29), it follows that for a reciprocal two-port where $z_{12} = z_{21}$ and $y_{12} = y_{21}$,

$$\Delta_T = AD - BC = 1 \tag{5-30}$$

Example 5-8 For the circuit of Fig. 5-7a,

$$\Delta_T = (sRC_2 + 1)(sRC_1 + 1) - R(s^2 RC_1 C_2 + sC_1 + sC_2) = 1$$

as expected.

Example 5-9 For the circuit of Fig. 5-10a, the chain parameters can be found from the admittance parameters calculated earlier, for example:

$$A = -\frac{y_{22}}{y_{21}} = -\frac{1/R_3 + 1/R_2}{-G/R_3 - 1/R_2} = \frac{R_2 + R_3}{GR_2 + R_3}$$

$$B = -\frac{1}{y_{21}} = -\frac{1}{-G/R_3 - 1/R_2} = \frac{R_2 R_3}{GR_2 + R_3}$$

$$C = -\frac{y_{11}y_{22} - y_{12}y_{21}}{y_{21}} = -\frac{y_{11}y_{22}}{y_{21}} + y_{12}$$

$$= -\frac{(1/R_1 + 1/R_2)(1/R_3 + 1/R_2)}{-G/R_3 - 1/R_2} - \frac{1}{R_2}$$

$$= \frac{(R_1 + R_2)(R_2 + R_3)}{(GR_2 + R_3)R_1 R_2} - \frac{1}{R_2}$$

$$D = -\frac{y_{11}}{y_{21}} = -\frac{1/R_1 + 1/R_2}{-G/R_3 - 1/R_2} = \frac{(R_1 + R_2)R_3}{R_1(GR_2 + R_3)}$$

Hence

$$AD - BC = \frac{\dfrac{(R_2 + R_3)(R_1 + R_2)R_3}{R_1} - R_2 R_3 \left[\dfrac{(R_1 + R_2)(R_2 + R_3)}{R_1 R_2} - \dfrac{GR_2 + R_3}{R_2} \right]}{(GR_2 + R_3)^2}$$

or, after simplifications,

$$AD - BC = \frac{R_3}{GR_2 + R_3}$$

Hence, $AD - BC = 1$ holds only if $R_3 \to \infty$ or $G \to 0$, that is, only if the controlled source is removed from the circuit.

Since Eq. (5-4) can be rearranged six different ways with two parameters on the left-hand side and two on the right-hand side, we can define six different sets of two-port parameters. The reader should consult Ref. 1, table 17-1, for a listing of these parameters and for the formulas needed to convert from one set to another.

5-2 REALIZABILITY CONDITIONS FOR THE Z AND Y OF *RLCM* TWO-PORTS

There is a close analogy between the impedance Z of an *RLCM* one-port and the open-circuit impedance matrix $\mathbf{Z}$ of an *RLCM* two-port. In fact, as will be shown in this section, the PR properties of $Z(s)$ can readily be generalized to $\mathbf{Z}(s)$.

Consider again the circuit of Fig. 5-4a. To restore the associated directions (see Fig. 2-4b) in branches 1 and 2, i.e., the input and output branches, we introduce the branch currents $J_1 \triangleq -I_1$ and $J_2 \triangleq -I_2$. Then by Tellegen's theorem, written in the form of (2-41),

$$\sum_{k=1}^{N} V_k(s)J_k^*(s) = V_1 J_1^* + V_2 J_2^* + \sum_{k=3}^{N} V_k(s)J_k^*(s) = 0 \tag{5-31}$$

From the definition of I_1 and I_2, therefore,

$$V_1 I_1^* + V_2 I_2^* = \sum_{\substack{\text{all internal} \\ \text{branches}}} V_k(s) J_k^*(s) \tag{5-32}$$

Using (5-9), we can rewrite the left-hand side as

$$V_1 I_1^* + V_2 I_2^* = \begin{bmatrix} V_1 \\ V_2 \end{bmatrix}^T \begin{bmatrix} I_1 \\ I_2 \end{bmatrix}^* = \mathbf{V}^T \mathbf{I}^* = \mathbf{I}^T \mathbf{Z}^T \mathbf{I}^* \tag{5-33}$$

Assume now that the two-port contains only *RLCM* elements. Then $\mathbf{Z}^T = \mathbf{Z}$; also, as shown in Sec. 2-3, the right-hand side in (5-32) can be written

$$\sum_{\substack{\text{all internal} \\ \text{branches}}} V_k(s) J_k^* = F_0(s) + \frac{V_0(s)}{s} + s M_0(s) \tag{5-34}$$

Here the energy functions $F_0(s)$, $V_0(s)$, and $M_0(s)$ are nonnegative real quantities for all values of s, real or complex. From this fact, we have shown in Sec. 2-3 that the right-hand side of (5-34) is a PR function of s. By (5-33), therefore, $\mathbf{I}^T \mathbf{Z}^T \mathbf{I}^*$ is a PR function of s. If the real and imaginary parts of I_1 and I_2 are denoted by a and b, respectively, so that

$$I_1 = a_1 + jb_1 \qquad I_2 = a_2 + jb_2 \tag{5-35}$$

then

$$\mathbf{I}^T \mathbf{Z}^T \mathbf{I}^* = (a_1 + jb_1 \quad a_2 + jb_2) \begin{bmatrix} z_{11} & z_{12} \\ z_{12} & z_{22} \end{bmatrix} \begin{bmatrix} a_1 - jb_1 \\ a_2 - jb_2 \end{bmatrix}$$

$$= z_{11}(a_1^2 + b_1^2) + 2z_{12}(a_1 a_2 + b_1 b_2) + z_{22}(a_2^2 + b_2^2) \tag{5-36}$$

This expression must be a PR function† of s for all real a_1, a_2, b_1, and b_2. This requires that

$$z_{11} a_1^2 + 2z_{12} a_1 a_2 + z_{22} a_2^2 \tag{5-37}$$

be a PR function† of s for all real a_1 and a_2. The necessity of this follows for $b_1 = b_2 = 0$; the sufficiency is obvious since the sum of two PR functions is again PR.

If we now also recall our discussions in Sec. 2-1 about the rational character of transfer impedances,‡ we can summarize the necessary conditions for the open-circuit impedance matrix $\mathbf{Z}$ of an *RLCM* two-port as follows:

(a) All elements z_{ij} of $\mathbf{Z}$ must be real rational functions of s, with $z_{12} = z_{21}$.
(b) The expression

$$Z(s) \triangleq z_{11} a_1^2 + 2z_{12} a_1 a_2 + z_{22} a_2^2 \tag{5-38}$$

must be a PR function for all real a_1 and a_2.

---

† Here s enters through $z_{ij}(s)$.
‡ See, in particular, Eqs. (2-16) to (2-17) and the accompanying discussions.

A matrix satisfying the above conditions will be called a *PR rational matrix*.

Our two conditions are entirely analogous to the Brune conditions given in Sec. 2-3 for the impedance Z of an *RLCM* one-port.

A conceptually simpler proof for condition *b* is implied in Prob. 5-8*a*.

The derivation performed for **Z** can be carried out in a similar manner for **Y**. The proof is based here on Eqs. (2-44) and (2-60); it is left as an exercise for the reader. The result shows that all y_{ij} are real rational functions of s, that $y_{12} = y_{21}$, and that

$$Y(s) \triangleq y_{11} a_1^2 + 2 y_{12} a_1 a_2 + y_{22} a_2^2 \tag{5-39}$$

is a PR function for all real a_1 and a_2. A conceptually simpler proof is implied in Prob. 5-8*b*.

One immediate conclusion from (5-38) and (5-39) can be obtained by setting $a_1 = 0$ or $a_2 = 0$. This shows that all four parameters z_{11}, z_{22}, y_{11}, and y_{22} are PR functions of s. This conclusion is not really surprising, since these parameters represent driving-point immittances measured at one of the ports, with the other port open- or short-circuited.

Using the equivalent PR conditions given in Sec. 2-4, we can also obtain a set of equivalent PR matrix conditions as follows:

(*a'*) Identical to condition *a*.

(*b'*) For $s = j\omega$, ω real, with

$$r_{ij} \triangleq \operatorname{Re} z_{ij} \tag{5-40}$$

(5-38) gives

$$\operatorname{Re} Z(j\omega) = r_{11} a_1^2 + 2 r_{12} a_1 a_2 + r_{22} a_2^2 \geq 0 \tag{5-41}$$

(*c'*) All poles of the right-hand side of (5-38) are either in the closed LHP of the s plane or on the $j\omega$ axis. Those on the $j\omega$ axis are simple, and their residues are real positive.

Next, (5-41) will be replaced by a more useful set of inequalities. For $a_2 = 0$, we have $r_{11} \geq 0$; for $a_1 = 0$, $r_{22} \geq 0$. Hence, the first and third terms in $\operatorname{Re} Z(j\omega)$ are nonnegative. $\operatorname{Re} Z(j\omega) < 0$ can therefore occur only if $2 r_{12} a_1 a_2$ is negative and its absolute value is greater than $r_{11} a_1^2 + r_{22} a_2^2$:

$$-2 r_{12} a_1 a_2 > r_{11} a_1^2 + r_{22} a_2^2 > 0 \tag{5-42}$$

But

$$(a_1 \sqrt{r_{11}} - a_2 \sqrt{r_{22}})^2 = a_1^2 r_{11} - 2 a_1 a_2 \sqrt{r_{11} r_{22}} + a_2^2 r_{22} \geq 0$$

$$r_{11} a_1^2 + r_{22} a_2^2 \geq 2 a_1 a_2 \sqrt{r_{11} r_{22}} \tag{5-43}$$

Hence, if

$$2 a_1 a_2 \sqrt{r_{11} r_{22}} \geq -2 r_{12} a_1 a_2 \qquad r_{11} r_{22} \geq r_{12}^2 \tag{5-44}$$

holds, (5-42) cannot be valid. Therefore, the conditions

$$r_{11}(j\omega) \geq 0 \qquad r_{22}(j\omega) \geq 0 \qquad r_{11}(j\omega)r_{22}(j\omega) - r_{12}^2(j\omega) \geq 0 \qquad (5\text{-}45)$$

are equivalent to (5-41). The inequalities of (5-45) are often called the *real-part conditions*. An alternative proof of (5-45) is implied in Prob. 5-9.

Turning to condition c', it is clear from (5-38) that z_{12} is not allowed to have any poles in the RHP since this would appear in $Z(s)$; it cannot be canceled by a corresponding pole term in z_{11} and z_{22} since those are themselves PR functions. A $j\omega$-axis pole at, say, $s_1 = j\omega_1$, must, by condition c', be simple. Therefore, partial-fraction expansion gives†

$$Z(s) = \frac{k}{s - j\omega_1} + \text{other terms}$$

$$z_{ij}(s) = \frac{k_{ij}}{s - j\omega_1} + \text{other terms} \qquad i, j = 1, 2$$

(5-46)

Substituting into (5-38) and multiplying by $s - j\omega_1$ gives

$$k + (s - j\omega_1)(\text{other terms}) = a_1^2 k_{11} + 2a_1 a_2 k_{12} + a_2^2 k_{22}$$
$$+ (s - j\omega_1)(\text{other terms}) \quad (5\text{-}47)$$

Since the other terms do not have $s - j\omega_1$ in their denominators, for $s = j\omega_1$ we obtain

$$k = a_1^2 k_{11} + 2a_1 a_2 k_{12} + a_2^2 k_{22} \geq 0 \qquad (5\text{-}48)$$

Here, the last part of the inequality follows from condition c'.

A comparison reveals that (5-48) is in the same form as (5-41). Hence, using the same argument as when we derived (5-45), we obtain the *residue conditions*

$$k_{11} \geq 0 \qquad k_{22} \geq 0 \qquad k_{11}k_{22} - k_{12}^2 \geq 0 \qquad (5\text{-}49)$$

which must hold at all $j\omega$-axis poles of the z_{ij}.

Naturally, all conditions can be proved, using identical steps, for the admittance parameters y_{ij}.

All arguments given above prove merely the *necessity* of conditions a and b (or, equivalently, conditions a', b', and c'). Their *sufficiency* can be shown by actually realizing a two-port whose **Z** or **Y** satisfies these conditions. This process, which is analogous to the Brune synthesis, was discovered by Gewertz. Since it is complicated and of little practical importance, it will not be described here.

---

† We assume here for generality that all three z_{ij} have this pole. If the pole is absent in, say, z_{12}, we can set $k_{12} = 0$.

5-3 REALIZABILITY CONDITIONS FOR REACTANCE TWO-PORT PARAMETERS

Among *RLCM* two-ports, the class of reactance two-ports, that is, *LCM* two-ports, is of the greatest practical importance for the following reasons:†

1. A reactance two-port can transmit a signal without dissipating any of its energy. Hence, it can maintain favorable signal-to-noise conditions.
2. A reactance two-port can exhibit a loss-vs.-frequency response which varies very rapidly with frequency. Hence, it is useful for realizing filters.
3. A reactance two-port can transmit signals and modify only their phase or delay characteristics. Hence, it can be used to realize delay circuits or phase equalizers.
4. As will be shown in Sec. 10-4, a reactance two-port terminated in resistors at both ports can be made inherently insensitive to variations of its elements.

These and other less important advantages made reactance two-ports the most widely used passive networks. For this reason, their design theory is well developed; we shall devote the remainder of this chapter and also Chap. 6 to this topic.

For a reactance two-port, $F_0(s)$, which in (5-34) represented the contribution of internal branches containing resistors, is identically zero. Hence, an argument similar to that leading to (5-38) gives again

$$Z(s) = z_{11}a_1^2 + 2z_{12}a_1a_2 + z_{22}a_2^2 \tag{5-50}$$

where now $Z(s)$ is a positive real odd function, i.e., a *reactance*. Hence, it can be written in the form of Eq. (3-9):‡

$$Z(s) = k^\infty s + \frac{k^0}{s} + \sum_{i=1}^{n} \frac{k^i s}{s^2 + \omega_i^2} \tag{5-51}$$

Since z_{11} and z_{22} are driving-point impedances measured at the input and output ports, respectively, they too are reactances and can be expanded into partial fractions

$$z_{jj}(s) = k_{jj}^\infty s + \frac{k_{jj}^0}{s} + \sum_{i=1}^{n} \frac{k_{jj}^i s}{s^2 + \omega_i^2} \qquad j = 1, 2 \tag{5-52}$$

Now z_{12} can be expressed from (5-50) in the form (for $a_1, a_2 \neq 0$)

$$z_{12}(s) = \frac{1}{2a_1a_2} Z(s) - \frac{a_1}{2a_2} z_{11}(s) - \frac{a_2}{2a_1} z_{22}(s) \tag{5-53}$$

from which the following conclusions can be drawn:

1. $z_{12}(s)$ is an odd rational function of s, since $Z(s)$, $z_{11}(s)$, and $z_{22}(s)$ all are such functions.

---

† Some of these are only valid for *resistively terminated* reactance two-ports (see Chap. 6).
‡ For convenience, the notation for the residues has been slightly changed here.

2. All poles of $z_{12}(s)$ are simple $j\omega$-axis poles, since they must be present in some of the reactance functions $Z(s)$, $z_{11}(s)$, and $z_{22}(s)$. The residue of $z_{12}(s)$ at any pole ω_i is given by (5-48)

$$k_{12}^i = \frac{k^i}{2a_1 a_2} - \frac{a_1 k_{11}^i}{2a_2} - \frac{a_2 k_{22}^i}{2a_1} \tag{5-54}$$

for a_1, $a_2 \neq 0$. Hence, k_{12}^i is real but may be negative. Its size is restricted; by (5-49)

$$(k_{12}^i)^2 \leq k_{11}^i k_{22}^i \tag{5-55}$$

Hence, we can write

$$z_{12}(s) = k_{12}^\infty s + \frac{k_{12}^0}{s} + \sum_{i=1}^n \frac{k_{12}^i s}{s^2 + \omega_i^2} \tag{5-56}$$

where all k_{12} are real, positive or negative, satisfying (5-55).

The above discussions can be summed up in the following theorem (due to W. Cauer):

Theorem The necessary and sufficient condition for a symmetric matrix

$$\mathbf{Z} = \begin{bmatrix} z_{11} & z_{12} \\ z_{12} & z_{22} \end{bmatrix} \tag{5-57}$$

to be the impedance matrix of a reactance two-port is that its elements be given by the partial-fraction expansions (5-52) and (5-56), where the residues are all real and at each pole satisfy (5-49).

The necessity of these conditions has been demonstrated; their sufficiency will be shown by carrying out Cauer's two-port synthesis in the next section.

From (5-56)

$$z_{12}(s) = \frac{N(s)}{D(s)} = \frac{N(s)}{s \prod_{i=1}^n (s^2 + \omega_i^2)} \tag{5-58}$$

where $N(s)$ is an even polynomial† in s with real coefficients. The zeros of $N(s)$ can be anywhere in the s plane; however, as shown in Sec. 2-5, the real coefficients of $N(s)$ assure that for any complex zero $z_i = \sigma_i + j\omega_i$ its conjugate value $z_i^* = \sigma_i - j\omega_i$ is also present as a zero. Furthermore, since $N(s)$ is even (or odd), if z_i is a zero of $N(s)$, then

$$N(-z_i) = \pm N(z_i) = 0 \tag{5-59}$$

and hence $-z_i$ is also a zero.

---

† $N(s)$ may contain a factor s^{2k}. Then $z_{12}(s) = N'(s)/\prod_{i=1}^n (s^2 + \omega_i^2)$, $N'(s)$ odd.

From conditions a and b, it follows that A, B, C, and D are all real rational functions of s. Concerning condition b, it can be shown through the same argument which we used to prove the sufficiency of condition c' for $\mathbf{Z}$ [see Eqs. (5-60) to (5-64)] that all four of the ratios A/C, D/C, D/B, and A/B are PR odd functions if any three are.

Hence, our three conditions for $\mathbf{T}$ imply conditions a, b, and c' for $\mathbf{Z}$. They are therefore both necessary and sufficient for the realizability of a chain matrix in the form of a reactance two-port.

Example 5-12 Test the realizability of†

$$\mathbf{T} = \frac{1}{s^4 + 3s^2 + 4} \begin{bmatrix} s^4 + 8s^2 + 8 & 3s^3 + 8s \\ 2s^3 + 3s & s^4 + 4s^2 + 2 \end{bmatrix}$$

Here

$$C = \frac{2s^3 + 3s}{s^4 + 3s^2 + 4}$$

clearly satisfies condition a. The ratios

$$\frac{A}{C} = \frac{s^4 + 8s^2 + 8}{2s^3 + 3s} = \frac{\frac{8}{3}}{s} + \frac{\frac{7}{12}s}{s^2 + \frac{3}{2}} + \frac{1}{2}s$$

$$\frac{D}{C} = \frac{s^4 + 4s^2 + 2}{2s^3 + 3s} = \frac{\frac{2}{3}}{s} + \frac{\frac{7}{12}s}{s^2 + \frac{3}{2}} + \frac{1}{2}s$$

and

$$\frac{D}{B} = \frac{s^4 + 4s^2 + 2}{3s^3 + 8s} = \frac{\frac{1}{4}}{s} + \frac{\frac{7}{36}s}{s^2 + \frac{8}{3}} + \frac{1}{3}s$$

are clearly PR odd functions. Finally

$$AD - BC = \frac{(s^4 + 8s^2 + 8)(s^4 + 4s^2 + 2) - (3s^3 + 8s)(2s^3 + 3s)}{(s^4 + 3s^2 + 4)^2}$$

$$= \frac{s^8 + 6s^6 + 17s^4 + 24s^2 + 16}{s^8 + 6s^6 + 17s^4 + 24s^2 + 16} = 1$$

All conditions are met; hence $\mathbf{T}$ is the chain matrix of a realizable reactance two-port.

5-4 REACTANCE TWO-PORT SYNTHESIS USING PARTIAL-FRACTION EXPANSION (CAUER'S TWO-PORT SYNTHESIS)

We still owe the reader the sufficiency proof of the realizability conditions given in Sec. 5-3 for $\mathbf{Z}$, $\mathbf{Y}$, and $\mathbf{T}$. In other words, we have to show that from a given $\mathbf{Z}$ meeting the realizability conditions described above we can always design a reactance two-port containing only positive elements. A straightforward proof, based on the actual construction of such a two-port, has been given by Cauer and will be described next. Since it utilizes the partial-fraction expansion of the z_{ij}, it can be regarded as a generalization of Foster's reactance synthesis, described in Sec. 3-2.

---

† The scalar factor of a matrix multiplies all elements of the matrix.

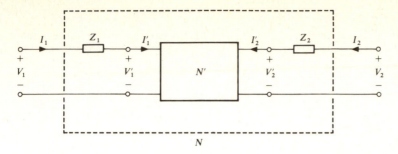

Figure 5-13 The realization of the private poles of z_{11} and z_{22}.

Assume that the partial-fraction expansions (5-52) and (5-56) of the z_{ij} are given, with the residue conditions (5-49) satisfied. We begin by collecting those terms which correspond to the private poles of z_{11} into an impedance Z_1, and similarly for z_{22}. Hence we write

$$z_{11} = Z_1 + k_{11}^\infty s + \frac{k_{11}^0}{s} + \sum_{i=1}^{n} \frac{k_{11}^i s}{s^2 + \omega_i^2}$$

$$z_{22} = Z_2 + k_{22}^\infty s + \frac{k_{22}^0}{s} + \sum_{i=1}^{n} \frac{k_{22}^i s}{s^2 + \omega_i^2} \tag{5-65}$$

Clearly, Z_1 and Z_2 are realizable reactances.

Consider now the circuit of Fig. 5-13. From the definition of the z_{ij}, the two-port parameters of the augmented two-port N are given by

$$z_{11} \triangleq \left[\frac{V_1}{I_1}\right]_{I_2=0} = \left(\frac{I_1 Z_1 + V_1'}{I_1}\right)_{I_2=0} = Z_1 + \left(\frac{V_1'}{I_1'}\right)_{I_2'=0} \tag{5-66}$$

$$z_{11} = Z_1 + z_{11}'$$

where $z_{11}' \triangleq (V_1'/I_1')_{I_2'=0}$ is the impedance parameter of the inside two-port N'. A similar argument gives

$$z_{22} = Z_2 + z_{22}' \tag{5-67}$$

Furthermore

$$z_{12} \triangleq \left(\frac{V_2}{I_1}\right)_{I_2=0} = \left(\frac{V_2'}{I_1'}\right)_{I_2'=0} = z_{12}' \tag{5-68}$$

We conclude that the series impedances Z_1 and Z_2 contribute only to z_{11} and z_{22}, respectively. They can therefore be used to represent the private poles in these impedances.† Hence, we need only concern ourselves with realizing those terms in (5-52) and (5-56) which are common in all four elements of **Z**.

---

† Note, however, that the presence of the series impedances Z_1 or Z_2 does *not* necessarily indicate that private poles exist. Nor does their absence mean that there are no private poles. See Probs. 5-40 and 5-41 for a detailed discussion of these phenomena.

Next, we can write for the impedance matrix formed from these terms

$$
\mathbf{Z}' =
\begin{bmatrix}
k_{11}^\infty s + \dfrac{k_{11}^0}{s} + \displaystyle\sum_{i=1}^{n} \dfrac{k_{11}^i s}{s^2 + \omega_i^2} & k_{12}^\infty s + \dfrac{k_{12}^0}{s} + \displaystyle\sum_{i=1}^{n} \dfrac{k_{12}^i s}{s^2 + \omega_i^2} \\[3ex]
k_{12}^\infty s + \dfrac{k_{12}^0}{s} + \displaystyle\sum_{i=1}^{n} \dfrac{k_{12}^i s}{s^2 + \omega_i^2} & k_{22}^\infty s + \dfrac{k_{22}^0}{s} + \displaystyle\sum_{i=1}^{n} \dfrac{k_{22}^i s}{s^2 + \omega_i^2}
\end{bmatrix}
$$

$$
\mathbf{Z}' =
\begin{bmatrix} k_{11}^\infty s & k_{12}^\infty s \\ k_{12}^\infty s & k_{22}^\infty s \end{bmatrix}
+
\begin{bmatrix} \dfrac{k_{11}^0}{s} & \dfrac{k_{12}^0}{s} \\[2ex] \dfrac{k_{12}^0}{s} & \dfrac{k_{22}^0}{s} \end{bmatrix}
+
\sum_{i=1}^{n}
\begin{bmatrix} \dfrac{k_{11}^i s}{s^2 + \omega_i^2} & \dfrac{k_{12}^i s}{s^2 + \omega_i^2} \\[2ex] \dfrac{k_{12}^i s}{s^2 + \omega_i^2} & \dfrac{k_{22}^i s}{s^2 + \omega_i^2} \end{bmatrix}
\tag{5-69}
$$

$$
\mathbf{Z}' = \mathbf{Z}^\infty + \mathbf{Z}^0 + \sum_{i=1}^{n} \mathbf{Z}^i
$$

Thus, as in the Foster reactance synthesis, we have expanded $\mathbf{Z}'$ as a sum of simple terms. Consider next the circuit of Fig. 5-14, where all mutually coupled inductors are ideal transformers with $1:1$ turn ratios. The currents in the secondary windings of these transformers clearly satisfy $I_2^\infty = I_2^{\infty'}$, $I_2^0 = I_2^{0'}$, etc. The primary currents J_2^∞, $J_2^{\infty'}$, etc., are also pairwise equal. Since by KCL

$$
I_1^\infty - I_1^{\infty'} + J_2^\infty - J_2^{\infty'} = 0 \tag{5-70}
$$

we get also

$$
I_1^\infty = I_1^{\infty'} \tag{5-71}
$$

and similar relations for I_1^0 and $I_1^{0'}$, I_1^1 and $I_1^{1'}$, etc. Therefore, all four-terminal circuits operate as two-ports, and we can write

$$
V_1' = V_1^\infty + V_1^0 + \sum_{i=1}^{n} V_1^i = (I_1^\infty z_{11}^\infty + J_2^\infty z_{12}^\infty) + (I_1^0 z_{11}^0 + J_2^0 z_{12}^0)
$$

$$
+ \sum_{i=1}^{n} (I_1^i z_{11}^i + J_2^i z_{12}^i)
$$

and since obviously $I_1^\infty = I_1^0 = I_1^1 = \cdots = I_1'$ and $J_2^\infty = I_2^\infty = J_2^0 = I_2^0 = \cdots = I_2'$,

$$
V_1' = \left(z_{11}^\infty + z_{11}^0 + \sum_{i=1}^{n} z_{11}^i \right) I_1' + \left(z_{12}^\infty + z_{12}^0 + \sum_{i=1}^{n} z_{12}^i \right) I_2' \tag{5-72}
$$

A similar derivation gives

$$
V_2' = \left(z_{12}^\infty + z_{12}^0 + \sum_{i=1}^{n} z_{12}^i \right) I_1' + \left(z_{22}^\infty + z_{22}^0 + \sum_{i=1}^{n} z_{22}^i \right) I_2' \tag{5-73}
$$

It follows that the overall impedance matrix $\mathbf{Z}'$ of the two-port satisfies

$$
\mathbf{Z}' = \mathbf{Z}^\infty + \mathbf{Z}^0 + \sum_{i=1}^{n} \mathbf{Z}^i \tag{5-74}
$$

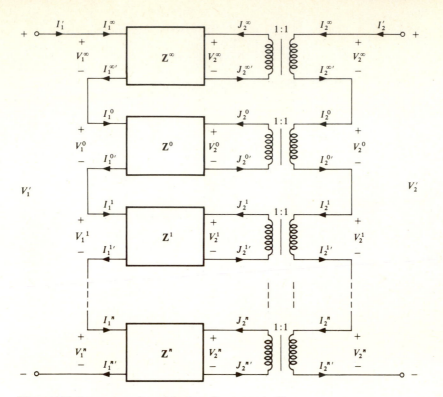

Figure 5-14 Cauer expansion of the impedance matrix $\mathbf{Z}'$.

Comparison of (5-74) and (5-69) reveals that N' can be realized in the configuration of Fig. 5-14, where each elementary two-port corresponds to an elementary impedance matrix on the right-hand side of (5-69).

To show how the impedance matrix

$$\mathbf{Z}^\infty \triangleq \begin{bmatrix} k_{11}^\infty s & k_{12}^\infty s \\ k_{12}^\infty s & k_{22}^\infty s \end{bmatrix} \tag{5-75}$$

can be realized, consider the circuit of Fig. 5-15a. Its impedance parameters are

$$z_{11}^\infty = \left(\frac{V_1}{I_1}\right)_{I_2=0} = (L_a + L_b)s$$

$$z_{12}^\infty = \left(\frac{V_2}{I_1}\right)_{I_2=0} = tL_b s \tag{5-76}$$

$$z_{22}^\infty = \left(\frac{V_2}{I_2}\right)_{I_1=0} = t^2 L_b s$$

Hence, this circuit realizes the impedance matrix (5-75) if we choose

$$L_a + L_b = k_{11}^\infty \qquad tL_b = k_{12}^\infty \qquad t^2 L_b = k_{22}^\infty \tag{5-77}$$

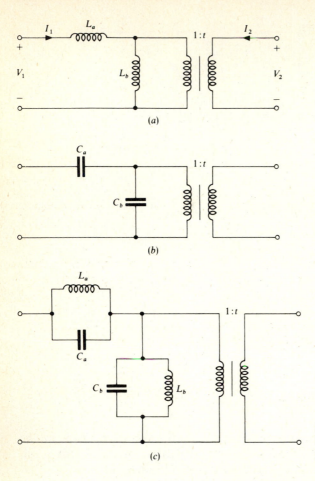

(a)

(b)

(c)

Figure 5-15 Elementary two-ports for Cauer's two-port synthesis process.

Solving (5-77) leads to

$$t = \frac{k_{22}^{\infty}}{k_{12}^{\infty}} \qquad L_b = \frac{k_{12}^{\infty}}{t} = \frac{(k_{12}^{\infty})^2}{k_{22}^{\infty}}$$

$$L_a = k_{11}^{\infty} - L_b = \frac{k_{11}^{\infty} k_{22}^{\infty} - (k_{12}^{\infty})^2}{k_{22}^{\infty}} = \frac{\Delta_k^{\infty}}{k_{22}^{\infty}} \tag{5-78}$$

Hence, t, L_a, and L_b are all real; furthermore, if the residue conditions (5-49) hold, then $L_a \geq 0$ and $L_b > 0$.

At this stage we introduce a definition. We shall call a shared (nonprivate) pole of the z_{ij} *compact* if the residues k_{ij} at that pole meet the residue condition with the equal sign, i.e., if

$$\Delta_k = k_{11} k_{22} - k_{12}^2 = 0 \tag{5-79}$$

Then we can state that $\mathbf{Z}^{\infty}$ in (5-69) can be realized by the circuit of Fig. 5-15a. All element values are real; $L_a \geq 0$ and $L_b > 0$, while $t < 0$ or $t > 0$ according to the sign of k_{12}^∞. $L_a = 0$ if the pole at infinity is compact; otherwise, $L_a > 0$.

An analogous derivation reveals that $\mathbf{Z}^0$ can be realized by the circuit of Fig. 5-15b, with the element values

$$t = \frac{k_{22}^0}{k_{12}^0} \qquad C_a = \frac{k_{22}^0}{k_{11}^0 k_{22}^0 - (k_{12}^0)^2} = \frac{k_{22}^0}{\Delta_k^0} \qquad C_b = \frac{k_{22}^0}{(k_{12}^0)^2} \qquad (5\text{-}80)$$

All element values are real; $0 < C_a \leq \infty$ and $C_b > 0$. If the pole at zero is compact, $C_a \to \infty$; that is, it is replaced by a short circuit.

Finally, consider the circuit of Fig. 5-15c. If we postulate $L_a C_a = L_b C_b = \omega_i^{-2}$, the impedance parameters of the two-port are

$$z_{11}^i = \frac{s/C_a}{s^2 + \omega_i^2} + \frac{s/C_b}{s^2 + \omega_i^2} \qquad z_{12}^i = t \frac{s/C_b}{s^2 + \omega_i^2} \qquad z_{22}^i = t^2 \frac{s/C_b}{s^2 + \omega_i^2} \quad (5\text{-}81)$$

Comparison with $\mathbf{Z}^i$ in (5-69) gives then

$$k_{11}^i = \frac{1}{C_a} + \frac{1}{C_b} \qquad k_{12}^i = \frac{t}{C_b} \qquad k_{22}^i = \frac{t^2}{C_b} \qquad\qquad (5\text{-}82)$$

Solving (5-82) gives

$$t = \frac{k_{22}^i}{k_{12}^i} \qquad C_b = \frac{k_{22}^i}{(k_{12}^i)^2} \qquad L_b = \frac{(k_{12}^i)^2}{\omega_i^2 k_{22}^i}$$

$$C_a = \frac{k_{22}^i}{k_{11}^i k_{22}^i - (k_{12}^i)^2} = \frac{k_{22}^i}{\Delta_k^i} \qquad L_a = \frac{\Delta_k^i}{\omega_i^2 k_{22}^i} \qquad (5\text{-}83)$$

Again, all elements are real; $t \lessgtr 0$, $L_b > 0$ and $C_b > 0$. If ω_i is a compact pole, $L_a \to 0$ and $C_a \to \infty$; otherwise $0 < L_a < \infty$ and $0 < C_a < \infty$.

In conclusion, all elementary matrices on the right-hand side of (5-69) are realizable by the elementary two-ports of Fig. 5-15. If a pole is compact, the series impedance Z_a is replaced by a short circuit.

Combining the partial results obtained thus far, we obtain the two-port realization shown in Fig. 5-16. Note that the $1:1$ turn-ratioed ideal transformers included in Fig. 5-14 have been combined with the $1:t$ ideal transformers of Fig. 5-15. Note also that the elements L_a^∞, C_a^0, C_a^1, L_a^1, etc., can be moved into the series branch and combined with Z_1.

The circuit of Fig. 5-16 needs many more reactances than the degree of $\mathbf{Z}$ justifies. It also needs, in general, $n + 2$ ideal transformers. Hence, it is not suitable for practical realization. Its importance lies in its *feasibility*, as defined in Sec. 3-2; i.e., *the Cauer two-port is always realizable from a $\mathbf{Z}$ meeting the realizability criteria given in Sec. 5-3.* This follows from the real values of all transformer ratios t and the nonnegative real values of all elements C_a, L_a, C_b, and L_b of Fig. 5-15 proved above, as well as the realizability of Z_1 and Z_2.

A dual derivation, which results in a different realization for the two-port, can be carried out from $\mathbf{Y}$. The resulting circuit[3] requires a multiwinding ideal transformer and is hence no more practical than that of Fig. 5-16.

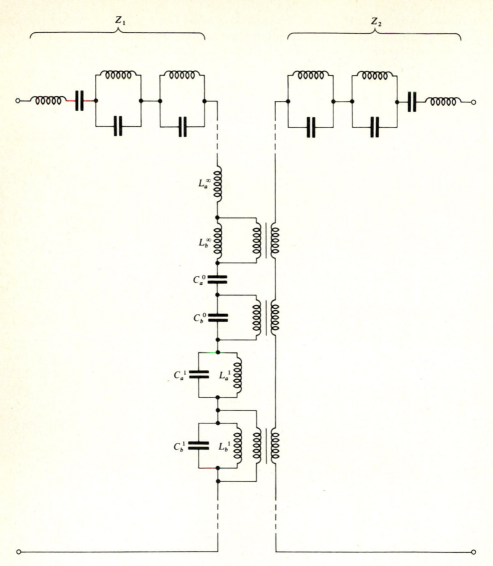

Figure 5-16 Cauer's two-port.

Example 5-13 Consider the z_{ij} tested for realizability in Sec. 5-3:

$$z_{11} = \frac{s^3 + 5s}{s^4 + 10s^2 + 9} = \frac{\frac{1}{2}s}{s^2 + 1} + \frac{\frac{1}{2}s}{s^2 + 9}$$

$$z_{12} = \frac{s}{s^4 + 10s^2 + 9} = \frac{\frac{1}{8}s}{s^2 + 1} + \frac{-\frac{1}{8}s}{s^2 + 9}$$

$$z_{22} = \frac{s^3 + 4s}{s^4 + 10s^2 + 9} = \frac{\frac{3}{8}s}{s^2 + 1} + \frac{\frac{5}{8}s}{s^2 + 9}$$

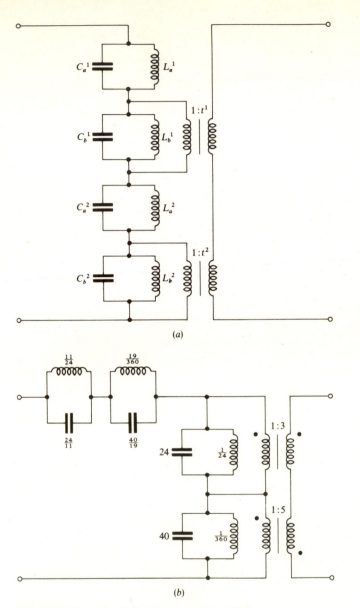

(a)

(b)

Figure 5-17 Example of Cauer's two-port synthesis.

There are no private poles nor any poles at 0 or ∞. Hence, the circuit is that shown in Fig. 5-17a, with the element values given by (5-83):

$$t^1 = \frac{\frac{3}{8}}{\frac{1}{8}} = 3 \qquad C_b^1 = \frac{\frac{3}{8}}{(\frac{1}{8})^2} = 24 \qquad L_b^1 = \frac{1}{(24)(1^2)} = \frac{1}{24}$$

$$C_a^1 = \frac{\frac{3}{8}}{(\frac{1}{2})(\frac{3}{8}) - (\frac{1}{8})^2} = \frac{24}{11} \qquad L_a^1 = \frac{1}{[(\frac{24}{11})(1^2)]} = \frac{11}{24}$$

$$t^2 = \frac{\left(\frac{5}{8}\right)}{-\frac{1}{8}} = -5 \qquad C_b^2 = \frac{\frac{5}{8}}{\left(-\frac{1}{8}\right)^2} = 40 \qquad L_b^2 = \frac{1}{(40)(3^2)} = \frac{1}{360}$$

$$C_a^2 = \frac{\frac{5}{8}}{\left(\frac{1}{2}\right)\left(\frac{5}{8}\right) - \left(-\frac{1}{8}\right)^2} = \frac{40}{19} \qquad L_a^2 = \frac{1}{\left(\frac{40}{19}\right)(3^2)} = \frac{19}{360}$$

When the $L_a^1 - C_a^1$ and $L_a^2 - C_a^2$ circuits are moved into the series input branch, the circuit of Fig. 5-17b results.

5-5 TRANSFER FUNCTIONS

When the two-port is excited by a generator and terminated by a load, as illustrated in Fig. 5-1, the signal-transfer properties of the *complete* circuit can be described by an appropriately chosen *transfer function*. We define the transfer function as the ratio of an output variable (voltage or current) to a known input quantity (generator voltage or current).† Since most practical generator and load impedances are essentially resistive, we shall restrict our discussions to resistor-terminated reactance two-ports.

In the simplest (and least useful) situation, both terminations are zero or infinite. Then we have one of the four configurations depicted in Fig. 5-18. These circuits are called *unterminated (unloaded)* two-ports. The proper choice of a transfer function for any of these circuits is obvious and unique. For example, for the circuit of Fig. 5-18a, the output variable must be V_2 since $I_2 \equiv 0$; the known input quantity is the generator voltage E. Hence, we must choose the *voltage ratio* A_V, defined by

$$A_V(s) \triangleq \frac{V_2(s)}{E(s)} \tag{5-84}$$

[The reader should keep in mind the dual interpretation of the variable s. Thus, for $s = j\omega$, $A_V(j\omega)$ may represent the ratio of the steady-state sine-wave voltage phasors at the output and input. In general, however, $A_V(s)$ is the ratio of the Laplace-transformed output signal $v_2(t)$ and generator signal $e(t)$ for a two-port initially free of stored energy.]

Similarly, for the circuit in Fig. 5-18b the transfer function must be the *transfer admittance*

$$Y_T(s) \triangleq \frac{I_2(s)}{E(s)} \tag{5-85}$$

For the circuit of Fig. 5-18c, the transfer function is the *transfer impedance*

$$Z_T(s) \triangleq \frac{V_2(s)}{I(s)} \tag{5-86}$$

---

† Here all quantities are assumed to be functions of s, not t.

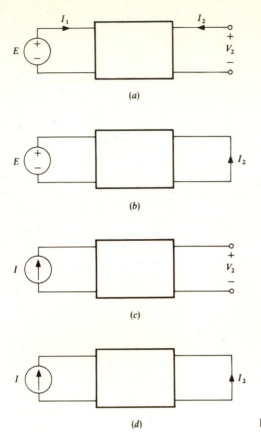

(a)

(b)

(c)

(d)　　　　　　　　　　**Figure 5-18** Unterminated two-ports.

Finally, for the circuit of Fig. 5-18d, the transfer function is the *current ratio*

$$A_I(s) \triangleq \frac{I_2(s)}{I(s)} \tag{5-87}$$

The transfer functions can readily be calculated from the two-port parameters **Z** or **Y** or **T**. For example, for the circuit of Fig. 5-18a, from (5-9)

$$V_1 = z_{11}I_1 + z_{12}I_2 = z_{11}I_1 \qquad V_2 = z_{12}I_1 + z_{22}I_2 = z_{12}I_1 \tag{5-88}$$

Hence
$$A_V = \frac{V_2}{E} = \frac{V_2}{V_1} = \frac{z_{12}}{z_{11}} \tag{5-89}$$

Alternatively, from (5-10)

$$I_2 = y_{12}V_1 + y_{22}V_2 = 0 \tag{5-90}$$

which gives

$$A_V = \frac{V_2}{V_1} = -\frac{y_{12}}{y_{22}} \tag{5-91}$$

Or, from (5-24),

$$V_1 = AV_2 - BI_2 = AV_2 \tag{5-92}$$

so that

$$A_V = \frac{V_2}{V_1} = \frac{1}{A} \tag{5-93}$$

For the circuit of Fig. 5-18b, from (5-9),

$$V_1 = z_{11}I_1 + z_{12}I_2 = E \qquad V_2 = z_{12}I_1 + z_{22}I_2 = 0 \tag{5-94}$$

Solving (5-94) for I_2 gives

$$I_2 = -\frac{Ez_{12}}{z_{11}z_{22} - z_{12}^2} = -E\frac{z_{12}}{\Delta_z} \tag{5-95}$$

so that

$$Y_T \triangleq \frac{I_2}{E} = -\frac{z_{12}}{\Delta_z} \tag{5-96}$$

Alternatively, from (5-10)

$$I_2 = y_{12}V_1 + y_{22}V_2 = y_{12}E \tag{5-97}$$

so that

$$Y_T = \frac{I_2}{E} = y_{12} \tag{5-98}$$

Or, from (5-24),

$$V_1 = E = AV_2 - BI_2 = -BI_2 \tag{5-99}$$

which gives

$$Y_T = \frac{I_2}{E} = -\frac{1}{B} \tag{5-100}$$

directly.

Exactly analogous manipulations give

$$Z_T = \frac{V_2}{I} \triangleq z_{12} = -\frac{y_{12}}{\Delta_Y} = \frac{1}{C} \tag{5-101}$$

for the circuit of Fig. 5-18c and

$$A_I = -\frac{z_{12}}{z_{22}} = \frac{y_{12}}{y_{11}} = -\frac{1}{D} \tag{5-102}$$

for the circuit of Fig. 5-18d.

If the two-port has a single resistive termination, it is called a *singly terminated* or *singly loaded* two-port. Four possible circuits for such a two-port are illustrated in Fig. 5-19. Notice that two other possible circuits exist which may be obtained by replacing the generator and its internal impedance R by its Norton equivalent in the circuit of Fig. 5-19b or by its Thevenin equivalent in Fig. 5-19d. Their transfer functions differ only by a factor R from those of Fig. 5-19b and d, and hence they do not merit separate treatment.

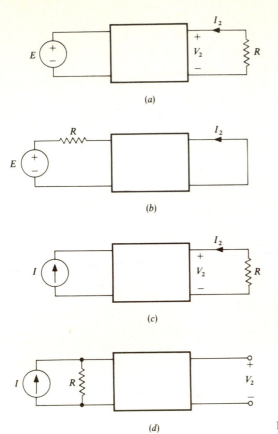

(d)   **Figure 5-19** Singly terminated two-ports.

For the circuit of Fig. 5-19a, we can choose $A_V = V_2/E$ as the transfer function. (Again, a trivially different choice is to select $Y_T \triangleq I_2/E$; since $I_2 = -V_2/R$, here $Y_T = -A_V/R$.)

For the circuit of Fig. 5-19b, the transfer function may be selected as $Y_T = I_2/E$ [or as $A_I = I_2/I = I_2/(E/R) = RY_T$ if the Norton generator model is substituted].

For the circuit of Fig. 5-19c, we can use $A_I = I_2/I$ or, as trivial variant, $Z_T = V_2/I = -I_2 R/I = -RA_I$. For the circuit of Fig. 5-19d, the transfer function can be $Z_T = V_2/I$ or if the Thevenin equivalent is used for the generator, $A_V = V_2/E = V_2/(IR) = Z_T/R$.

The transfer functions of the singly loaded two-port can also easily be found in terms of the two-port parameters and R, as will be shown next. For the circuit of Fig. 5-19a, combining the branch relations

$$V_1 = E \qquad V_2 = -RI_2 \qquad (5\text{-}103)$$

and the two-port relations (5-9), we get

$$z_{11}I_1 + z_{12}I_2 = E \qquad z_{12}I_1 + (z_{22} + R)I_2 = 0 \qquad (5\text{-}104)$$

which gives

$$I_2 = \frac{z_{12} E}{-z_{11} z_{22} + z_{12}^2 - z_{11} R_r} \tag{5-105}$$

so that

$$A_V = \frac{V_2}{E} = \frac{-I_2 R}{E} = \frac{z_{12} R}{\Delta_z + z_{11} R} \tag{5-106}$$

Alternatively, from (5-103) and (5-10)

$$I_2 = y_{12} V_1 + y_{22} V_2 = y_{12} E - y_{22} R I_2 \tag{5-107}$$

which gives

$$I_2 = \frac{y_{12} E}{1 + y_{22} R} \qquad A_V = \frac{-I_2 R}{E} = \frac{-y_{12} R}{1 + y_{22} R} \tag{5-108}$$

Finally, from (5-24), using (5-103), we have

$$V_1 = E = AV_2 - BI_2 = -ARI_2 - BI_2$$

$$I_2 = \frac{-E}{AR + B} \qquad A_V = \frac{-I_2 R}{E} = \frac{R}{AR + B} \tag{5-109}$$

Similar calculations performed for the circuit of Fig. 5-19*b* give

$$Y_T \triangleq \frac{I_2}{E} = \frac{-z_{12}}{\Delta_z + z_{22} R} = \frac{y_{12}}{1 + y_{11} R} = \frac{-1}{B + DR} \tag{5-110}$$

For the circuit of Fig. 5-19*c*,

$$A_I \triangleq \frac{I_2}{I} = \frac{-z_{12}}{z_{22} + R} = \frac{y_{12}}{\Delta_Y R + y_{11}} = \frac{-1}{CR + D} \tag{5-111}$$

Finally, for the circuit of Fig. 5-19*d*,

$$Z_T \triangleq \frac{V_2}{I} = \frac{z_{12} R}{z_{11} + R} = \frac{-y_{22} R}{\Delta_Y R + y_{22}} = \frac{R}{A + RC} \tag{5-112}$$

The most important and most widely used circuit is the doubly terminated (or doubly loaded) reactance two-port, illustrated in Fig. 5-20. Depending on whether Thevenin or Norton model is used for the generator and whether V_2 or I_2 is used as output variable, any one of the four transfer functions A_V, A_I, Z_T, and Y_T can be used to describe the transmission properties of the circuit. If the circuit of Fig. 5-20*a* is chosen, for example, i.e., a Thevenin generator model, and V_2 as output variable, A_V is the proper transfer function. Now the branch relations are

$$V_1 = E - R_G I_1 \qquad V_2 = -I_2 R_L \tag{5-113}$$

They can be combined with the two-port relations (5-9) to give

$$(z_{11} + R_G)I_1 + z_{12} I_2 = E \qquad z_{12} I_1 + (z_{22} + R_L)I_2 = 0 \tag{5-114}$$

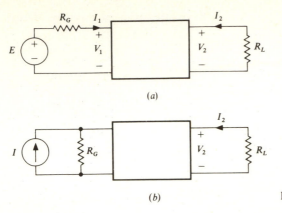

(a)

(b)

Figure 5-20 Doubly terminated two-ports.

Solving (5-114) for I_2 gives

$$I_2 = \frac{-z_{12}E}{\Delta_Z + z_{11}R_L + z_{22}R_G + R_GR_L} \qquad (5\text{-}115)$$

Hence

$$A_V = \frac{V_2}{E} = \frac{-I_2R_L}{E} = \frac{z_{12}R_L}{\Delta_Z + z_{11}R_L + z_{22}R_G + R_GR_L} \qquad (5\text{-}116)$$

Carrying out the calculations in terms of the y_{ij}, that is, combining and solving (5-113) and (5-10), gives

$$A_V = \frac{-y_{12}R_L}{\Delta_Y R_GR_L + y_{11}R_G + y_{22}R_L + 1} \qquad (5\text{-}117)$$

Finally, to express A_V in terms of the chain parameters, we combine and solve Eqs. (5-113) and (5-24). This results in

$$A_V = \frac{R_L}{AR_L + B + CR_GR_L + DR_G} \qquad (5\text{-}118)$$

As will be shown in Chap. 6, the design of a doubly terminated two-port is expediently performed using a different transfer function $H(s)$, which is related to $A_V(s)$ by

$$H(s) \triangleq \frac{1}{2}\sqrt{\frac{R_L}{R_G}\frac{E}{V_2}} = \frac{\sqrt{R_L/R_G}}{2A_V} \qquad (5\text{-}119)$$

5-6 THE CALCULATION OF Z, Y, AND T FROM THE TRANSFER FUNCTION OF AN UNTERMINATED OR SINGLY TERMINATED TWO-PORT

In the preceding section, we showed how the transfer function can be calculated from the two-port parameters z_{ij}, y_{ij} or A, B, C, and D. In actual design, as mentioned in Sec. 1-2, the procedure is reversed: we first derive the transfer function directly from the specifications, then the two-port parameters from the transfer function, and finally the element values from the two-port parameters. Hence, we must answer two questions:

1. In order to find appropriate transfer functions, we have to know the realizability conditions on them, i.e., when is a transfer function realizable by a resistively terminated two-port?
2. Given such a realizable transfer function, how can we obtain the two-port parameters from it?

We shall next answer these questions for unloaded and singly loaded two-ports. The discussion of doubly loaded two-ports will be given in Chap. 6.

5-6.1 Realizability Conditions

For *unloaded two-ports*, the transfer functions are, from (5-89) to (5-102),

$$A_V = \frac{z_{12}}{z_{11}} \quad \text{or} \quad Y_T = y_{12} \quad \text{or} \quad Z_T = z_{12} \quad \text{or} \quad A_I = \frac{y_{12}}{y_{11}} \tag{5-120}$$

Consider $A_V(s)$. Since for a reactance two-port both z_{12} and z_{11} are real rational odd functions of s, $A_V(s)$ must be a real rational even function of s, that is, a ratio of polynomials in s^2. The zeros of $A_V(s)$ consist of the zeros of z_{12} (which may be anywhere in the s plane but must have quadrantal symmetry, as illustrated in Fig. 5-12) and of the private poles of z_{11} (which must be on the $j\omega$ axis). The poles of $A_V(s)$ are more restricted. They consist of those zeros of z_{11} which are not also zeros of z_{12}. Since z_{11} is a reactance function, its zeros and thus the poles of $A_V(s)$ are restricted to the $j\omega$ axis. Furthermore, $A_V(s)$ cannot have a pole at $s = 0$ or $s \to \infty$. For if $z_{11}(s)$ has a zero at $s = 0$, then $z_{12}(s)$ must have a zero there too: it cannot have a pole (since that would be a private pole not allowed for z_{12}) and as an odd rational function z_{12} cannot have a finite nonzero value at $s = 0$. If $z_{11}(s)$ and $z_{12}(s)$ both have a zero at $s = 0$, that is, both have a factor s, that factor cancels in $A_V(s)$ which therefore will be finite† at $s = 0$.

A similar argument rules out the possibility of a pole of $A_V(s)$ for $s \to \infty$.

Since $A_I(s)$ has the same form in terms of the y_{ij} as $A_V(s)$ in terms of the z_{ij}, and since the properties of y_{ij} and z_{ij} are identical, we conclude that $A_I(s)$ and $A_V(s)$ are real rational even functions of s, with poles restricted to the finite nonzero portion of the $j\omega$ axis.

Consider now $Y_T(s)$ and $Z_T(s)$ as given in (5-120). From the known properties of y_{12} and z_{12}, it follows that $Y_T(s)$ and $Z_T(s)$ must be real rational odd functions of s, with only simple poles along the $j\omega$ axis (including $s = 0$ and $s \to \infty$).

Example 5-14 The functions $1/(s^2 + 1)$, $s^2/(s^2 + 2)$, and $(s^2 - 4)/(s^2 + 1)(s^2 + 3)$ are realizable $A_V(s)$ or $A_I(s)$ functions, while $(s^3 + s)/(s^4 + s^2 + 1)$, $s^4/(s^2 + 1)$, and $(s^2 - 4)/(s^4 + s^2)$ are not. (Why?)

† Note that z_{12} (but not z_{11}) can have a multiple zero at $s = 0$; then $A_V(s)$ will have a zero there.

Example 5-15 The functions $(s^4 + 3s^2 + 4)/(2s^3 + 3s)$, $1/(s^3 + 2s)$, and $s/(s^4 + 4s^2 + 2)$ are realizable Z_T or Y_T functions, while $(s + 1)/(s^3 + s)$, $s/(s^4 + 2s^2 + 1)$, and $1/(s^2 - 4)$ are not. (Why?)

For *singly loaded two-ports*, from (5-108) to (5-112), the transfer functions can be expressed as

$$A_V = -\frac{y_{12}}{y_{22} + 1/R} \quad \text{or} \quad Y_T = \frac{y_{12}}{y_{11} + 1/R}$$

or
$$A_I = -\frac{z_{12}}{z_{22} + R} \quad \text{or} \quad Z_T = \frac{Rz_{12}}{z_{11} + R}$$

(5-121)

Since all functions are in the same general form, it is sufficient to analyze the properties of, say, $A_V(s)$. Writing the y_{ij} as ratios of even or odd polynomials in s

$$-y_{12} = \frac{N_{12}(s)}{D_{12}(s)} \qquad y_{22}(s) = \frac{N_{22}(s)}{D_{22}(s)}$$

(5-122)

and impedance-normalizing to make $R = 1$, we have

$$A_V(s) = \frac{N_{12}/D_{12}}{N_{22}/D_{22} + 1} = \frac{N_{12}D_{22}/D_{12}}{N_{22} + D_{22}} = \frac{N(s)}{D(s)}$$

(5-123)

Here, the zeros of D_{22} are the finite poles of y_{22}; those of D_{12} are the finite poles of y_{12}. Since y_{12} cannot have any poles *not* shared with y_{22}, all root factors of D_{12} are also present in D_{22}. Hence, D_{22}/D_{12} becomes a *polynomial* after all common factors are canceled; therefore, $N(s) \triangleq N_{12}D_{22}/D_{12}$ is also a polynomial in s. $N(s)$ is even or odd since N_{12}, D_{22}, and D_{12} are all even or odd.

The denominator $D(s) \triangleq N_{22} + D_{22}$ is the sum of the numerator and denominator of the reactance function $y_{22}(s)$. Hence, by the theorem given in Sec. 4-1, it is a strictly Hurwitz polynomial.

Hence, $A_V(s)$ *is the ratio of an even or odd polynomial* $N(s)$ *and a strictly Hurwitz polynomial* $D(s)$. The zeros of $A_V(s)$ can be anywhere in the s plane (in quadrantal symmetry); the poles must be inside the LHP. There can be no poles at $s = 0$ or at $s \to \infty$. This last statement follows since such a pole of $-y_{12}/(y_{22} + 1)$ would imply a private pole of y_{12} at $s = 0$ or ∞; $y_{22} + 1$ cannot be zero at either frequency.

The above statements also hold, of course, for the Y_T, A_I, and Z_T functions of the singly terminated reactance two-port.

Example 5-16 $(s^2 + 1)/(s^3 + 2s^2 + 2s + 1)$ and $(s^2 - 8)/[(s + 1)(s^2 + 2s + 4)]$ are realizable transfer functions of a singly loaded reactance two-port; $(s^3 + s)/[(s^2 + 1)(s + 4)]$ and $(s + 4)/(s^2 + 2s + 1)$ are not. (Why?)

5-6.2 Calculation of the Two-Port Parameters from the Transfer Function

For an *unloaded two-port*, if a realizable $Y_T = y_{12}$ is given, we are free to choose any realizable reactance functions as y_{11} and y_{22} as long as the residue condition

(5-55) is satisfied at all poles of y_{12}. A simple way to achieve this is to expand y_{12} into partial fractions as in (5-56) and then choose

$$y_{11} = y_{22} = |k_{12}^\infty|s + \frac{|k_{12}^0|}{s} + \sum_{i=1}^{n} \frac{|k_{12}^i|s}{s^2 + \omega_i^2} \tag{5-124}$$

This choice results in realizable reactance functions y_{11}, y_{22} which satisfy (5-55) with the equal sign at all poles. As Probs. 5-16 and 5-17 illustrate, this can result in an economical realization of the circuit.

Example 5-17 For

$$z_{12} = \frac{s}{s^4 + 10s^2 + 9} = \frac{\frac{1}{8}s}{s^2 + 1} + \frac{-\frac{1}{8}s}{s^2 + 9}$$

this process gives

$$z_{11} = z_{22} = \frac{\frac{1}{8}s}{s^2 + 1} + \frac{\frac{1}{8}s}{s^2 + 9} = \frac{s^3 + 5s}{4(s^4 + 10s^2 + 9)}$$

The same process, of course, applies to finding $\mathbf{Z}$ from a given $Z_T = z_{12}$.

For a given $A_V = z_{12}/z_{11}$ of an unloaded two-port, there is again considerable arbitrariness in choosing z_{12} and z_{11}. A possible method is as follows. A realizable $A_V(s)$ is in the form

$$A_V(s) = \frac{N(s)}{D(s)} \tag{5-125}$$

where $N(s)$ and $D(s)$ are even polynomials and all zeros of $D(s)$ are simple, lying on the finite nonzero part of the $j\omega$ axis. Let these zeros be at $\pm\omega_1, \pm\omega_2, \ldots, \pm\omega_n$. Then choosing a polynomial

$$P(s) = s(s^2 + \omega_1'^2)(s^2 + \omega_2'^2) \cdots (s^2 + \omega_{n-1}'^2) \tag{5-126}$$

where the ω_i and ω_i' interlace, i.e.,

$$\omega_1^2 < \omega_1'^2 < \omega_2^2 < \omega_2'^2 < \cdots < \omega_{n-1}'^2 < \omega_n^2 \tag{5-127}$$

we can set

$$z_{12} = \frac{N(s)}{P(s)} \qquad z_{11} = \frac{D(s)}{P(s)} \tag{5-128}$$

Then z_{12}/z_{11} clearly gives the $A_V(s)$ of (5-125); it is also assured that $z_{11}(s)$ is a realizable reactance (recall the discussions of Sec. 3-1). Furthermore, z_{12} has the correct form and no poles which are not shared with z_{11}. Hence, no realizability condition is violated by z_{11} and z_{12}. If necessary, z_{22} can be constructed from the partial-fraction expansions of z_{11} and z_{12}, by choosing appropriate residues $k_{22}^i \geq (k_{12}^i)^2/k_{11}^i$ at all poles of z_{12}.

The above results also hold, of course, for the $A_I(s)$ of an unloaded two-port.

Example 5-18 Let

$$A_V = \frac{1}{(s^2 + 1)(s^2 + 3)}$$

Choosing arbitrarily

$$P(s) = s(s^2 + 2)$$

which satisfies (5-126) and (5-127), we see that Eq. (5-128) now gives

$$z_{12} = \frac{N(s)}{P(s)} = \frac{1}{s(s^2 + 2)} \qquad z_{11} = \frac{D(s)}{P(s)} = \frac{(s^2 + 1)(s^2 + 3)}{s(s^2 + 2)}$$

which are clearly realizable.

For a *singly loaded two-port*, with $R = 1$,

$$A_V(s) = \frac{N(s)}{D(s)} = \frac{-y_{12}}{y_{22} + 1} \tag{5-129}$$

where $N(s)$ is even or odd and $D(s)$ is a strictly Hurwitz polynomial. Assume first that $N(s)$ *is even*. Then we can write

$$A_V(s) = \frac{N(s)}{D_e(s) + D_o(s)} = \frac{N(s)/D_o(s)}{D_e(s)/D_o(s) + 1} \tag{5-130}$$

where, as before, D_e denotes the even part and D_o the odd part of $D(s)$. Choosing now

$$-y_{12} = \frac{N(s)}{D_o(s)} \qquad y_{22} = \frac{D_e(s)}{D_o(s)} \tag{5-131}$$

we conclude that $y_{22}(s)$, given by the ratio of the even to odd parts of a strictly Hurwitz polynomial, is a realizable reactance. Also, y_{12} is an odd rational function with the same denominator as that of y_{22}. Since a realizable $A_V(s)$ cannot have a pole for $s \to \infty$, it follows that the degree of $N(s)$ is at most equal to that of $D_e(s)$.† Hence, y_{12} cannot have a private pole at $s \to \infty$. Thus, y_{12} and y_{22} are realizable admittance parameters for a reactance two-port.

Example 5-19 For

$$A_V(s) = \frac{s^2 + 1}{s^3 + 2s^2 + 2s + 1}$$

we have, from (5-131),

$$-y_{12} = \frac{s^2 + 1}{s^3 + 2s} \qquad y_{22} = \frac{2s^2 + 1}{s^3 + 2s}$$

which form a realizable pair of two-port parameters. As before, y_{11} can be constructed by partial-fraction expansion, using the residue condition (5-55).

---

† Recall that $N(s)$ is even and that since $D(s)$ is strictly Hurwitz, the degrees of $D_e(s)$ and $D_o(s)$ differ by 1.

If $N(s)$ *is odd*, similarly

$$A_V(s) = \frac{N(s)}{D_e(s) + D_o(s)} = \frac{N(s)/D_e(s)}{D_o(s)/D_e(s) + 1} \tag{5-132}$$

which leads to the realizable parameters

$$-y_{12} = \frac{N(s)}{D_e(s)} \qquad y_{22} = \frac{D_o(s)}{D_e(s)} \tag{5-133}$$

Example 5-20 For

$$A_V(s) = \frac{s^3 + s}{s^3 + 2s^2 + 2s + 1}$$

(5-133) gives

$$-y_{12} = \frac{s^3 + s}{2s^2 + 1} \qquad y_{22} = \frac{s^3 + 2s}{2s^2 + 1}$$

The above derivations for the singly loaded two-port were carried out in terms of $A_V(s)$. However, since the other transfer functions have the same realizability properties, the preceding relations can equally well be applied to find y_{11} and y_{12} from Y_T, or z_{12} and z_{22} from A_I, or z_{11} and z_{12} from Z_T.

It is also noteworthy that the calculations performed in Sec. 5-6.2 show the realizability conditions of Sec. 5-6.1 to be not only *necessary* but also *sufficient* conditions.

5-7 LADDER-NETWORK REALIZATION OF AN UNLOADED OR SINGLY LOADED TWO-PORT

In Sec. 3-3 we introduced ladder networks and showed how they could be realized by performing alternating pole removals from the immittances Z, Y', Z'', ... A general form of the resulting circuit is shown in Fig. 5-21.

The ladder is the most widely used passive-circuit configuration, since it has two important advantages:

1. The generator E and the load impedance Z_n have a common (lower) terminal; hence, both can be grounded at the same time. This is not true for the *lattice*

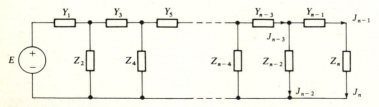

Figure 5-21 Ladder network.

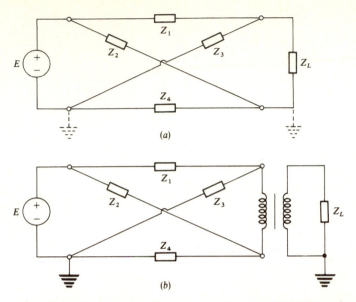

Figure 5-22 (a) Lattice network; (b) lattice network with output transformer.

network shown in Fig. 5-22a, for example, where simultaneously grounding both generator and load short-circuits Z_4 and thus changes the behavior of the circuit. Hence, if both generator and load must be grounded, a transformer must be included to isolate galvanically the load (as in Fig. 5-22b) or the generator from the rest of the circuit.

2. A ladder network has only one path (that going through Y_1, Y_3, ..., Y_{n-1} in Fig. 5-21) between the generator and the load. Hence, it is easy to interrupt the signal flow, either by causing one of the series Y_i to be zero or, alternatively, by making one of the shunt Z_j equal to zero. The latter short-circuits the signal path to ground. Again, the lattice behaves differently. There, two signal paths lead to both terminals of Z_L. Hence, the output can be made zero either by setting *two* admittances (say Y_1 and Y_2 or Y_1 and Y_3) equal to zero or by causing the two voltages at the two terminals of Z_L to be the same. The latter can be achieved by satisfying the bridge-balance condition $Z_1/Z_3 = Z_2/Z_4$, since the lattice is just a Wheatstone bridge with Z_L in the meter arm.

To make this last property more specific, we define a *transmission zero* as a zero of the transfer function defined for the terminated two-port. Note that some arbitrariness is implied here. For example, for the circuit shown in Fig. 5-23 we can choose as the transfer function either $A_V \triangleq V_2/E = 1/(s^2LC + 1)$ or $Y_T \triangleq I_2/E = -sC/(s^2LC + 1)$. For the former choice, the circuit has no transmission zero for $s = 0$; for the latter it does. With the former choice, the circuit has two transmission zeros for $s \to \infty$; with the latter, only one. (Other such examples are given in Probs. 5-25 and 5-26.) We can now state the following theorem.

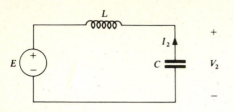

Figure 5-23 Simple *LC* two-port.

Theorem A ladder has a transmission zero only if either a series admittance Y_i or a shunt impedance Z_j equals zero.

Note that the above condition is a *necessary* but not a *sufficient* one. For example, the circuit of Fig. 5-43b (shown later) with $A_V = V_2/E$ as the transfer function, has no transmission zero at $\omega = \frac{1}{2}$ even though $Y_1 = 0$ there; nor does it have transmission zeros at $\omega = 0$ or $\omega \to \infty$, where $Z_2 = 0$. This is because Z_1 and Z_2 form a voltage divider in which both arm impedances simultaneously go to zero or infinity.

PROOF The proof of the theorem is straightforward. Referring to the ladder in Fig. 5-21, let us denote the current through the immittance Z_k or Y_k by J_k and the voltage across it by V_k. Next, we assume that the theorem is *false* so that a transmission zero exists at some frequency ω_z even though none of the series admittances $Y_1, Y_3, \ldots, Y_{n-1}$ or shunt impedances $Z_2, Z_4, \ldots, Z_n$ is zero there. Then at the output both V_n and $J_n = V_n/Z_n$ equal zero.† Clearly then also J_{n-1} and hence $V_{n-1} = J_{n-1}/Y_{n-1}$ are zero.† Since $Z_{n-2} \neq 0$, $J_{n-2} = (V_{n-1} + V_n)/Z_{n-2} = 0$. But $J_{n-3} = J_{n-2} + J_{n-1}$ also is zero then, and so is $V_{n-3} = J_{n-3}/Y_{n-3}$, etc. Working our way back to the input, we find $V_2 + V_1 = E = 0$. Hence, the output can only be zero at ω_z if the generator voltage is; therefore, no transmission zero is possible. This proves the theorem.

Since in a reactance ladder all branch immittances are reactance functions with zeros and poles located interlaced on the $j\omega$ axis, *all transmission zeros of a reactance ladder must be on the $j\omega$ axis, including $\omega = 0$ and $\omega \to \infty$*. Since in applications such as spectral filtering we want to block sine-wave signals from propagating to the output, we do want all transmission zeros on the $j\omega$ axis. Hence, in such applications this restriction of the reactance ladder circuit does not constitute a drawback; in fact, it is helpful.

According to our theorem, the transmission zeros of a ladder network can always be identified with the zero of one (sometimes several) of its branch immittances. If the zero is at a finite nonzero frequency ω_z, the branch causing it is almost always in one of the two forms shown in Fig. 5-24a and b. A zero at $\omega = 0$ corresponds to one or more branches shown in Fig. 5-24c and d, and a zero at

† Since we assumed $Z_n, Z_{n-2}, \ldots \neq 0$ and $Y_{n-1}, Y_{n-3}, \ldots \neq 0$, all divisions are permissible.

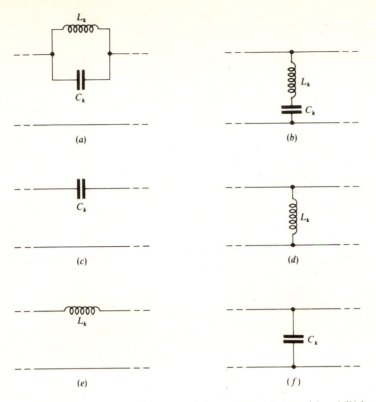

Figure 5-24 Branches realizing transmission zeros in a ladder: (*a*) and (*b*) branches realizing a zero at $\omega_z = (L_k C_k)^{-1/2}$; (*c*) and (*d*) branches realizing a zero at $\omega = 0$; (*e*) and (*f*) branches realizing a zero at $\omega \to \infty$.

$\omega \to \infty$ to branches of the form shown in Fig. 5-24*e* and *f*. In the actual construction of the circuit, it is usual to make one of the elements (say C_k) of a branch like that in Fig. 5-24*a* or *b* as close as possible to its prescribed value. The other (say L_k) is then carefully *tuned* so that it resonates with C_k at the prescribed frequency $\omega_z = (L_k C_k)^{-1/2}$. This process results in a compensation of the errors of L_k and C_k, and, as will be shown in Sec. 10-4, in a greatly reduced sensitivity to such errors for all frequencies close to ω_z.

Consider now the transfer function Z_T of the singly terminated two-port shown in Fig. 5-19*d*. From (5-112),

$$Z_T \triangleq \frac{V_2}{I} = \frac{z_{12} R}{z_{11} + R} = \frac{N(s)}{D(s)} \tag{5-134}$$

where $N(s)$ and $D(s)$ are polynomials. It is clear that Z_T will be zero at the zeros of z_{12} and at the private poles of z_{11}.

We shall now show that by synthesizing z_{11} in the form of a ladder and by choosing the zeros of the ladder immittances appropriately we can obtain a circuit for which the transfer function Z_T is a scaled version of the prescribed one. To be more specific, the realized transfer function $(Z_T)_{act}$, where the subscript stands for "actual," differs from the specified one only by a scale factor K so that $(Z_T)_{act} = KZ_T(s)$. The proof is based on the properties of the zeros and poles of $Z_T(s)$, which will therefore be examined next.

The zeros of $N(s)$, as defined in (5-134), are the transmission zeros.† The zeros s_i of $D(s)$, as shown in Eqs. (2-12) to (2-18), are the zeros of the circuit determinant $\Delta(s)$ formed from the coefficients $Y_{ij}(s)$ of Eq. (2-14). From this, it follows (Ref. 1, pp. 600–603) that the s_i are the *natural frequencies*‡ of the singly terminated network; i.e., they are the exponents in its zero-input response

$$x(t) = \sum_{i=1}^{n} K_i e^{s_i t} \qquad t \geq 0 \qquad (5\text{-}135)$$

[This shows again that $D(s)$ must be strictly Hurwitz, a conclusion we reached earlier.]

It follows from (5-134) that the natural modes are given by the zeros of $z_{11} + R$. Note that the poles of z_{12} are *not* poles of Z_T, since, by the residue condition, any pole of z_{12} is also a pole of z_{11}.

We can now make the following statements. In realizing Z_T for the circuit of Fig. 5-19d, the natural modes are determined solely by z_{11} and R. If the two-port realization is in the form of a reactance ladder, the transmission zeros are solely determined by the zeros of the series Y_i and shunt Z_j. Hence, if the reactance ladder has the correct z_{11} for the given natural frequencies and the Y_i and Z_j the correct zeros, i.e., the transmission zeros, then *both* the zeros and poles of Z_T will be realized by the circuit. Hence, the actual ratio of V_2 and I can only differ from the prescribed Z_T by a constant K, the scale factor.

The above statements form the basis of the design of reactance ladders.

Example 5-21 Let the prescribed transfer function be

$$Z_T = \frac{2}{s^3 + 2s^2 + 2s + 1} = \frac{z_{12}}{z_{11} + 1}$$

with $R = 1$. This is in the form given in (5-130); hence now

$$z_{11} = \frac{2s^2 + 1}{s^3 + 2s} \qquad z_{12} = \frac{2}{s^3 + 2s}$$

All transmission zeros are at $s \to \infty$; since z_{11} has no private poles which would cause Z_T to be zero, all transmission zeros are due to the zeros of z_{12}. The three transmission zeros must be realized by three branches of the form shown in Fig. 5-24e and f. Hence, our ladder has either the configuration shown in Fig. 5-25a or the one in Fig. 5-25b. Since here

$$z_{11} = \frac{2s^2 + 1}{s^3 + 2s} \quad \to \quad \begin{cases} \infty & \text{for} \quad s \to 0 \\ 0 & \text{for} \quad s \to \infty \end{cases}$$

evidently, the circuit of Fig. 5-25b is the correct one. In this circuit, the structure guarantees

---

† If the degree of $N(s)$ is lower than that of $D(s)$, there will be transmission zeros at infinite frequency as well.

‡ Also often called *natural modes*.

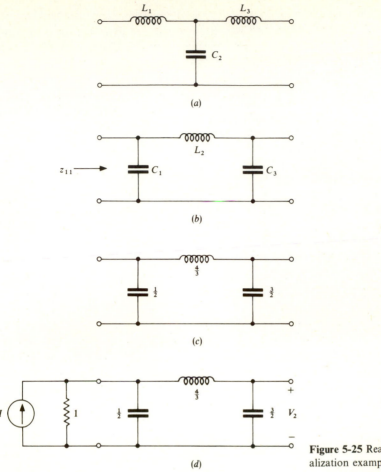

Figure 5-25 Reactance-ladder realization example.

the transmission zeros.† To obtain the correct natural modes, the $z_{11}(s)$ of the circuit must equal the prescribed function. This can be achieved by finding the element values $C_1, L_2,$ and C_3 through a Cauer 1 expansion of z_{11}:

$$2s^2 + 1) \quad s^3 + 2s) \quad \dfrac{s}{2} \swarrow^{C_1 s}$$
$$\underline{-s^3 - \dfrac{s}{2}}$$
$$\tfrac{3}{2}s) \quad 2s^2 + 1) \quad \tfrac{4}{3}s \swarrow^{L_2 s}$$
$$\underline{-2s^2}$$
$$1) \quad \tfrac{3}{2}s) \quad \tfrac{3}{2}s \swarrow^{C_1 s}$$
$$\underline{-\tfrac{3}{2}s}$$
$$0$$

† As long as all element values satisfy $0 < L < \infty, 0 < C < \infty.$

Hence, the two-port of Fig. 5-25c results. The full circuit is shown in Fig. 5-25d.

Next, we recall that while our realization procedure guarantees the correct zeros and poles of Z_T, the actual and specified transfer functions may differ by a constant scale factor K. It is clear from Fig. 5-25d that for the realized circuit

$$Z_T(s) = \frac{V_2(s)}{I(s)} \to 1 \qquad \text{for} \quad s \to 0$$

since the series inductor becomes a short circuit, while the shunt capacitors open circuits at dc. The specified Z_T, on the other hand, is $2/1 = 2$ for $s = 0$. Hence, our circuit realizes a $Z_T(s)$ which is one-half of the prescribed one, that is, $K = 1/2$. In many applications, such a discrepancy is acceptable; if it is not, it can be remedied by cascading a transformer with a turns ratio $1 : 2$ or a voltage amplifier of gain 2 with the two-port.

Example 5-22 Let $R = 1$ and

$$Z_T = \frac{s}{s^3 + 3s^2 + 3s + 1}$$

Now (5-132) applies, and hence

$$z_{11} = \frac{s^3 + 3s}{3s^2 + 1} \qquad z_{12} = \frac{s}{3s^2 + 1}$$

Z_T has three zeros, one at $s = 0$ and two for $s \to \infty$. Of these, two are due to zeros of z_{12}, one at $s = 0$ and one for $s \to \infty$. The third transmission zero, occurring for $s \to \infty$, is due to the private pole which z_{11} has for $s \to \infty$.

In the design, the private pole of z_{11} can be realized as a series impedance Z_1 (remember the discussions given in Sec. 5-4, in connection with Figs. 5-13 and 5-16?). Since

$$z_{11} = \frac{s}{3} + \frac{\frac{8}{3}s}{3s^2 + 1}$$

here $Z_1 = s/3$; that is, Z_1 is an inductor of $\frac{1}{3}$ H. Z_1 is of the form of Fig. 5-24e; it realizes one transmission zero at $s \to \infty$.

The remaining two transmission zeros at $s = 0$ and $s \to \infty$ can be obtained by branches of the form of Fig. 5-24d and f, respectively. The two-port then has the configuration shown in Fig. 5-26a. This corresponds to a Foster 1 expansion of z_{11}; the element values are $L_2 = 8/3$ and $C_3 = 9/8$. The complete circuit is therefore that of Fig. 5-26b.

The application of this process to transfer functions other than Z_T is equally straightforward, as illustrated below.

Example 5-23 Let the prescribed transfer function of the singly terminated two-port shown in Fig. 5-19a be

$$A_V \triangleq \frac{V_2}{E} = \frac{-y_{12}}{y_{22} + 1/R} = \frac{3s^2}{18s^2 + 3s + 1}$$

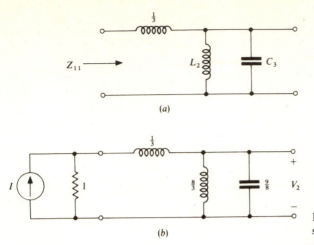

$Z_{11} \longrightarrow$

L_2

C_3

(a)

I

1

L_2

V_2

(b)

Figure 5-26 Reactance-ladder design example.

Choosing $R = 1$, we see that Eqs. (5-130) and (5-131) lead to

$$-y_{12} = \frac{3s^2}{3s} = s \qquad y_{22} = \frac{18s^2 + 1}{3s}$$

Now y_{22}, illustrated in Fig. 5-27a, must be expanded into a ladder which at the same time realizes the two transmission zeros (both at $s = 0$) of $A_V(s)$. By Fig. 5-24c and d, the circuit must be one of the two shown in Fig. 5-27b and c. The first of these circuits is clearly inappropriate, since L is short-circuited and hence $y_{22} = sC$ which cannot correspond to the specified function. On the other hand, the circuit of Fig. 5-27c has the admittance $y_{22} = sC + 1/sL$, which is of the required form. This circuit represents a Cauer 2 expansion of y_{22}; hence, the element values can be obtained by continued-fraction expansion

$$3s) \quad 1 + 18s^2) \quad \frac{1}{3s} \sqrt{\frac{1}{sL}}$$

$$\frac{-1}{18s^2) \quad 3s) \quad \frac{1}{6s} \sqrt{\frac{1}{sC}}$$

$$\frac{-3s}{0}$$

Thus, $L = 3$, $C = 6$, and the circuit has the form shown in Fig. 5-27d. For $s \to \infty$, the capacitor becomes a short circuit and the inductor an open circuit. Hence, $A_V \to 1$ for $s \to \infty$ in the actual circuit. Since the specified function $A_V(s) \to \frac{1}{6}$ for $s \to \infty$, we have six times as much voltage gain as required, that is, $K = 6$. This can be corrected, if necessary, by cascading a transformer with the circuit or by changing the 1-Ω resistor into a voltage divider.

All previous examples dealt with the realization of *singly terminated* ladders. The next example illustrates the design of unterminated reactance ladders.

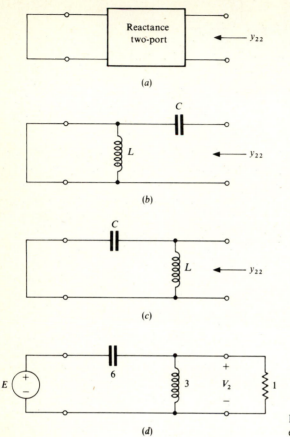

(a)

(b)

(c)

(d)

Figure 5-27 Reactance-ladder design example.

Example 5-24 Design a reactance ladder such that when it is connected in the configuration of Fig. 5-18d, the current ratio is

$$A_I \triangleq \frac{I_2}{I} = -\frac{z_{12}}{z_{22}} = \frac{1}{(s^2 + 1)(s^2 + 3)}$$

Using the process of Eqs. (5-125) to (5-128) and choosing $P(s) = s(s^2 + 2)$, we can easily derive the impedance parameters

$$z_{12} = \frac{-1}{s(s^2 + 2)} \qquad z_{22} = \frac{(s^2 + 1)(s^2 + 3)}{s(s^2 + 2)}$$

for the two-port.

Since there are four transmission zeros, all at infinite frequency, we immediately deduce that the circuit of either Fig. 5-28a or b is suitable. In the ladder of Fig. 5-28a, L_4 is open-circuited; furthermore, the behavior of $z_{22}(s)$ for $s \to \infty$ is inappropriate. The ladder of Fig. 5-28b, on the other hand, has a $z_{22}(s)$ of the correct form. Clearly, it represents a

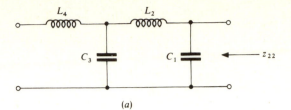

(a)

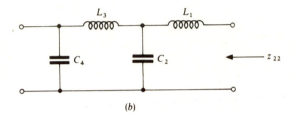

(b)

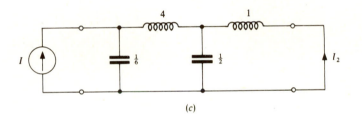

(c)

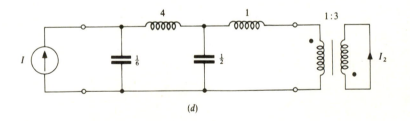

(d)

Figure 5-28 Reactance-ladder design example.

Cauer 1 expansion of z_{22}; hence, we can write

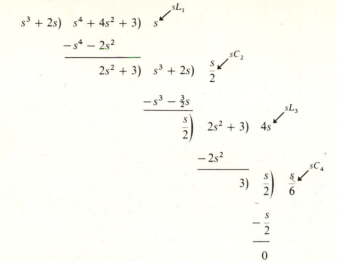

The element values are $L_1 = 1$, $C_2 = \frac{1}{2}$, $L_3 = 4$, and $C_4 = \frac{1}{6}$; the complete circuit is shown in Fig. 5-28c. For $s \to 0$, the circuit gives $A_I(0) = -1$, while for the specified function the dc value is $+\frac{1}{3}$. Hence, $K = -\frac{1}{3}$. If this scale factor is objectionable, an inverting transformer of turns ratio $1:3$ can be used at the output of the ladder, as shown in Fig. 5-28d.

On the basis of our derivations and design examples, we can state the following conclusions for this design process:†

1. The process realizes the expanded driving-point impedance z_{11} exactly.
2. The poles and zeros of the transfer function are exactly realized; however, the transfer function of the resulting circuit may differ from the prescribed function by a constant scale factor. This factor can often be found simply by comparing the values of the two functions for $s = 0$ or $s \to \infty$.
3. It follows from the previous two conclusions and from the expression $Z_T = z_{12}/(z_{11} + 1)$ that the z_{12} of the *actual* circuit differs from the z_{12} obtainable from the *specified* Z_T at most by a constant scale factor K. This factor is the same as that for the transfer functions.

5-8 PARTIAL POLE REMOVAL, ZERO SHIFTING

In the design examples given in Sec. 5-7, the transmission zeros of the ladder networks were all located at zero or infinite frequency. Consider, by contrast, the design of an unterminated ladder in the circuit of Fig. 5-18a, with the transfer function

$$A_V \triangleq \frac{V_2}{E} = \frac{s^2 + 4}{(s^2 + 1)(s^2 + 3)} = \frac{z_{12}}{z_{11}}$$

† To be specific, these conclusions refer to the transfer function $Z_T \triangleq V_2/I = z_{12}/(z_{11} + 1)$. They apply equally to other transfer functions.

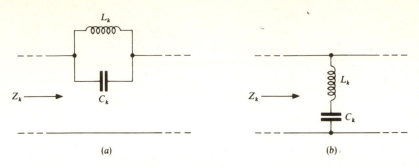

Figure 5-29 Ladder branches which produce finite nonzero-frequency transmission zeros: (a) $Z_k \to \infty$ for $s \to j/\sqrt{L_k C_k}$; (b) $Z_k = 0$ for $s = j/\sqrt{L_k C_k}$.

Choosing in the usual way $P(s) = s(s^2 + 2)$, we obtain

$$z_{11} = \frac{(s^2 + 1)(s^2 + 3)}{s(s^2 + 2)} \qquad z_{12} = \frac{s^2 + 4}{s(s^2 + 2)}$$

There are four transmission zeros; two are located at $s \to \infty$ and two at $s = \pm j2$. The new feature, of course, is the appearance of the latter. By the theorem given in Sec. 5-7, these transmission zeros must be realized by the zeros of some series $Y(s)$ or shunt $Z(s)$ in the two-port. As mentioned earlier, these zero-producing branch immittances almost always take the simple forms shown in Fig. 5-24a and b. However, as illustrated in Fig. 5-29a, a series branch of the form of Fig. 5-24a can be realized only if the remainder impedance Z_k at its left has a pole at $\omega_z = (L_k C_k)^{-1/2}$. Similarly, a shunt branch of the form of Fig. 5-24b implies that $Z_k(j\omega_z) = 0$ (Fig. 5-29b). To bring these conditions about, we can apply the strategy used in Brune's impedance synthesis (Sec. 4-3). There, to obtain a remainder impedance $Z_2(s)$ which was zero at a prescribed $s = j\omega_1$, a series impedance $Z_s(s)$ was first realized which satisfied $Z_s(j\omega_1) = Z_1(j\omega_1)$, as shown in Fig. 5-30$a$. This then left a remainder $Z_2(s)$ which was zero at $s = \pm j\omega_1$. Hence, $1/Z_2(s)$ had a pole at $\pm j\omega_1$ which could be removed by realizing the shunt L_2-C_2 branch. Since in the Brune synthesis $Z_1(s)$ was a minimum function, $Z_s(s)$ had to be realized by a nonphysical (negative) element; in the present case, however, all our immittances are reactance functions, and hence no such problems arise.

Returning to our example, we note that z_{11} has a private pole (and correspondingly A_V a transmission zero) at $s \to \infty$. Hence, it is legitimate to begin the realization with a series inductance L_1 (Fig. 5-30b). As in the Brune synthesis, we choose L_1 to satisfy

$$j\omega_1 L_1 = z_{11}(j\omega_1) \qquad (5\text{-}136)$$

Then the remainder

$$Z_2(s) = z_{11}(s) - sL_1 \qquad (5\text{-}137)$$

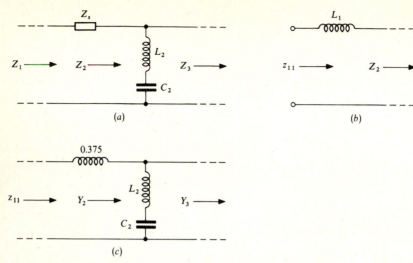

Figure 5-30 Steps in ladder network design: (a) $Z_s(j\omega_1) = Z_1(j\omega_1), Z_2(j\omega_1) = 0$; (b) $j\omega_1 L_1 = z_{11}(j\omega_1)$, $Z_2(j\omega_1) = 0$.

is, for $s = j\omega_1$,

$$Z_2(j\omega_1) = z_{11}(j\omega_1) - j\omega_1 L_1 = 0 \tag{5-138}$$

Consequently, $Y_2(s) = 1/Z_2(s)$ has a pole at $j\omega_1$.

In the given example, we want to introduce the transmission zero at $j2$; hence, $\omega_1 = 2$, and

$$L_1 = \frac{z_{11}(j2)}{j2} = \frac{j0.75}{j2} = 0.375$$

The remainder is (Fig. 5-30c)

$$Z_2(s) = z_{11}(s) - 0.375s = \frac{5s^4 + 26s^2 + 24}{8s^3 + 16s} = \frac{(s^2 + 4)(5s^2 + 6)}{8s^3 + 16s}$$

Clearly, $Z_2(s)$ has the required zeros at $\pm j2$; hence, $Y_2(s)$ has poles there. Therefore writing

$$Y_2(s) = \frac{K_1 s}{s^2 + 4} + Y_3(s)$$

gives $K_1 = \frac{8}{7}$ and hence (Fig. 5-30c) $L_2 = 1/K_1 = \frac{7}{8}$, $C_2 = 1/\omega_1^2 L_2 = \frac{2}{7}$. Also

$$Y_3(s) = \frac{8s^3 + 16s}{5s^4 + 26s^2 + 24} - \frac{\frac{8}{7}s}{s^2 + 4} = \frac{\frac{16}{7}s}{5s^2 + 6}$$

[Notice that the pole of $Y_2(s)$ at $s^2 = -4$ is no longer present in $Y_3(s)$.] At this stage, we have realized the transmission zero at $\pm j\omega_1$. The presence of the series inductor L_1 also leads us to believe that a transmission zero at $s \to \infty$ has been realized. Surprisingly, this is not the case. To illustrate this point, Fig. 5-31 shows

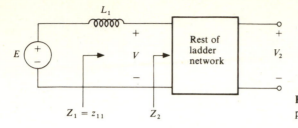

$Z_1 = z_{11}$ Z_2

Figure 5-31 Illustration of the partial pole-removal process.

the complete circuit after the realization of L_1. If the frequency of the generator is very high, i.e., if $s \to \infty$, the voltage V is given by

$$V = \lim_{s \to \infty} \left(E \frac{Z_2}{sL_1 + Z_2} \right) \tag{5-139}$$

Assume now that the partial-fraction expansion of $Z_1 = z_{11}$ is

$$z_{11} = k_{11}^\infty s + \frac{k_{11}^0}{s} + \sum_{i=1}^{n} \frac{k_{11}^i s}{s^2 + \omega_i^2} \tag{5-140}$$

Then clearly

$$Z_2 = z_{11} - sL_1 = (k_{11}^\infty - L_1)s + \frac{k_{11}^0}{s} + \sum_{i=1}^{n} \frac{k_{11}^i s}{s^2 + \omega_i^2} \tag{5-141}$$

Recall that the value of L_1, as given by (5-136), is determined by the transmission zero ω_1. Hence, in general $L_1 \neq k_{11}^\infty$, and therefore Z_2 still has a pole at infinite frequency with a residue $k_{11}^\infty - L_1$. Hence, $Z_2 \to (k_{11}^\infty - L_1)s$ for $s \to \infty$, and by (5-139) for $s \to \infty$

$$V \to E \frac{(k_{11}^\infty - L_1)s}{sL_1 + (k_{11}^\infty - L_1)s} = E \frac{k_{11}^\infty - L_1}{k_{11}^\infty} \tag{5-142}$$

Therefore in general $V \neq 0$. Hence, by the theorem of Sec. 5-7, unless a series Y or a shunt Z is zero for $s \to \infty$ in the rest of the ladder network (Fig. 5-31), V_2 will not be zero for $s \to \infty$. Thus, the realization of L_1 did *not* create a transmission zero at infinite frequency.

Notice that the removal of sL_1 from z_{11} did not eliminate the pole which z_{11} had at $s \to \infty$; it merely "weakened" it in the sense of decreasing the residue at that pole. For this reason, the operation is called *partial pole removal*, in contrast to the *full pole removal*, which the Foster expansion of a reactance employs, as discussed in Sec. 3-2.

Of course, if $L_1 = k_{11}^\infty$ had been chosen so that Z_2 no longer had a pole at infinite frequency, then by (5-142) or (5-139) $V \to 0$ for $s \to \infty$ and hence also $V_2 \to 0$. Hence, we can conclude that *for our unterminated two-port a full pole removal creates a transmission zero; a partial pole removal, however, does not.* A similar conclusion can be reached also for singly terminated two-ports. The situation is different, however, for doubly terminated reactance two-ports. This topic will be investigated in more detail in Chap. 6.

Returning to Eqs. (5-140) and (5-141), it is important to note that if the elements following L_1 in the ladder are all to be positive, Z_2 must be PR. This requires that the residue $k_{11}^\infty - L_1$ of Z_2 be nonnegative, i.e., that

$$L_1 = \frac{z_{11}(j\omega_1)}{j\omega_1} \le k_{11}^\infty \qquad (5\text{-}143)$$

Note that (5-143) *is not always satisfied*. In our circuit, $L_1 = 0.375$, while k_{11}^∞ is clearly 1 for the given z_{11}. Hence (5-143) holds. However, if we change the numerator of A_V to, say, $s^2 + 1.96$, so that the transmission zeros lie at $s = \pm j1.4$, then (5-136) gives

$$L_1 \approx 12.735 > k_{11}^\infty$$

and $Z_2(s)$ will not be PR.

Therefore, a ladder obtained using partial pole removals is not always realizable; it is *not* a feasible circuit as we defined the term in Sec. 3-2. By contrast, ladders obtained by full pole removals only are simply Cauer expansions of a driving-point reactance and hence are always feasible.

Returning to our example, after removing the shunt L_2-C_2 branch (Fig. 5-30c), we are left with the remainder $Y_3(s) = \frac{16}{7}s/(5s^2 + 6)$ and two yet-to-be-realized transmission zeros at $s \to \infty$. Following the process described in Sec. 5-7, we obtain the full circuit by the Foster 1 expansion of

$$Z_3(s) = \frac{1}{Y_3(s)} = \frac{7}{16}\frac{5s^2 + 6}{s} = \frac{35}{16}s + \frac{42}{16s}$$

Removing the pole of Z_3 at $s \to \infty$ gives $L_3 = \frac{35}{16}$ (Fig. 5-32) and leaves a remainder $Z_4 = 42/16s = 1/\frac{8}{21}s$. This corresponds to a shunt capacitor with a value $C_4 = 8/21$.

Thus, we obtain the circuit of Fig. 5-32. The element values can be quickly checked by noting that for $s \to \infty$ the impedances of C_2 and C_4 become negligible compared with those of L_2 and L_3, with which they are in series. Hence, for the realized circuit as $s \to \infty$

$$z_{11}(s) \to s\left(L_1 + \frac{L_2 L_3}{L_2 + L_3}\right) = s$$

which agrees with the asymptotic value of the prescribed z_{11} for $s \to \infty$. Similarly, for $s \to 0$, $z_{11}(s) \to 1/[(C_2 + C_4)s] = 1/\frac{2}{3}s$, which also agrees with the value obtainable from the prescribed z_{11}.

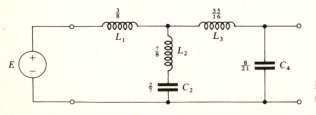

Figure 5-32 Ladder network obtained using partial pole removal.

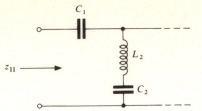

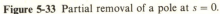

Figure 5-33 Partial removal of a pole at $s = 0$.

To find the scale factor of A_V, we note that for the realized circuit $A_V(0) = 1$ since the inductors become short circuits and the capacitors open circuits at dc. By contrast, for the prescribed voltage ratio $A_V(0) = 4/3$. Hence, the scale factor is $K = 3/4$.

In the design process which gave the circuit of Fig. 5-32 we started with a partial removal of the pole which $z_{11}(s)$ had at $s \to \infty$. But $z_{11}(s)$ also has a pole at $s = 0$. Hence, we may be tempted to remove a part of that pole, realizing the circuit of Fig. 5-33. The value of C_1 is then determined by a relation analogous to (5-136)

$$\frac{1}{j\omega_1 C_1} = z_{11}(j\omega_1) \qquad C_1 = \frac{1}{j\omega_1 z_{11}(j\omega_1)} = \frac{1}{(j2)(j0.75)} \approx -0.667$$

Hence, the circuit of Fig. 5-33 is not realizable.

A convenient way to anticipate which operations lead to realizable ladders is to plot the pole-zero patterns of all immittances encountered during the design. For the circuit of Fig. 5-32, designed earlier in this section, these patterns are shown in Fig. 5-34. A detailed explanation of the patterns will be given next.

The top line of the figure shows the location of the zeros of A_V on a non-linearly scaled ω axis. The scale is arbitrary and unimportant since only the relative position of the various zeros and poles is of concern. The zeros of A_V (which are the transmission zeros) are denoted by small triangles; the double triangle at $\omega = \infty$ denotes a double transmission zero.

The next line shows, on the same ω scale, the zeros (small circles) and poles (crosses) of $z_{11}(j\omega)$. The last six lines illustrate the zeros and poles of the remainder immittances Z_2, Y_2, Y_3, Z_3, Z_4, and Y_5. These functions, as can be found in the earlier part of this section, were

$$z_{11} = \frac{(s^2 + 1)(s^2 + 3)}{s(s^2 + 2)} \qquad Z_2 = \frac{(s^2 + 4)(5s^2 + 6)}{8s(s^2 + 2)}$$

$$Y_2 = \frac{1}{Z_2} \qquad Y_3 = \frac{\frac{16}{7}s}{5s^2 + 6} \qquad Z_3 = \frac{1}{Y_3} \qquad Z_4 = \frac{21}{8s}$$

It is easy to check that the zeros and poles of these functions are those shown in Fig. 5-34. The admittance Y_5 left after C_4 has been removed is identically zero. Hence, it has no poles or zeros shown in the figure.

On the extreme right of the figure, the symbol of L_1 indicates that Z_2 was obtained from z_{11} by extracting the impedance of L_1. Similarly, the L_2-C_2 branch indicates how Y_3 was obtained from Y_2, etc.

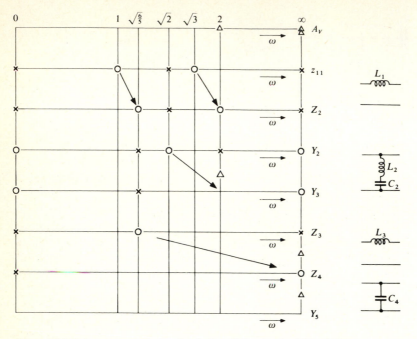

Figure 5-34 Pole-zero patterns for the design of the circuit of Fig. 5-32.

On the vertical line at $\omega = 2$ between the Y_2 and Y_3 patterns, the triangle indicates that the transmission zero at $\omega = 2$ is being realized at this stage. Similarly, the triangles at $\omega = \infty$, between the Z_3 and Z_4, as well as the Z_4 and Y_5 patterns indicate the realization of the two transmission zeros at infinite frequency.

Consider now the effect of the partial pole removal from z_{11}. As Eqs. (5-140) and (5-141) make clear, the poles of z_{11} reappear at the same locations in Z_2;

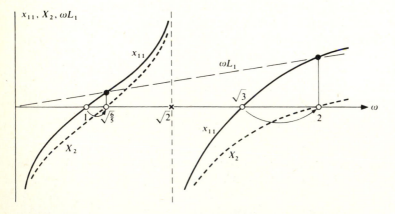

Figure 5-35 Reactance diagram for partial pole removal at infinite frequency.

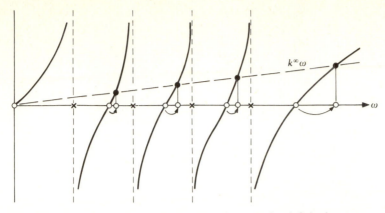

Figure 5-36 Reactance diagram for partial pole removal at infinite frequency.

however, *the zeros of z_{11} appear shifted toward higher frequencies in Z_2.* To show why this must be so, Fig. 5-35 illustrates the reactances $x_{11} \triangleq z_{11}(j\omega)/j$, ωL_1, and $X_2 \triangleq Z_2(j\omega)/j = x_{11} - \omega L_1$. The zeros of X_2 are at the intersections of the $x_{11}(\omega)$ curve and the ωL_1 line. They clearly are at higher frequencies than the zeros of $x_{11}(\omega)$. Furthermore, it follows from the monotone increasing character of all reactance functions that this conclusion is generally true, i.e., that any removal from a pole at infinity which entails subtraction of a $k^\infty \omega$ term from the reactance function must shift its zeros upward. This point is illustrated for a higher-degree reactance in Fig. 5-36, which also shows that a zero of the reactance function at $\omega = 0$ does not shift; this can also be concluded from Eqs. (5-140) and (5-141) if $k_{11}^0 = 0$ is substituted.

The preceding discussion explains why in Fig. 5-34 the zeros of Z_2 are higher in frequency than the corresponding zeros of z_{11}. It also points out that the purpose of the partial pole removal was to *shift the zero of z_{11} located at $\omega = \sqrt{3}$ to $\omega = 2$,* where we require a transmission zero. This process is often called *zero shifting.*

In going from Y_2 to Y_3, the pole of Y_2 at $\omega = 2$ is fully removed. As explained earlier, such a full removal creates a transmission zero, which is what we intended to achieve. The pole removal does not affect the other poles of Y_2; we simply eliminated one of the terms in the partial-fraction expansion without altering the other terms, which represent the remaining poles. Also, the zeros of Y_2 at $\omega = 0$ and $\omega = \infty$ remain unchanged. However, the zero at $\omega = \sqrt{2}$ shifts to $\omega = 2$ and cancels the pole there. The remainder of the calculations can be similarly analyzed.

To illustrate how partial pole removal at $\omega = 0$ can be performed, we consider the following design problem.

Example 5-25 Design a reactance ladder such that in the circuit of Fig. 5-37a (which is the same as that of Fig. 5-19a) the transfer function is

$$A_V = \frac{V_2}{E} = \frac{-y_{12}}{y_{22} + 1} = \frac{3s(4s^2 + 1)}{12s^3 + 16s^2 + 5s + 6}$$

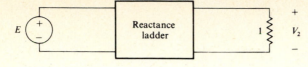

(a)

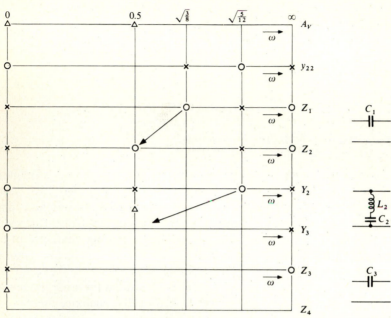

(b)

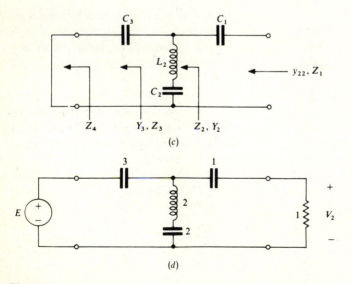

(c)

(d)

Figure 5-37 Ladder-network design example.

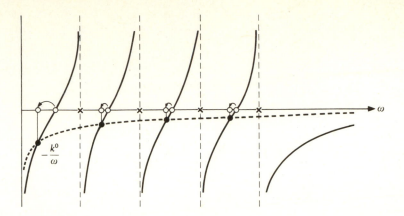

Figure 5-38 Reactance diagram illustrating the effects of a partial pole removal at zero frequency.

The transmission zeros are at $\omega = 0$ and $\omega = \pm\frac{1}{2}$. Finding y_{22} the usual way, we have

$$y_{22} = \frac{12s^3 + 5s}{16s^2 + 6} = \frac{3s(s^2 + \frac{5}{12})}{4(s^2 + \frac{3}{8})}$$

The transmission zeros and the zero-pole pattern of y_{22} are shown in the first two lines of the chart of Fig. 5-37b. We could try to perform a partial pole removal at $\omega = \infty$ from y_{22} to shift a zero of y_{22} to the transmission zero at $\omega_1 = 1/2$, but since the only movable, i.e., finite- and nonzero-frequency, zero of y_{22} lies at $\omega = \sqrt{5/12} \approx 0.645 > \omega_1$, and since a partial pole removal at infinite frequency would shift all zeros *up*, this approach will fail. We use instead

$$Z_1 \triangleq \frac{1}{y_{22}} = \frac{4}{3} \frac{s^2 + \frac{3}{8}}{s(s^2 + \frac{5}{12})}$$

and partially remove its pole at $\omega = 0$ by finding C_1 such that

$$\frac{1}{j\omega_1 C_1} = \frac{1}{j0.5C_1} = Z_1(j\omega_1) = -2j \qquad C_1 = 1$$

This causes the zero of Z_1 at $\omega = \sqrt{3/8} \approx 0.612$ to shift *down* to $\omega_1 = 1/2$. As Fig. 5-38 illustrates, zeros always shift down when we perform a partial pole removal at $\omega = 0$ from a reactance function, i.e., whenever we subtract from a reactance function a function $-k^0/\omega$, where† $k^0 > 0$. As the same figure also illustrates, a zero at $\omega \to \infty$ (if any) will not shift.

The rest of the calculations is routine. The remainder after C_1 is extracted is (Fig. 5-37c)

$$Z_2 = Z_1(s) - \frac{1}{sC_1} = \frac{4s^2 + 1}{12s^3 + 5s}$$

so that

$$Y_2 = \frac{1}{Z_2} = \frac{12s^3 + 5s}{4s^2 + 1} = \frac{\frac{1}{2}s}{s^2 + \frac{1}{4}} + 3s$$

† For simpler notation, we do not use a new symbol for the removed residue k^0.

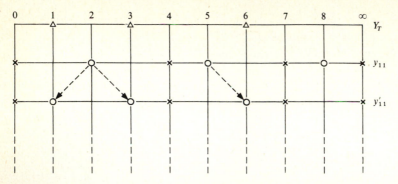

Figure 5-39 A pole-zero pattern illustrating possible choices in ladder-network design.

This gives the element values $L_2 = 2$ and $C_2 = 2$ and the remainder $Y_3 = 3s$. Hence, the last element C_3 is a 3-F capacitor. In establishing the location of C_3 in the circuit, we must keep in mind that we are expanding y_{22}; hence the ladder has a short circuit at the generator end, and C_3 must be in the last *series* branch.† The complete circuit is shown in Fig. 5-37*d*.

For more complicated circuits, there may be several possible pole-removal sequences. Consider, for example, the pole-zero patterns shown in Fig. 5-39. There, by partial pole removal at $\omega = 0$ or $\omega \to \infty$, we can achieve any one of the indicated three zero-shifting operations. It is possible, of course, that some or all of these manipulations will result in negative element values later on if the removed residue is more than what y_{11} possessed at the affected pole.

We have shown that a partial (or, for that matter, also a full) pole removal at $\omega = 0$ or ∞ shifts all zeros located at finite nonzero frequencies toward the pole being partially removed. As Fig. 5-40 demonstrates, this conclusion holds also for

† For this circuit, it is not possible to identify the transmission zero at $\omega = 0$ with either C_1 or C_3. We can easily show, however, that at dc the ladder degenerates into a capacitive T network composed of C_1, C_2, and C_3, which represents a *single* transmission zero at $\omega = 0$.

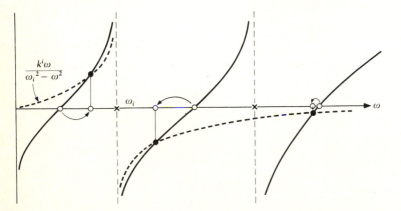

Figure 5-40 Reactance diagram showing the effects of partial pole removal at a finite nonzero frequency ω_i.

the removal of a pole located at a finite nonzero frequency. Zeros located above the pole shift down; those below the pole shift up in frequency.

In conclusion, we can make the following observations:

1. A partial pole removal shifts all zeros (except those located at $\omega = 0$ or ∞) toward the affected pole. In general, the closer a zero is to the pole, the more it moves.
2. The more residue is removed, the more the zeros shift. If *all* the residue which the reactance function possessed is removed, the nearest zero vanishes along with the pole; effectively, the zero moves on top of the pole and cancels it.† This property can be seen, for example, in Fig. 5-36, by drawing a second $k^\infty \omega$ line with a larger k^∞, that is, larger slope, and observing the increase in the shifts of the zeros. Similar observations can be made in Figs. 5-38 and 5-40.
3. No zero can be shifted beyond an adjacent pole. This is evident from Figs. 5-36, 5-38, and 5-40.

The operation shown in Fig. 5-40 is seldom needed. However, in rare cases the last-mentioned limitation of zero shifting makes it convenient, as will be illustrated next.

Example 5-26 For the circuit of Fig. 5-41a, let

$$A_V = \frac{-y_{12}}{1 + y_{22}} = \frac{s^2 + 1}{s^3 + 2s^2 + 2s + 1}$$

The usual process gives

$$y_{22} = \frac{2s^2 + 1}{s^3 + 2s}$$

and the pole-zero patterns of Fig. 5-41b. Considering the first three patterns, it is obvious that the zero of y_{22} at $\omega = 1/\sqrt{2}$ should be shifted *up* to $\omega = 1$ for the realization of the transmission zero. However, y_{22} has no pole at $\omega = \infty$ which could be partially removed to accomplish the shifting. The situation is similar for $Z_1 = 1/y_{22}$; its zero at $\omega = \sqrt{2}$ should be shifted *down* to $\omega = 1$; however, it lacks a pole at $\omega = 0$.

The solution is to use the partial removal of a finite nonzero pole. Reducing the residue of the pole of y_{22} at $\omega = \sqrt{2}$ can be used to shift the zero at $\omega = 1/\sqrt{2}$ to $\omega_1 = 1$. Thus, k is found from

$$\frac{kj\omega_1}{-\omega_1^2 + 2} = y_{22}(j\omega_1) \qquad k = 1$$

Hence, the ladder begins with a shunt branch admittance

$$\frac{s}{s^2 + 2} = \frac{s/L_1}{s^2 + 1/L_1 C_1}$$

---

† In a process reminiscent of Act III of "Tristan and Isolde." If the removed pole is at $\omega = 0$ or $\omega \to \infty$, however, the zero replaces the pole rather than cancels it.

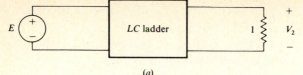

(a)

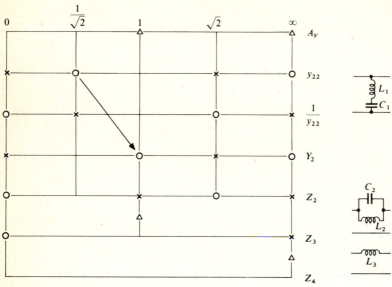

(b)

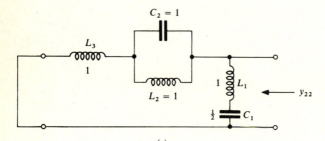

(c)

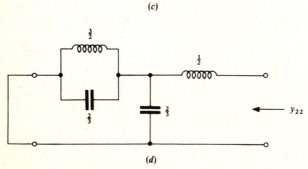

(d)

Figure 5-41 Ladder-network design illustrating partial pole removal at a finite nonzero frequency: (a) circuit configuration; (b) pole-zero diagram; (c) ladder circuit; (d) equivalent ladder.

which gives $L_1 = 1$, $C_1 = 1/2$. The rest of the realization is conventional; it is illustrated by the pole-zero patterns of Fig. 5-41b and leads to the circuit of Fig. 5-41c.

Note that an alternative sequence of pole removals gives the ladder of Fig. 5-41d *without* the partial removal of a finite nonzero pole. For more complicated circuits, such alternatives may not exist, and the designer may be forced to resort to partial pole removal at a finite nonzero frequency.

Finally, it should be made clear that the discussions of Secs. 5-7 and 5-8 represent a highly simplistic description of the theory of ladder-circuit synthesis. The exact theory (due to Darlington, Bader, and Piloty) is beyond the scope of this book. The interested reader should consult Refs. 7 and 8 for further information.

5-9 SUMMARY

The main subjects discussed in this chapter were the following:

1. The definition of a two-port as a special four-terminal circuit whose terminal currents and voltages satisfy the constraints of Eq. (5-1); the definition of the impedance, admittance, and chain parameters for two-ports (Sec. 5-1).
2. Realizability conditions (analogous to the PR impedance conditions) for the immittance parameters of passive two-ports (Sec. 5-2).
3. Cauer's realizability conditions for reactance two-ports, as well as some equivalent conditions which are less compact but easier to apply. The use of these conditions was illustrated by numerical examples (Sec. 5-3).
4. Cauer's partial-fraction-expansion technique for the realization of reactance two-ports (Sec. 5-4).
5. The definition of transfer functions (voltage ratio, current ratio, transfer impedance, and transfer admittance) for a two-port excited by a generator and terminated by a load. The derivation of the appropriate transfer function for unloaded, singly loaded, and doubly loaded two-ports in terms of the immittance and chain parameters of the two-port (Sec. 5-5).
6. Realizability conditions on the transfer functions and calculation of the two-port immittance parameters from a given transfer function for unloaded and singly loaded two-ports (Sec. 5-6).
7. The properties of ladder networks and their transmission zeros and the realization of unloaded or singly loaded reactance two-ports in the form of ladder networks (Sec. 5-7).
8. The design of ladder two-ports with finite nonzero transmission zeros, using partial pole removal (zero shifting) and pole-zero diagrams (Sec. 5-8).

The content of this chapter is of great theoretical importance since it establishes the realizability properties of the various two-port parameters and transfer functions; however, the design processes given are suitable only for unterminated or singly terminated two-ports, which are of limited practical usefulness. It will be shown in the next chapter how these concepts and design procedures can be extended to the much more useful class of doubly loaded reactance two-ports.

PROBLEMS

5-1 Calculate all z_{ij} parameters for the circuit of Fig. 5-7a, using the scheme of Fig. 5-4.

5-2 Calculate all y_{ij} parameters of the circuit of Fig. 5-8 using the scheme of Fig. 5-5.

5-3 Calculate $\mathbf{Z}$ and $\mathbf{Y}$ for the two-port shown in Fig. 5-9.

5-4 Calculate **Z** for the circuit of Fig. 5-10a.

5-5 (a) Find the value of G in Fig. 5-10a for which $\Delta_Y = 0$.

(b) For this value of G, **Z** does not exist. What is the physical reason for this?

5-6 Find the chain parameters of the circuit of Fig. 5-7a using (5-26) and the results obtained in Sec. 5-1 for the **Z** and **Y** of this circuit. Compare with the values obtained using Fig. 5-11.

5-7 Find the chain parameters of the circuit of Fig. 5-10a using the scheme shown in Fig. 5-11. Compare your results with those obtained in Sec. 5-1.

5-8 (a) From the circuit of Fig. 5-42a containing two ideal transformers with turns ratios a_1, a_2, and an RLCM two-port, show that the $Z(s)$ function defined in Eq. (5-38) must be a PR function for all real a_1 and a_2.

(b) From the circuit of Fig. 5-42b, show that the $Y(s)$ function defined in (5-39) is a PR function for all real a_1 and a_2 values.

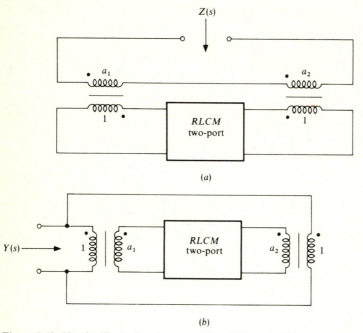

(a)

(b)

Figure 5-42 Circuits illustrating the properties of $Z(s)$ and $Y(s)$ given in Eqs. (5-38) and (5-39).

5-9 (a) From (5-41) of Sec. 5-2, if $a_1 = 0$, $r_{22}(j\omega) \geq 0$ follows; if $a_2 = 0$, $r_{11}(j\omega) \geq 0$. Show that for $a_1 \neq 0$, $a_2 \neq 0$, (5-44) is a necessary and sufficient condition. *Hint:* Dividing (5-41) by $a_2^2 > 0$ gives $F(x) = r_{11} x^2 + 2r_{12} x + r_{22} \geq 0$, where $x = a_1/a_2$.

(b) Find the minimum value of $F(x)$ and draw the obvious conclusions.

5-10 Show that the functions

$$y_{11} = y_{22} = \frac{4s^4 + 6s^2 + 1}{4s^3 + 4s} \qquad y_{12} = -\frac{4s^4 + 4s^2 + 1}{4s^3 + 4s}$$

are realizable admittance parameters for a reactance two-port.

5-11 What is the maximum value C for which the impedance parameters

$$z_{11} = z_{22} = \frac{2s^2 + 1}{s^3 + 3s} \qquad z_{12} = C \frac{2s^2 + 1}{s^3 + 3s}$$

are realizable?

5-12 Which of the Z matrices listed below are realizable as the open-circuit impedance parameters of an *LCM* two-port? Why?

(a) $z_{11} = \dfrac{s^4 + 8s^2 + 8}{2s^3 + 3s}$ $\qquad z_{12} = \dfrac{s^4 + 3s^2 + 4}{2s^3 + 3s}$ $\qquad z_{22} = \dfrac{s^4 + 4s^2 + 2}{2s^3 + 3s}$

(b) $z_{11} = \dfrac{\frac{2}{3}s^2 + 1}{s^3 + 2s}$ $\qquad z_{12} = \dfrac{1}{s^3 + 2s}$ $\qquad z_{22} = \dfrac{2s^2 + 1}{s^3 + 2s}$

(c) $z_{11} = \dfrac{s^2 + 3}{s(s^2 + 4)}$ $\qquad z_{12} = \dfrac{s^3 + 4s}{s^2 + 4}$ $\qquad z_{22} = \dfrac{s^3 + 8s}{s^2 + 4}$

5-13 Using (5-26), calculate the chain matrices which correspond to the impedance parameters given in Prob. 5-12. Test the chain matrices for realizability; compare your results with those obtained in Prob. 5-12.

5-14 Find **T** from the admittance parameters given in Prob. 5-10. Show that **T** is realizable.

5-15 Prove the relations (5-80) for the Cauer two-port of Fig. 5-15b.

5-16 Using the maximum permissible value C_{max} of C, find the Cauer two-port realization of the z_{ij} given in Prob. 5-11. Repeat with $C = C_{max}/2$. What is the effect of reducing C on the two-port?

5-17 Carry out the Cauer two-port synthesis for all **Z** given in Prob. 5-12. Which ones have negative elements? Why?

5-18 Derive the relations (5-101) and (5-102) giving the transfer functions of the circuits of Fig. 5-18c and d in terms of the two-port parameters.

5-19 Derive the relations (5-110) to (5-112) for the transfer functions of the singly loaded circuits shown in Fig. 5-19b, c, and d.

5-20 Derive the relations (5-117) and (5-118) for the doubly terminated two-port shown in Fig. 5-20a.

5-21 Decide which of the functions $s/(s + 2)$, $s^2/(s^4 + 8s^2 + 8)$, $(\frac{2}{3}s^2 + 1)/(s^3 + 2s)$, $1/(s^3 + 4s)$, $1/s^2$, and $(s^4 + 4s^2)/(s^2 + 4)$ are realizable as the $A_V(s)$ or $Z_T(s)$ of an unloaded reactance two-port.

5-22 Find a realizable $z_{22}(s)$ for the given impedance parameters

$$z_{12} = \frac{1}{s(s^2 + 2)} \qquad z_{11} = \frac{(s^2 + 1)(s^2 + 3)}{s(s^2 + 2)}$$

such that the residue condition is satisfied with the equals sign at all poles of z_{12}.

5-23 Find y_{11} and y_{12} from the following $A_I(s)$ functions of an unloaded reactance two-port:

$$\frac{1}{s^2 + 1} \qquad \frac{s^2}{s^2 + 2} \qquad \frac{s^2 - 4}{(s^2 + 1)(s^2 + 9)}$$

Determine y_{22} such that the residue condition is met with the equals sign at all poles of y_{12}.

5-24 Give a *physical* interpretation of the fact that for the circuit of Fig. 5-23 we have:

(a) Two transmission zeros for $s \to \infty$ and none for $s = 0$ if the transfer function is chosen to be $A_V = V_2/E$.

(b) One transmission zero at $s = 0$ and one for $s \to \infty$ if we choose $Y_T = I_2/E$ as the transfer function.

5-25 Find all transmission zeros of the circuits shown in Fig. 5-43 if the transfer function chosen is:

(a) $A_V = \dfrac{V_2}{E}$ $\qquad$ (b) $Y_T = \dfrac{I_2}{E}$

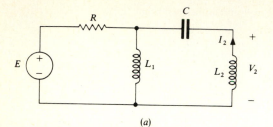

(a)

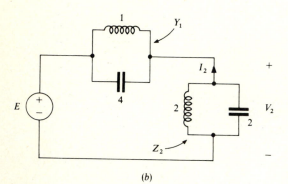

(b)   **Figure 5-43** Ladder circuits.

5-26 Find the transmission zeros for the ladder circuit shown in Fig. 5-44, with the transfer function chosen to be:

(a) $A_V = \dfrac{V_2}{E}$ (b) $Y_T = \dfrac{I_2}{E}$

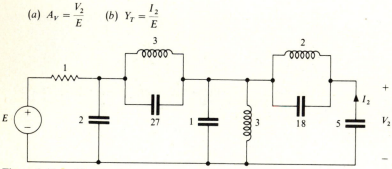

Figure 5-44 Ladder circuit.

5-27 Calculate z_{11} and z_{12} for the two-port of Fig. 5-25c. Show that z_{11} is equal to the one specified in the example of Sec. 5-7, while z_{12} is one-half of the prescribed function.

5-28 Calculate the value of the constant scale factor K between the Z_T of the circuit shown in Fig. 5-26b and the prescribed function:

$$Z_T = \frac{s}{s^3 + 3s^2 + 3s + 1}$$

5-29 Design an unterminated ladder for the circuit of Fig. 5-18d with the transfer function

$$A_I = \frac{Ks^2}{(s^2 + 1)(s^2 + 5)}$$

Use $P(s) = s(s^2 + 3)$.
  What is K for the realized circuit?

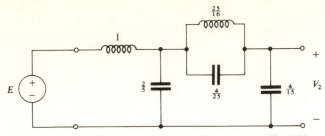

Figure 5-45 Ladder circuit equivalent to that shown in Fig. 5-32.

5-30 Show that the transfer function

$$A_V = \frac{z_{12}}{z_{11}} = \frac{s^2 + 4}{(s^2 + 1)(s^2 + 3)}$$

and the corresponding impedance parameters

$$z_{11} = \frac{(s^2 + 1)(s^2 + 3)}{s(s^2 + 2)} \qquad z_{12} = \frac{s^2 + 4}{s(s^2 + 2)}$$

which were realized by the circuit of Fig. 5-32, can also be realized in the form shown in Fig. 5-45.

5-31 Which, if any, of the three zero-shifting operations indicated in Fig. 5-39 can result in a realizable ladder?

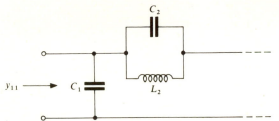

Figure 5-46 Reactance-ladder realization.

5-32 The short-circuit admittance parameter y_{11} is to be developed into the ladder shown in Fig. 5-46. For

$$y_{11} = \frac{(s^2 + 1)(s^2 + 3)}{3s(s^2 + 2)}$$

what are the possible values of the transmission zero $\omega_1 = 1/\sqrt{L_2 C_2}$ if the ladder is to be realizable?

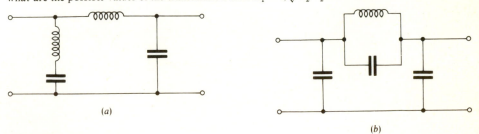

Figure 5-47 Equivalent ladders.

5-33 Design two ladders in the configurations shown in Fig. 5-47a and b for the transfer function

$$Z_T = \frac{z_{12}}{z_{22} + 1} = \frac{s^2 + 4}{s^3 + s^2 + 2s + 1}$$

5-34 Carry out the design steps giving the circuits of Fig. 5-41c and d from

$$A_V = \frac{-y_{12}}{1 + y_{22}} = \frac{s^2 + 1}{s^3 + 2s^2 + 2s + 1}$$

5-35 Design a ladder network from

$$Z_T = \frac{z_{12}}{z_{11} + 1} = \frac{Ks^2}{s^3 + 2s^2 + 2s + 1}$$

What is K for the realized circuit?

5-36 Design a ladder circuit from

$$A_I = \frac{-z_{12}}{z_{22} + 1} = \frac{K(s^2 + 4)}{s^3 + 2s^2 + 2s + 1}$$

What is K for the realized circuit?

5-37 Design a ladder from

$$Z_T = \frac{z_{12}}{z_{11} + 1} = \frac{K(s^2 + 4)(s^2 + 9)}{s^5 + s^4 + 7s^3 + 4s^2 + 10s + 3}$$

What is the scale factor K for the realized circuit?

5-38 Design a ladder from

$$z_{11} = \frac{(s^2 + 1)(s^2 + 9)}{s(s^2 + 4)(s^2 + 16)}$$

if the circuit is to have transmission zeros at $\omega = \pm 5, \pm 6$, and ∞. How many possible configurations can this circuit have?

5-39 Design a ladder for

$$A_I = \frac{-z_{12}}{z_{22} + 1} = \frac{K(s^2 + 1)}{s^3 + 2s^2 + 3s + 1}$$

5-40 Referring to the circuit shown in Fig. 5-13, show that:

(a) All private poles of z_{11} can be extracted in the form of a realizable Z_1; similarly, all private poles of z_{22} can be realized by Z_2.

(b) In a dual manner, all private poles of y_{11} and y_{22} can be realized by shunt branches connected across the input ports.

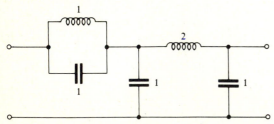

Figure 5-48 Two-port for which z_{11} does *not* have any private pole.

(c) The presence of Z_1 (or Z_2) does not necessarily indicate private poles in z_{11} (or z_{22}). For example, examine z_{11} and z_{12} for the circuit of Fig. 5-48. Show why z_{11} does *not* have a private pole.

5-41 (a) Referring to the circuit of Fig. 5-13, show that it is possible for a ladder network to begin with a shunt branch, that is, $Z_1(s) \equiv 0$, but z_{11} to have some private poles. For instance, consider the circuit of Fig. 5-49a; find its z_{ij} parameters. Show that it is equivalent to the circuit of Fig. 5-49b where $Z_1(s) \neq 0$.

(b) How can you generate such networks? *Hint:* What happens if the transmission zeros coincide with the poles of z_{11}?

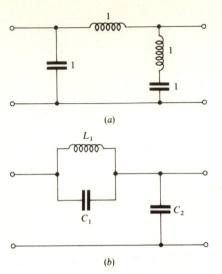

(a)

(b)

Figure 5-49 Equivalent two-ports for which z_{11} does have private poles.

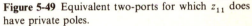

REFERENCES

1. Desoer, C. A., and E. S. Kuh: "Basic Circuit Theory," McGraw-Hill, New York, 1969.
2. Rupprecht, W.: "Netzwerksynthese," Springer, Berlin, 1972.
3. Cauer, W.: "Synthesis of Linear Communication Networks," McGraw-Hill, New York, 1958.
4. Guillemin, E. A.: "Synthesis of Passive Networks," Wiley, New York, 1957.
5. Van Valkenburg, M. E.: "Introduction to Modern Network Synthesis," Wiley, New York, 1960.
6. Balabanian, N.: "Network Synthesis," Prentice Hall, Englewood Cliffs, N.J., 1958.
7. Unbehauen, R.: "Synthese elektrischer Netzwerke," Oldenbourg, Munich, 1972.
8. Darlington, S.: Synthesis of Reactance 4-Poles Which Produce Prescribed Insertion Loss Characteristics, *J. Math. Phys.*, September 1939, pp. 257–353.

CHAPTER SIX

THE SYNTHESIS OF DOUBLY TERMINATED REACTANCE TWO-PORTS

6-1 INTRODUCTION

The synthesis of two-ports which are resistively terminated at both ports is an *important and difficult topic*. Its *importance* lies partly in the fact that all physical generators have a finite nonzero internal impedance and all loads are at least partly resistive. Hence, in practice all two-ports are doubly loaded. In addition, doubly loaded two-ports offer the possibility of matching both generator and load simultaneously to the rest of the transmission system. This, in turn, brings several benefits. One is the lack of reflected waves on the connecting wires, cables, and other coupling elements. This may be of great importance, for example, in signal transmission systems. The other is the resulting low sensitivity of the system performance to the variations of its elements. To demonstrate this last property, let us refer to Fig. 6-1. Assume that generator, two-port, and load are all matched to each other, i.e., that $R_G = Z_1$ and $Z_2 = R_L$. Then, as is well known, the maximum amount of power available from the generator, $P_{max} = E_G^2/4R_G$, enters the two-port at port 1. Since, as indicated in Fig. 6-1, the two-port contains only reactive components, no power is lost there and the total power P_{max} emerges at port 2, where it is dissipated in R_L. Assume now that any element of the two-port, say the inductor L_i, is varied from its nominal value. Whether L_i is decreased or increased, the perfect match between the parts of the system will be disturbed and the output power P_2 decreases from P_{max} to a lower value. The situation is schematically illustrated in Fig. 6-2. The most important conclusion that can be drawn from Fig. 6-2 is that in the immediate vicinity of its nominal value, L_i has very little effect on the output power. More explicitly, the sensitivity of P_2 to

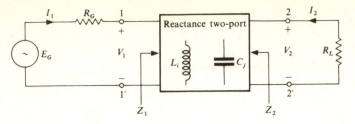

Figure 6-1 Doubly terminated reactance two-port.

changes in L_i, $\partial P_2/\partial L_i$, is zero for the nominal value of L_i since the $P_2(L_i)$ curve has a maximum there. Thus, a matched system consisting of a reactive two-port and its terminations has a natural immunity against the variations of any of its elements. (Needless to say, the above derivation could be duplicated for variations in any C_j or even R_G or R_L.) Singly terminated or unterminated networks do not have this property and are thus inherently more sensitive.

The difficulty in designing doubly terminated reactance two-ports arises because the voltages and currents at both ports are in general nonzero. As a consequence, the transfer function depends on all three immittance parameters (z_{ij} or y_{ij}), rather than just the one or two involved in the design of singly loaded or unloaded two-ports. This makes the circuit equations much more complicated and necessitates the introduction of auxiliary functions, such as the reflection and characteristic functions, into the calculation. Finally, because of the finite power available in the circuit, it is advantageous to specify the power, rather than voltage or current, relations of the two-port. It turns out, as we shall see, that this necessitates the extensive manipulation, including factoring, of polynomials. These additional complications represent the price the designer pays for the increased flexibility, decreased sensitivity, and other practical advantages (described above) of doubly terminated two-ports.

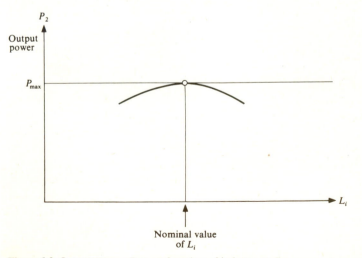

Figure 6-2 Output power P_2 as a function of inductance L_i.

6-2 TRANSDUCER PARAMETERS

The transmission properties of the doubly terminated two-ports can conveniently be described using the voltage ratio (Fig. 6-1)

$$A_V = \frac{V_2(s)}{E_G} \tag{6-1}$$

since E_G is known. It is more expedient, however, to start out by considering the *power-transmission* properties of the two-port. Note that for singly or unterminated two-ports power-transfer relations were meaningless, since either the generator was a pure voltage or current source, with (theoretically) unlimited power-supplying capabilities or the load was an open or short circuit which required zero power for sustaining a nonzero output voltage or current, respectively. For doubly terminated two-ports, however, the generator can only supply a finite amount of power. This power, as is easy to prove, is at most

$$P_{max} = \frac{E_G^2}{4R_G} \tag{6-2}$$

P_{max} is obtained only for a matched load, i.e., for $Z_1 = R_G$ (Fig. 6-1).† At the same time, maintaining a voltage V_2 across the resistor R_L requires a power input

$$P_2 = \frac{|V_2|^2}{R_L} \tag{6-3}$$

which is transformed into heat or radiant energy. Obviously, the relation

$$\frac{P_{max}}{P_2} = \left| \frac{1}{2} \sqrt{\frac{R_L}{R_G}} \frac{E_G}{V_2} \right|^2 \geq 1 \tag{6-4}$$

must hold. The ratio P_{max}/P_2 therefore provides a well-defined measure of the power-transmission efficiency of the terminated two-port. For $P_{max}/P_2 = 1$, all the power which the generator can supply is transmitted to the load; for $P_{max}/P_2 \to \infty$, none of it is. It seems therefore logical to define and use, instead of A_V, the *transducer factor*

$$H(s) = \frac{1}{2} \sqrt{\frac{R_L}{R_G}} \frac{E_G}{V_2(s)} = \frac{1}{2} \sqrt{\frac{R_L}{R_G}} \frac{1}{A_V(s)} \tag{6-5}$$

to characterize the two-port. It should be noticed that $H(s)$, in contrast to previously defined transfer functions, is an input-quantity/output-quantity ratio. By (6-4), for a passive two-port

$$|H|^2 = \frac{P_{max}}{P_2} \geq 1 \tag{6-6}$$

---

† Note that all voltages and currents in the discussions of this chapter refer to *effective* values. Thus, a voltage signal $V_0 \cos(\omega t + \varphi)$ is denoted by its phasor $V = (V_0/\sqrt{2})e^{j\varphi}$.

Let the *transducer constant* Γ be defined as the natural logarithm of H:

$$\Gamma \triangleq \ln H = \alpha + j\beta \qquad \alpha \triangleq \ln |H| \qquad \beta \triangleq \underline{/H} \qquad (6\text{-}7)$$

Here α is the *transducer loss* (in nepers), and β is the *phase lag* of the output voltage V_2 behind E_G; β, which is measured in radians, will simply be called the *phase* of the two-port.

It is usual to measure the loss in decibels. This unit is defined by

$$\alpha(\text{in dB}) = 20 \log |H| \qquad (6\text{-}8)$$

Hence
$$1 \text{ Np} = 20 \log e \text{ dB} \approx 8.686 \text{ dB} \qquad (6\text{-}9)$$

where Np is the abbreviation for nepers.

Other useful parameters of the doubly terminated two-port are the *reflection factors*. Referring to Fig. 6-1, we see that these are defined by the relations[†]

$$\rho_1 = \frac{R_G - Z_1}{R_G + Z_1} \qquad (6\text{-}10)$$

and
$$\rho_2 = \frac{R_L - Z_2}{R_L + Z_2} \qquad (6\text{-}11)$$

Here, Z_1 is the impedance seen at the primary port (terminals 1-1'), with the secondary port terminated in R_L

$$Z_1 = \frac{V_1}{I_1} \qquad (6\text{-}12)$$

Similarly, Z_2 is the driving-point impedance at the secondary port when the primary port is terminated by R_G (but with E_G set to zero and R_L removed). Note that with E_G at the primary port, as shown in Fig. 6-1,

$$\frac{V_2}{-I_2} = R_L \neq Z_2 \qquad (6\text{-}13)$$

To give a physical meaning to the reflection coefficients, consider the power flow through the network. Although the generator is capable of supplying a maximum amount of power P_{max}, in fact it supplies only

$$P_1 = \text{Re } V_1 I_1^* \leq P_{max} \qquad (6\text{-}14)$$

to the input of the two-port. Since the two-port is lossless, P_1 travels through the two-port undiminished and eventually emerges as the power flow into the load:

$$P_2 = \frac{|V_2|^2}{R_L} = P_1 \qquad (6\text{-}15)$$

[†] The alternative definitions $\rho_1 = (Z_1 - R_G)/(Z_1 + R_G)$ and $\rho_2 = (Z_2 - R_L)/(Z_2 + R_L)$ have some conceptual advantages and are also often used.

This realistic picture can be replaced by a hypothetical one as follows: the generator is assumed *always* to supply its maximum power P_{max} to the primary port of the two-port. Because of the mismatch ($R_G \neq Z_1$) a part P_r of P_{max} is reflected from the two-port, and only the remainder

$$P_2 = P_{max} - P_r \qquad (6\text{-}16)$$

passes on to the load. The *reflected power* P_r leaves the two-port at port 1 and is eventually reabsorbed by the generator. The power flow is schematically illustrated in Fig. 6-3. It is readily recognized that in the steady state the two concepts described give the same results for the net power flow at any point in the network and hence the second interpretation (borrowed from the physics of transmission-line systems) is permissible.

By Eqs. (6-14) to (6-16),

$$P_r = P_{max} - P_1 = \frac{E_G^2}{4R_G} - \operatorname{Re} I_1 I_1^* Z_1 \qquad (6\text{-}17)$$

We denote

$$Z_1 = R_1 + jX_1 \qquad (6\text{-}18)$$

Therefore for $s = j\omega$

$$P_r = \frac{E_G^2}{4R_G}\left(1 - \frac{4R_G}{E_G^2}|I_1|^2 R_1\right) = P_{max}\left(1 - \frac{4R_G R_1}{|R_G + Z_1|^2}\right)$$

$$= P_{max}\frac{(R_G + R_1)^2 + X_1^2 - 4R_G R_1}{|R_G + Z_1|^2} \qquad (6\text{-}19)$$

since $I_1 = E_G/(R_G + Z_1)$. After some simple calculation, we obtain

$$P_r = P_{max}\left|\frac{R_G - Z_1}{R_G + Z_1}\right|^2 = P_{max}|\rho_1|^2 \qquad (6\text{-}20)$$

Hence, $|\rho_1|^2$ is the ratio of the reflected power to the available generator power. Evidently, $|\rho_1|^2 \leq 1$.

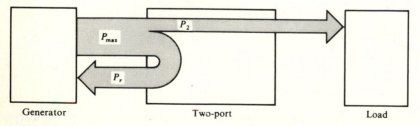

Generator   Two-port   Load

Figure 6-3 Power flow in a doubly terminated two-port.

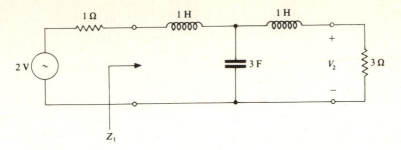

Figure 6-4 Two-port example.

Example 6-1 Find P_{max}, P_r, ρ_1, P_2, and V_2 for the circuit of Fig. 6-4 at $\omega = 1$ rad/s.

By elementary calculations

$$P_{max} = \frac{2^2}{(4)(1)} = 1 \text{ W} \qquad Z_1 = \frac{3s^3 + 9s^2 + 2s + 3}{3s^2 + 9s + 1}$$

and

$$\rho_1 = \frac{1 - Z_1}{1 + Z_1} = \frac{-3s^3 - 6s^2 + 7s - 2}{3s^3 + 12s^2 + 11s + 4}$$

Hence

$$|\rho_1|^2 = \frac{(6\omega^2 - 2)^2 + \omega^2(3\omega^2 + 7)^2}{(-12\omega^2 + 4)^2 + \omega^2(-3\omega^2 + 11)^2}$$

For $\omega = 1$, $|\rho_1|^2 = \frac{29}{32}$. Hence

$$P_r = |\rho_1|^2 P_{max} = \frac{29}{32} \text{ W} \qquad P_2 = P_{max} - P_r = \frac{3}{32} \text{ W}$$

On the other hand, by analyzing the network, we find

$$V_2 = \frac{6}{3s^3 + 12s^2 + 11s + 4}$$

and hence, for $s = j1$,

$$|V_2|^2 = \frac{36}{(-12\omega^2 + 4)^2 + \omega^2(-3\omega^2 + 11)^2} = \frac{9}{32}$$

Therefore

$$P_2 = \frac{|V_2|^2}{R_2} = \frac{3}{32}$$

and

$$P_r = P_{max} - P_2 = 1 - \frac{3}{32} = \frac{29}{32}$$

as obtained via $|\rho_1|$.

Next, we introduce the *characteristic function* $K(s)$ of the terminated two-port by the definition

$$K(s) = \rho_1(s)H(s) \qquad (6\text{-}21)$$

Then, for $s = j\omega$,

$$|K|^2 = |\rho_1|^2|H|^2 = \frac{P_r}{P_{max}} \frac{P_{max}}{P_2} = \frac{P_r}{P_2} \qquad (6\text{-}22)$$

Thus, $|K|^2$ gives the ratio of the reflected and transmitted powers. Hence $|K|^2$ may take on any positive value between zero and infinity for $s = j\omega$.

The power relation

$$P_{max} = P_r + P_2 \qquad (6\text{-}23)$$

now gives

$$\frac{P_{max}}{P_2} = \frac{P_r}{P_2} + 1 \qquad (6\text{-}24)$$

or, by Eqs. (6-6) and (6-22), for $s = j\omega$,

$$|H|^2 = |K|^2 + 1 \qquad (6\text{-}25)$$

This equality (often called the *Feldtkeller equation†*) plays a central role in the design of doubly terminated two-ports.

By their definitions, $H(s)$, $\rho_1(s)$, and $K(s)$ are all rational functions of s, with real coefficients. Denoting their even and odd parts in s by the subscripts e and o, respectively, we have

$$H(s) = H_e(s) + H_o(s) \qquad (6\text{-}26)$$

$$H(-s) = H_e(s) - H_o(s) \qquad (6\text{-}27)$$

On the $j\omega$ axis $H_e(j\omega)$ is real and $H_o(j\omega)$ is imaginary. Hence

$$|H(j\omega)|^2 = [H_e(j\omega)]^2 + \left[\frac{H_o(j\omega)}{j}\right]^2 \qquad (6\text{-}28)$$

$$|H(j\omega)|^2 = H_e^2(j\omega) - H_o^2(j\omega) = H(j\omega)H(-j\omega)$$

When we rewrite $|K(j\omega)|^2$ this same way and replace $j\omega$ by s, the Feldtkeller equation becomes

$$H(s)H(-s) = K(s)K(-s) + 1 \qquad (6\text{-}29)$$

This form is much more convenient for numerical calculations than Eq. (6-25).

Note that the argument used in deriving Eq. (6-29) from Eq. (6-25) is valid only for $s = j\omega$. But Eq. (6-25) *was* only valid for $s = j\omega$ to start with, since it was derived from steady-state power considerations. Hence, this restriction is harmless.

Next, some of the properties of $H(s)$, $\rho_1(s)$, $\rho_2(s)$, and $K(s)$ will be discussed from a physical viewpoint. After writing $H(s)$ in the form

$$H(s) = \frac{E(s)}{P(s)} \qquad (6\text{-}30)$$

where $E(s)$ and $P(s)$ are polynomials, the properties of $E(s)$ will be investigated. First of all, it is clear that the degree of $E(s)$ must be equal to or greater than that

---

† From Eqs. (6-22) and (6-25), $|H|^{-2} + |\rho_1|^2 = 1$. Historically, it is this relation which is due to Feldtkeller.

of $P(s)$, since otherwise $|H(j\omega)|$ would become less than 1 for sufficiently large ω, in violation of Eq. (6-6). Therefore, $H(s)$ can be zero only if $E(s)$ equals zero. Next, upon writing

$$|H|^2 = \frac{P_{max}}{P_2} = \frac{P_{max}}{P_1} = \frac{E_G^2/4R_G}{|E_G/(R_G + Z_1)|^2 R_1} \tag{6-31}$$

or $$|H|^2 = \frac{|R_G + Z_1|^2}{4R_G R_1} \tag{6-32}$$

it becomes evident that $H(s)$ and thus $E(s)$ is zero if and only if

$$Z_1(s) = -R_G \tag{6-33}$$

Now since $Z_1(s)$ is a physical impedance of a (passive) network, it is PR. It cannot have a negative real part in the RHP or on the $j\omega$ axis. Hence, Eq. (6-33) can be satisfied only for s values in the (open) LHP. Therefore, in conclusion, $E(s)$ *is a strictly Hurwitz polynomial*.

From the equation

$$|\rho_1|^{-2} = \left| \frac{R_G + Z_1}{R_G - Z_1} \right|^2 = \frac{P_{max}}{P_r} \geq 1 \tag{6-34}$$

it follows in an exactly analogous way that the numerator of $1/\rho_1$ is of a degree at least as high as its denominator. Therefore again the numerator contains all zeros of $1/\rho_1$, and, furthermore, these zeros are the same as those of $H(s)$, as a comparison of Eqs. (6-34) and (6-32) shows. Hence, $\rho_1(s)$ can be written in the form

$$\rho_1(s) = \frac{F(s)}{E(s)} \tag{6-35}$$

where $E(s)$ is the strictly Hurwitz numerator polynomial of $H(s)$ and the degree of $F(s)$ is at most that of $E(s)$.

Returning to the original definition (6-5) of $H(s)$, we see also that the equation $E(s) = 0$ is equivalent to the simultaneous physical conditions of zero generator voltage

$$E_G \equiv 0 \tag{6-36}$$

and nonzero output

$$|V_2(s)| > 0 \tag{6-37}$$

The zeros of $E(s)$ are therefore the finite nonzero complex *natural frequencies* s_i of the network (which includes the two-port and its terminations R_G, R_L). This again shows that all s_i must lie in the left half of the s plane.†

Consider now the numerator polynomial $F(s)$ of $\rho_1(s)$. At the solutions of the equation

$$F(s) = 0 \tag{6-38}$$

---

† We could also base our proof on derivations similar to those presented in Secs. 5-6 and 5-7.

ρ_1 vanishes. This, by Eq. (6-10), shows that at these frequencies [and, if the degree of $E(s)$ is higher than that of $F(s)$, also at infinite frequency]

$$Z_1 = R_G \tag{6-39}$$

Hence, if a zero of $F(s)$ lies on the $j\omega$ axis, there is a perfect match at the primary port; $P_1 = P_{max}$ and $P_r = 0$ at that frequency. From $P_1 = P_2$, furthermore, $\rho_2 = 0$ and $Z_2 = R_L$ also follow. Because of these properties, such a frequency is called a *reflection zero*. (The *reflection poles*, where ρ_1 and ρ_2 become infinite, coincide with the natural modes, as we have seen.)

Next, consider the characteristic function $K(s)$. By Eqs. (6-21), (6-30), and (6-35),

$$K(s) = \rho_1(s)H(s) = \frac{F(s)\,E(s)}{E(s)\,P(s)} = \frac{F(s)}{P(s)} \tag{6-40}$$

Hence, the zeros of $K(s)$ are the reflection zeros. Also, if $P(s)$ is of higher degree than $F(s)$, $K \to 0$ for $\omega \to \infty$.

The poles of $K(s)$ are the same as those of $H(s)$. At these poles, by Eq. (6-5), V_2 is zero although E_G is not. Also, both $|H|$ and the transducer loss α become infinite. For these reasons, the zeros of $P(s)$ are called *transmission zeros* or *loss poles*. Note that if $E(s)$ is of higher degree than $P(s)$, $\omega \to \infty$ is also a loss pole.

The natural frequencies, reflection zeros, and loss poles are called the *critical frequencies* of the terminated two-port. Together, they give a full qualitative description of the network's behavior.

The polynomials $E(s)$, $F(s)$, and $P(s)$ are not independent; by the Feldtkeller relation (6-25),

$$|E|^2 = |F|^2 + |P|^2 \tag{6-41}$$

or, in a more practical form, from Eq. (6-29),

$$E(s)E(-s) = F(s)F(-s) + P(s)P(-s) \tag{6-42}$$

From this equation, if any two of the polynomials are given, the third one can be obtained (although, as we shall see later, not always uniquely).

Example 6-2 Find $H(s)$, $K(s)$, and the critical frequencies of the circuit shown in Fig. 6-4 and analyzed in Example 6-1.

From the $V_2(s)$ function found there,

$$H(s) = \frac{E_G}{2}\sqrt{\frac{R_L}{R_G}}\frac{1}{V_2(s)} = 1\sqrt{\frac{3}{1}}\frac{3s^3 + 12s^2 + 11s + 4}{6}$$

$$H(s) = \frac{1}{\sqrt{12}}(3s^3 + 12s^2 + 11s + 4)$$

From $\rho_1(s)$, found also in Example 6-1,

$$K(s) = \rho_1(s)H(s) = \frac{-3s^3 - 6s^2 + 7s - 2}{3s^3 + 12s^2 + 11s + 4}\frac{3s^3 + 12s^2 + 11s + 4}{\sqrt{12}}$$

$$K(s) = \frac{1}{\sqrt{12}}(-3s^3 - 6s^2 + 7s - 2)$$

Note that the zeros of $H(s)$ and the poles of $\rho_1(s)$ indeed coincide, as predicted by theory: they are the zeros of the natural-mode polynomial

$$E(s) = 3s^3 + 12s^2 + 11s + 4$$

These zeros turn out to be duly in the LHP

$$s_1 \approx -2.8913 \qquad s_2, s_3 = -0.55433 \pm j0.39226$$

It is apparent that all loss poles [which are the poles of $H(s)$] are at $s \to \infty$. The reflection zeros are the zeros of

$$F(s) = -3s^3 - 6s^2 + 7s - 2$$

These are

$$s_1' = -2.8880 \qquad s_2', s_3' = +0.44395 \pm j0.18372$$

Evidently, $F(s)$ is neither Hurwitz (all zeros in LHP) nor anti-Hurwitz (all zeros in RHP). In fact, there are no physical reasons for it to be either.

Example 6-3 Find the functions $H(s)$, $\rho_1(s)$, and $Z_1(s)$ from a prescribed characteristic function

$$K(s) = \frac{s^2 + 1}{s}$$

Assume that $R_G = 1\ \Omega$. Realize the corresponding two-port.

Here $F(s)$ is $s^2 + 1$ and $P(s)$ is simply s; hence, by the Feldtkeller equation (6-42),

$$E(s)E(-s) = (s^2 + 1)^2 - s^2 = s^4 + s^2 + 1$$

The roots of this equation are

$$s_1, s_2 = -0.5 \pm j0.86603 \qquad s_3, s_4 = +0.5 \pm j0.86603$$

Upon plotting these roots in the s plane (Fig. 6-5) it becomes evident that they show a

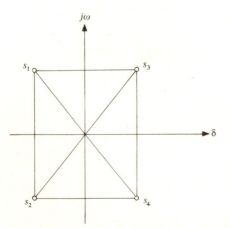

Figure 6-5 Roots of the equation $E(s)E(-s) = 0$.

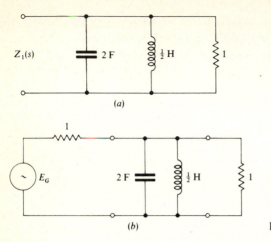

(a)

(b)

Figure 6-6 Design example.

quadrantal symmetry. Now $E(s)$ is a Hurwitz polynomial, while $E(-s)$ [whose zeros are the negatives of those of $E(s)$] is anti-Hurwitz. This makes it quite clear that in the equation

$$E(s)E(-s) = (s - s_1)(s - s_2)(s - s_3)(s - s_4)$$

the first two factors form $E(s)$ while the last two belong to $E(-s)$. Hence

$$E(s) = s^2 + s + 1 \qquad H(s) = \frac{E(s)}{P(s)} = \frac{s^2 + s + 1}{s} \qquad \rho_1(s) = \frac{F(s)}{E(s)} = \frac{s^2 + 1}{s^2 + s + 1}$$

From $\rho_1(s)$, by Eqs. (6-10) and (6-35)

$$Z_1(s) = R_G \frac{1 - \rho_1(s)}{1 + \rho_1(s)} = R_G \frac{E(s) - F(s)}{E(s) + F(s)} \qquad Z_1(s) = \frac{s}{2s^2 + s + 2}$$

Using the methods of Chap. 4 (or by recognition), we can develop $Z_1(s)$ into the circuit of Figure 6-6a. Hence, the complete network giving the specified characteristic function is that of Fig. 6-6b. A consideration of the behavior of $|H(j\omega)|$ for $\omega = 0$, $\omega = 1$, and $\omega \to \infty$ easily shows that the network just obtained is a simple bandpass filter with its passband centered around $s = j1$.

6-3 RELATIONS BETWEEN THE TRANSDUCER AND IMMITTANCE PARAMETERS

The transmission properties of a doubly terminated reactance two-port are usually specified in terms of the transducer parameters $H(s)$ or $K(s)$. The realization, on the other hand, is performed using some of the two-port immittance parameters as a starting point. Hence, it is essential to bridge the gap and be able to calculate the immittance parameters from a given $H(s)$ and $K(s)$. The formulas making this calculation possible will be developed in this section.

The derivation of the formulas will be performed in a roundabout way. We shall first find out how to express $H(s)$ and $K(s)$ in terms of the immittance parameters; then the resulting equations will be solved for the immittance parameters. The derivations will be carried out in terms of the impedance parameters; it could equally well be done on an admittance basis, using dual equations.

With reference to the circuit of Fig. 6-1 (reproduced here as Fig. 6-7 for convenience), evidently

$$V_1 = E_G - I_1 R_G \tag{6-43}$$

and
$$V_2 = -I_2 R_L \tag{6-44}$$

Substituting these values into Eqs. (5-4), which defined the impedance parameters, gives

$$z_{11} I_1 + z_{12} I_2 = E_G - I_1 R_G \qquad z_{12} I_1 + z_{22} I_2 = -I_2 R_L \tag{6-45}$$

Solving these two linear equations, we obtain the port currents

$$I_1 = E_G \frac{z_{22} + R_L}{(z_{11} + R_G)(z_{22} + R_L) - z_{12}^2} \tag{6-46}$$

and
$$I_2 = -E_G \frac{z_{12}}{(z_{11} + R_G)(z_{22} + R_L) - z_{12}^2} \tag{6-47}$$

Hence, the transducer factor is

$$H(s) = \frac{1}{2}\sqrt{\frac{R_L}{R_G}} \frac{E_G}{-I_2 R_L} = \frac{(z_{11} + R_G)(z_{22} + R_L) - z_{12}^2}{2\sqrt{R_G R_L}\, z_{12}} \tag{6-48}$$

Also, the primary driving-point impedance satisfies

$$Z_1 = \frac{V_1}{I_1} = \frac{E_G - I_1 R_G}{I_1} = \frac{E_G}{I_1} - R_G \tag{6-49}$$

By Eq. (6-46), therefore,

$$Z_1 = \frac{(z_{11} + R_G)(z_{22} + R_L) - z_{12}^2}{z_{22} + R_L} - R_G \tag{6-50}$$

and
$$Z_1 = z_{11} - \frac{z_{12}^2}{z_{22} + R_L} \tag{6-51}$$

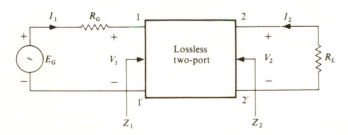

Figure 6-7 Doubly terminated two-port.

Obviously, Z_2 can be obtained by simply interchanging the subscripts 1 and 2 as well as G and L in Eq. (6-51):

$$Z_2 = z_{22} - \frac{z_{12}^2}{z_{11} + R_G} \tag{6-52}$$

The reflection coefficients can now be obtained from Eqs. (6-50) to (6-52):

$$\rho_1 = \frac{R_G - Z_1}{R_G + Z_1} = \frac{(R_G - z_{11})(R_L + z_{22}) + z_{12}^2}{(R_G + z_{11})(R_L + z_{22}) - z_{12}^2} \tag{6-53}$$

and

$$\rho_2 = \frac{R_L - Z_2}{R_L + Z_2} = \frac{(R_G + z_{11})(R_L - z_{22}) + z_{12}^2}{(R_G + z_{11})(R_L + z_{22}) - z_{12}^2} \tag{6-54}$$

It should be noticed that the denominators of ρ_1 and ρ_2 are both the same as the numerator of $H(s)$. In view of the discussions of the preceding section, this should not be particularly surprising.

Next, the characteristic function can be obtained from Eqs. (6-48) and (6-53):

$$K = \rho_1 H = \frac{(R_G - z_{11})(R_L + z_{22}) + z_{12}^2}{2\sqrt{R_G R_L}\, z_{12}} \tag{6-55}$$

The first stage of our program has been accomplished: in Eqs. (6-48) and (6-55), we have derived formulas giving $H(s)$ and $K(s)$ in terms of the z parameters. Now these equations should be solved for z_{11}, z_{12}, and z_{22}. This is seemingly quite a task, since it involves the solution of the *two* equations (6-48) and (6-55) for their *three* unknowns. We have, however, additional information about the unknown z_{ij}: they are all *odd* functions of s, since they represent the impedance parameters of a reactance two-port. On the basis of this information, we are able to separate $H(s)$ and $K(s)$ into their even and odd parts. The even part of the transducer factor is evidently

$$H_e = \frac{R_G z_{22} + R_L z_{11}}{2\sqrt{R_G R_L}\, z_{12}} \tag{6-56}$$

while the odd part is

$$H_o = \frac{(z_{11} z_{22} - z_{12}^2) + R_G R_L}{2\sqrt{R_G R_L}\, z_{12}} \tag{6-57}$$

Similarly, for the characteristic function,

$$K_e = \frac{R_G z_{22} - R_L z_{11}}{2\sqrt{R_G R_L}\, z_{12}} \tag{6-58}$$

and

$$K_o = \frac{-(z_{11} z_{22} - z_{12}^2) + R_G R_L}{2\sqrt{R_G R_L}\, z_{12}} \tag{6-59}$$

Evidently, we overshot our goal: instead of the earlier *two* equations, now we have *four* equations (6-56) to (6-59) for our three unknown z parameters. These equations are, however, compatible, as will be shown later.

Adding and subtracting the equations giving H_e, H_o, K_e, and K_o, we get

$$H_o + K_o = \frac{\sqrt{R_G R_L}}{z_{12}} \tag{6-60a}$$

$$H_o - K_o = \frac{z_{11} z_{22} - z_{12}^2}{\sqrt{R_G R_L} \, z_{12}} \tag{6-60b}$$

$$H_e + K_e = \sqrt{\frac{R_G z_{22}}{R_L z_{12}}} \tag{6-60c}$$

$$H_e - K_e = \sqrt{\frac{R_L z_{11}}{R_G z_{12}}} \tag{6-60d}$$

From Eq. (6-60a),

$$z_{12} = \frac{\sqrt{R_G R_L}}{H_o + K_o} \tag{6-61}$$

Substituting this into the remaining three equalities of Eq. (6-60) gives

$$z_{11} = R_G \frac{H_e - K_e}{H_o + K_o} \tag{6-62}$$

$$z_{22} = R_L \frac{H_e + K_e}{H_o + K_o} \tag{6-63}$$

$$\Delta_z \triangleq z_{11} z_{22} - z_{12}^2 = R_G R_L \frac{H_o - K_o}{H_o + K_o} \tag{6-64}$$

The compatibility of these four equations can now be checked by substituting the formulas for z_{11}, z_{22}, and z_{12} into that for Δ_z and showing that this results in a valid equation, namely the Feldtkeller equation. This task is given as Prob. 6-6.

The derivation of the y parameters can be performed on a dual basis, or use can simply be made of the formulas connecting the z and y parameters. In either case, the result is

$$y_{11} = \frac{1}{R_G} \frac{H_e + K_e}{H_o - K_o} \tag{6-65}$$

$$y_{12} = \frac{1}{\sqrt{R_L R_G}} \frac{1}{H_o - K_o} \tag{6-66}$$

$$y_{22} = \frac{1}{R_L} \frac{H_e - K_e}{H_o - K_o} \tag{6-67}$$

$$\Delta_y \triangleq y_{11} y_{22} - y_{12}^2 = \frac{1}{R_G R_L} \frac{H_o + K_o}{H_o - K_o} = \frac{1}{\Delta_z} \tag{6-68}$$

The driving-point impedances are even easier to obtain. From

$$\rho_1(s) = \frac{K(s)}{H(s)} \tag{6-69}$$

we get directly

$$Z_1 = R_G \frac{1 - \rho_1}{1 + \rho_1} = R_G \frac{H - K}{H + K} \tag{6-70}$$

Next, it can be deduced from Eqs. (6-53) and (6-54) that $\rho_1(s)$ and $\rho_2(s)$ have the same denominators but that their numerators differ. The only difference, however, is evidently that z_{11}/R_G and z_{22}/R_L are interchanged in the two numerators. By Eqs. (6-62) and (6-63), this is equivalent to interchanging $K_e(s)$ and $-K_e(s)$. Therefore†

$$\rho_2(s) = \frac{-K(-s)}{H(s)} = \frac{-K_e(s) + K_o(s)}{H(s)} \tag{6-71}$$

and hence

$$Z_2 = R_L \frac{1 - \rho_2(s)}{1 + \rho_2(s)} = R_L \frac{H + K_e - K_o}{H - K_e + K_o} \tag{6-72}$$

These equations can be written in a more directly usable form if we note from Eqs. (6-48) and (6-55) that the common denominator $P(s)$ of $H(s)$ and $K(s)$ must be either a pure even or a pure odd function of s unless common factors of $E(s)$ and $P(s)$ or $F(s)$ and $P(s)$ cancel. This can be seen by noting again that z_{11}, z_{12}, and z_{22} are all *odd* rational functions of s; hence, multiplying through by their common denominator in $H(s)$ and $K(s)$ must leave an even or odd denominator $P(s)$. Now if $P(s)$ is even, then

$$H_e(s) = \frac{E_e(s)}{P(s)} \qquad H_o(s) = \frac{E_o(s)}{P(s)}$$

$$K_e(s) = \frac{F_e(s)}{P(s)} \qquad K_o(s) = \frac{F_o(s)}{P(s)} \tag{6-73}$$

while for an odd $P(s)$,

$$H_e(s) = \frac{E_o(s)}{P(s)} \qquad H_o(s) = \frac{E_e(s)}{P(s)}$$

$$K_e(s) = \frac{F_o(s)}{P(s)} \qquad K_o(s) = \frac{F_e(s)}{P(s)} \tag{6-74}$$

Substituting these expressions in Eqs. (6-61) to (6-72), we obtain the results given in Table 6-1.

Example 6-4 Assuming $R_L = R_G = 1 \ \Omega$, find the impedance and admittance parameters of the bandpass filter discussed in Example 6-3, using Table 6-1. Since $P(s)$ is now odd,

$$z_{11} = R_G \frac{E_o - F_o}{E_e + F_e} = \frac{s}{2(s^2 + 1)}$$

$$z_{12} = \sqrt{R_G R_L} \frac{P}{E_e + F_e} = \frac{s}{2(s^2 + 1)}$$

† If this line of thought seems too involved, the reader should work out Prob. 6-7, which leads to a conceptually simpler, although more tedious, proof.

$$z_{22} = R_L \frac{E_o + F_o}{E_e + F_e} = \frac{s}{2(s^2 + 1)}$$

$$y_{11} = \frac{1}{R_G} \frac{E_o + F_o}{E_e - F_e} \to \infty$$

$$y_{12} = \frac{1}{\sqrt{R_G R_L}} \frac{P}{E_e - F_e} \to \infty$$

$$y_{22} = \frac{1}{R_L} \frac{E_o - F_o}{E_e - F_e} \to \infty$$

These results can easily be checked by inspection from Fig. 6-6.

Table 6-1

Function	$P(s)$ even	$P(s)$ odd
$\dfrac{z_{11}}{R_G}$	$\dfrac{E_e - F_e}{E_o + F_o}$	$\dfrac{E_o - F_o}{E_e + F_e}$
$\dfrac{z_{12}}{\sqrt{R_G R_L}}$	$\dfrac{P}{E_o + F_o}$	$\dfrac{P}{E_e + F_e}$
$\dfrac{z_{22}}{R_L}$	$\dfrac{E_e + F_e}{E_o + F_o}$	$\dfrac{E_o + F_o}{E_e + F_e}$
$y_{11} R_G$	$\dfrac{E_e + F_e}{E_o - F_o}$	$\dfrac{E_o + F_o}{E_e - F_e}$
$y_{12}\sqrt{R_G R_L}$	$\dfrac{P}{E_o - F_o}$	$\dfrac{P}{E_e - F_e}$
$y_{22} R_L$	$\dfrac{E_e - F_e}{E_o - F_o}$	$\dfrac{E_o - F_o}{E_e - F_e}$
$\dfrac{\Delta_z}{R_G R_L} = \dfrac{1}{\Delta_y R_G R_L}$	$\dfrac{E_o - F_o}{E_o + F_o}$	$\dfrac{E_e - F_e}{E_e + F_e}$
ρ_1	$\dfrac{F}{E}$	$\dfrac{F}{E}$
ρ_2	$\dfrac{-F_e + F_o}{E}$	$\dfrac{F_e - F_o}{E}$
$\dfrac{Z_1}{R_G}$	$\dfrac{E - F}{E + F}$	$\dfrac{E - F}{E + F}$
$\dfrac{Z_2}{R_L}$	$\dfrac{E + F_e - F_o}{E - F_e + F_o}$	$\dfrac{E - F_e + F_o}{E + F_e - F_o}$

6-4 REALIZABILITY CONDITIONS ON THE TRANSDUCER PARAMETERS

It is important to analyze the conditions under which the transducer parameters $H(s)$, $K(s)$ and $\rho_1(s)$, $\rho_2(s)$ can lead to a realizable doubly terminated reactance two-port. These conditions will be listed next. Afterward, they will be proved and their necessity and sufficiency analyzed.

Conditions on the transducer factor $H(s)$

1. $H(s)$ can be written in the form $H(s) = E(s)/P(s)$, where $E(s)$ and $P(s)$ are polynomials with real coefficients.
2. $E(s)$ is a strictly Hurwitz polynomial.
3. $P(s)$ is a purely even or odd polynomial unless common factors of $E(s)$ and $P(s)$ cancel in $H(s)$.
4. The modulus of $H(s)$ satisfies $|H(s)| \geq 1$ on the $s = j\omega$ axis.
5. The degree of $E(s)$ is greater than or equal to that of $P(s)$.

PROOF Condition 1 is obvious from the defining equation (6-5) of $H(s)$. Condition 2 has been proved in Sec. 6-2. Condition 3 follows easily from Eq. (6-48), which expressed $H(s)$ in terms of the z parameters: multiplying through by the common denominator of z_{11}, z_{12}, and z_{22} must leave an even or odd denominator for $H(s)$, since all z parameters contain even or odd numerators and denominators. Conditions 4 and 5 were proved earlier, from the definition of $H(s)$, in Sec. 6-2.

Conditions on the characteristic function $K(s)$

1. $K(s)$ can be written in the form

$$K(s) = \frac{F(s)}{P(s)} \tag{6-75}$$

where $F(s)$ and $P(s)$ are polynomials with real coefficients.
2. $P(s)$ is a purely even or odd polynomial.

PROOF Condition 1 follows from the defining equation of $K(s)$

$$K(s) = \rho_1(s)H(s) = \frac{R_G - Z_1(s)}{R_G + Z_1(s)} H(s) \tag{6-76}$$

Condition 2 is identical to condition 3 for the transducer factor and can be proved in exactly the same way.

Conditions on simultaneously given $H(s)$ and $K(s)$

In addition to the conditions described above for $H(s)$ and $K(s)$ individually, if H and K are given simultaneously for the same network, they must also satisfy the following restrictions:

1. $H(s)$ and $K(s)$ must have the same denominator $P(s)$, barring cancellations.
2. The polynomials forming $H(s)$ and $K(s)$ must satisfy

$$E(s)E(-s) = F(s)F(-s) + P(s)P(-s) \tag{6-77}$$

for $s = j\omega$.
3. The degree of $E(s)$ must be at least as high as that of either $F(s)$ or $P(s)$.

PROOF Condition 1 can be deduced directly from Eqs. (6-48) and (6-55), which show that in terms of the z parameters, $H(s)$ and $K(s)$ have the same denominator. Condition 2 (the Feldtkeller relation) and condition 3 were proved in Sec. 6-2.

Conditions on the reflection factors $\rho_1(s)$ and $\rho_2(s)$

1. The modulus of both reflection factors satisfies

$$|\rho_i(s)| \leq 1 \qquad i = 1 \text{ or } 2 \qquad \text{for } s = j\omega \tag{6-78}$$

2. The reflection factors can be written in the forms

$$\rho_1(s) = \frac{F(s)}{E(s)} \qquad \rho_2(s) = \mp \frac{F(-s)}{E(s)} \tag{6-79}$$

where $F(s)$ and $E(s)$ are polynomials with real coefficients. The minus sign holds for ρ_2 if $P(s)$ is even; the plus sign if $P(s)$ is odd.
3. The common denominator of $\rho_1(s)$ and $\rho_2(s)$, $E(s)$, is a strictly Hurwitz polynomial, of a degree higher than or equal to that of $F(s)$.

PROOF All these conditions were proved before. In particular, the relation between the numerators of $\rho_1(s)$ and $\rho_2(s)$ follows from

$$\rho_1(s) = \frac{K(s)}{H(s)} \qquad \rho_2(s) = \frac{-K(-s)}{H(s)} \tag{6-80}$$

which were derived as Eqs. (6-69) and (6-71), respectively. Assume now that $P(s)$ is *even*. Then

$$-K(-s) = -\frac{F(-s)}{P(-s)} = \frac{-F(-s)}{P(s)} \tag{6-81}$$

and

$$\rho_2(s) = \frac{-F(-s)/P(s)}{E(s)/P(s)} = \frac{-F(-s)}{E(s)} \tag{6-82}$$

For *odd* $P(s)$, an analogous derivation gives

$$\rho_2(s) = \frac{+F(-s)}{E(s)} \tag{6-83}$$

This completes the discussion on the *necessity* of the above given realizability conditions. To complete the analysis, an argument about the *sufficiency* of these conditions will be given next.

To this end, we shall establish that the impedance $Z_1(s)$ is always realizable as is the impedance-parameter set z_{ij} if they are all calculated from some $H(s)$ and $K(s)$ satisfying the conditions described above.

First the admittance formed by the series combination of R_G and Z_1

$$Y = \frac{1}{R_G + Z_1} = \frac{1}{R_G + R_G(H - K)/(H + K)} = \frac{1}{2R_G} \frac{H + K}{H} \tag{6-84}$$

or

$$Y = \frac{1}{2R_G} \frac{E/P + F/P}{E/P} = \frac{1}{2R_G} \frac{E + F}{E} \tag{6-85}$$

will be analyzed. Y is a PR function, since (1) it is a ratio of real polynomials, (2) its denominator is strictly Hurwitz, and (3) its real part G is nonnegative along the $j\omega$ axis:

$$G \triangleq \text{Re } Y(j\omega) = \frac{1}{2R_G} \text{Re } \frac{(E + F)(E_e - E_o)}{|E|^2}$$

$$G = \frac{1}{2R_G|E|^2} (|E|^2 + F_e E_e - F_o E_o) \tag{6-86}$$

$$G = \frac{1}{2R_G|E|^2} \left(\frac{|E + F|^2}{2} + \frac{|E|^2 - |F|^2}{2} \right) > 0$$

The last inequality follows, since by the Feldtkeller equation

$$|E|^2 - |F|^2 = |P|^2 \geq 0 \tag{6-87}$$

The positive reality of $Y(s)$ establishes the strictly Hurwitz character of $E(s) + F(s)$. Analysis of the impedance formed from the *parallel* combination of R_G and Z_1 yields the fact that $E(s) - F(s)$ is also strictly Hurwitz. Now it follows that

$$Z_1 = R_G \frac{H - K}{H + K} = R_G \frac{E - F}{E + F} \tag{6-88}$$

is positive real, since its numerator and denominator are both strictly Hurwitz while its real part is

$$R_1 = R_G \text{ Re } \frac{(E - F)(E_e - E_o + F_e - F_o)}{|E + F|^2}$$

$$= R_G \frac{|E|^2 - |F|^2}{|E + F|^2} = R_G \frac{|P|^2}{|E + F|^2} = \frac{R_G}{|H + K|^2} \geq 0 \tag{6-89}$$

Similar analysis of the admittance $1/(R_L + Z_2)$ and the impedance $R_L Z_2/(R_L + Z_2)$ for both even and odd $P(s)$ reveals that both polynomials $E + F_e - F_o = E(s) + F(-s)$ and $E - F_e + F_o = E(s) - F(-s)$ are also strictly Hurwitz and that $Z_2(s)$ is PR.

From the Hurwitz character of the four polynomials $E(s) \pm F(\pm s)$ it follows that all six immittances z_{11}, z_{12}, z_{22} and y_{11}, y_{12}, y_{22} are separately realizable. To show that the z parameters are compatible, i.e., realizable simultaneously by a reactance two-port, we must also prove that they satisfy the residue conditions given in Eq. (5-49). To this end, consider the quantity z_{11}/y_{22}. By Eqs. (6-62) and (6-67),

$$\frac{z_{11}}{y_{22}} = \frac{R_G(H_e - K_e)/(H_o + K_o)}{(1/R_L)(H_e - K_e)/(H_o - K_o)} \tag{6-90}$$

and hence, using Eq. (6-64),

$$\frac{z_{11}}{y_{22}} = R_G R_L \frac{H_o - K_o}{H_o + K_o} = z_{11}z_{22} - z_{12}^2 \tag{6-91}$$

Now let $j\omega_i$ be an arbitrary pole of z_{12}. At that pole the realizable impedances z_{11} and $1/y_{22}$ both have nonnegative residues. Hence, z_{11}/y_{22} has a double pole at $s = j\omega_i$, and the partial-fraction expansion of z_{11}/y_{22} will contain a term $Cs^2/(s^2 + \omega_i^2)^2$. Here C is the product of the residues of z_{11} and $1/y_{22}$, and thus $C \geq 0$. But, by Eq. (6-91), $C = k_{11}k_{22} - k_{12}^2$, where k_{11}, k_{22}, and k_{12} are the residues of z_{11}, z_{22}, and z_{12}, respectively, at $s = j\omega_i$. Therefore, $k_{11}k_{22} - k_{12}^2$ at $j\omega_i$ is nonnegative, and the residue condition is thus satisfied. An exactly dual derivation proves the compatibility of the y parameters.

6-5 THE REALIZATION OF DOUBLY TERMINATED REACTANCE LADDERS

The ladder expansion described in Chap. 5 can also be utilized for doubly terminated two-ports. There are, however, some important differences. The first concerns partial pole removals. As already stated in Chap. 5, this operation does not create a transmission zero for singly loaded or unloaded reactance two-ports, but it does for doubly loaded networks. To illustrate and verify this property, consider the circuit of Fig. 6-8a, where Z_k represents an impedance which we used to accomplish a partial pole removal, from, say, a primary driving-point impedance. The output voltage V_2 can be found by first replacing the one-port consisting of two-port I and the generator by its Thevenin equivalent, as shown in Fig. 6-8b. The elements of this equivalent circuit can easily be found from the impedance-parameter equations of two-port I:

$$E^{(I)} = E_G \frac{z_{12}^{(I)}}{R_G + z_{11}^{(I)}} \tag{6-92}$$

and

$$Z_2^{(I)} = z_{22}^{(I)} - \frac{(z_{12}^{(I)})^2}{R_G + z_{11}^{(I)}} \tag{6-93}$$

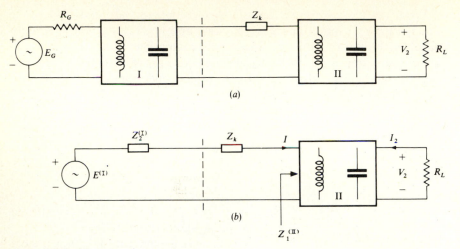

Figure 6-8 Ladder expansion.

By Eq. (6-92), $E^{(I)}$ cannot become infinite for any purely imaginary $s = j\omega$, since all poles of $z_{12}^{(I)}$ are also poles of $z_{11}^{(I)}$ and since $z_{11}^{(I)} = -R_G$ cannot occur for $s = j\omega$.

The driving-point impedance of two-port II is, by Eq. (6-51),

$$Z_1^{(II)} = z_{11}^{(II)} - \frac{(z_{12}^{(II)})^2}{R_L + z_{22}^{(II)}} \tag{6-94}$$

Also, by Fig. 6-8b, the input current I of two-port II is

$$I = \frac{E^{(I)}}{Z_2^{(I)} + Z_k + Z_1^{(II)}} \tag{6-95}$$

Next, from the impedance equations of two-port II,

$$V_2 = z_{12}^{(II)} I + I_2 z_{22}^{(II)} = -I_2 R_L \tag{6-96}$$

and so

$$V_2 = I \frac{z_{12}^{(II)}}{1 + z_{22}^{(II)}/R_L} \tag{6-97}$$

Combining Eqs. (6-94) to (6-97), we get the output voltage

$$V_2 = \frac{E^{(I)} R_L z_{12}^{(II)}}{(Z_2^{(I)} + Z_k + z_{11}^{(II)})(z_{22}^{(II)} + R_L) - (z_{12}^{(II)})^2} \tag{6-98}$$

At the $j\omega$-axis pole which Z_k partially removes, Z_k, and hence the denominator, tends to infinity. $E^{(I)}$ remains finite, as shown above; $z_{12}^{(II)} \to \infty$ is possible, but even then V_2 tends to zero. Hence, the partial pole removal has created a transmission zero at the frequency of the pole.

When we let $R_L \to \infty$, however, Eq. (6-98) becomes

$$V_2 = \frac{E^{(I)} z_{12}^{(II)}}{Z_2^{(I)} + Z_k + z_{11}^{(II)}} \tag{6-99}$$

Now since Z_k represented a *partial* pole removal, $z_{11}^{(II)}$, and thus in general also $z_{12}^{(II)}$, has a pole at the same frequency. Therefore V_2 can be nonzero, and no transmission zero is generated.

The second new consideration concerns the fact that in the case of doubly terminated reactance two-ports all three immittance parameters are prescribed. Hence, we must make sure that the ladder development leads to a network actually realizing all three parameters simultaneously. Assume, for definiteness, that z_{11} is developed. Let the ladder development be performed so that the transmission zeros are also all realized, as discussed in Chap. 5. Then, all poles of z_{12} are automatically obtained, since they are among the poles of z_{11}. Similarly, the zeros of z_{12} are included amongst the transmission zeros; hence, those too are taken care of. Since the zeros and poles of z_{12} are thus both represented by the network, the only difference between the prescribed z_{12} and that of the synthesized two-port can be a constant factor. This factor can easily be found from the network's behavior at, say, zero or infinite frequency. The discrepancy represented by the constant scale factor can then be corrected, as we have seen in Chap. 5, by cascading an ideal transformer with the circuit.

The above considerations show that z_{11} and z_{12} can be simultaneously realized by only one ladder development. But what about z_{22}? No steps have been taken to assure that the circuit has the prescribed z_{22} realized; nor do we know how to remedy the situation if the prescribed and actually realized z_{22} do not agree.

The key to this situation lies in the residue theorem. As described in Chap. 5, the residue theorem states that at any pole of the z_{ij} parameters their residues satisfy the condition

$$k_{11}k_{22} - k_{12}^2 \geq 0 \qquad (6\text{-}100)$$

It is convenient at this state to recall some definitions given in Chap. 5. We called a pole *shared* if it was present in all three z_{ij} with nonzero residue; we called it a *private pole* of z_{11} (or z_{22}) if it was present only in z_{11} (or z_{22}). Finally, we called a shared pole *compact* if the residue condition was satisfied at this pole with the equality sign:[†]

$$k_{11}k_{22} - k_{12}^2 = 0 \qquad (6\text{-}101)$$

noncompact if it was satisfied as an inequality

$$k_{11}k_{22} - k_{12}^2 > 0 \qquad (6\text{-}102)$$

In the course of the ladder development, we alternate between partial and full pole removals. The connection between the way these operations are performed and the properties of the actual z_{22} of the network thus obtained can be summarized in the following theorem, due to Bader.[‡]

[†] Note that a *private* pole always satisfies Eq. (6-101). By our definition, however, only shared poles may qualify as compact poles.

[‡] W. Bader, Kopplungsfreie Kettenschaltungen, *Telegr. Fernsprech. Technol.*, vol. 31, pp. 177–189; July 1942. The proof of this theorem is somewhat lengthy and will therefore be omitted here.

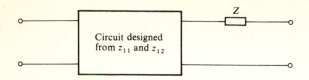

Figure 6-9 Correcting z_{22} for a two-port designed from z_{11} and z_{12}.

Theorem If each partial removal in the ladder development is followed later by a full removal of a pole at the same frequency, then the actual z_{22} of the realized circuit will only have shared compact poles.

Notice, for example, that in the example of Fig. 5-41 this condition was *not* satisfied; the partial removal of the pole at $\omega = \sqrt{2}$ was *not* followed by a full pole removal at this same frequency. However, if all partial pole removals are performed exclusively at zero and/or infinite frequency, the condition will be often satisfied.

The advantage of realizing a circuit which has a z_{22} with only shared compact poles is evident. It means that the actually realized z_{22} has no private poles and satisfies the residue condition at all its shared poles with the equal sign. On the other hand, the prescribed z_{22} is presumably realizable and hence satisfies the residue condition at all its poles with the $\geq$ sign. Hence, at each pole, the residue of the *prescribed* z_{22} is at least as large as that of the *actually* synthesized z_{22}. Let the difference between the residues of the prescribed and actual z_{22} at the ith pole $j\omega_i$ be $\Delta k_{22}^{(i)} \geq 0$. Then

$$z_{22_{\text{pres}}} = z_{22_{\text{act}}} + Z \tag{6-103}$$

where

$$Z = \sum_{(i)} \frac{\Delta k_{22}^{(i)}}{s - j\omega_i} \tag{6-104}$$

is a realizable reactance. Hence, merely connecting Z in series with the secondary terminals of the realized two-port will result in obtaining the desired z_{22} while leaving z_{11} and z_{12} unaffected (Fig. 6-9).

In summary, the design of the doubly terminated network can be carried out in the following steps:

1. z_{11} is expanded into a reactance ladder which has the required transmission zeros as the zeros of its shunt impedances and series admittances. Partial pole removals are performed from poles located at zero and infinite frequencies only and followed eventually by full removals.
2. The discrepancy between actual and prescribed z_{12} (which amounts to a constant scale factor) is corrected for by cascading an ideal transformer with the ladder network obtained in step 1.
3. The actual z_{22} of the cascade $z_{22_{\text{act}}}$ is next found, and the PR reactance

$$Z = z_{22} - z_{22_{\text{act}}} \tag{6-105}$$

is series-connected at the secondary port. The resulting two-port is shown in Fig. 6-10a.

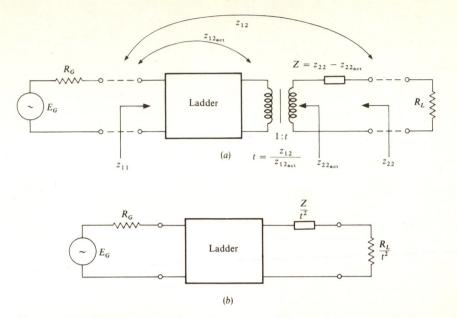

Figure 6-10 Design of a two-port from given z_{11}, z_{12}, and z_{22}.

4. The ideal transformer can be eliminated, if desired, by bringing both Z and R_L to its primary side. In order to keep the impedance seen by the rest of the circuit the same, the values of these impedances must of course be multiplied by $1/t^2$. The resulting network is shown in Fig. 6-10b. Here the open-circuited (and thus unimportant) transformer is omitted. This only results in a change of the constant multiplier of V_2 and of the impedance level at port 2, both of which are inconsequential in most applications.

Example 6-5 Realize a doubly loaded two-port from the prescribed impedance parameters

$$z_{11} = \frac{24s^3 + 36s}{8s^2 + 3} \qquad z_{12} = \frac{12s^3 + 6s}{8s^2 + 3} \qquad z_{22} = \frac{6s^4 + 21s^2 + 7}{8s^3 + 3s}$$

Assume $R_G = R_L = 1$.

An examination of Eq. (6-48) shows that transmission zeros result at those frequencies where z_{12} is zero and where z_{11} or z_{22} has either a private or a noncompact pole. Here, $z_{12} = 0$ for $s = 0$ and $s = \pm j/\sqrt{2}$; z_{22} has a private pole at $s = 0$. The shared pole at $s = \pm j\sqrt{3/8}$ is compact, as a calculation of the residues shows; so is the pole at infinite frequency. We conclude therefore that there are two transmission zeros at $\omega = 0$ and one each at $\omega = \pm 1/\sqrt{2}$. The pole-zero pattern for the development of the ladder from z_{11} is shown in Fig. 6-11a; the circuit itself is given in Fig. 6-11b. The reader should by now be able to work out the details without difficulty. The resulting element values are $L_1 = 12$, $C_2 = \frac{1}{4}$, $L_3 = 4$, and $C_3 = \frac{1}{2}$. The transformer turns ratio t can be found from the behavior of z_{12} and $z_{12_{act}}$ as $s \to \infty$. Clearly, for the prescribed function $z_{12} \to 12s^3/8s^2 = \frac{3}{2}s$, while from the circuit, for large ω, $1/\omega C_2 \ll \omega L_3$ and $1/\omega C_3 \ll \omega L_3$; hence $z_{12_{act}} \to sL_1 sL_3/(sL_1 + sL_3) = 3s$. Hence, $t = z_{12}/z_{12_{act}} = \frac{1}{2}$.

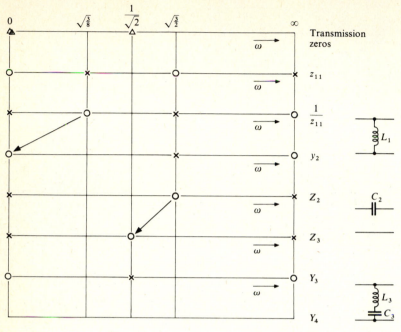

(a)

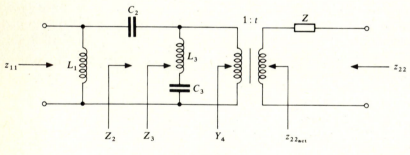

(b)

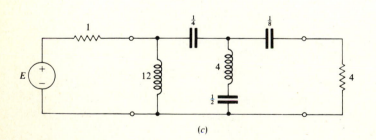

(c)

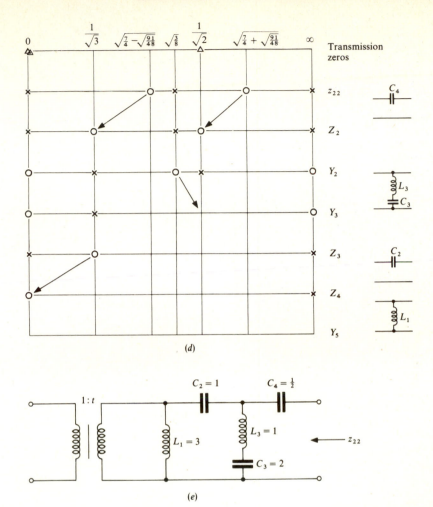

(d)

(e)

Figure 6-11 Design example.

For the actual circuit,† from Fig. 6-11b,

$$z_{22_{act}} = t^2 \frac{1}{\dfrac{1}{sL_3 + 1/sC_3} + \dfrac{1}{sL_1 + 1/sC_2}} = \frac{6s^4 + 5s^2 + 1}{8s^3 + 3s}$$

Hence

$$z_{22} - z_{22_{act}} = \frac{16s^2 + 6}{8s^3 + 3s} = \frac{2}{s}$$

Therefore, Z is a capacitor of $\frac{1}{2}$ F. Using the circuit of Fig. 6-10b, we then obtain the circuit of Fig. 6-11c.

---

† Note that Bader's condition is violated here for the pole at $s = 0$. Hence, $z_{22_{act}}$ does have a private pole at dc.

For a complex circuit, the calculation of $z_{12_{act}}$ and $z_{22_{act}}$ may be difficult. Then a different procedure is expedient for the calculation of t and Z. This consists of developing ladder networks from both z_{11} and z_{22} independently but in the same circuit configurations. For our example, the development of z_{11} gave the circuit of Fig. 6-11b as well as the values of L_1, C_2, L_3, and C_3. Developing z_{22} in the configuration of Fig. 6-11c results in the pole-zero patterns of Fig. 6-11d and the circuit of Fig. 6-11e. Consider now the values of L_3 as obtained from z_{11} (denoted $L_3^{(1)}$) and z_{22} (denoted $L_3^{(2)}$). As a comparison of Fig. 6-11b and e indicates, these values must be related by $L_3^{(2)} = L_3^{(1)}t^2$. Similar considerations give $C_3^{(2)} = C_3^{(1)}/t^2$, $L_2^{(2)} = L_2^{(1)}t^2$, and $C_1^{(2)} = C_1^{(1)}/t^2$. Hence, we can find the transformer turns ratio from

$$t = \sqrt{\frac{L_k^{(2)}}{L_k^{(1)}}} = \sqrt{\frac{C_i^{(1)}}{C_i^{(2)}}} \tag{6-106}$$

where L_k and C_i are arbitrary elements obtainable from both z_{11} and z_{22}.

For some circuits, y_{11} and y_{22} rather than z_{11} and z_{22} must be used in the process; for others, z_{11} and y_{22} or y_{11} and z_{22}. For example, Fig. 6-12a indicates a circuit where the ladder cannot be developed from z_{22} since L_3 and C_3 must both be found from partial pole removal, which cannot be done uniquely (Fig. 6-12b). By contrast, the development of y_{22} is possible, since L_3 is now obtained by full pole removal at infinite frequency. The reader is encouraged to work out this example in detail. The immittance parameters are

$$z_{11} = \frac{2s^2 + 5}{2s} \qquad z_{12} = \frac{2s^2 + 1}{s} \qquad z_{22} = \frac{16s^2 + 6}{s}$$

and

$$y_{22} = \frac{2s^3 + 5s}{24s^4 + 84s^2 + 28}$$

The process of developing the ladder network twice, from both the input and the output ports, requires an a priori knowledge of the circuit configuration. This configuration can be deduced in advance from the loss poles, since we know the manipulations (partial and full pole removals) needed to produce these poles.[†] The process then gives all element values as well as t. In addition, for larger networks, there are usually many elements which are obtained from both developments. Their ratio must be the same, namely t^2. Hence, calculation and comparison of these ratios provides a very useful check on the accuracy and correctness of all preceding computations. The process will be illustrated on another example, this time in terms of the y parameters.

Example 6-6 Let the parameters

$$y_{11}(s) = \frac{21s^2 + 1}{3s} \qquad y_{22}(s) = \frac{72s^2 + 3}{s}$$

as well as the attenuation poles $\omega_1 = 1/3\sqrt{2}$ and $\omega_2 \to \infty$ be prescribed. Find the doubly terminated two-port.

---

† See Ref. 5 for a detailed description of this procedure.

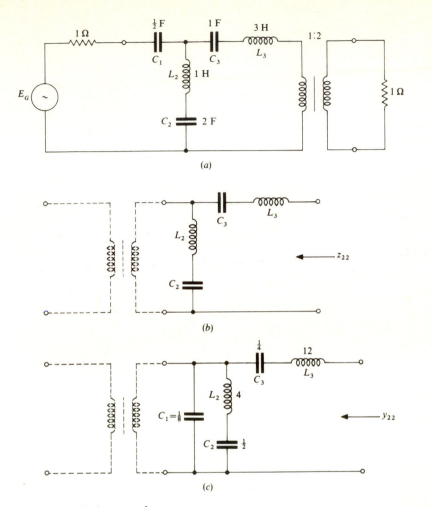

Figure 6-12 Design example.

Here, y_{11}, $y_{22} \to \infty$ for $s \to \infty$, and also there is a loss pole for $s \to \infty$. Hence, the obvious procedure is the following:

1. Remove a part of the pole that y_{11} has at $s \to \infty$ in the form of a shunt capacitor. The remainder of y_{11} must have a zero at ω_1.
2. Inverting the remainder gives an impedance which has a pole at ω_1. Full removal of this pole then creates the required transmission zero and terminates the development of y_{11}.
3. Repeat steps 1 and 2 for y_{22}.

The details of the actual calculations follow:

Step 1:
$$C_1^{(1)} = \left[\frac{y_{11}(s)}{s}\right]_{s=j\omega_1} = \left(7 + \frac{1}{3s^2}\right)_{s^2 = -1/18} = 1\ \mathrm{F}$$

Step 2:
$$Y_2 = y_{11} - sC_1^{(1)} = \frac{18s^2 + 1}{3s}$$

$$\frac{1}{Y_2} = \frac{3s}{18s^2 + 1} = \frac{s/6}{s^2 + 1/18} = \frac{s/C_2^{(1)}}{s^2 + 1/L_2^{(1)}C_2^{(1)}}$$

$$C_2^{(1)} = 6 \text{ F} \qquad L_2^{(1)} = 3 \text{ H}$$

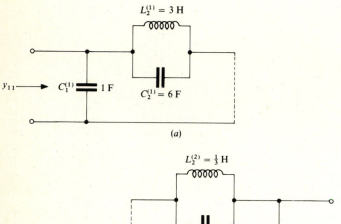

(a)

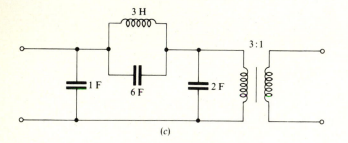

(b)

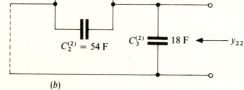

(c)

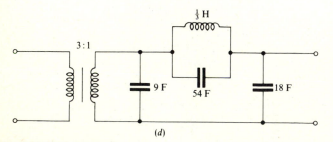

(d)

Figure 6-13 Design example.

The circuit elements obtained in steps 1 and 2 are shown in Fig. 6-13a.

Step 3:
$$C_3^{(2)} = \left[\frac{y_{22}(s)}{s}\right]_{s=j\omega_1} = \left(72 + \frac{3}{s^2}\right)_{s^2 = -1/18} = 18 \text{ F}$$

Step 4:
$$Y_2 = y_{22} - sC_3^{(2)} = \frac{54s^2 + 3}{s}$$

$$\frac{1}{Y_2} = \frac{s}{54s^2 + 3} = \frac{s/54}{s^2 + 1/18} = \frac{s/C_2^{(2)}}{s^2 + 1/L_2^{(2)}C_2^{(2)}}$$

$$C_2^{(2)} = 54 \text{ F} \qquad L_2^{(2)} = \tfrac{1}{3} \text{ H}$$

The elements obtained from y_{22} are shown in Fig. 6-13b. The transformer ratio can now be found from

$$t = \sqrt{\frac{L_2^{(2)}}{L_2^{(1)}}} = \sqrt{\frac{C_2^{(1)}}{C_2^{(2)}}} = \frac{1}{3}$$

Hence the complete network may be represented in either of the forms shown in Fig. 6.13c and d.

6-6 THE REALIZATION OF SYMMETRICAL REACTANCE LATTICES

By definition, a two-port will be called *symmetric* if its ports are electrically indistinguishable. This means that if the two-port is connected at both its ports to a larger electrical system, all voltages and currents in this outside system remain the same if the input and output ports of the two-port are interchanged, i.e., the two-port is turned around. Note that if a two-port is built up from its elements symmetrically, this so-called structural symmetry guarantees the electrical symmetry as defined above; but a two-port may be structurally quite unsymmetric and still possess electrical symmetry. Thus the two-ports of Fig. 6-14 are both electrically symmetrical although only that of Fig. 6-14a exhibits structural symmetry.

By definition, the external performance of a symmetrical two-port is invariant with respect to the exchange of its ports; hence interchanging the subscripts 1 and 2 in any driving-point or transfer function leaves the function unchanged. Thus, for example,

$$z_{11} = z_{22} \qquad y_{11} = y_{22} \qquad A = D \qquad Z_1 = Z_2 \qquad \rho_1 = \rho_2 \quad (6\text{-}107)$$

for a symmetric two-port.† Hence, a symmetric two-port is completely described by only two of its immittance parameters, say z_{11} and z_{12} or y_{11} and y_{12}.

A circuit configuration of great importance for symmetric two-ports is the *symmetric lattice*. This circuit, shown in Fig. 6-15a, has two parameters z_a and z_b, the lattice impedances, which may be used instead of z_{11} and z_{12} to describe the

---

† Of course, the last two of these conditions hold only if also $R_L = R_G$.

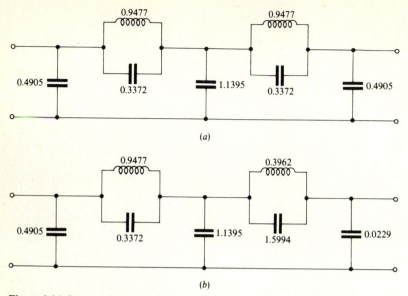

(a)

(b)

Figure 6-14 Symmetrical two-ports.

two-port (a symbolic diagram, often used to denote the full circuit of Fig. 6-15a, is shown in Fig. 6-15b).

Analysis of the circuit shows directly that

$$z_{11} = \frac{V_1}{I_1}\bigg|_{I_2=0} = \frac{V_1}{I_a + I_b}\bigg|_{I_2=0} = \tfrac{1}{2}(z_a + z_b) \tag{6-108}$$

$$z_{12} = \frac{V_2}{I_1}\bigg|_{I_2=0} = \frac{I_a z_b - I_b z_a}{I_a + I_b}\bigg|_{I_2=0} = \tfrac{1}{2}(z_b - z_a) \tag{6-109}$$

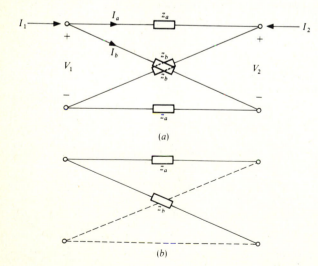

(a)

(b)

Figure 6-15 (a) Lattice network; (b) symbol used for lattice network.

since, for $I_2 = 0$,

$$I_a = I_b = \frac{I_1}{2} = \frac{V_1}{z_a + z_b} \tag{6-110}$$

Solving for z_a and z_b gives

$$z_a = z_{11} - z_{12} \qquad z_b = z_{11} + z_{12} \tag{6-111}$$

On the basis of these relations, it can easily be shown that the lattice is a feasible realization of a symmetrical two-port in the sense that if a symmetrical two-port is at all realizable with positive elements, it is certainly realizable as a symmetric lattice. For a simple proof in the general case of *RLC* two-ports, the reader is referred to Prob. 6-9. For the restricted class of reactance two-ports considered here, it follows from Eqs. (6-111) that z_a and z_b have only $j\omega$-axis poles with real residues, since the same is true for any realizable z_{11} and z_{12}. Furthermore, by the residue theorem, at any pole of z_a or z_b the residues k_{ij} satisfy

$$k_{11} \geq 0 \tag{6-112}$$

and

$$k_{11}^2 - k_{12}^2 = k_{11}^2 - |k_{12}|^2 \geq 0 \tag{6-113}$$

Therefore, $k_{11} \geq |k_{12}|$, and hence both $k_{11} - k_{12}$ and $k_{11} + k_{12}$ are nonnegative. Hence, z_a as well as z_b have positive (or zero) real residues. These conditions then guarantee that z_a and z_b are both realizable reactances whenever z_{11} and z_{12} are realizable. Note, however, that the lattice is not a canonical realization from the viewpoint of economy; in fact, it uses twice as many elements as the degree necessitates. Also, it precludes the simultaneous grounding of both input and output ports, since this would short-circuit the lower branch containing z_a.

Both the larger number of elements and the lack of common ground terminal for the ports can be helped by using the so-called half-lattice networks,† two of which are shown in Fig. 6-16. The reader is urged to show that both networks have the same *z* parameters as the lattice of Fig. 6-15 and are therefore equivalent to it. (A detailed discussion of these and other equivalent circuits will be given in Sec. 6-7 and in the problems at the end of this chapter.)

The complete design process for lattice networks will now be illustrated by way of a simple but fairly representative example.

Example 6-7 Design a symmetrical lattice from the absolute value of its characteristic function

$$|K| = \omega^3$$

A comparison‡ of Eqs. (6-62) and (6-63) giving z_{11} and z_{22} shows that these impedances will be equal to each other only if $R_L = R_G = R$ and $K_e = 0$. The latter condition, which demands that $K(s)$ be a pure *odd* function, is satisfied for our specification; by

---

† A more proper but lengthier name is *unbalanced transformer-coupled lattice equivalents.*
‡ The solution of this problem will be worked out in some detail to permit its generalization afterward.

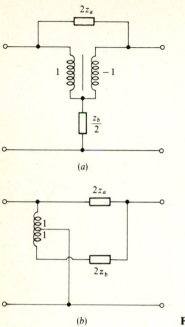

(a)

(b)

Figure 6-16 Half-lattice two-ports.

choosing $R_L = R_G = 1 \; \Omega$, the first condition is also valid, and the requirements of symmetry are thus met. Next, by the Feldtkeller equation, with $F(s) = s^3$ and $P(s) \equiv 1$,

$$E(s)E(-s) = F(s)F(-s) + P(s)P(-s)$$

$$E(s)E(-s) = s^3(-s^3) + 1 = -s^6 + 1$$

This leads in the usual way to

$$E(s) = (s + 1)(s^2 + s + 1) = s^3 + 2s^2 + 2s + 1$$

The open-circuit impedances are then, by Table 6-1,

$$z_{11} = z_{22} = \frac{2s^2 + 1}{2s^3 + 2s} \qquad \text{and} \qquad z_{12} = \frac{1}{2s^3 + 2s}$$

Next, by Eq. (6-111),

$$z_a = z_{11} - z_{12} = \frac{2s^2}{2s^3 + 2s} = \frac{s}{s^2 + 1} \qquad z_b = z_{11} + z_{12} = \frac{2s^2 + 2}{2s^3 + 2s} = \frac{1}{s}$$

The circuit is therefore that shown in Fig. 6-17a. Note that the same z parameters lead to the circuit of Fig. 6-17b if a ladder development is used. The 50 percent saving in element values is clearly visible.

Another interesting comparison can be drawn between the lattice and ladder realizations of these low-pass specifications. For high frequencies, the ladder circuit achieves the required high attenuation by virtue of the high shunt admittance of its capacitors and the high series impedance of its coil. This keeps the output voltage and current low. The lattice, on the other hand, achieves its small output

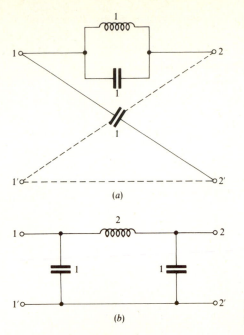

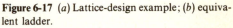

Figure 6-17 (a) Lattice-design example; (b) equivalent ladder.

for high frequencies by acting as a nearly balanced bridge circuit containing a 1-F capacitor in each arm. Obviously, this balanced condition is a sensitive one, easily upset by small deviations in the value of any of the capacitors.

One final item of interest is the cancellation of common factors in the numerator and denominator of both z_a and z_b. This turns out to be the rule rather than a coincidence. In fact, a somewhat lengthier derivation reveals that simplified formulas can be found for z_a and z_b in which all cancellations have already been performed. The resulting design procedure is then as follows:

1. Factor $F(s) + P(s)$ into the form

$$F(s) + P(s) = h(s)a(s) \tag{6-114}$$

where $h(s)$ is a strictly Hurwitz and $a(s)$ a strictly anti-Hurwitz polynomial.
2. If $P(s)$ is *even* [and hence $F(s)$ is odd, since $K(s)$ must be purely odd], then

$$z_a(s) = R\,\frac{-a_o(s)}{a_e(s)} \qquad z_b(s) = R\,\frac{h_e(s)}{h_o(s)} \tag{6-115}$$

If $P(s)$ is *odd* [and hence $F(s)$ is even], then

$$z_a(s) = R\,\frac{-a_o(s)}{a_e(s)} \qquad z_b(s) = R\,\frac{h_o(s)}{h_e(s)} \tag{6-116}$$

Here R is the common value of R_L and R_G, and the subscripts e and o denote even and odd parts, as before.

These relations can be proved simply by expressing E, F, and P in terms of $a(s)$ and $h(s)$ and then substituting into the expressions

$$z_a = z_{11} - z_{12} = R\frac{H_e - 1}{H_o + K_o} \qquad z_b = z_{11} + z_{12} = R\frac{H_e + 1}{H_o + K_o} \qquad (6\text{-}117)$$

After canceling factors common to both numerator and denominator in these expressions, the simplified formulas are obtained. The reader should work out the lengthy but straightforward details as an exercise in polynomial manipulation.

Example 6-8 For the problem solved above, $F(s) = s^3$ and $P(s) \equiv 1$; that is, P is even. Also

$$F + P = s^3 + 1 = (s + 1)(s^2 - s + 1)$$

so that

$$h(s) = s + 1 \qquad a(s) = s^2 - s + 1$$

Hence, with $R = 1$,

$$z_a = R\frac{-a_o}{a_e} = \frac{s}{s^2 + 1} \qquad z_b = R\frac{h_e}{h_o} = \frac{1}{s}$$

as before.

For complicated networks, where the common factors are not as evident as in our simple example, the simplified formulas are especially useful.

6-7 EQUIVALENTS OF SYMMETRICAL LATTICES

Important conclusions can be drawn from the formula giving the transducer factor $H(s)$ of a symmetric lattice as a function of its lattice impedances. From Eq. (6-48)

$$H(s) = \frac{(z_{11} + R)^2 - z_{12}^2}{2Rz_{12}} = \frac{(z_{11} + z_{12} + R)(z_{11} - z_{12} + R)}{2Rz_{12}} \qquad (6\text{-}118)$$

Hence

$$H(s) = \frac{(z_b + R)(z_a + R)}{R(z_b - z_a)} = \frac{(z_a/R + 1)(Z_b/R + 1)}{z_b/R - z_a/R} \qquad (6\text{-}119)$$

Equation (6-119) can also be rewritten in the form

$$H(s) = \frac{(R/z_a + 1)(R/z_b + 1)}{R/z_a - R/z_b} \qquad (6\text{-}120)$$

It can be seen from these equations that interchanging z_a and z_b results merely in a change of the sign of $H(s)$ (this is also obvious from Fig. 6-15). Less evidently, changing z_a into R^2/z_a and z_b into R^2/z_b has the same effect, as Eq. (6-120) shows. Hence, doing *both*, as illustrated in Fig. 6-18a for the circuit of Example 6-8, leaves $H(s)$ unchanged. As far as the value of $|H(j\omega)|$, or equivalently the loss, is concerned, the four lattice impedance pairs (z_a, z_b), (z_b, z_a), $(R^2/z_a, R^2/z_b)$, and

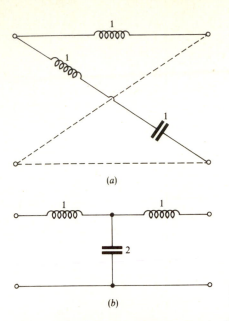

(a)

(b)

Figure 6-18 (a) Lattice equivalent to that shown in Fig. 6-17a; (b) equivalent ladder.

$(R^2/z_b, R^2/z_a)$ are all perfectly equivalent. Hence, from any known lattice network with a specified loss response, three more can be found directly with the same performance. For example, in addition to the circuit of Fig. 6-18, the circuits of Fig. 6-19 also have the same loss response as the original circuit given in Fig. 6-17. Their phase, of course, differs by 180°.

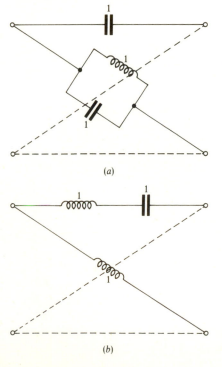

(a)

(b)

Figure 6-19 Additional equivalent lattices with 180° phase shift.

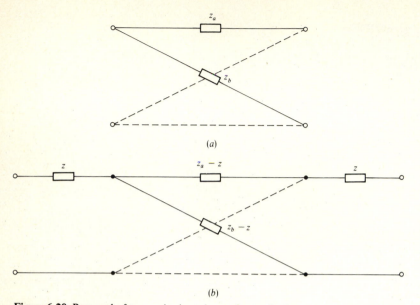

(a)

(b)

Figure 6-20 Removal of two series impedances from a lattice.

Next, consider the two circuits of Fig. 6-20. Calculation of the impedance parameters for the first circuit gives

$$z_{11} = z_{22} = \tfrac{1}{2}(z_a + z_b) \qquad z_{12} = \tfrac{1}{2}(z_b - z_a) \tag{6-121}$$

as before. For the second circuit, the corresponding parameters are

$$z_{11} = z_{22} = z + \tfrac{1}{2}(z_a - z + z_b - z) = \tfrac{1}{2}(z_a + z_b)$$

$$z_{12} = \tfrac{1}{2}[(z_b - z) - (z_a - z)] = \tfrac{1}{2}(z_b - z_a) \tag{6-122}$$

i.e., the same. Hence, these circuits are completely equivalent electrically.

In a perfectly dual derivation, the admittance parameters of the circuits of Fig. 6-21 can be obtained. For the lattice of Fig. 6-21a, by Fig. 6-22,

$$y_{11} = y_{22} = \frac{I_1}{E}\bigg|_{V_2=0} = \frac{I_a + I_b}{E}\bigg|_{V_2=0} = \frac{y_a(E/2) + y_b(E/2)}{E}$$

$$y_{11} = y_{22} = \tfrac{1}{2}(y_a + y_b) \tag{6-123}$$

and
$$y_{12} = \frac{I_2}{E}\bigg|_{V_2=0} = \frac{I_b - I_a}{E}\bigg|_{V_2=0} = \tfrac{1}{2}(y_b - y_a) \tag{6-124}$$

For the circuit of Fig. 6-21b, the admittance parameters are

$$y_{11} = y_{22} = y + \tfrac{1}{2}[(y_a - y) + (y_b - y)] = \tfrac{1}{2}(y_a + y_b)$$

$$y_{12} = \tfrac{1}{2}[(y_b - y) - (y_a - y)] = \tfrac{1}{2}(y_b - y_a) \tag{6-125}$$

i.e., unchanged. Hence again the two circuits are equivalent.

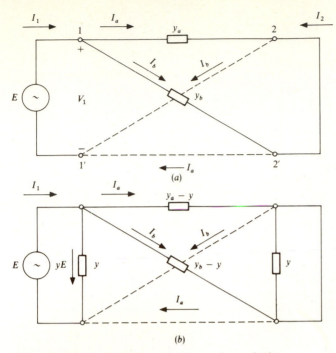

(a)

(b)

Figure 6-21 Removal of shunt admittances from a lattice.

Example 6-9 The two equivalences derived above sometimes enable us to move elements out of the lattice and into series or shunt ladder branches. For example, the original lattice circuit, obtained in Example 6-8 and shown in Fig. 6-17, contains a shunt capacitance of 1 F in both branches. Regarding this capacitor as the admittance y of Fig. 6-21b and utilizing the equivalence of the circuits given in Fig. 6-21a and b, we can go through the steps shown in Fig. 6-23 and obtain the ladder network already exhibited in Fig. 6-17b. The last step indicated in Fig. 6-23 is valid because the admittances $y_0 = 0$ may be left out and because the balanced two-port consisting of the two 1-H inductors is fully equivalent to an unbalanced two-port containing the single 2-H inductor† as long as we refrain from some special operations forbidden for two-ports anyway; e.g., connecting an impedance or a generator, say, between terminals 1' and 2'.

† This equivalence can be checked by comparing the z or y parameters of the two two-ports.

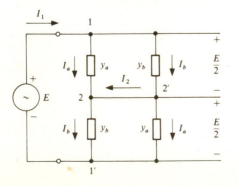

Figure 6-22 Finding the admittance parameters of a lattice.

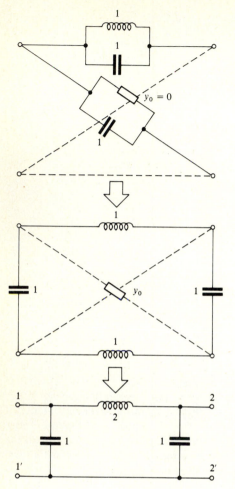

Figure 6-23 Ladder-to-lattice transformation.

Example 6-10 As an exercise in using the equivalence of Fig. 6-20, consider the lattice shown in Fig. 6-18a. Here, the 1-H inductor may be regarded as z. Then, the equivalent ladder of Fig. 6-18b is obtained directly.

A particularly interesting circuit type is obtained if we constrain the lattice impedances by the relation

$$z_a z_b = R^2 \tag{6-126}$$

Substituting $z_b = R^2/z_a$ into Eq. (6-119), we obtain

$$H(s) = \frac{(z_a/R + 1)(R/z_a + 1)}{R/z_a - z_a/R} = \frac{(z_a + R)^2}{R^2 - z_a^2} \tag{6-127}$$

or

$$H(s) = \frac{R + z_a}{R - z_a} \tag{6-128}$$

The driving-point impedances can be written, by Eq. (6-51), for any symmetric lattice in the form

$$Z_1 = Z_2 = \frac{(z_{11} + z_{12})(z_{11} - z_{12}) + Rz_{11}}{R + z_{11}} = \frac{z_a z_b + (R/2)(z_a + z_b)}{R + \frac{1}{2}(z_a + z_b)} \quad (6\text{-}129)$$

and for $z_a z_b = R^2$

$$Z_1 = Z_2 = R \frac{R + \frac{1}{2}(z_a + z_b)}{R + \frac{1}{2}(z_a + z_b)} = R \quad (6\text{-}130)$$

Hence, irrespective of the value of z_a (and thus of z_b), both driving-point impedances become constants, equal to the terminating resistances. For this reason, these circuits are called *constant-impedance lattices*. In addition, for the reactance two-ports discussed in this chapter, the lattice impedances are restricted to be both purely reactive

$$z_a = jx_a \qquad z_b = jx_b \quad (6\text{-}131)$$

Then, by Eq. (6-128),

$$H(s) = \frac{R + jx_a}{R - jx_a} \quad (6\text{-}132)$$

and

$$|H(j\omega)|^2 = \frac{R^2 + x_a^2}{R^2 + x_a^2} = 1 \quad (6\text{-}133)$$

By Eq. (6-8), $|H|^2 = 1$ is equivalent to zero transducer loss. Now since Eq. (6-133) is valid for any ω value, such a network will pass sine waves of all frequencies without attenuation. For this reason, the constant-impedance reactance two-port is often called an *allpass* network. The phase shift of these networks is easily obtained from Eq. (6-132):

$$\beta = 2 \tan^{-1} \frac{x_a}{R} \quad (6\text{-}134)$$

Example 6-11 Design an allpass lattice from the specified transducer factor

$$H(s) = \frac{(1 + s)(1 + s + s^2)}{(1 - s)(1 - s + s^2)}$$

SOLUTION 1 Choosing $R = 1$, we can identify z_a from the rewritten form

$$H(s) = \frac{1 + 2s + 2s^2 + s^3}{1 - 2s + 2s^2 - s^3} = \frac{1 + (s^3 + 2s)/(2s^2 + 1)}{1 - (s^3 + 2s)/(2s^2 + 1)}$$

and from Eq. (6-128), as

$$z_a = \frac{s^3 + 2s}{2s^2 + 1} = \frac{s}{2} + \frac{\frac{3}{4}s}{s^2 + \frac{1}{2}}$$

Then

$$y_b = \frac{z_a}{R^2} = \frac{s}{2} + \frac{\frac{3}{4}s}{s^2 + \frac{1}{2}}$$

Hence, the lattice of Fig. 6-24 is obtained.

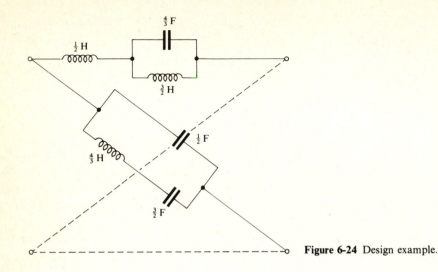

Figure 6-24 Design example.

SOLUTION 2 A different solution can be obtained on the basis of the following observations. For cascaded constant-impedance two-ports (Fig. 6-25), each two-port is working between constant impedances. Hence, they can be designed, independently of each other, to operate between terminations $R_G = R_L = R$; thereafter, cascading them as in Fig. 6-25 will not affect their terminations or their performance. The transducer factor of the overall two-port is therefore

$$H(s) = \frac{1}{2} \frac{E}{V_2} = \frac{E/2}{V_{2A}} \frac{V_{1B}}{V_{2B}} \frac{V_{1C}}{V_{2C}} \cdots \tag{6-135}$$

or

$$H(s) = H_A(s)H_B(s) \cdots \tag{6-136}$$

Hence, a constant-impedance realization can be obtained by first factoring $H(s)$ into a product of lower-degree allpass factors $H_A(s)$, $H_B(s)$, ..., next realizing each factor as a separate allpass two-port, and finally by cascading these elementary two-ports. For our example, the factors are evidently

$$H_A(s) = \frac{1+s}{1-s} \qquad H_B(s) = \frac{1+s+s^2}{1-s+s^2} = \frac{1+s/(1+s^2)}{1-s/(1+s^2)}$$

Hence, $z_{aA} = y_{bA} = s$ and $z_{aB} = y_{bB} = s/(1+s^2)$ may be chosen. The circuit is then that shown in Fig. 6-26.

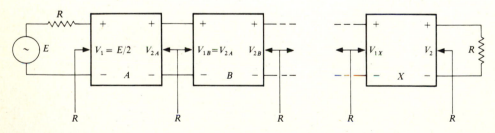

Figure 6-25 Cascaded constant-impedance two-ports.

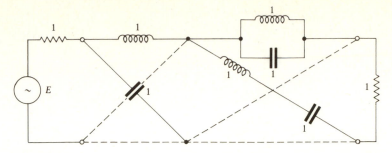

Figure 6-26 Design example.

A comparison of the two solutions (Figs. 6-24 and 6-26) reveals that the number of circuit elements is the same for the two design techniques. However, the fact that the simple elementary two-ports of the second circuit can be built, adjusted, and tested individually, represents a great practical advantage which becomes more significant for higher degrees of $H(s)$.

We have investigated the problem of finding equivalent circuits for given lattices. The reverse problem of finding a lattice equivalent for a given symmetric two-port is also of some importance. This task is simplified in the case of structurally symmetrical two-ports by using the *bisection theorem* of Bartlett and Brune. This theorem, illustrated in Fig. 6-27, shows that the arm impedances for the lattice equivalent of a structurally symmetrical two-port can be obtained in the following steps:

1. The network is cut into two symmetrical halves $N/2$ along its axis of symmetry.
2. The input impedance of either half is then equal to z_b.
3. All wires cut by the bisection† are shorted to each other; the input impedances now become z_a.

> PROOF To prove the bisection theorem, consider first the circuit of Fig. 6-28a. By symmetry, the two-port inputs are $V_1 = V_2 = E$ and $I_1 = I_2 = I$. Substituting into the impedance equations (5-4) gives
>
> $$E = (z_{11} + z_{12})I \qquad (6\text{-}137)$$
>
> and hence, using Eq. (6-111),
>
> $$\frac{E}{I} = z_{11} + z_{12} = z_b \qquad (6\text{-}138)$$
>
> (Equation 6-138 follows even more directly from the circuit shown in Fig. 6-28c.)
>   But because of the complete symmetry of the two-port and its generator circuit, there can be no current flowing in the wires crossing the line of

† It is assumed here that these wires do not cross each other in the symmetry axis prior to bisection.

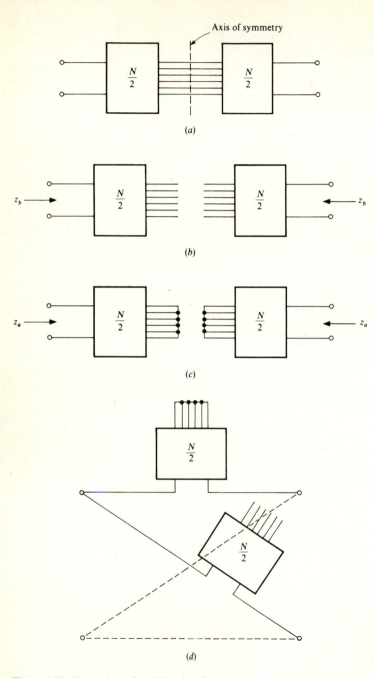

Figure 6-27 Illustration of the bisection theorem.

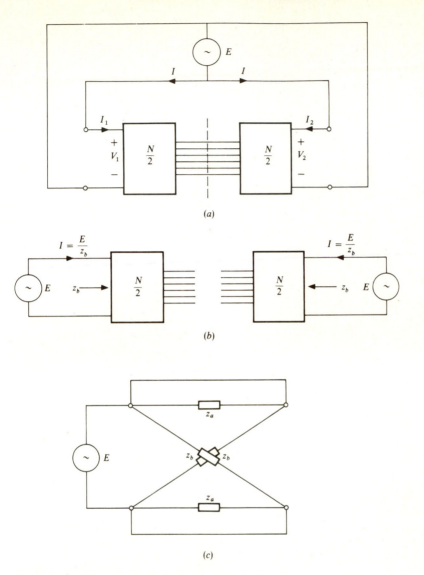

Figure 6-28 Proving the bisection theorem.

symmetry. Hence, the currents I at both ports will remain undisturbed if these wires are *cut* (Fig. 6-28b). Therefore, the driving-point impedances also remain z_b. This proves the validity of the process used to obtain z_b.

To prove the second half of the bisection theorem, consider the circuit of Fig. 6-29a. Now the inputs are $V_1 = -V_2 = E$ and $I_1 = -I_2 = I$, due to the skew symmetry of the circuit. By Eq. (5-4),

$$E = (z_{11} - z_{12})I \qquad (6\text{-}139)$$

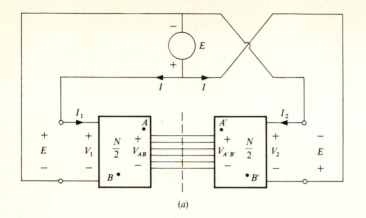

(a)

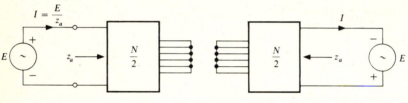

(b)

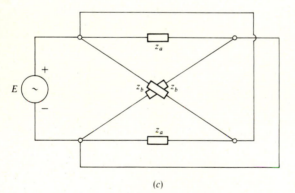

(c)

Figure 6-29 Proving the bisection theorem (continued).

and, using Eq. (6-111),†

$$\frac{E}{I} = z_{11} - z_{12} = z_a \qquad (6\text{-}140)$$

Also due to the skew symmetry, if pairs of points are chosen symmetrically with respect to the axis of structural symmetry, such as A, B and A', B'

† Equation (6-140) also follows directly from Fig. 6-29c.

in Fig. 6-29a, their potential differences will be equal in size but of opposite sign

$$V_{AB} = -V_{A'B'} \tag{6-141}$$

Hence, points which are *on* the symmetry axis must be at the same potential and therefore can be connected together without affecting I_1 or I_2. Hence, the relations indicated by Fig. 6-29b are valid. This proves that the process shown in Fig. 6-27c for finding z_a is justified.

This completes the proof of the bisection theorem. Note that strictly speaking, the generator E should have been coupled via $1:1$ and $1: -1$ ideal transformers, rather than directly, to the two-ports in all circuits utilized in the proof. This would guarantee that the currents in the two wires of each port are equal but of opposite directions. It is easy to see, however, that this does not affect the arguments presented (although it makes the circuits unnecessarily complicated).

Example 6-12 Give the relations between the impedances for the equivalence indicated in Fig. 6-30.

Using the bisection theorem (Fig. 6-31), for given z, z_1, and z_2, we have

$$z_b = z + 2z_2 \qquad z_a = \frac{zz_1/2}{z + z_1/2} = \frac{zz_1}{2z + z_1}$$

and, for given z, z_a, and z_b,

$$z_2 = \frac{z_b - z}{2} \qquad z_1 = 2\frac{zz_a}{z - z_a} = 2\frac{z_a(-z)}{z_a - z} = \frac{1}{\frac{1}{2}(1/z_a - 1/z)}$$

Obviously, for given PR functions z, z_1, and z_2, the lattice impedances are also PR. Hence, the bridged T always has an equivalent lattice (this follows also from the canonical nature of the lattice, proved in Sec. 6-6). On the other hand, for given PR z_a and z_b, the impedances z_1 and z_2 will be PR only if an impedance z can be found which is a series element of z_b and at the same time a shunt element in z_a. (This can be seen both from the equations and from Fig. 6-31.)

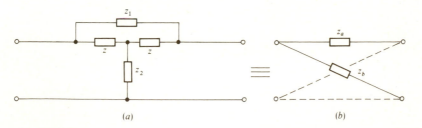

(a)   (b)

Figure 6-30 Equivalent two-ports.

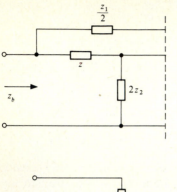

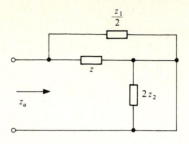

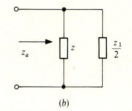

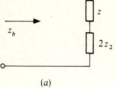

(a)

(b)

Figure 6-31 Using the bisection theorem.

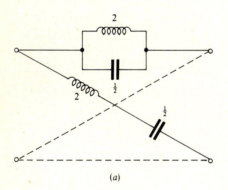

(a)

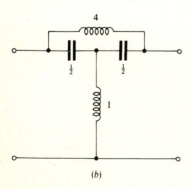

(b)

Figure 6-32 Lattice–to–bridged-T transformation.

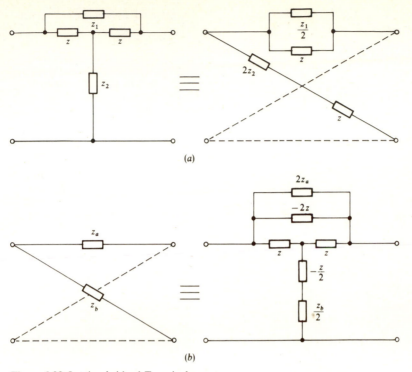

Figure 6-33 Lattice–bridged-T equivalences.

For example, for the allpass lattice shown in Fig. 6-32a, $z = 2/s$ may be chosen. The bridged-T equivalent is then that shown in Fig. 6-32b.

The equivalence illustrated in Figs. 6-30 and 6-31 can be summarized as shown in Fig. 6-33. The equivalence is valid for any finite nonzero z.

6-8 SUMMARY

The subjects discussed in this chapter were the following:

1. The advantages of doubly terminated reactance two-ports and the difficulties involved in their analysis and design (Sec. 6-1).
2. The transducer parameters (transducer factor, transducer constant, transducer loss, and phase), the characteristic function, and the reflection factors of a doubly loaded reactance two-port (Sec. 6-2).
3. Power relations of the two-port and its terminations; the Feldtkeller equation; the critical frequencies of the doubly loaded reactance two-port (Sec. 6-2).
4. The calculation of the transducer parameters from given two-port immittance parameters, and vice versa (Sec. 6-3).
5. Realizability conditions on the transducer factor, the characteristic function, and the reflection factors (Sec. 6-4).
6. Ladder realization of a doubly loaded reactance two-port from given immittance parameters and transmission zeros; the calculation of the turns ratio of the ideal transformer at its output (Sec. 6-5).
7. The properties of symmetrical lattices; the realization of reactance lattices; half-lattice networks (Sec. 6-6).

8. Equivalent lattice networks; extraction of series or shunt branches from a lattice; constant-impedance lattices; allpass networks (Sec. 6-7).
9. Finding the lattice equivalent of a symmetric two-port: the bisection theorem and its applications (Sec. 6-7).

The material contained in this chapter is directly applicable to the design of passive filters, delay lines, and delay equalizers, etc., and is therefore of great practical importance. It is, however, restricted to circuits containing only passive lumped components. In the next two chapters, we shall discuss the design of two-ports containing active elements (Chap. 7) and distributed components (Chap. 8).

PROBLEMS

6-1 Prove that the maximum power obtainable in the variable resistor R of Fig. 6-34 for given E_G and R_G is $E_G^2/4R_G$. How much is R for this power? How much can R change in either direction before the dissipated power in it is halved?

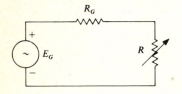

Figure 6-34 Maximum-power-transfer example.

6-2 Prove that for a rational function $F(s)$, with real coefficients, the even and odd parts can be written

$$F_e(s) = \tfrac{1}{2}[F(s) + F(-s)] \qquad F_o(s) = \tfrac{1}{2}[F(s) - F(-s)]$$

Show that, for $s = j\omega$, F_e is real and F_o is imaginary.

6-3 Find $H(s)$, $\rho_1(s)$, $\rho_2(s)$, and $K(s)$ for the circuit of Fig. 6-35. Calculate the loss and phase at $\omega = 2$ rad/s.

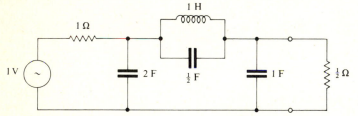

Figure 6-35 Two-port example.

6-4 Let a reactive two-port T consist of a cascade combination of two two-ports T_a and T_b, as shown in Fig. 6-36a.

(a) Show that the portion of the circuit to the left of the broken vertical line can be replaced by the Thevenin equivalent shown in Fig. 6-36b as far as its effect on the remainder of the circuit is concerned; here

$$E_a = E_G \frac{z_{12_a}}{R_G + z_{11_a}} \qquad Z_a = z_{22_a} - \frac{z_{12_a}^2}{R_G + z_{11_a}}$$

and the z_{ij_a} are the impedance parameters of T_a.

(b) Show that the remainder of the circuit presents an impedance

$$Z_b = z_{11_b} - \frac{z_{12_b}^2}{R_L + z_{22_b}}$$

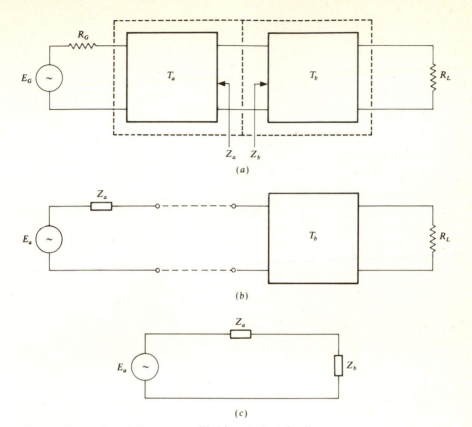

Figure 6-36 (a) Cascaded two-ports; (b), (c) equivalent circuits.

(c) From the resulting equivalent circuit (Fig. 6-36c), show that the transducer power ratio of T can be written

$$\frac{P_{\max}}{P_2} = \frac{|Z_a + Z_b|^2}{4R_a R_b}$$

where $R_a = \mathrm{Re}\, Z_a$ and $R_b = \mathrm{Re}\, Z_b$.

(d) Show that, equivalently,

$$\frac{P_{\max}}{P_2} = \frac{|Y_a + Y_b|^2}{4G_a G_b}$$

where $Y_a = 1/Z_a$, $Y_b = 1/Z_b$, $G_a = \mathrm{Re}\, Y_a$, and $G_b = \mathrm{Re}\, Y_b$.

6-5 From Eqs. (6-53) and (6-54), prove that the zeros of $\rho_1(s)$ and $\rho_2(s)$ are negatives of each other, i.e., that if

$$\rho_1 = \frac{F(s)}{E(s)}$$

then

$$\rho_2 = \pm \frac{F(-s)}{E(s)}$$

Hint: Make use of the odd character of the $z_{ij}(s)$:

$$z_{ij}(-s) = -z_{ij}(s) \qquad \text{for } i, j = 1, 2$$

6-6 Substitute formulas (6-61) to (6-63) giving z_{11}, z_{12}, and z_{22} into Eq. (6-64) giving

$$\Delta_z = z_{11}z_{22} - z_{12}^2$$

in terms of H and K. Show that this results in a simple identity and that therefore these equations are compatible. *Hint:* Use the Feldtkeller equation.

6-7 Substitute relations (6-61) to (6-63), giving the z parameters in terms of H and K, in the expression (6-54) of $\rho_2(s)$ to obtain Eq. (6-71), which expresses $\rho_2(s)$ via H and K.

6-8 Analyze the circuit of Fig. 6-14b to prove that it is symmetrical.

6-9 From the circuit of Fig. 6-37 show that the lattice impedances of any realizable symmetrical two-port T are realizable impedances. *Hint:* Prove that the input impedance Z is $2z_a$ $(2z_b)$ if the ideal transformed ratio is $+1$ (-1).

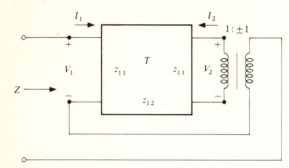

Figure 6-37 Circuit proving the realizability of the lattice impedances.

6-10 Design a symmetrical lattice from its characteristic function $K(s) = s/(1 + s^2)$.

6-11 (a) Find the immittance parameters of the circuits shown in Fig. 6-38. *Hint:* Treat them as parallel combinations of simpler two-ports.

  (b) Design symmetrical lattices which are equivalent to these two-ports.

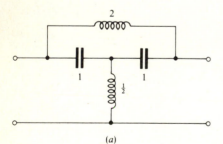

(a)

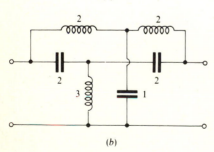

(b)

Figure 6-38 (a) Bridged-T circuit, (b) twin-T circuit.

6-12 Prove that the lattice and ladder networks shown in Fig. 6-39 are equivalent.

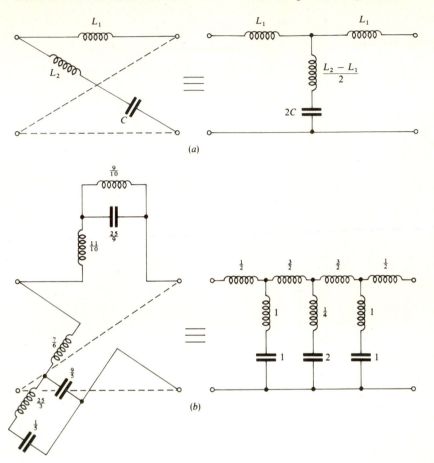

Figure 6-39 Lattice equivalents. (*From J. E. Storer, "Passive Network Synthesis," p. 127, McGraw-Hill, New York, 1957.*)

6-13 Prove the equivalences shown in Fig. 6-40. *Hint:* Treat the two-port of Fig. 6-40a as a series combination of two simpler two-ports and derive its z parameters; use the dual process on the circuit of Fig. 6-40b. Use the equivalence of Fig. 6-40a in proving that of Fig. 6-40c; the equivalence of Fig. 6-40b in proving that of Fig. 6-40d. Use the equivalences of Fig. 6-40a and b in proving Fig. 6-40f.

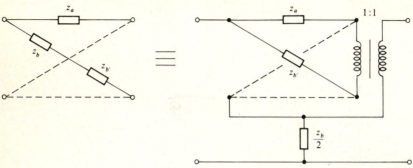

(a)

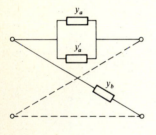

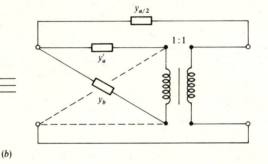

(b)

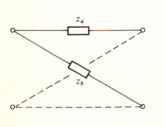

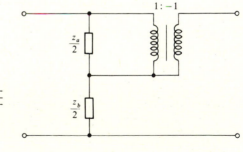

(c)

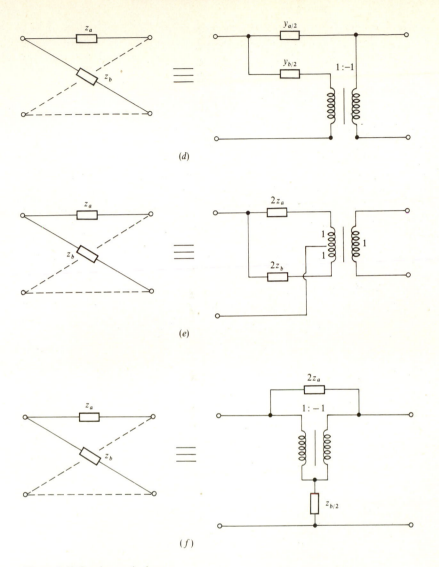

(d)

(e)

(f)

Figure 6-40 Lattice equivalences.

6-14 Realize in the form of single constant-impedance reactance lattices each of the transducer factors given below. Thereafter, obtain the equivalent realization in the form of cascaded elementary lattices.

(a) $H(s) = \dfrac{(s + 2)(s + 5)(s^2 + 3s + 5)}{(-s + 2)(-s + 5)(s^2 - 3s + 5)}$

(b) $H(s) = \dfrac{(s^2 + s + 3)(s^2 + 2s + 4)}{(s^2 - s + 3)(s^2 - 2s + 4)}$

(c) $H(s) = \dfrac{s^3 + 3s^2 + 6s + 4}{-s^3 + 3s^2 - 6s + 4}$

6-15 (*a*) Compare the transfer functions $H(s)$ of the lattices shown in Fig. 6-41. Prove that apart from a constant factor they are identical. What practical applications can you visualize for the equivalence indicated?† *Hint:* Use Eqs. (6-48) and (6-119).

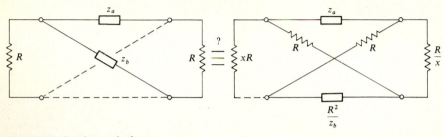

Figure 6-41 Lattice equivalence.

(*b*) For a constant-impedance lattice, $z_b = R^2/z_a$, and the equivalent lattice derived in part (*a*) becomes balanced, i.e., symmetric with respect to the horizontal axis. Show that repeated application of the equivalence of part (*a*) then results in the transformations shown in Fig. 6-42 for an initially constant-impedance lattice.

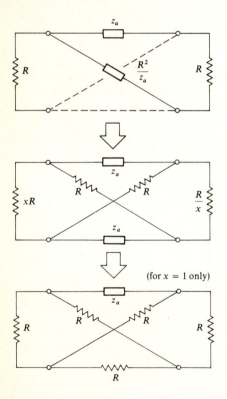

Figure 6-42 Lattice transformation.

† Due to S. Darlington.

6-16 Show the validity of the equivalence illustrated in Fig. 6-43. When are the T and Π networks realizable?

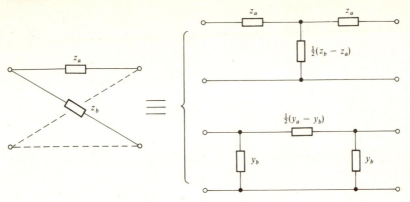

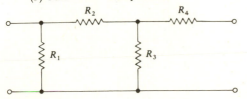

Figure 6-43 Lattice equivalences.

6-17 Show that the characteristic function and reflection factors of a symmetric lattice are given by

$$K(s) = \frac{R^2 - z_a z_b}{R(z_b - z_a)} \quad \text{and} \quad \rho_1(s) = \rho_2(s) = \frac{R^2 - z_a z_b}{(R + z_a)(R + z_b)}$$

Analyze the effects (a) of interchanging z_a and z_b; (b) of replacing z_a by R^2/z_a and z_b by R^2/z_b; (c) of performing both operations upon $H(s)$, $K(s)$, and $\rho(s)$.

6-18 (a) For an allpass lattice network, $|H(j\omega)| \equiv 1$, and we can write

$$H(j\omega) = e^{j\beta(\omega)}$$

where $\beta(\omega)$ is a real function, the phase function, of ω. Show that the lattice impedances can then be written as

$$z_a = jR \tan \frac{\beta}{2} \qquad z_b = -jR \cot \frac{\beta}{2}$$

(b) Find $\beta(\omega)$, $z_a(\omega)$, and $z_b(\omega)$ for the $H(s)$ functions of Prob. 6-14.

6-19 (a) Give the condition for the electrical symmetry for the two-port shown in Fig. 6-44.
(b) Find the lattice equivalent.

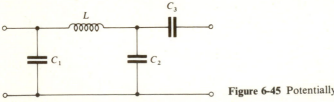

Figure 6-44 Potentially symmetric two-port.

6-20 Can the reactance two-port shown in Fig. 6-45 be symmetric? Why or why not?

Figure 6-45 Potentially symmetric two-port.

6-21 The reactance two-port shown in Fig. 6-46 is antimetric. What relations must hold between its element values?

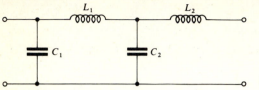

Figure 6-46 Antimetric two-port.

6-22 Find the lattice equivalent of the twin-T circuit shown in Fig. 6-47.

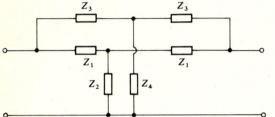

Figure 6-47 Twin-T circuit.

6-23 Show from Eqs. (5-4) and (5-10) of Sec. 5-1 that for an arbitrary two-port the relation

$$\Delta_z = \frac{z_{22}}{y_{11}} = \frac{z_{11}}{y_{22}}$$

holds. Using an argument similar to that presented at the end of Sec. 6-4, give an alternative proof of the residue theorem.

REFERENCES

1. Darlington, S.: Synthesis of Reactance 4-Poles Which Produce Prescribed Insertion Loss Characteristics, *J. Math. Phys.*, September 1939, pp. 257–353.
2. Cauer, W.: "Synthesis of Linear Communication Networks," McGraw-Hill, New York, 1958.
3. Storer, J. E.: "Passive Network Synthesis," McGraw-Hill, New York, 1957.
4. Rupprecht, W.: "Netzwerksynthese," Springer, Berlin, 1972.
5. Saal, R., and E. Ulbrich: On the Design of Filters by Synthesis, *IRE Trans. Circuit Theory*, December 1958, pp. 284–327.

CHAPTER SEVEN

LINEAR ACTIVE TWO-PORTS†

7-1 INDUCTORLESS TWO-PORTS

As discussed in Sec. 6-1, doubly loaded reactance ladder two-ports have attractive features, the most important being their low sensitivity to variations of all element values. However, the recent trend in electronics has been toward reducing the size of circuits, a trend which culminated in the development of integrated circuits. It proved to be relatively simple to reduce greatly the dimensions of resistors and capacitors; unfortunately, it proved impractical to achieve a comparable reduction in the size of inductors. The main reasons for this are the following:

1. No ferromagnetic effect can be obtained in semiconductors, which provide the building material of integrated circuits. Hence, both the magnetic material forming the core and the conductors forming the windings of the inductor must be deposited on the semiconductor surface. This arrangement results in inductors of very low inductance L and poor quality factor Q_L.

2. Even more fundamental is the inherent relation between the physical size of an inductor and its quality factor. To understand this relation, let us consider a solenoid (Fig. 7-1a). It is well known that the inductance of the solenoid is given by

$$L = \mu n^2 F(d/l)d \tag{7-1}$$

† The contents of this chapter have been greatly influenced by useful discussions with H. J. Orchard.

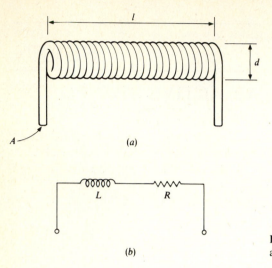

(a)

(b)

Figure 7-1 (a) Solenoid inductor; (b) equivalent circuit.

where μ = permeability of core material
n = number of turns
l = length
d = diameter
$F(d/l)$ = form factor
As indicated, F depends only on the ratio of the coil diameter to the coil length.

The imperfect conductivity of the wire forming the winding means that the inductor also has a resistance associated with it, given by

$$R = \rho \frac{l_w}{A} \tag{7-2}$$

where $l_w \approx n\pi d$ = total length of wire
A = area of cross section of wire
ρ = resistivity of wire material
An approximate equivalent circuit of the inductor is therefore that given in Fig. 7-1b. The inductor is lossy; the amount of loss at radian frequency ω is characterized by the quality factor

$$Q_L = \frac{\omega L}{R} \tag{7-3}$$

Assume now that the size of the inductor is reduced by replacing every linear dimension d, l, etc., by xd, xl, ..., where $0 < x < 1$. By (7-1) L will then become

$$L_x = \mu n^2 F(xd/xl)xd = xL \tag{7-4}$$

while R becomes, by (7-2),

$$R_x = \rho \frac{xl_w}{x^2 A} = \rho \frac{l_w}{xA} = \frac{R}{x} \tag{7-5}$$

(Note that A is scaled down to $x^2 A$, since it represents an area.) Therefore, the quality factor of the scaled-down inductor is

$$Q_{Lx} \triangleq \frac{\omega L_x}{R_x} = \frac{\omega x L}{R/x} = x^2 \frac{\omega L}{R} = x^2 Q_L \tag{7-6}$$

Hence, decreasing the physical size of an inductor by, say, a factor of 10 reduces Q_L by a factor of 100. Since good full-sized inductors have Q's of the order of 100 to 1000, a scaling down by 10 of their size would result in inductors with quality factors of 1 to 10. Such low Q values are in most applications unacceptable. Hence, tiny inductors are usually not practical. Since on the other hand the full-sized inductors are heavy and bulky, they have no place in a subminiature circuit.

By contrast, consider a planar capacitor such as shown in Fig. 7-2a. Its capacitance is

$$C = \varepsilon \frac{A}{d} \tag{7-7}$$

and the value of its leakage conductance is

$$G = \sigma \frac{A}{d} \tag{7-8}$$

where ε is the dielectric permittivity and $\sigma = 1/\rho$ is the conductivity of the dielectric material between the electrodes. Now the quality factor is

$$Q_c = \frac{\omega C}{G} = \omega \varepsilon \rho \tag{7-9}$$

which is independent of the physical dimensions. Hence, the size of a capacitor can be greatly reduced without significant reduction of Q_c.

In addition to the large size and weight of inductors, they also have the following drawbacks:

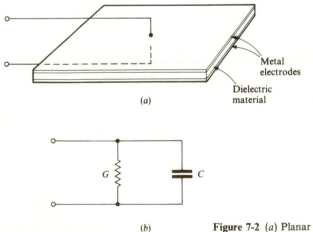

Metal electrodes

Dielectric material

(a)

(b)

Figure 7-2 (a) Planar capacitor; (b) equivalent circuit.

1. Even large inductors are quite lossy. The best attainable Q_L is about 1000; by contrast, capacitors with Q_c values of 5000 to 10,000 can be obtained.
2. For frequencies below, say, 20 Hz, the size and weight of inductors become preposterous; also, Q_L becomes very low. Hence, inductors are seldom used at such low frequencies.
3. Inductors are not off-the-shelf items. If a circuit designer requires, for example, an inductor of 8.16 μH to operate at 2 MHz with $Q_L \geq 200$, he must have the component specially built for his specifications. By contrast, resistors and capacitors are readily available for all usual requirements.
4. Inductors using ferromagnetic materials are basically nonlinear elements. Hence, unless the amplitude of the signal which they handle is kept small and direct currents are avoided, they generate harmonic distortion.
5. Inductors tend to act as small antennas, radiating as well as picking up electromagnetic waves. This can result in undesirable noise and coupling of signals in circuits containing inductors.

   In many applications, these disadvantages can be tolerated, or the low cost and low sensitivity of $RLCM$ circuits outweigh them. Hence, such passive networks still dominate the field as this book is being written. However, circuits in which inductors are replaced by active elements are now used in increasing numbers.

   To understand why active elements are usually needed in inductorless circuits, we must examine the limitations of passive RC two-ports. Consider the circuit of Fig. 7-3. If the two-port is described by its open-circuit impedance parameters, the port voltages and currents are related by

$$V_1 = z_{11} I_1 + z_{12} I_2 \qquad V_2 = z_{12} I_1 + z_{22} I_2 \tag{7-10}$$

Since $V_1 = E$ and $I_2 = 0$, the voltage ratio now is

$$A_V = \frac{V_2}{E} = \frac{z_{12}}{z_{11}} \tag{7-11}$$

To find the properties of z_{11} and z_{12}, we must carry out a derivation closely parallel to that done in Secs. 5-2 and 5-3. From (5-32) to (5-34), with $M_0(s) \equiv 0$,

$$\mathbf{I} \mathbf{Z}^T \mathbf{I}^* = F_0(s) + \frac{V_0(s)}{s} \tag{7-12}$$

where the right-hand side is an RC impedance function; from (5-35) to (5-38), it then follows that the auxiliary function

$$Z(s) \triangleq z_{11} a_1^2 + 2 z_{12} a_1 a_2 + z_{22} a_2^2 \tag{7-13}$$

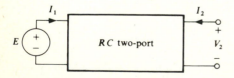

Figure 7-3 RC two-port with voltage-source input and open-circuit output.

must also be an RC impedance function. Since $Z(s)$, $z_{11}(s)$, and $z_{22}(s)$ are thus all RC impedances, they can be written in the partial-fraction form given in Eq. (3-39)

$$Z(s) = k_\infty + \frac{k_0}{s} + \sum_{i=1}^{n} \frac{k_i}{s + \sigma_i}$$

$$z_{jj}(s) = k_{jj}^\infty + \frac{k_{jj}^0}{s} + \sum_{i=1}^{n} \frac{k_{jj}^i}{s + \sigma_i} \qquad j = 1, 2$$

(7-14)

Here all constants (K, k, σ) are nonnegative.

It follows then, as in Sec. 5-3, that $z_{12}(s)$ must be in the form

$$z_{12}(s) = k_{12}^\infty + \frac{k_{12}^0}{s} + \sum_{i=1}^{n} \frac{k_{12}^i}{s + \sigma_i}$$

(7-15)

where all $\sigma_i > 0$ but where the real residues k_{12}^∞, k_{12}^0, and the k_{12}^i may be either positive, zero, or negative.

Substituting (7-14) and (7-15) into (7-13), multiplying by $s + \sigma_1$ (where σ_1 is an arbitrary pole), and substituting $s = -\sigma_1$ gives

$$K_1 = a_1^2 k_{11}^1 + 2a_1 a_2 k_{12}^1 + a_2^2 k_{22}^1$$

(7-16)

in close analogy to (5-46) to (5-48).

Since $K_1 \geq 0$, the right-hand side must be nonnegative for all real values of a_1 and a_2. Then it follows, as in the derivation of (5-41) to (5-45), that the residue condition (5-49) is valid for RC two-ports as well, i.e., that

$$k_{11}^1 \geq 0 \qquad k_{22}^1 \geq 0 \qquad k_{11}^1 k_{22}^1 - (k_{12}^1)^2 \geq 0$$

(7-17)

Similar results can be obtained for the k_{ij}^0 and k_{ij}^∞.

We conclude that, by analogy to Cauer's theorem given in Sec. 5-3 for RC two-ports, the following theorem holds.

Theorem The necessary conditions for a symmetric matrix

$$\mathbf{Z} = \begin{bmatrix} z_{11} & z_{12} \\ z_{12} & z_{22} \end{bmatrix}$$

(7-18)

to be the impedance matrix of an RC two-port are that its elements be given by the partial-fraction expansions (7-14) to (7-15), where all $\sigma_i > 0$, and that all residues be real and satisfy (7-17) at all poles.†

These conditions are sufficient as well; however, all we are concerned with here is their necessity. Returning to (7-11), we note that all poles of A_v are due to the zeros of z_{11}, since the poles of z_{12} are, by (7-17), all shared by z_{11}. Since z_{11} is

---

† Similar conditions can, of course, be proved for the y_{ij}. The only difference is in the form of the partial-fraction expansion, which is $y_{ij} = k_{ij}^0 + k_{ij}^\infty s + \sum_{m=1}^{n} k_{ij}^m s/(s + \sigma_m)$.

an RC impedance, all its zeros are simple and lie on the negative real axis; also $z_{11} \neq 0$ for $s = 0$, as (7-14) shows. Hence, all poles of A_v lie on the negative real axis ($-\sigma$ axis) of the s plane, excluding the origin. Furthermore, if $z_{11} \to 0$ for $s \to \infty$, then, by (7-14), $k_{11}^{\infty} = 0$. But, as (7-17) shows, then also $k_{12}^{\infty} = 0$, and $z_{12} \to 0$ for $s \to \infty$ as well. Hence, $A_v = z_{12}/z_{11}$ cannot have a pole for $s \to \infty$. We conclude that *all poles of A_v are simple and are located on the finite nonzero part of the $-\sigma$ axis* (Fig. 7-4).

By contrast, the zeros of $A_v(s)$ may lie anywhere, with the usual restriction that any complex zero must be accompanied by its conjugate. This follows from the real character of the residues of $z_{12}(s)$.

Three alternative loading conditions of the RC two-port are illustrated in Fig. 7-5. The transfer function for the circuit of Fig. 7-5a is

$$Y_T = \frac{I_2}{E} = y_{12} \tag{7-19}$$

For the circuit of Fig. 7-5b it is

$$Z_T = \frac{V_2}{I} = z_{12} \tag{7-20}$$

and for the circuit of Fig. 7-5c it is

$$A_I = \frac{I_2}{I} = -\frac{z_{12}}{z_{22}} \tag{7-21}$$

From the properties of the z_{ij} and the y_{ij}, it follows that all poles of all these transfer functions also lie on the negative σ axis, although $Y_T = y_{12}$ can have poles at $s \to \infty$ and $Z_T = z_{12}$ at $s = 0$.

Resistive terminations at either or both ports do not change the above properties, since they can be absorbed into the two-port without altering its RC character.

We conclude that *the poles of all transfer functions of an RC two-port are restricted to the negative σ axis. The zeros may lie anywhere in the complex plane.*

This restriction on the poles of the transfer function is a very serious one for filters or phase correctors which are used to change the amplitude and/or the

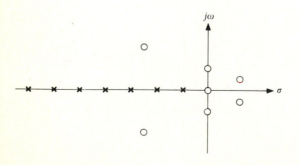

Figure 7-4 Zeros and poles of the voltage ratio $A_v(s)$ of an RC two-port.

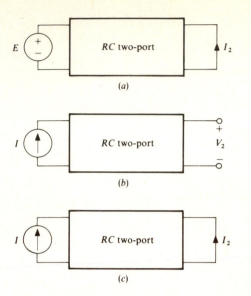

E $\quad$ RC two-port $\quad$ I_2

(a)

I $\quad$ RC two-port $\quad$ $+$ V_2 $-$

(b)

I $\quad$ RC two-port $\quad$ I_2

(c)

Figure 7-5 *RC* two-port under different input-output conditions.

phase response of an input signal along the $j\omega$ axis. To illustrate this point, we note that $A_v(s)$ is in the form

$$A_v(s) = \frac{N(s)}{\displaystyle\prod_{i=1}^{n}(s+\sigma_i)} \tag{7-22}$$

We can define the loss α_v and the phase β_v by

$$\alpha_v \triangleq -20 \log \left| \frac{V_2(j\omega)}{E(j\omega)} \right| = -20 \log |N(j\omega)| + \sum_{i=1}^{n} 20 \log |j\omega + \sigma_i| \tag{7-23}$$

and

$$\beta_v \triangleq -\underline{/V_2(j\omega)} + \underline{/E(j\omega)} = -\underline{/N(j\omega)} + \sum_{i=1}^{n} \underline{/(j\omega + \sigma_i)}$$

$$= -\underline{/N(j\omega)} + \sum_{i=1}^{n} \tan^{-1} \frac{\omega}{\sigma_i} \tag{7-24}$$

Hence, a pole at $s = -\sigma_i$ contributes the term $\alpha_i = 20 \log |j\omega + \sigma_i| = 10 \log (\omega^2 + \sigma_i^2)$ to the loss and the term $\beta_i = \tan^{-1} (\omega/\sigma_i)$ to the phase. As an illustration, these functions are shown, for $\sigma_i = 1$, in Fig. 7-6. Clearly, both curves are smooth and vary only slowly with frequency.

By contrast, a complex pole at $s = -\sigma_i + j\omega_i$ would contribute a factor $s + \sigma_i - j\omega_i$ to the denominator of $A_v(s)$. This factor, in turn, would give rise to a term

$$\alpha_i' = 20 \log |\sigma_i + j(\omega - \omega_i)| = 10 \log [\sigma_i^2 + (\omega - \omega_i)^2] \tag{7-25}$$

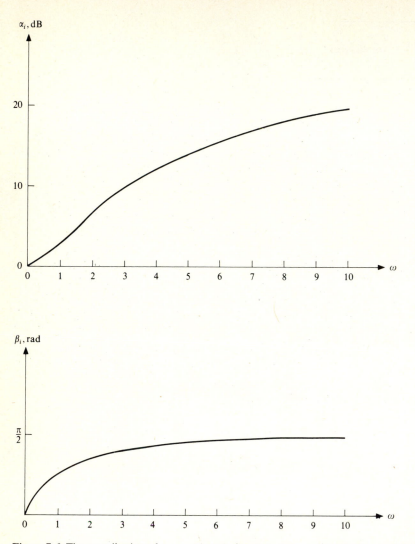

Figure 7-6 The contribution of a $-\sigma$ axis pole (located at $s = -1$) to the loss and the phase.

in the loss and a term

$$\beta'_i = \underline{/\,[\sigma_i + j(\omega - \omega_i)]} = \tan^{-1} \frac{\omega - \omega_i}{\sigma_i} \qquad (7\text{-}26)$$

in the phase. The corresponding responses are shown, for $\sigma_i = 0.1$ and $\omega_i = 1$, in Fig. 7-7. As comparison of the curves of Figs. 7-6 and 7-7 shows, complex poles are much more efficient than real ones in producing rapid variation of either loss or phase. Hence, it requires a large number of real-axis poles to achieve the result obtained by a few complex poles. Consequently, RC two-ports require a much larger number of elements than RLC ones to perform a similar filtering or phase-

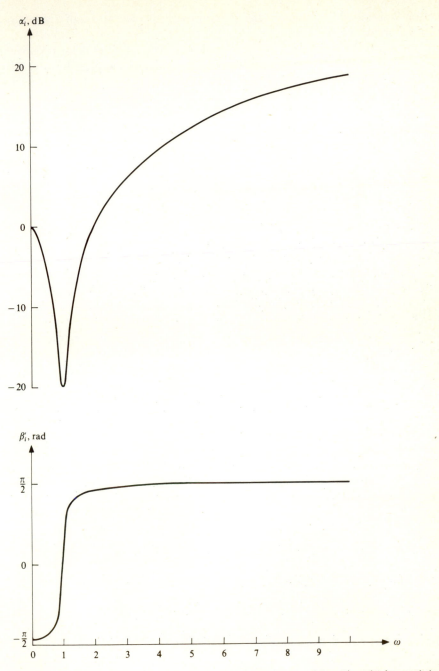

Figure 7-7 The contribution of a complex pole (located at $s = -0.1 + j1$) to the loss and the phase.

correcting function. Also, much of the signal power gets lost through dissipation in the resistors of the RC two-port.

We conclude that RC two-ports are inefficient as replacements for resistively terminated LC two-ports. As will be shown later in this chapter, however, by permitting active elements in the circuit, in addition to resistors and capacitors, we can obtain performance comparable to that of RLC circuits. Thus, in a loose manner of speaking, we can swap inductors for active components.

The active elements most useful in linear-circuit applications will be discussed in the next two sections.

7-2 OPERATIONAL AMPLIFIERS

The basic off-the-shelf active element available for linear-circuit design is the *operational amplifier*,† a complex integrated circuit which usually contains the equivalent of 10 to 30 transistors. However, for our purposes we can replace it by the simple equivalent circuit shown in Fig. 7-8a. Although the two-port appears to be resistive and frequency-independent, in fact its open-circuit voltage gain μ varies greatly with frequency; its amplitude response for a typical amplifier (the 741) is shown schematically in Fig. 7-8b. As the figure illustrates, the voltage gain is very large ($|\mu| > 100,000$) for very low frequencies; thereafter, the gain is divided by about a factor of 10 for an increase by a factor of 10 in frequency.‡ It reaches the value 1 around 1 MHz. The phase of $\mu(j\omega)$ is shown in Fig. 7-8c.

The input resistance R_i ranges from a few tens of kilohms to many megohms for various types of amplifiers ($R_i \approx 2$ MΩ for the 741); the output resistance from a few tens to a few hundred ohms ($R_0 \approx 75$ Ω for the 741). The values of R_i, R_0, and μ vary greatly from unit to unit even for the same type of amplifier. Fortunately, as will be shown below, in the usual applications of op-amps the exact values of these quantities are unimportant as long as $|\mu|$ is large enough.

Note that the lower terminal of the output port is shown grounded. This is because this terminal is usually chosen as the negative side of the dc power supply of the amplifier, and hence, unless a floating (ungrounded) power supply is permissible, it must be at ac ground potential.§ Note also that the structure of the circuit of Fig. 7-8a guarantees that $I_1 = I'_1$ and $I_2 = I'_2$ and hence that the circuit operates as a two-port as long as the model is valid.

Consider now the circuit of Fig. 7-9, representing an op-amp connected to a linear network. In the figure, the linear network N may contain active and passive elements but no generators. It is assumed that all natural modes of the complete circuit are inside the LHP of the s plane and hence the circuit is stable. Clearly, at each terminal pair of N the condition $I_i = I'_i$ holds; hence, N operates as a four-

---

† The name will be abbreviated to op-amp in our discussions.

‡ This corresponds to a slope of -20 dB/decade when gain and frequency are both expressed in logarithmic units.

§ The lower output terminal is also usually connected to the metal case of the device, which therefore must be insulated from ground if the output "floats," i.e., has no grounded terminal.

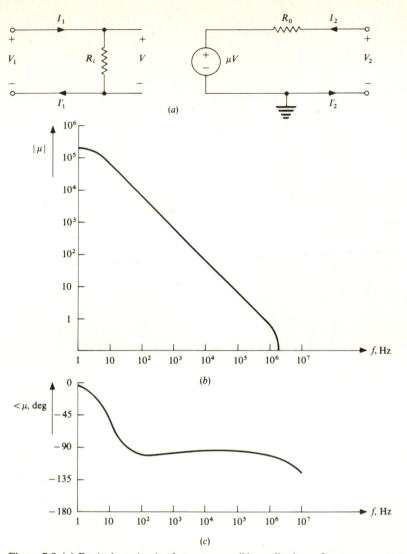

Figure 7-8 (a) Equivalent circuit of an op-amp; (b) amplitude vs. frequency response of the open-circuit gain μ; (c) phase vs. frequency response of the open-circuit gain μ.

port network. Since it is linear, we can obtain by superposition [in the same manner as we obtained Eqs. (5-4) and (5-10)]

$$
\begin{aligned}
I_3 &= y_{31} E_1 + y_{32} E_2 + y_{33} V_3 + y_{34} V_4 \\
I_4 &= y_{41} E_1 + y_{42} E_2 + y_{43} V_3 + y_{44} V_4
\end{aligned}
\tag{7-27}
$$

In addition, the branch relations of the op-amp connected between ports 3 and 4 are

$$
V_3 = -I_3 R_i \qquad V_4 = V_0 = \mu V_3 - I_4 R_0
\tag{7-28}
$$

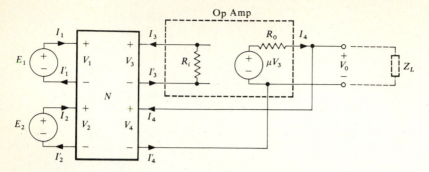

Figure 7-9 Linear circuit connected to an op-amp.

Expressing V_3 and I_3 in terms of V_4 and I_4 from (7-28) and substituting into (7-27) gives

$$\left[-\left(\frac{1}{R_i} + y_{33}\right)\delta - y_{34}\right]V_4 - R_0\delta\left(\frac{1}{R_i} + y_{33}\right)I_4 = I_{30}$$

$$(-\delta y_{43} - y_{44})V_4 + (1 - \delta R_0 y_{43})I_4 = I_{40}$$

(7-29)

where we use the abbreviated notations

$$\delta \triangleq \frac{1}{\mu} \qquad I_{30} \triangleq y_{31}E_1 + y_{32}E_2 \qquad I_{40} \triangleq y_{41}E_1 + y_{42}E_2 \qquad (7\text{-}30)$$

Solving (7-29) gives

$$V_4 = \frac{I_{30}(1 - \delta R_0 y_{43}) + I_{40}\delta(1/R_i + y_{33})}{-y_{34} + \delta R_0 y_{34} y_{43} - \delta R_0(1/R_i + y_{33})(y_{44} + 1/R_0)} \qquad (7\text{-}31)$$

We now make a number of plausible restrictions, shown below along with their physical interpretations:

1. $|I_{30}| < \infty$, $|I_{40}| < \infty$: the generators E_1, E_2 are not wired directly to ports 3 and 4.
2. $R_0 < \infty$, $R_i > 0$: the operational amplifier is not defective.
3. $|y_{33}| < \infty$, $|y_{43}| < \infty$, $|y_{44}| < \infty$: ports 3 and 4 are not short-circuited or wired directly to each other or to any other port.
4. $|y_{34}| > 0$: a voltage V_4 at port 4 will result in some nonzero current at port 3; that is, N provides a feedback path for the op-amp.
5. $\delta \to 0$: the open-circuit voltage gain is so high that all terms containing the factor $1/\mu$ become negligible compared with all other terms.

Under these assumptions, (7-31) becomes

$$V_4 \approx -\frac{I_{30}}{y_{34}} = -\frac{y_{31}E_1 + y_{32}E_2}{y_{34}} \qquad (7\text{-}32)$$

Substituting into the first equation of (7-27) leads to

$$I_3 = y_{31} E_1 + y_{32} E_2 + y_{33} V_3 - (y_{31} E_1 + y_{32} E_2)$$

$$I_3 - y_{33} V_3 = 0 \qquad\qquad\qquad (7\text{-}33)$$

Since $V_3 = -I_3 R_i$ by (7-28), this gives

$$(1 + y_{33} R_i) I_3 = 0 \qquad\qquad\qquad (7\text{-}34)$$

Hence, unless $y_{33} = -1/R_i$ (which indicates that a natural mode exists at the frequency of E_1 and E_2), we get

$$I_3 = 0 \qquad V_3 = 0 \qquad\qquad\qquad (7\text{-}35)$$

Thus, as Eqs. (7-32) to (7-35) indicate, *the output voltage $V_0 = V_4$ of the amplifier adjusts to a value which causes the input voltage and current of the amplifier to vanish.* This conclusion is independent of the exact values of R_i, R_0, and μ as long as $R_i > 0$, $R_0 < \infty$, $|\mu|$ is extremely large, and the phase angle of μ is such that the circuit remains stable. The coupling network N must also satisfy the obvious constraints 1 to 4 listed above.

The derivation was carried out for two independent sources E_1 and E_2 but can clearly be extended to any number of such sources. Also, N may contain other op-amps; of course the above conclusions hold for them as well.

Next, assume that an arbitrary load impedance Z_L is added to the circuit (Fig. 7-9) and $|Z_L| > 0$; then Z_L can be considered as an added component part of N. This addition clearly has the effect of replacing y_{44} by $y_{44} + 1/Z_L$. However, y_{31}, y_{32}, y_{33}, and y_{34} remain unchanged, as can be seen from (7-27). For example,

$$y_{31} = \left(\frac{I_3}{E_1}\right)_{E_2 = V_3 = V_4 = 0} \qquad\qquad (7\text{-}36)$$

which means that y_{31} is given by the current flowing through a short circuit placed across port 3 when $E_1 = 1$ V and ports 2 and 4 are short-circuited. Clearly, Z_L is then shunted by a short circuit and hence has no effect on y_{31}. Similar reasoning applies to y_{32} and y_{34}. By (7-32), therefore, $V_4 = V_0$, which depends on y_{31}, y_{32}, y_{34}, E_1, and E_2 only, does not change when Z_L is connected across the output port. We conclude that *the output terminals of the op-amp act as a voltage source under the conditions which led to (7-32).* Again, the result holds for any number of voltage generators driving N and is valid for the output terminals of any operational amplifier inside N.

Example 7-1 To illustrate how closely a physical device approaches the ideal situation described above, consider the integrator circuit shown in Fig. 7-10. Note that in the circuit the positive input terminal of the op-amp is grounded and the negative input terminal receives the input signal. The op-amp equivalent circuit uses the values of the 741 amplifier for $f = 1$ kHz, where $|\mu| \approx 1000$ and $\underline{/\mu} \approx -90°$, and so $\mu \approx -j1000$ (Fig. 7-8).

Analyzing the circuit for $R_L = 10$ kΩ gives $V_i \approx 8.011 \times 10^{-4} + j6.681 \times 10^{-6}$ V, $I_i \approx 4.01 \times 10^{-10} + j3.34 \times 10^{-12}$ A, and $V_0 \approx 8.06 \times 10^{-4} + j0.795$ V. Thus, $|V_i| \approx 0.8$ mV and $|I_i| \approx 400$ pA, both negligible compared with the other voltages and currents in the circuit. Changing R_L to 10 Ω still leaves the inputs V_i and I_i negligibly small; the

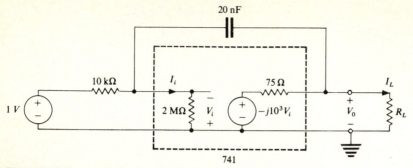

Figure 7-10 Integrator circuit using 741 op-amp.

output voltage V_0 changes to $6.68 \times 10^{-3} + j0.790$ V. Hence, reducing R_L by a factor of 10^{-3} reduces $|V_0|$ by less than 1 percent.

We conclude that (at least for this example) the behavior of the op-amp is close to being ideal.

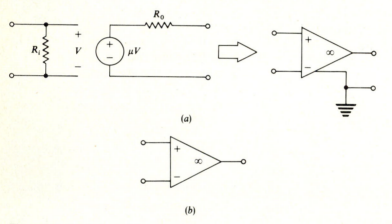

(a)

(b)

Figure 7-11 Symbols used for op-amps.

Since under the conditions stated before the behavior of the op-amp is very nearly independent of the values of R_i, R_0, and μ, in the rest of this chapter we shall use the simple symbol shown in Fig. 7-11a for the op-amp. Often the lower (grounded) output terminal will also be omitted and the symbol of Fig. 7-11b will be used.

The use of op-amps to build more elaborate and useful active devices will be discussed in the next section.

7-3 ACTIVE BUILDING BLOCKS

In this section, we show how op-amps can be used to construct controlled sources as well as impedance converters and inverters, circuits which are directly useful as building blocks in the design of linear active networks. We shall begin with a discussion of controlled-source realizations.

Controlled voltage sources can easily be realized using op-amps, as illustrated in Fig. 7-12. Figure 7-12a shows a positive-gain voltage-controlled voltage source (VCVS). By our earlier results, $V \approx V_1$ and $I = I' \approx 0$. Since R_1 and R_2 form a voltage divider,

$$V_1 \approx V \approx V_2 \frac{R_1}{R_1 + R_2} \tag{7-37}$$

and hence†

$$V_2 \approx \left(\frac{R_2}{R_1} + 1\right) V_1 = \mu V_1 \tag{7-38}$$

† The gain μ of the VCVS realized by the circuit should not be confused with the gain of the VCVS in the op-amp model of Fig. 7-8a.

Figure 7-12 VCVS and CCVS circuits using op-amps.

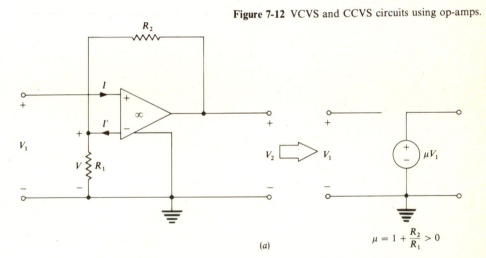

$$\mu = 1 + \frac{R_2}{R_1} > 0$$

(a)

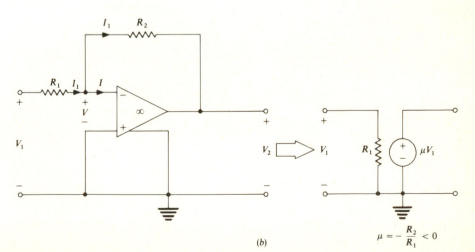

$$\mu = -\frac{R_2}{R_1} < 0$$

(b)

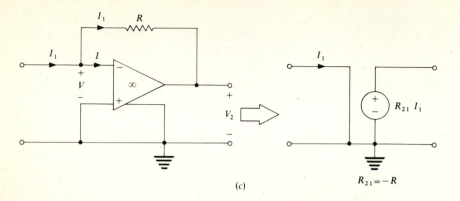

(c)

$R_{21} = -R$

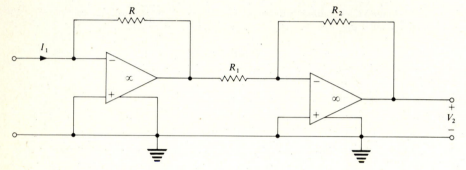

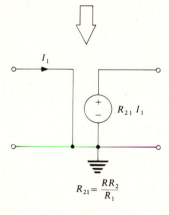

$$R_{21} = \frac{R R_2}{R_1}$$

(d)

Figure 7-12 (*continued*) VCVS and CCVS circuits using op-amps.

where $\mu > 0$, so that the circuit is a noninverting VCVS. An inverting VCVS is shown in Fig. 7-12b. Since $I \approx 0$, the currents through R_1 and R_2 have the same value I_1. Also, since $V \approx 0$,

$$V_1 - V = I_1 R_1 \approx V_1 \qquad V - V_2 = I_1 R_2 \approx -V_2 \qquad (7\text{-}39)$$

so that
$$V_2 \approx -I_1 R_2 \approx -\frac{R_2}{R_1} V_1 \qquad (7\text{-}40)$$

Hence this circuit is also a VCVS with $\mu = -R_2/R_1 < 0$.

Figure 7-12c shows a current-controlled voltage source (CCVS). Here

$$V_2 - V \approx V_2 \approx -I_1 R \qquad (7\text{-}41)$$

Hence the gain is $R_{21} = -R$. If a positive gain is required, the inverting circuit of Fig. 7-12b can be cascaded with that of Fig. 7-12c. The resulting circuit is shown in Fig. 7-12d; the gain is clearly

$$R_{21} \triangleq \frac{V_2}{I_1} = \frac{R_2 R}{R_1} \qquad (7\text{-}42)$$

The realization of controlled current sources using op-amps tends to be more difficult, since the op-amp is by nature a voltage source. Some circuits are shown in Fig. 7-13. A voltage-controlled current source (VCCS) is shown in Fig. 7-13a; clearly,

$$V_1 \approx I_2 R \qquad (7\text{-}43)$$

and hence $I_2 \approx (1/R)V_1$, so that the circuit is a VCCS with a gain $G = 1/R > 0$. Negative gain can be obtained by using a voltage inverter (Fig. 7-12b) at the input of the circuit. Note that the load Z_L connected to the current source must be floating (not grounded). The potentials of the output terminals are V_1 and $(1 + Z_L/R)V_1$, as a comparison with Fig. 7-12a shows.

If a grounded load must be driven by the VCCS, the circuit of Fig. 7-13b can be used. This has the disadvantage that the lower output terminal of the op-amp must float. As explained in the previous section, this necessitates a floating power supply and insulated case for the op-amp, which are seldom permissible. Since Eq. (7-43) holds here as well, $G = 1/R$ is the gain of this circuit.

Figure 7-13c illustrates a current-controlled current source (CCCS). From the figure,

$$V_1 - V \approx -V = I_1 R_1 \qquad V \approx (I_1 + I_2)R_2 \qquad (7\text{-}44)$$

Hence
$$\frac{(I_1 + I_2)R_2}{I_1 R_1} = \left(1 + \frac{I_2}{I_1}\right)\frac{R_2}{R_1} \approx -1 \qquad I_2 = -\left(\frac{R_1}{R_2} + 1\right)I_1 \qquad (7\text{-}45)$$

Thus, the current gain is $-(R_1/R_2 + 1) < 0$.

Again, the output of the op-amp floats in the circuit, with the described disadvantages. A CCCS circuit which has grounded op-amp output but floating load is shown in Fig. 7-13d. Equations (7-44) and (7-45) remain valid for this circuit, and therefore the current gain is again $-(R_1/R_2 + 1)$.

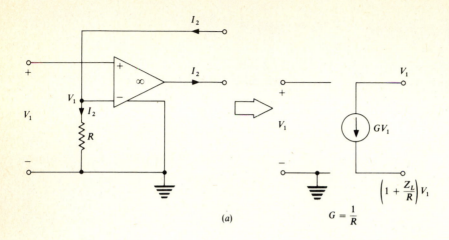

$$(a) \qquad G = \frac{1}{R}$$

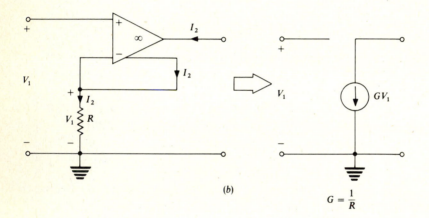

$$(b) \qquad G = \frac{1}{R}$$

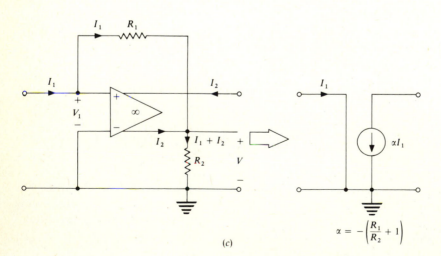

$$(c) \qquad \alpha = -\left(\frac{R_1}{R_2} + 1\right)$$

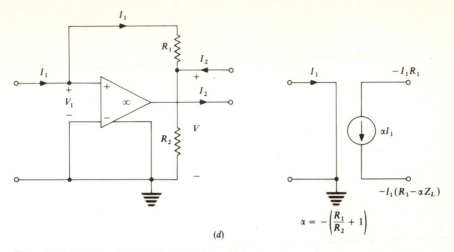

$$\alpha = -\left(\frac{R_1}{R_2} + 1\right)$$

(d)

Figure 7-13 VCCS and CCCS circuits using op-amps.

An important building block for linear active circuits is the *impedance converter* (Fig. 7-14), also often called *generalized impedance converter* (GIC). This is a two-port characterized by the property that if the secondary port is terminated by an impedance Z_{L_2}, the impedance Z_1 seen at its primary port is proportional to Z_{L_2}:

$$Z_1 = f(s)Z_{L_2} \qquad (7\text{-}46)$$

Here, as indicated, the proportionality factor† $f(s)$ is in general a function of s.

Let the GIC be described by its chain parameters. By Eq. (5-24),

$$V_1 = AV_2 - BI_2 \qquad I_1 = CV_2 - DI_2 \qquad (7\text{-}47)$$

If Z_{L_2} is connected to the secondary port, then $V_2 = -Z_{L_2}I_2$ and

$$Z_1 = \frac{V_1}{I_1} = \frac{-AZ_{L_2}I_2 - BI_2}{-CZ_{L_2}I_2 - DI_2} = \frac{AZ_{L_2} + B}{CZ_{L_2} + D} \qquad (7\text{-}48)$$

For (7-46) to hold, we must have

$$B = 0 \qquad C = 0 \qquad Z_1 = \frac{A}{D}Z_{L_2} \qquad (7\text{-}49)$$

† $f(s)$, often called the *conversion factor*, is a dimensionless quantity.

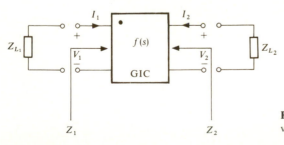

Figure 7-14 Generalized impedance converter.

Hence, $f(s) = A/D$. For a prescribed $f(s)$, we may choose $A = 1$ and $D = 1/f(s)$, or $A = f(s)$ and $D = 1$. In the former case $V_1 = V_2$ and $I_1 = -I_2/f(s)$; in the latter, $I_1 = -I_2$ and $V_1 = f(s)V_2$. The circuit described by the former is called a current-conversion GIC (CGIC); the latter, a voltage-conversion GIC (VGIC).

An interesting special case of the GIC is the *negative-impedance converter* (NIC). Here, $f(s) \equiv -1$, and hence $Z_1 = -Z_{L_2}$. Again, we call it a VNIC if $I_1 = -I_2$ and $V_1 = -V_2$; we call it a CNIC if $I_1 = I_2$ and $V_1 = V_2$. A realization of the CNIC using op-amps is shown in Fig. 7-15a. Since $V \approx 0$ at the op-amps input, clearly $V_1 \approx V_2$; also,

$$V_1 - V_0 = I_1 R_1 \qquad V_2 - V_0 = I_2 R_2 \qquad \frac{I_1}{I_2} = \frac{R_2}{R_1} \qquad (7\text{-}50)$$

Hence, for $R_1 = R_2$ we obtain the CNIC relations.

A VNIC realization is shown in Fig. 7-15b. From the circuit, we can write the following relations. For the op-amp

$$V \approx 0 \qquad I \approx 0$$

From the KCL for node A

$$-I_1 - I_0 - I_2 + I_0 = 0 \qquad I_1 = -I_2 \qquad (7\text{-}51)$$

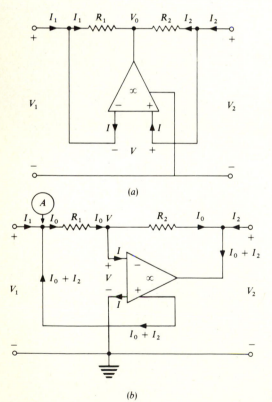

(a)

(b)

Figure 7-15 CNIC and VNIC realizations.

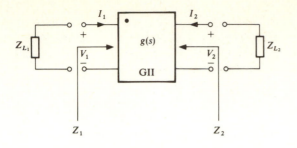

Figure 7-16 Generalized impedance inverter.

and from the KVL, for $R_1 = R_2 = R$,

$$V_1 - I_0 R_1 = V \approx 0 \qquad V_2 + I_0 R_2 = V \approx 0 \qquad V_1 + V_2 \approx 0 \qquad V_1 = -V_2$$
$$(7\text{-}52)$$

Hence, the VNIC relations are obtained. Note that this circuit requires a floating op-amp output or (if the ground is moved to the lower output terminal of the op-amp) a floating generator and load.

A circuit closely related to the GIC is the *generalized impedance inverter* (GII). The GII is a two-port characterized by the property that if its secondary port is terminated by Z_{L_2}, the impedance at its primary port is inversely proportional to Z_{L_2}. Hence (Fig. 7-16)

$$Z_1 = \frac{g(s)}{Z_{L_2}} \tag{7-53}$$

where $g(s)$, the *inversion factor*, has the dimension ohms^2.

From (7-48) and (7-53), it follows that

$$A = 0 \qquad D = 0 \qquad Z_1 = \frac{B}{C Z_{L_2}} \tag{7-54}$$

Hence, $g(s) = B/C$.

An important special case is obtained for $g(s) = R^2$, where R is a positive real constant. Then choosing $B = R_1$, $C = R_1/R^2 = 1/R_2$, where R_1 and $R_2 \triangleq R^2/R_1$ are positive real constants, we see that (7-47) becomes

$$V_1 = -R_1 I_2 \qquad I_1 = \frac{1}{R_2} V_2 \tag{7-55}$$

The two-port described by (7-55) is called a *gyrator*. It is usually represented by the symbol shown in Fig. 7-17a.

The total power entering the gyrator is

$$P = I_1 V_1 + I_2 V_2 = -I_1 R_1 I_2 + I_2 R_2 I_1 = I_1 I_2 (R_2 - R_1) \tag{7-56}$$

Since the two-port is passive if and only if $P \geq 0$ for all values of I_1 and I_2, and since $I_1 I_2$ may be either positive or negative, we conclude that the gyrator is passive only if $R_1 = R_2 = R$. Then the power entering the gyrator is always zero,

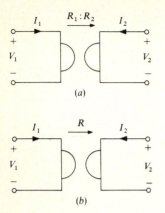

(a)

(b)

Figure 7-17 Symbols for (a) general and (b) lossless gyrator.

and by the definition given in Sec. 1-1 the gyrator is *lossless*.† From Eqs. (5-28) and (5-29) the immittance parameters of the lossless gyrator are

$$z_{11} = z_{22} = 0 \qquad z_{12} = -B = -R \qquad z_{21} = \frac{1}{C} = R \qquad (7\text{-}57)$$

and

$$y_{11} = y_{22} = 0 \qquad y_{12} = C = \frac{1}{R} \qquad y_{21} = -\frac{1}{B} = -\frac{1}{R} \qquad (7\text{-}58)$$

Since $z_{12} = -z_{21}$ and $y_{12} = -y_{21}$, the gyrator is a nonreciprocal two-port. Note also that all four driving-point immittance parameters z_{11}, z_{22}, y_{11}, and y_{22} of the gyrator are zero.

The symbol for the lossless gyrator is shown in Fig. 7-17b.

An important application for the gyrator can immediately be deduced from (7-53) for $g(s) = R^2$. Let the gyrator be loaded by a capacitor, so that $Z_{L_2} = 1/sC$. Then, by (7-53),

$$Z_1 = \frac{R^2}{1/sC} = sR^2C = sL_{eq} \qquad (7\text{-}59)$$

where the equivalent inductance L_{eq} is given by $L_{eq} = R^2C$. We conclude that *the input impedance of a capacitively loaded gyrator is equivalent to the impedance of an inductor*, and, vice versa, if the gyrator is terminated in an inductor, the input impedance will be capacitive. This last property is seldom needed, however; in practical applications we are trying to get rid of inductors and replace them by capacitors, for the reasons discussed in Sec. 7-1.

A conceptually simple gyrator realization, which still introduces several practical problems, can be deduced directly from (7-58). Since

$$I_1 = y_{11}V_1 + y_{12}V_2 = \frac{1}{R}V_2$$

$$I_2 = y_{21}V_1 + y_{22}V_2 = -\frac{1}{R}V_1 \qquad (7\text{-}60)$$

† A lossless gyrator is also called sometimes an *ideal gyrator*.

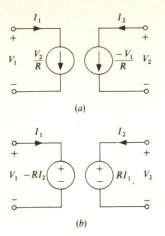

(a)

(b)

Figure 7-18 Ideal gyrator circuits.

the circuit of Fig. 7-18a is an ideal gyrator. The circuit requires two VCCSs, which necessitate either floating loads or floating op-amp outputs when realized with the single op-amp circuits shown in Fig. 7-13a and b.

Another simple gyrator circuit is shown in Fig. 7-18b. It is based on the relations

$$V_1 = z_{11}I_1 + z_{12}I_2 = -RI_2$$
$$V_2 = z_{21}I_1 + z_{22}I_2 = RI_1 \tag{7-61}$$

obtainable from (7-57). Unfortunately, in this circuit either the controlling currents must be in a floating branch or the controlled voltage sources must be floating. Hence, the simple op-amp realizations shown in Fig. 7-12c and d for the CCVSs cannot be used in the gyrator realization.

A circuit which uses two grounded-output op-amps and is useful for the realization of either GICs or GIIs is shown in Fig. 7-19a.† The input impedance Z can easily be found, as follows. When we recall that the input voltage of an op-amp is very nearly zero,

$$V \approx V_2 \approx V_4 \tag{7-62}$$

is obtained. Also, if we denote the current through Z_1 by I_1 (with the reference direction pointing left to right), the current through Z_2 by I_2, etc., clearly

$$
\begin{aligned}
I_1 &\approx I & V - V_1 &= I_1 Z_1 \approx V_2 - V_1 = -I_2 Z_2 \\
I_3 &\approx I_2 & V_2 - V_3 &= I_3 Z_3 \approx V_4 - V_3 = -I_4 Z_4 \\
I_5 &\approx I_4 & V &\approx V_4 = I_5 Z_5
\end{aligned}
\tag{7-63}
$$

Here we assumed, as usual, that the current in the input leads of the op-amps is zero.

---

† This circuit, a modified version of that described in Ref. 8, is called *Riordan's circuit.*

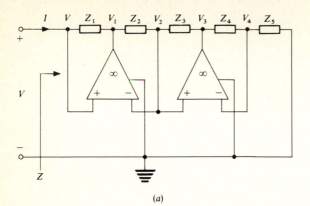

(a)

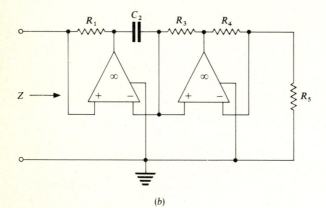

(b)

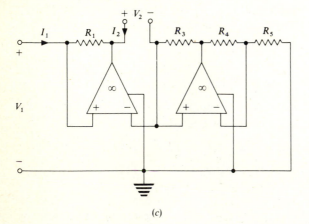

(c)

Figure 7-19 The Riordan circuit: (a) basic circuit; (b) used as an inductor; (c) used as a gyrator.

Working backward in (7-63) leads to

$$V \approx I_5 Z_5 \approx I_4 Z_5 \approx -I_3 \frac{Z_3}{Z_4} Z_5 \approx -I_2 \frac{Z_3}{Z_4} Z_5 \approx I_1 \frac{Z_1 Z_3}{Z_2 Z_4} Z_5 \approx I \frac{Z_1 Z_3 Z_5}{Z_2 Z_4}$$

$$(7\text{-}64)$$

Hence
$$Z = \frac{V}{I} \approx \frac{Z_1 Z_3 Z_5}{Z_2 Z_4} \tag{7-65}$$

If Z_5 is regarded as a load impedance, the circuit behaves like a GIC; (7-46) takes the form

$$Z(s) = f(s) Z_5(s) \qquad f(s) \equiv \frac{Z_1(s) Z_3(s)}{Z_2(s) Z_4(s)} \tag{7-66}$$

If, for example, $Z_1 = R_1$, $Z_2 = 1/sC_2$, $Z_3 = R_3$, $Z_4 = R_4$, and $Z_5 = R_5$ (Fig. 7-19b), then $f(s) = R_1 R_3 / [(1/sC_2)R_4]$ and

$$Z = \frac{R_1 R_3}{(1/sC_2)R_4} R_5 = s \frac{R_1 C_2 R_3 R_5}{R_4} \tag{7-67}$$

Hence, the input impedance is that of an *inductor*, with an equivalent inductance value $L_{eq} = R_1 C_2 R_3 R_5 / R_4$.

As (7-67) suggests, and as can be directly verified from (7-65), the two-port formed by regarding the terminals of Z_2 as an output port is a *gyrator* if all other impedances are purely resistive (Fig. 7-19c). More generally, if the terminals of Z_5 (or Z_1 or Z_3) constitute the output port, the circuit of Fig. 7-19a is a GIC; if the terminals of Z_2 (or Z_4) form the output port, the resulting two-port is a GII.

Assume now that we choose Z_2 and Z_4 as capacitive and Z_1, Z_3, and Z_5 as resistive impedances. Then (7-65) gives, for $s = j\omega$,

$$Z(j\omega) = \frac{R_1 R_3 R_5}{(1/j\omega C_2)(1/j\omega C_4)} = -\omega^2 R_1 C_2 R_3 C_4 R_5 \tag{7-68}$$

We note that $Z(j\omega)$ is pure real, negative, and a function of ω. Such an impedance† is called a *frequency-dependent negative resistance* (FDNR). A slightly different form of FDNR can be obtained, e.g., by choosing Z_1 and Z_3 as capacitors and Z_2, Z_4, and Z_5 as resistors. Then

$$Z(j\omega) = -\frac{R_5}{C_1 R_2 C_3 R_4} \frac{1}{\omega^2} \tag{7-69}$$

As we shall see later, FDNRs are very useful for the design of active filters.

Finally, it should be noted that the task of the op-amps in the circuit of Fig. 7-19a is to assure (1) that V, V_2, and V_4 are equal to each other and (2) that I

---

† Naturally, the impedance functions realized with these active networks are not PR functions; here, Re $Z(j\omega) < 0$.

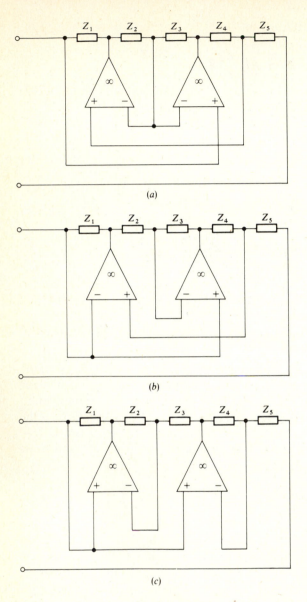

(a)

(b)

(c)

Figure 7-20 Circuits equivalent to the network of Fig. 7-19a.

and I_1, I_2 and I_3, and I_4 and I_5 are pairwise equal. This can also be achieved by connecting the op-amps in different ways; as long as the resulting circuit is stable, the operation of the circuit will be the same as that of Fig. 7-19a. Three such equivalent circuits are shown in Fig. 7-20.†

In the next sections, we discuss active-circuit design techniques which utilize the building blocks developed in this section.

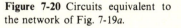

† The circuit of Fig. 7-20c is the original Riordan's circuit.[8] It should be noted that the stability properties of these circuits differ; see Ref. 24 for a thorough discussion.

7-4 LINEAR ACTIVE-CIRCUIT DESIGN USING INDUCTANCE SIMULATION

In the next sections, we shall describe briefly some methods for the design of active linear networks. While the number of design techniques described in the literature is very large, only a few will be discussed here; they have proved to be practically useful and illustrate important concepts. Specifically, the following methods are considered:

1. Inductance simulation
2. LC ladder simulation using impedance scaling or analog computation
3. Cascade realization

The first of these will be discussed in this section. Conceptually, inductance simulation is perhaps the most obvious method for designing inductorless filters. Its theory[9] is based on the observations made in Sec. 6-1 about the low sensitivity of doubly terminated reactance two-ports to element-value variations. If each inductor in the circuit is replaced by a gyrator terminated by a capacitor, this low-sensitivity property is extended to the variations of the gyration resistance R and the terminating capacitor C.

In practice, the open-circuit impedances of the gyrator will not exactly satisfy Eq. (7-57). In particular, z_{11} and z_{22} will not be exactly zero. Assume that

$$z_{11} = z_{22} = \varepsilon R \qquad 0 < \varepsilon \ll 1 \tag{7-70}$$

What will be the effect on the simulated inductor?

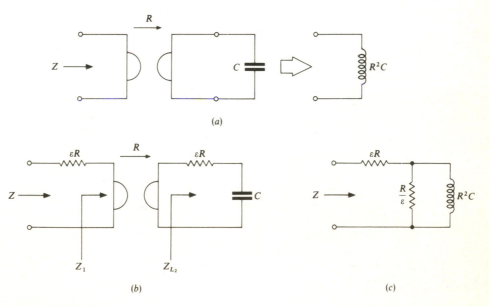

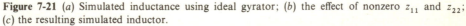

Figure 7-21 (a) Simulated inductance using ideal gyrator; (b) the effect of nonzero z_{11} and z_{22}; (c) the resulting simulated inductor.

As shown in Chap. 5 (see Fig. 5-13), an increment of z_{11} (or z_{22}) can be represented as a series impedance at the input (or output) port of the two-port. Hence, the effect of the nonzero z_{11} and z_{22} is to replace the ideal circuit of Fig. 7-21a by that of Fig. 7-21b. Since the new load impedance

$$Z_{L_2} = \varepsilon R + \frac{1}{sC} \tag{7-71}$$

is transformed into

$$Z_1 = \frac{R^2}{Z_{L_2}} = \frac{R^2}{\varepsilon R + 1/sC} = \frac{1}{\varepsilon/R + 1/sR^2C} \tag{7-72}$$

the equivalent impedance realized is that shown in Fig. 7-21c. Now

$$Z = \varepsilon R + \frac{sR^2C}{1 + \varepsilon sRC} \approx \varepsilon R + sR^2C(1 - \varepsilon sRC) \tag{7-73}$$

Hence, for $s = j\omega$,

$$Z(j\omega) \approx \varepsilon R[1 + (\omega RC)^2] + j\omega R^2C \tag{7-74}$$

Thus $Z(j\omega)$ has a real part due to the incremental impedances $z_{11} = z_{22} = \varepsilon R$. This indicates that the simulated inductor is *lossy*; its quality factor is given by

$$Q \triangleq \frac{\operatorname{Im} Z(j\omega)}{\operatorname{Re} Z(j\omega)} = \frac{1}{\varepsilon} \frac{\omega RC}{1 + (\omega RC)^2} \tag{7-75}$$

The $Q(\omega)$ curve is plotted in Fig. 7-22; the maximum value $1/2\varepsilon$ is obtained at $\omega RC = 1$.

A second imperfection of practical gyrators is that the gyration resistance R may be complex. If it is in the form

$$R_\varphi = Re^{j\varphi} \qquad |\varphi| \ll \pi/2 \tag{7-76}$$

the simulated impedance becomes

$$Z_\varphi = \frac{R_\varphi^2}{1/j\omega C} = j\omega R^2 C e^{j2\varphi} \qquad Z_\varphi \approx j\omega R^2 C(1 + j2\varphi)$$
$$Z_\varphi \approx -2\omega\varphi R^2 C + j\omega R^2 C \tag{7-77}$$

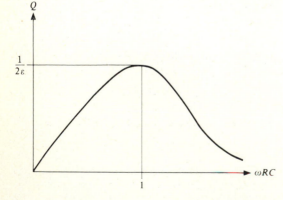

Figure 7-22 $Q(\omega)$ curve for simulated inductor.

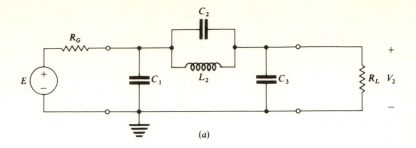

(a)

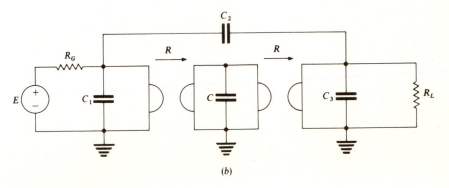

(b)

Figure 7-23 Floating-inductor simulation using grounded gyrators.

Hence, the quality factor is

$$Q = -\frac{1}{2\varphi} \tag{7-78}$$

which is negative for positive φ. Thus, the effects of the nonzero z_{11}, z_{22} and the complex gyration resistance may, to some extent, cancel.

Another difficulty is presented by the fact that for practical gyrators (like the one in Fig. 7-19c) the input port is grounded. Hence, if the inductance to be simulated does not have a grounded terminal, e.g., L_2 in the circuit of Fig. 7-23a, a direct simulation in the form of Fig. 7-21a is not feasible.

An ingenious scheme[10] to overcome this problem is illustrated in Fig. 7-24a. The chain matrix of the two lossless gyrators is, by (7-47) and (7-55),

$$\mathbf{T}_g = \begin{bmatrix} 0 & R \\ \dfrac{1}{R} & 0 \end{bmatrix} \tag{7-79}$$

The chain matrix of the two-port containing only the shunt capacitor C, as can readily be verified from Eq. (5-27), is

$$\mathbf{T}_c = \begin{bmatrix} 1 & 0 \\ sC & 1 \end{bmatrix} \tag{7-80}$$

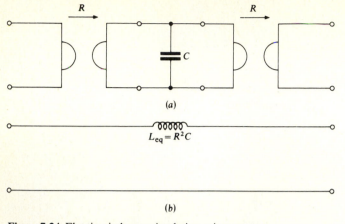

(a)

$L_{eq} = R^2 C$

(b)

Figure 7-24 Floating-inductor simulation using gyrators.

Since the chain matrices of cascaded two-ports are multiplied together to obtain the chain matrix of the cascade,† the two-port of Fig. 7-24a has the chain matrix

$$\mathbf{T} = \begin{bmatrix} 0 & R \\ \dfrac{1}{R} & 0 \end{bmatrix} \begin{bmatrix} 1 & 0 \\ sC & 1 \end{bmatrix} \begin{bmatrix} 0 & R \\ \dfrac{1}{R} & 0 \end{bmatrix} \qquad \mathbf{T} = \begin{bmatrix} 1 & sR^2C \\ 0 & 1 \end{bmatrix} \qquad (7\text{-}81)$$

It is easy to verify that the two-port containing the floating inductor $L_{eq} = R^2C$ shown in Fig. 7-24b has this same chain matrix **T**. Hence, the two two-ports of Fig. 7-24 are equivalent. With this process a floating inductor can therefore be replaced by two gyrators and a capacitor.

Example 7-2 For the filter circuit shown in Fig. 7-23a, L_2 can be replaced using the equivalence of Fig. 7-24. The resulting circuit is shown in Fig. 7-23b. If we choose R as the geometric mean of R_G and R_L, then

$$R = \sqrt{R_G R_L} \qquad L_{eq} = L_2 = R^2 C = R_G R_L C \qquad C = \frac{L_2}{R_G R_L} \qquad (7\text{-}82)$$

The circuit of Fig. 7-24a requires *two* gyrators to replace *one* floating inductor. In addition, both ports of both gyrators must have a grounded terminal. This rules out the use of the Riordan gyrator of Fig. 7-19c in this application. More complicated gyrators can be constructed which do have grounded ports.[12] Alternatively, it is possible to use the Riordan circuit as a GIC by omitting the load R_5 (Fig. 7-19b). The two-port thus obtained has the conversion factor

$$f(s) = s\frac{R_1 C_2 R_3}{R_4} = Ks \qquad (7\text{-}83)$$

† See, for example, pp. 746–748 of Ref. 11.

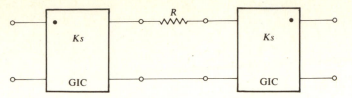

Figure 7-25 Floating-inductor simulation using GICs.

as shown earlier in Eqs. (7-66) and (7-67). Consider now the circuit of Fig. 7-25. Deriving and multiplying together the chain matrices of the individual two-ports, we easily find the overall chain matrix to be

$$\mathbf{T} = \begin{bmatrix} 1 & sKR \\ 0 & 1 \end{bmatrix} \tag{7-84}$$

Hence, choosing

$$KR = \frac{R_1 C_2 R_3 R}{R_4} = L_{eq} \tag{7-85}$$

we again obtain the chain matrix of the two-port of Fig. 7-24*b*.

Other methods are also available to simulate a floating inductor, the conceptually simplest one being the use of a floating gyrator terminated by a capacitor and fed by a separate floating power supply.[4] This method requires some auxiliary bias-control circuitry and puts restrictions on the op-amps which can be used in the gyrators. Nevertheless, it is a practical technique.

7-5 *LC* LADDER SIMULATION

As discussed in Chaps. 5 and 6, doubly terminated *LC* ladder networks have some very desirable properties. In the frequency regions where they transmit all (or almost all) input power, they have very low sensitivities, as the discussions of Sec. 6-1 showed. Furthermore, since each of their transmission zeros is usually produced by a single branch, these zeros are easy to tune and also quite stable. Thus, such networks tend to give reliable response in both their low-loss (passband) and high-loss (stopband) frequency regions. This is the motivation for the methods discussed in this section, which simulate such networks with equivalent inductorless circuits.

The first method, due to Bruton,[13] is based on a transformation (scaling) of all impedances in the ladder. Consider the ladder network of Fig. 7-26, in which Z_1, $Z_2, \ldots, Z_N$ are *LC* impedances. Let all currents in the circuit be calculated using, for example, mesh analysis (Ref. 11, pp. 461–462). The loop currents J_i satisfy

$$Z_{11}J_1 + Z_{12}J_2 + \cdots + Z_{1l}J_l = E$$

$$Z_{21}J_1 + Z_{22}J_2 + \cdots + Z_{2l}J_l = 0 \tag{7-86}$$

$$\cdots\cdots\cdots\cdots\cdots\cdots\cdots\cdots\cdots\cdots\cdots$$

$$Z_{l1}J_1 + Z_{l2}J_2 + \cdots + Z_{ll}J_l = 0$$

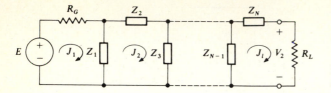

Figure 7-26 Doubly terminated *LC* ladder.

Here, $l = N/2 + 1$ is the number of loops in the circuit (Fig. 7-26), $Z_{11} = R_G + Z_1$ the total impedance in loop 1, $Z_{12} = -Z_1$ the negative of the total impedance common to loops 1 and 2, etc. The output voltage is clearly $V_2 = J_l R_L$.

Assume now that all impedances in the circuit (including R_G and R_L) are divided by s. Then all Z_{ij} are similarly divided by s; since the right-hand side of (7-86) remains unchanged, this means that all J_i must be multiplied by s to obtain the new loop currents J_i'. The new output voltage is therefore

$$V_2' = J_l' \frac{R_L}{s} = sJ_l \frac{R_L}{s} = J_l R_L = V_2 \tag{7-87}$$

We conclude that *the output voltage V_2 (and hence the voltage ratio $A_V \triangleq V_2/E$) remains unchanged when all impedances are scaled by the factor $1/s$ in the circuit.*

The change in the character of the circuit elements due to this scaling is illustrated in Fig. 7-27. Clearly, resistors and inductors become capacitors and resistors, respectively, in the scaled circuit. Capacitors become nonphysical elements† whose impedance, for $s = j\omega$, is

$$Z' = \frac{1}{s^2 D'} = -\frac{1}{\omega^2 D'} \tag{7-88}$$

Comparison with Eq. (7-69) shows that this new element is a *frequency-dependent negative resistor* (FDNR). We have already shown in Sec. 7-3 how such

† That is, not realizable with passive devices.

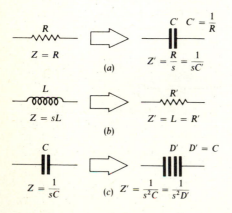

Figure 7-27 The effect of scaling by $1/s$ on circuit elements.

an FDNR can be realized using the Riordan circuit of Fig. 7-19a with $Z_1 = 1/sC_1$, $Z_3 = 1/sC_3$, $Z_2 = R_2$, $Z_4 = R_4$, and $Z_5 = R_5$. From (7-69), then,

$$D' = \frac{C_1 R_2 C_3 R_4}{R_5} \tag{7-89}$$

Note the double-capacitor symbol used in Fig. 7-27c for the FDNR.

Example 7-3 Consider again the circuit of Fig. 7-23. Analysis gives the voltage ratio

$$A_V = \frac{V_2}{E} \frac{s^2 L_2 C_2 + 1}{a_3 s^3 + a_2 s^2 + a_1 s + a_0} \tag{7-90}$$

where
$$a_3 = L_2 R_G(C_1 C_2 + C_1 C_3 + C_2 C_3)$$

$$a_2 = L_2 \left[(C_1 + C_2)\frac{R_G}{R_L} + C_2 + C_3 \right] \tag{7-91}$$

$$a_1 = \frac{L_2}{R_L} + R_G(C_1 + C_3) \qquad a_0 = 1 + \frac{R_G}{R_L}$$

The $a_3 s^3$ term can be rewritten

$$a_3 s^3 = L_2 R_G(C_1 C_2 + C_1 C_3 + C_2 C_3)s^3$$
$$= (sL_2)(R_G)[(sC_1)(sC_2) + (sC_1)(sC_3) + (sC_2)(sC_3)] \tag{7-92}$$

which, as Fig. 7-27 shows, is transformed by the scaling into

$$R'_2 \frac{1}{sC'_G}[(s^2 D'_1)(s^2 D'_2) + (s^2 D'_1)(s^2 D'_3) + (s^2 D'_2)(s^2 D'_3)]$$

$$= s^3 R'_2 \frac{1}{C'_G}(D'_1 D'_2 + D'_1 D'_3 + D'_2 D'_3) \tag{7-93}$$

Since $R'_2 = L_2$, $C'_G = 1/R_G$ and $D'_1 = C_1$, $D'_2 = C_2$, $D'_3 = C_3$, the transformed term equals $a_3 s^3$.

Similar calculations indicate that *all* terms in (7-90) remain unchanged by the scaling. Hence, $A_V(s)$ is the same for the transformed circuit as for the original *LCR* one. Furthermore, as a comparison of Eqs. (7-92) and (7-93) indicates, R'_2, $1/C'_G$, D'_1, D'_2, and D'_3 enter $a_3 s^3$ in exactly the same manner as L_2, R_G, C_1, C_2, and C_3, respectively, did for the *LCR* circuit. This conclusion can be readily generalized to the following statement: the element values R'_i, D'_i, and $1/C'_i$ enter the expression for the transfer function of the scaled circuit in exactly the same way as L_i, C_i, and R_i did for the transfer function of the *LCR* network. Since that function was insensitive to variations of L_i, C_i, R_i, it follows that the scaled function is just as insensitive to the variations of R'_i, D'_i, and $1/C'_i$.[†] This is what makes the impedance-scaling method such a useful one for inductorless filter design.

Returning now to our example, the scaling transforms the circuit of Fig. 7-23a into that shown in Fig. 7-28a. The latter, unfortunately, contains a floating FDNR. This makes its realization difficult, since the impedance Z of Fig. 7-19a (which we would normally use) must have a grounded terminal. The problem can be circumvented if the dual of the circuit, shown in Fig. 7-28b, is used. The scaled version of the dual circuit is shown in Fig. 7-28c. It has only one FDNR, which is grounded.

---

† Since $\partial A_V/\partial C'_i = -[\partial A_V/\partial(1/C'_i)]/C'^2_i$, it follows for $C'_i \neq 0$ that the sensitivity to C'_i is also low.

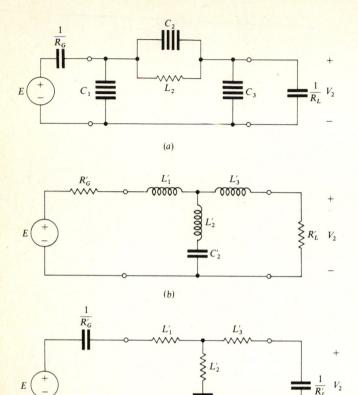

(a)

(b)

(c)

Figure 7-28 (*a*) The scaled version of the filter shown in Fig. 7-23*a*; (*b*) a circuit dual to that of Fig. 7-23*a*; (*c*) the scaled dual circuit.

The capacitive loads in the scaled dual circuit introduce two difficulties. (1) There is no dc path from the FDNR to the generator and to ground. This causes problems in biasing the op-amps in the FDNR. A brute-force solution is to shunt each load capacitor by a resistor which is so large that the frequency response remains unaffected. (2) The generator and load are often inherently resistive, and their internal impedances cannot be replaced by capacitors. Then a modification of the above procedure which is due to Panzer (described in Ref. 3, p. 518) can be used. The process is based on the equivalence illustrated in Fig. 7-29*a*, which can easily be proved. Let the *LC* two-port be described by its chain matrix

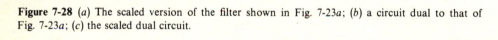

$$T = \begin{bmatrix} A & B \\ C & D \end{bmatrix} \tag{7-94}$$

Scaling all impedances by $1/Ks$ does not affect parameter A, which is the open-circuit voltage ratio, as a derivation similar to that given earlier in the section

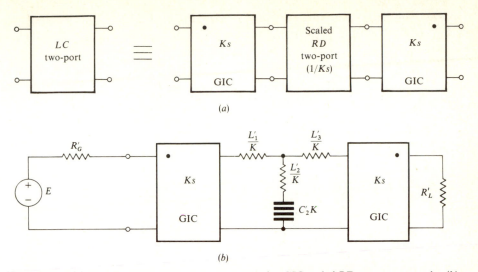

(a)

(b)

Figure 7-29 (a) Equivalence between *LC* two-port and a GIC-scaled-RD two-port cascade; (b) application of the equivalence.

proves. Similarly, the short-circuit current ratio D remains unaffected. However B, which is a transfer *impedance*, will be scaled by $1/Ks$, and C, which is a transfer *admittance*, by Ks. Hence, the scaled *RD* two-port's chain matrix is

$$\mathbf{T}_s = \begin{bmatrix} A & \dfrac{B}{Ks} \\ CKs & D \end{bmatrix} \tag{7-95}$$

If VGIC realization is chosen for both GICs, the chain matrix of the input GIC is

$$\mathbf{T}_1 = \begin{bmatrix} Ks & 0 \\ 0 & 1 \end{bmatrix} \tag{7-96}$$

For the output GIC, which has interchanged input and output ports, the chain relations are

$$V_2 = KsV_1 \qquad I_2 = -I_1 \tag{7-97}$$

and hence the chain matrix is

$$\mathbf{T}_2 = \begin{bmatrix} \dfrac{1}{Ks} & 0 \\ 0 & 1 \end{bmatrix} \tag{7-98}$$

The chain matrix of the cascade is therefore

$$\mathbf{T}_c = \mathbf{T}_1 \, \mathbf{T}_s \, \mathbf{T}_2 = \begin{bmatrix} Ks & 0 \\ 0 & 1 \end{bmatrix} \begin{bmatrix} A & \dfrac{B}{Ks} \\ CKs & D \end{bmatrix} \begin{bmatrix} \dfrac{1}{Ks} & 0 \\ 0 & 1 \end{bmatrix} = \begin{bmatrix} A & B \\ C & D \end{bmatrix} \tag{7-99}$$

as stated.

Using this equivalence, we can replace the circuit of Fig. 7-28b by that of Fig. 7-29b. This circuit has the required resistive generator and load impedances. The constant K should be chosen so as to make the element values practical.

It is worth reiterating that the usefulness of the impedance-scaling method lies in the following two properties:

1. The transfer function of the scaled network is identical to that of the doubly loaded LC two-port on which it is modeled.
2. The element values R', D', and $1/C'$ of the scaled circuit enter the transfer function exactly the same way as L, C, and R entered the transfer function of the model circuit. Hence, the low-sensitivity properties of the model are maintained for the scaled circuit with respect to variations of R', D', and $1/C'$.†

The above advantages are also obtained when an entirely different modeling approach[14,15] is used. Consider the circuit of Fig. 7-30a. Writing the KVL and KCL for the network, we obtain

$$I_1 = Y_1(E - V_2) = \frac{1}{R_1 + sL_1}(E - V_2)$$

$$V_2 = Z_2(I_1 - I_3) = \frac{1}{sC_2}(I_1 - I_3)$$

$$I_3 = Y_3(V_2 - V_4) = \frac{1}{sL_3}(V_2 - V_4)$$

$$V_4 = Z_4 I_3 = \frac{1}{sC_4 + G_4} I_3$$

(7-100)

Next we construct an active inductorless circuit in which the *voltages* will be described by equations analogous to Eqs. (7-100) describing the *voltages and currents* of the model of Fig. 7-30a. As a first step, we construct the block diagram shown in Fig. 7-30b.

In the diagram, the input of the first box is $E - V_2$. Since the transfer function of the box is $1/(R_1 + sL_1)$, its output is $(E - V_2)/(R_1 + sL_1)$, which, by (7-100), is numerically equal to I_1. The input of the second box is $I_1 - I_3$; hence its output is $(I_1 - I_3)/sC_2$, which, by (7-100), equals V_2. We can follow through this way the operation of the whole block diagram to verify that it represents the relations in (7-100) correctly.‡

In constructing a circuit from the block diagram, we note that the related building blocks shown in Fig. 7-31 can be built easily and reliably using

---

† Strictly speaking, we should consider the variations of the internal element values of the FDNR, not of D'. The low-sensitivity property, however, is valid for these variations as well. It is also of interest in some applications that FDNR filters are bilateral, while most other active filter types are not.

‡ The feedback arrangement of its block diagram caused this circuit to be nicknamed *leapfrog filter*. Its other, more dignified, names are *active-ladder* or *multiple-feedback filter*.[14,15]

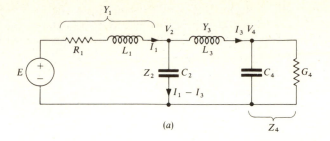

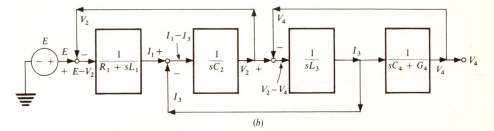

Figure 7-30 (a) Ladder-circuit model; (b) block diagram of an equivalent system.

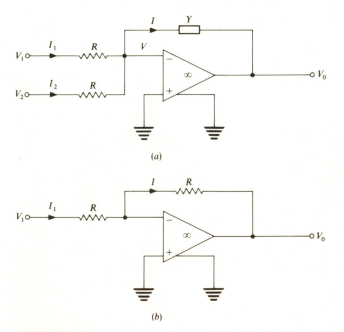

Figure 7-31 Grounded-op-amp active building blocks: (a) general two-input circuit; (b) voltage inverter.

grounded-input–grounded-output op-amps. Since the input voltage V as well as the input current of the op-amp is zero, for the circuit of Fig. 7-31a we obtain

$$I = I_1 + I_2 = \frac{V_1 + V_2}{R} \qquad V_0 = -\frac{I}{Y} = -\frac{1}{RY}(V_1 + V_2) \qquad (7\text{-}101)$$

If $Y = sC + G$, this becomes

$$V_0 = -\frac{1}{sRC + GR}(V_1 + V_2) \qquad (7\text{-}102)$$

If $Y = sC$,

$$V_0 = -\frac{1}{sRC}(V_1 + V_2) \qquad (7\text{-}103)$$

A similar analysis for the circuit of Fig. 7-31b gives

$$V_0 = -IR = -I_1 R = -\frac{V_1}{R}R = -V_1 \qquad (7\text{-}104)$$

Hence, this circuit is simply an inverter.

An economical system using the building blocks of Fig. 7-31 for realizing the block diagram of Fig. 7-30b is shown in Fig. 7-32, where voltage inverters are denoted by triangles marked -1. Following through the operation of this diagram reveals that it is equivalent to that of Fig. 7-30b, and hence it also models the relations of (7-100).

To illustrate the actual circuit realization for this system, consider the first box, reproduced in Fig. 7-33a. Comparing its transfer function

$$\frac{-R}{R_1 + sL_1} = \frac{-1}{s(L_1/R) + R_1/R} \qquad (7\text{-}105)$$

with that given in (7-102) reveals that the box can be realized by the circuit of Fig. 7-31a if we choose

$$Y = sC + G \qquad C = \frac{L_1}{R^2} \qquad G = \frac{R_1}{R^2} \qquad (7\text{-}106)$$

(Fig. 7-33b).

Similarly, the next part of the system is shown in Fig. 7-34a. When the circuits

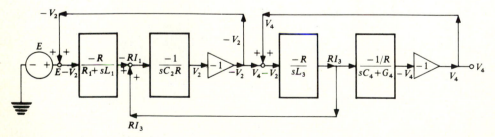

Figure 7-32 System equivalent to the ladder of Fig. 7-30a and suitable for the building blocks of Fig. 7-31.

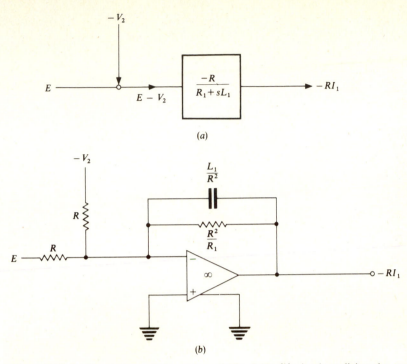

(a)

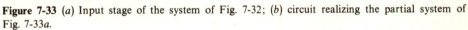

(b)

Figure 7-33 (a) Input stage of the system of Fig. 7-32; (b) circuit realizing the partial system of Fig. 7-33a.

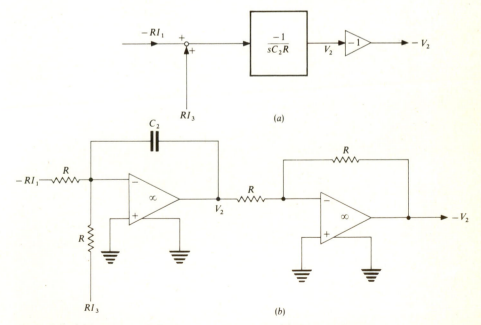

(a)

(b)

Figure 7-34 (a) Second stage of the system of Fig. 7-32; (b) circuit realization.

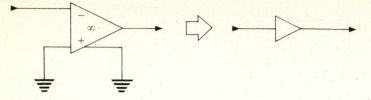

Figure 7-35 Simplified symbol for a grounded-input–grounded-output op-amp.

of Fig. 7-31 are used and Eq. (7-103) is applied, the equivalent circuit of Fig. 7-34b results.

The rest of the system of Fig. 7-32 can be treated similarly. Using the simplified symbol of Fig. 7-35 for the grounded-input–grounded-output op-amp, we get the overall circuit of Fig. 7-36a. (In the diagram, all unmarked resistors are of value R.) A somewhat more compact and ladderlike rearrangement of the same circuit is shown in Fig. 7-36b.

The principle was illustrated here only for the ladder circuit of Fig. 7-30a. It can be extended to other ladder types[14,15] but only at the cost of a more complicated design process and a more elaborate final circuit.

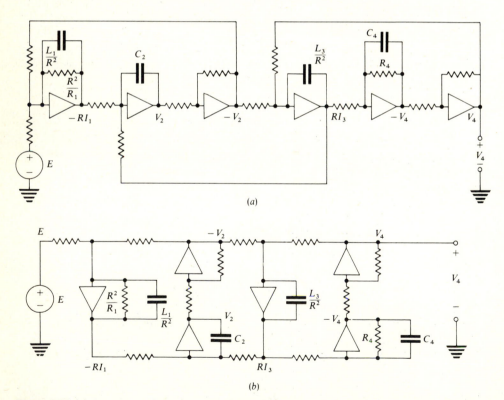

Figure 7-36 (a) Complete active equivalent of the *LCR* ladder shown in Fig. 7-30a; (b) rearranged circuit.

By construction, the final circuit shown in Fig. 7-36 has exactly the same transfer function as the model. Furthermore, if the two transfer functions are expressed symbolically in terms of element values, they will still have the same forms although R_1 in the transfer function of the circuit of Fig. 7-30a will be replaced by R^2/R_1' in that of Fig. 7-36, where $R_1' \triangleq R^2/R_1$ is the resistor in the feedback loop of the first op-amp. Similarly, L_1 will be replaced by $R^2 C_1'$, where $C_1' \triangleq L_1/R^2$, etc. Hence, if the sensitivity of the model transfer function to variations of, say, L_1 is low, the sensitivity of the actual transfer function to variations of $R^2 C_1'$ and hence also to the variation C_1' will be equally low. Thus, a one-to-one correspondence is maintained between both the transfer functions and the element values of the RLC model and the active-RC final circuit, as it was for the scaled-impedance design. Hence the low-sensitivity property of the doubly loaded LC ladder is preserved.

The above argument does not prove that the sensitivity is low also with respect to variations of the ubiquitous resistors of value R (unmarked in Fig. 7-36). However, a numerical investigation[16] showed this to be true as well.

7-6 CASCADE REALIZATION

The simulation methods described in Secs. 7-4 and 7-5 give reliable circuits with low sensitivities to component variations. They also have some disadvantages, however, which include the following:

1. A large number of op-amps is needed for the realization of a given transfer function. For example, the simulation of the circuit of Fig. 7-30a requires six op-amps. Its transfer function is of degree 4; hence, if the methods to be described in this section are used, only two op-amps are needed.
2. The described simulation methods require that a ladder-circuit model be available. Since the design of ladders is a difficult task (unless tabulated circuits are handy), this approach necessitates time-consuming and tedious calculations.
3. As explained in Chap. 5, ladder circuits represent a restricted class of passive two-ports. Hence, the variety of transfer functions which can be realized by ladder simulation is unnecessarily narrowed down to those for which a ladder can be designed with positive element values.

The design approach discussed next in this section is, by contrast, simple and leads to economical circuits. The price paid is less reliability and a greater susceptibility of the performance to element-value variations. The basic approach is the following. Let the prescribed transfer function be factored as follows:

$$A_V(s) = \frac{a_n s^n + a_{n-1} s^{n-1} + \cdots + a_0}{b_n s^n + b_{n-1} s^{n-1} + \cdots + b_0} = \prod_{i=1}^{n/2} \frac{n_{2i} s^2 + n_{1i} s + n_{0i}}{d_{2i} s^2 + d_{1i} s + d_{0i}} \qquad (7\text{-}107)$$

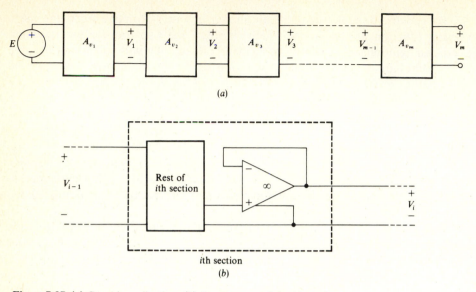

(a)

ith section
(b)

Figure 7-37 (a) Cascade realization; (b) the structure of the ith section.

where we assume for convenience that n is the common degree of the numerator and denominator and is even. We shall discard this restriction soon. Denoting $m = n/2$ and defining the biquadratic transfer functions

$$A_{V_i}(s) \triangleq \frac{n_{2i} s^2 + n_{1i} s + n_{0i}}{d_{2i} s^2 + d_{1i} s + d_{0i}} \qquad i = 1, 2, \ldots, m \qquad (7\text{-}108)$$

we have

$$A_V(s) = \prod_{i=1}^{m} A_{V_i}(s) \qquad (7\text{-}109)$$

Assume now that the circuit is realized as a cascade of m elementary two-ports (Fig. 7-37a). If the transfer function of the ith two-port is $A_{V_i}(s)$, and if connecting the two-ports does not change their transfer functions, then clearly

$$A_V(s) \triangleq \frac{V_m}{E} = \frac{V_1}{E} \frac{V_2}{V_1} \frac{V_3}{V_2} \cdots \frac{V_m}{V_{m-1}} = A_{V_1} A_{V_2} \cdots A_{V_m} \qquad (7\text{-}110)$$

As shown in Sec. 7-2, the output port of an op-amp acts as a voltage source under quite permissive conditions. Hence, if the elementary sections in the circuit have the general structure of Fig. 7-37b, the input impedance of the $(i + 1)$st section will have no effect on $A_V(s)$ and (7-110) will hold.

If n is odd, $n_{2m} = d_{2m} = 0$ will give a bilinear rather than biquadratic factor. Similarly, by setting $n_{2i} = n_{1i} = 0$ it is possible to represent functions where the numerator is of lower degree than the denominator.†

---

† The reverse situation cannot exist since the circuit must be stable for $\omega \to \infty$.

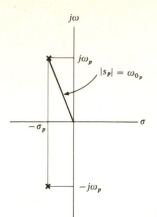

Figure 7-38 Complex pole frequency.

For $d_{2i} \neq 0$ and $n_{2i} \neq 0$, $A_{V_i}(s)$ can be rewritten

$$A_{V_i} = \frac{n_{2i}}{d_{2i}} \frac{s^2 + (n_{1i}/n_{2i})s + (n_{0i}/n_{2i})}{s^2 + (d_{1i}/d_{2i})s + (d_{0i}/d_{2i})}$$

$$A_{V_i} = K \frac{s^2 + (\omega_{0z}/Q_z)s + \omega_{0z}^2}{s^2 + (\omega_{0p}/Q_p)s + \omega_{0p}^2}$$

(7-111)

where in the second relation the subscript i was dropped for convenience on the right-hand side. Figure 7-38 shows a pair of complex poles $s_p = -\sigma_p \pm j\omega_p$ of $A_{V_i}(s)$ and illustrates the definitions of the *pole frequency* ω_{0p}

$$\omega_{0p} \triangleq \sqrt{\frac{d_0}{d_2}} = \sqrt{\sigma_p^2 + \omega_p^2} = |s_p|$$

(7-112)

as well as the *pole Q* (or *pole quality*)

$$Q_p \triangleq \frac{d_2}{d_1}\omega_{0p} = \frac{1}{d_1}\sqrt{d_0 d_2} = \frac{|s_p|}{2\sigma_p}$$

(7-113)

A high value of Q_p indicates that a pole of A_V, that is, a natural mode of the circuit, is relatively close to the $j\omega$ axis. This will be the case for some sections of a highly selective filter. As we shall see, such pole usually causes high sensitivity to element-value variations.

For a passive RC two-port, as we have seen in Sec. 7-1, all natural modes lie on the $-\sigma$ axis. Therefore for such circuit $\omega_p = 0$, $|s_p| = \sigma_p$, and $Q_p = \frac{1}{2}$ hold for all poles. To illustrate how an active element can raise this pole Q, consider the following example.

Example 7-4† Find the pole Q as a function of the VCVS gain K for the circuit of Fig. 7-39a.

---

† From W. J. Kerwin in L. P. Huelsman (ed.), "Active Filters," McGraw-Hill, New York, 1970.

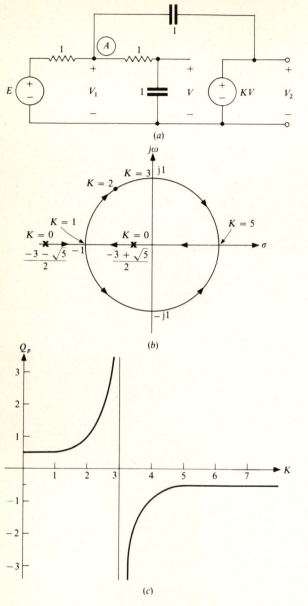

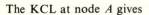

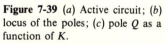

Figure 7-39 (a) Active circuit; (b) locus of the poles; (c) pole Q as a function of K.

The KCL at node A gives

$$\frac{E - V_1}{1} + \frac{V - V_1}{1} + \frac{KV - V_1}{1/s1} = 0 \qquad (7\text{-}114)$$

Also, by potential division

$$V = V_1 \frac{1/s1}{1/s1 + 1} = \frac{1}{s + 1} \qquad (7\text{-}115)$$

Solving (7-114) and (7-115) gives

$$V = \frac{E}{s^2 + (3 - K)s + 1} \tag{7-116}$$

and hence

$$A_V = \frac{V_2}{E} = \frac{KV}{E} = \frac{K}{s^2 + (3 - K)s + 1} \tag{7-117}$$

The poles are located at

$$s_{p1,\,2} = \frac{K - 3}{2} \pm j\sqrt{1 - \left(\frac{K - 3}{2}\right)^2} \tag{7-118}$$

The locus of the s_p is shown in Fig. 7-39b.

For $K = 0$, $s_{p1,2} = (-3 \pm \sqrt{5})/2$; for $K = 1$, $s_{p_1} = s_{p_2} = -1$; for $K = 2$, $s_{p1,2} = (-1 \pm j\sqrt{3})/2$; for $K = 3$, $s_{p1,2} = \pm j$. For $K \geq 3$, $-\sigma_p \geq 0$, and hence the circuit becomes unstable. For $1 \leq K \leq 5$, the pole frequency is $\omega_{0p} = |s_p| = 1$; hence, by (7-113),

$$Q_p = \frac{1}{2\sigma_p} = \frac{1}{3 - K} \tag{7-119}$$

For $K < 1$ and $K > 5$, $|\omega_{0p}| = |\sigma_p| = |s_p|$ and hence $|Q_p| = \frac{1}{2}$. The complete $Q_p(K)$ curve is illustrated in Fig. 7-39c.

Returning to Eq. (7-111), we note that *zero frequency*, $\omega_{0z} \triangleq |s_z|$, and *zero Q*, $Q_z \triangleq |s_z|/2\sigma_z$, are defined analogously to ω_{0p} and Q_p. Since, however, Q_z does not play the same important role as Q_p does ($Q_p \to \infty$ or $Q_p < 0$ indicates instability while $Q_z \to \infty$ or $Q_z < 0$ are of no consequence), these quantities are seldom used.

The realization of the biquadratic functions $A_V(s)$ can be carried out by using any of the hundreds of different circuits developed over the past decades. A few of the most important ones will be described. A more complete and detailed discussion of biquadratic filter sections will be found in Refs. 3, 5, 7, and 17.

One of the first filter types developed is the *Sallen-Key filter* (Fig. 7-40a). In this circuit, the active element is a VCVS; its widely used simplified notation is explained in Fig. 7-40b, which also shows a possible realization based on the op-amp circuit of Fig. 7-12a. The circuit of Fig. 7-40a is the general form of the example filter shown in Fig. 7-39a and can be analyzed the same way; compare Eqs. (7-114) to (7-117). The result is

$$A_V = \frac{V_2}{V_1} = \frac{K Y_1 Y_3}{(Y_1 + Y_2)(Y_3 + Y_4) + Y_3 Y_4 - K Y_2 Y_3} \tag{7-120}$$

Making any two of the Y_i conductive ($Y_i = G_i$) and the other two capacitive ($Y_i = sC_i$), we obtain the $(4)(\frac{3}{2}) = 6$ simple filter circuits shown in Fig. 7-41. The corresponding transfer functions are of the following general form. For circuit (a):

$$A_V = \frac{a_0}{b_2 s^2 + b_1 s + b_0} \tag{7-121}$$

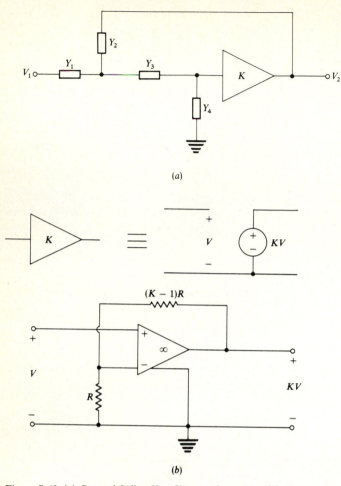

(a)

(b)

Figure 7-40 (a) General Sallen-Key filter configuration; (b) VCVS notation and realization.

For circuit (b):

$$A_V = \frac{a_2 s^2}{b_2 s^2 + b_1 s + b_0}$$

(7-122)

For circuits (c) and (d):†

$$A_V = \frac{a_1 s}{b_2 s^2 + b_1 s + b_0}$$

(7-123)

Circuits (e) and (f) have bilinear rather than biquadratic transfer functions. They are therefore of little practical interest, since bilinear transfer function could also be obtained using passive RC circuits, as illustrated, for example, by the two-ports shown in Fig. 7-42.

The general behavior of the logarithmic gain, defined as

† These two circuits require $K < 1$ for stability and hence the VCVS of Fig. 7-40b cannot be used here.

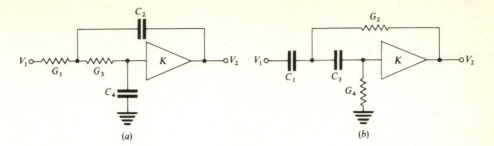

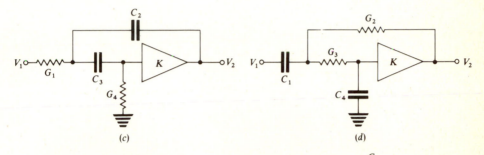

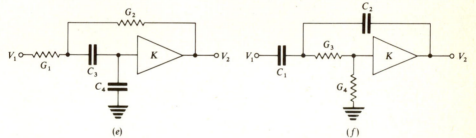

Figure 7-41 Sallen-Key filters $(K > 0)$.

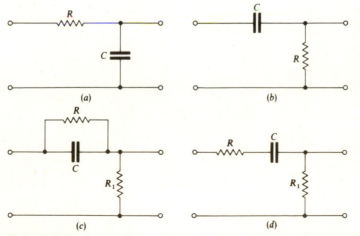

Figure 7-42 Passive-RC filter sections: (a) $A_V = 1/(sRC + 1)$; (b) $A_V = sRC/(sRC + 1)$; (c) $A_V = (1 + sRC)/(1 + R/R_1 + sRC)$; (d) $A_V = sR_1C/[1 + s(R + R_1)C]$.

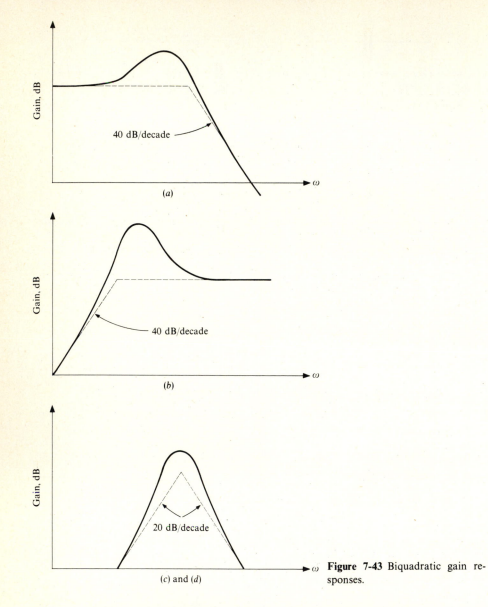

Figure 7-43 Biquadratic gain responses.

20 log $|V_2(j\omega)/V_1(j\omega)|$, is illustrated in Fig. 7-43 for the filter sections shown in Fig. 7-41a to d. Clearly, circuit (a) has the general character of a *low-pass* filter, circuit (b) the character of *high-pass* filter, and circuits (c) and (d) the gain response of a simple *bandpass* filter.

Example 7-5 Design a Sallen-Key filter from the transfer function

$$A_V(s) = \frac{s^3}{s^3 + 2s^2 + 2s + 1} \tag{7-124}$$

Factoring $A_V(s)$ gives

$$A_V(s) = \frac{s}{s+1} \frac{s^2}{s^2 + s + 1} \tag{7-125}$$

The first factor can be realized, for example, for $R = 1$ and $C = 1$, by the circuit of Fig. 7-42b. The second factor, as (7-122) indicates, can correspond to circuit (b) of Fig. 7-41. That circuit, from (7-120), has the transfer function

$$A_V = \frac{Ks^2 C_1 C_3}{(sC_1 + G_2)(sC_3 + G_4) + sC_3 G_4 - KsG_2 C_3}$$

$$= \frac{Ks^2}{s^2 + s[G_4/C_3 + (G_2 + G_4 - KG_2)/C_1] + G_2 G_4/C_1 C_3} \tag{7-126}$$

A comparison with (7-111) shows that $\omega_{0z}^2 = 0$, $\omega_{0p}^2 = G_2 G_4/C_1 C_3$, and $Q_p = \sqrt{G_2 G_4}/[G_4\sqrt{C_1/C_3} + (G_2 + G_4 - KG_2)\sqrt{C_3/C_1}]$.

The design equations are obtained by matching the coefficients in the denominator of $A_V(s)$:

$$\frac{G_4}{C_3} + \frac{(1 - K)G_2 + G_4}{C_1} = 1 \qquad \frac{G_2 G_4}{C_1 C_3} = 1 \tag{7-127}$$

We have two equations and five design parameters: C_1, C_3, G_2, G_4, and K. Arbitrarily choosing $C_1 = C_3 = 1$ and $K = 1$, we have from (7-127)

$$2G_4 = 1 \qquad G_4 = \tfrac{1}{2} \qquad G_2 = \frac{1}{G_4} = 2$$

As Fig. 7-40b indicates, the VCVS with $K = 1$ can be realized by the simple circuit of Fig. 7-44a. Hence the complete circuit is that shown in Fig. 7-44b. The biquadratic section is placed first in the cascade, so that the passive section is connected to the output of the op-amp and therefore its input impedance does not change the transfer properties of the preceding stage. If the order of the sections is to be changed, a second unity-gain VCVS must be inserted between the sections as a buffer (Fig. 7-44c).

An inspection of (7-121) to (7-123) shows that the transmission zeros of the Sallen-Key filters are either at zero or infinite frequency. Finite nonzero-frequency zeros can be created by adding more branches to the Sallen-Key circuit. Consider the circuit† of Fig. 7-45. If Y is realized by a resistor ($Y = G$), analysis gives the transfer function

$$A_V(s) = \frac{K(s^2 + a^{-2})}{s^2 + (G + 2 - K)2a^{-1}s + (1 + 2G)a^{-2}} \tag{7-128}$$

while for $Y = saC$

$$A_V(s) = \frac{[K/(2C + 1)](s^2 + a^{-2})}{s^2 + [2a^{-1}(C + 2 - K)/(2C + 1)]s + a^{-2}/(2C + 1)} \tag{7-129}$$

† Called *Kerwin's circuit*.

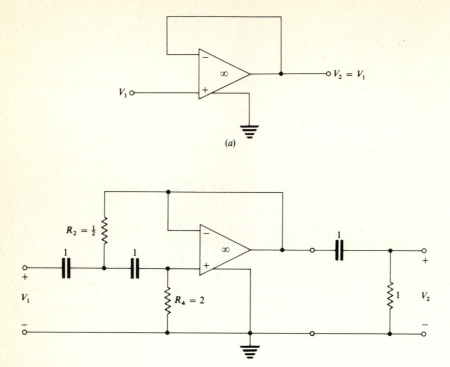

(a)

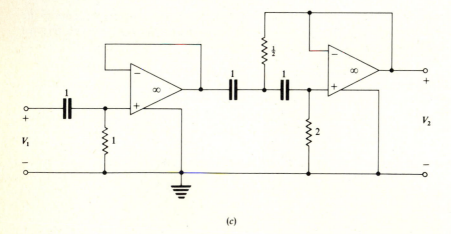

(b)

(c)

Figure 7-44 (a) Unity-gain VCVS circuit; (b) Sallen-Key filter example; (c) filter with its sections interchanged.

This gr
$Q_p = 10$), w
$K \approx 2.4293.$
order of 5 or
adjustment
next section

7-7 LOW

Several circ
more addit
circuits disc
terminated
active resor
The output
nodes A an
The cir
minals of a

Also, the cu
the KCL fo

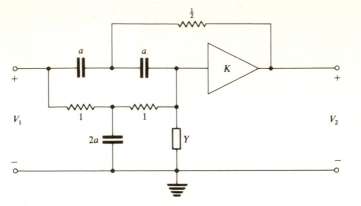

Figure 7-45 Kerwin's circuit. [*From W. J. Kerwin, in L. P. Huelsman (ed.), "Active Filters," p. 16, McGraw-Hill, New York, 1970.*]

Both transfer functions have zeros at $z_{1,2} = \pm j/a$. For complex poles $s_{p1,2}$, from (7-128),

$$\omega_{0p} \triangleq |s_{p1,2}| = \frac{\sqrt{1 + 2G}}{a} > |z_{1,2}| = \frac{1}{a} \tag{7-130}$$

For the function of (7-129), on the other hand,

$$|s_{p1,2}| = \frac{1}{\sqrt{1 + 2C}} \frac{1}{a} < |z_{1,2}| = \frac{1}{a} \tag{7-131}$$

This shows that $Y = G$ should be used if the poles are *farther* from the origin than the zeros and $Y = saC$ if the poles are *closer* to the origin than the zeros.†
The former represents a high-pass characteristic; the latter a low-pass one.

Example 7-6 Design a circuit for

$$A_V = \frac{s^2 + 2}{s^2 + 0.1s + 1} \tag{7-132}$$

Here, $\omega_{0z} = \sqrt{2}$, $\omega_{0p} = 1$, and $Q_p = 1/0.1 = 10$.
Since $|z_{1,2}| = \sqrt{2} > |s_{p1,2}| = 1$, we set $Y = saC$ and use (7-129). Comparing coefficients in the denominator and setting

$$a = \frac{1}{|z_{1,2}|} = \frac{1}{\sqrt{2}} \approx 0.7071$$

we get

$$\frac{2\sqrt{2}(C + 2 - K)}{2C + 1} = 0.1 \quad \text{and} \quad \frac{2}{2C + 1} = 1 \tag{7-133}$$

Introdu

This is the t
function of t

† This node
terminal of the
Z_0 and Z_5, Z_1

† $Y \equiv 0$ for $|s_{p_1}| = |z_1|$.

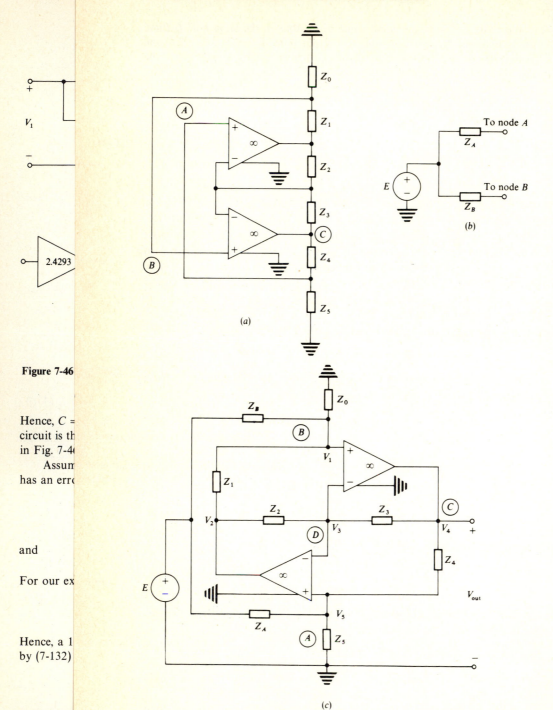

V_1

$+$

$-$

2.4293

Figure 7-46

Hence, $C =$
circuit is th
in Fig. 7-4

Assum
has an err

and

For our ex

Hence, a 1
by (7-132)

the magnit
VCVS gai

Figure 7-47 (a) Terminated GIC used as a resonator; (b) generator circuit; (c) complete GIC filter.

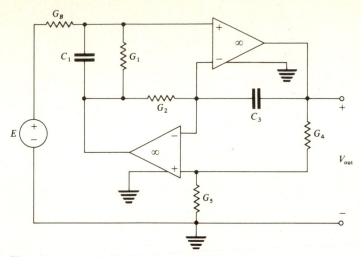

Figure 7-48 Low-pass-filter realization using the GIC.

This gives the circuit shown in Fig. 7-48 and, by (7-140), the transfer function

$$A_V = \frac{G_B G_2(G_4 + G_5)}{(G_1 + sC_1)sC_3 G_5 + G_2 G_4 G_B} = \frac{G_B G_2(G_4 + G_5)}{s^2 C_1 C_3 G_5 + sG_1 C_3 G_5 + G_2 G_4 G_B} \tag{7-142}$$

To obtain a *high-pass* filter, the admittances can be selected as follows:

$$Y_0 = G_0 \qquad Y_1 = G_1 \qquad Y_2 = sC_2 \qquad Y_3 = G_3$$

$$Y_4 = G_4 \qquad Y_5 = G_5 \qquad Y_A = 0 \qquad Y_B = sC_B \tag{7-143}$$

The corresponding transfer function is then, from (7-140),

$$A_V = \frac{s^2 C_B C_2(G_4 + G_5)}{G_1 G_3 G_5 + sC_2 G_4(sC_B + G_0)} = \frac{s^2 C_B C_2(G_4 + G_5)}{s^2 C_2 C_B G_4 + sC_2 G_0 G_4 + G_1 G_3 G_5}$$

$$= \frac{G_4 + G_5}{G_4} \frac{s^2}{s^2 + sG_0/C_B + G_1 G_3 G_5/C_2 C_B G_4} \tag{7-144}$$

The circuit is shown in Fig. 7-49. For a *bandpass* response, the choice

$$Y_0 = sC_0 \qquad Y_1 = G_1 \qquad Y_2 = sC_2 \qquad Y_3 = G_3$$

$$Y_4 = G_4 \qquad Y_5 = G_5 \qquad Y_A = 0 \qquad Y_B = G_B \tag{7-145}$$

may be made. This gives

$$A_V = \frac{G_B sC_2(G_4 + G_5)}{G_1 G_3 G_5 + sC_2 G_4(G_B + sC_0)}$$

$$= \frac{G_B(G_4 + G_5)}{G_4 C_0} \frac{s}{s^2 + sG_B/C_0 + G_1 G_3 G_5/C_2 G_4 C_0} \tag{7-146}$$

and the circuit of Fig. 7-50.

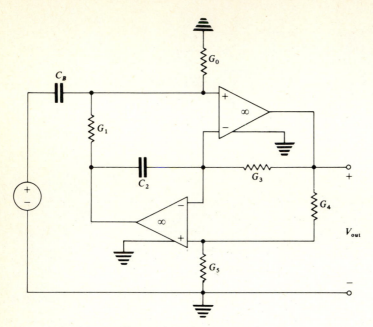

Figure 7-49 High-pass-filter section using the GIC.

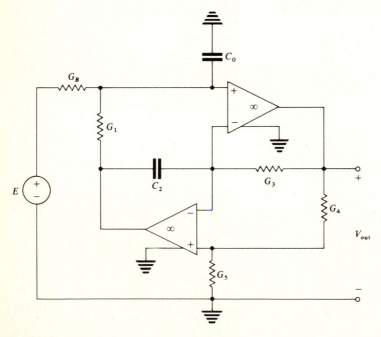

Figure 7-50 Bandpass-filter section using the GIC.

Example 7-7 Realize the transfer function

$$A_V = \frac{s^3}{s^3 + 2s^2 + 2s + 1} = \frac{s}{s+1}\frac{s^2}{s^2+s+1} \tag{7-147}$$

(earlier used in the design of the Sallen-Key filter of Fig. 7-44) using the GIC filter configuration of Fig. 7-47c.

Since the biquadratic factor is of a high-pass character, Eqs. (7-143) and (7-144) are used in the design. Equating the denominator coefficients gives

$$\frac{G_0}{C_B} = 1 \qquad \frac{G_1 G_3 G_5}{C_2 G_B G_4} = 1$$

Therefore $C_2 = C_B = 1$, and $G_0 = G_1 = G_3 = G_4 = G_5 = G_B = 1$ is a possible solution. Thus, a cascade of the biquadratic section of Fig. 7-49 and the bilinear section of Fig. 7-42b, with all unit element values, realizes the A_V of Eq. (7-147).

The circuit of Fig. 7-47c is also suitable for the construction of filter sections with finite nonzero $j\omega$-axis transmission zeros. If we choose

$$Y_0 = 0 \qquad Y_1 = sC_1 \qquad Y_2 = G_2 \qquad Y_3 = G_3$$
$$Y_4 = G_4 \qquad Y_5 = G_5 \qquad Y_A = sC_A \qquad Y_B = G_B \tag{7-148}$$

the transfer function becomes

$$A_V = \frac{s^2 C_A C_1 G_3 + G_B G_2(G_4 + G_5)}{sC_1 G_3(sC_A + G_5) + G_2 G_4 G_B} = \frac{s^2 + G_B G_2(G_4 + G_5)/C_A C_1 G_3}{s^2 + sG_5/C_A + G_2 G_4 G_B/C_1 C_A G_3} \tag{7-149}$$

By (7-111),

$$\omega_{0z}^2 = \frac{G_B G_2(G_4 + G_5)}{C_A C_1 G_3} \qquad \omega_{0p}^2 = \frac{G_B G_2 G_4}{C_A C_1 G_3} \qquad \omega_{0z}^2 > \omega_{0p}^2 \tag{7-150}$$

The circuit diagram is shown in Fig. 7-51a.

Another section with a finite nonzero $j\omega$-axis transmission zero is obtained for the choice

$$Y_0 = G_0 \qquad Y_1 = G_1 \qquad Y_2 = G_2 \qquad Y_3 = G_3$$
$$Y_4 = sC_4 \qquad Y_5 = 0 \qquad Y_A = G_A \qquad Y_B = sC_B \tag{7-151}$$

The corresponding transfer function is, by (7-140),

$$A_V = \frac{G_A(G_1 G_3 - G_0 G_2) + s^2 C_B G_2 C_4}{G_1 G_3 G_A + G_2 sC_4(sC_B + G_0)} = \frac{s^2 + G_A(G_1 G_3 - G_0 G_2)/C_B G_2 C_4}{s^2 + sG_0/C_B + G_1 G_3 G_A/C_B G_2 C_4} \tag{7-152}$$

This has $j\omega$-axis zeros only if $G_1 G_3 \geq G_0 G_2$. Then

$$\omega_{0z}^2 = \frac{G_A(G_1 G_3 - G_0 G_2)}{C_B G_2 C_4} \geq 0 \qquad \omega_{0p}^2 = \frac{G_1 G_3 G_A}{C_B G_2 C_4} \geq \omega_{0z}^2 \tag{7-153}$$

The circuit is shown in Fig. 7-51b.

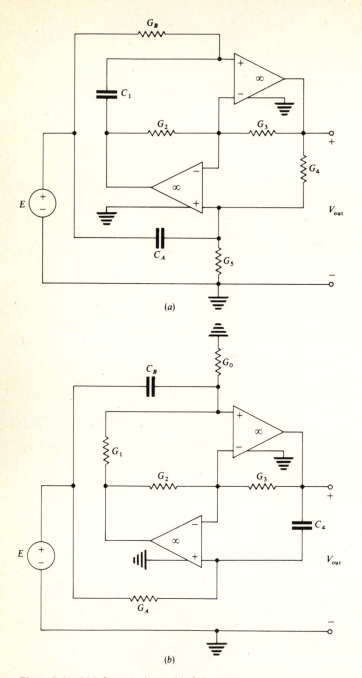

(a)

(b)

Figure 7-51 GIC-filter sections with finite nonzero $j\omega$-axis transmission zeros: (a) low-pass section; (b) high-pass section.

As Eqs. (7-150) and (7-153) indicate, the circuit of Fig. 7-51a is to be used for low-pass sections ($\omega_{0z}^2 > \omega_{0p}^2$) and that of Fig. 7-51b for high-pass sections.

Example 7-8 Design a GIC filter realizing the transfer function

$$A_V = \frac{s^2 + 2}{s^2 + 0.1s + 1} \tag{7-154}$$

used earlier in the design of the Kerwin filter.

Since $\omega_{0z}^2 = 2 > \omega_{0p}^2 = 1$, a low-pass section is needed. Hence, the circuit of Fig. 7-51a is selected. From Eqs. (7-149) and (7-154)

$$\frac{G_B G_2 (G_4 + G_5)}{C_A C_1 G_3} = 2 \qquad \frac{G_B G_2 G_4}{C_A C_1 G_3} = 1 \qquad \frac{G_5}{C_A} = 0.1 \tag{7-155}$$

Dividing the first two equations gives

$$\frac{G_4 + G_5}{G_4} = 1 + \frac{G_5}{G_4} = 2 \qquad G_4 = G_5 \tag{7-156}$$

Dividing the second two equations gives

$$\frac{G_B G_2 G_4}{G_5 C_1 G_3} = 10 \tag{7-157}$$

Hence, if we choose $C_1 = C_A = 1$, the last equation of (7-155) gives $G_5 = 0.1$. Equation (7-157) can be satisfied if we set $G_2 = G_4 = 2$, $G_B = 0.1$, and $G_3 = 0.4$. Then the complete set of element values of the circuit of Fig. 7-51a is

$$C_1 = C_A = 1 \text{ F} \qquad G_2 = G_4 = 2 \text{ ℧} \qquad G_3 = 0.4 \text{ ℧} \qquad G_B = G_5 = 0.1 \text{ ℧} \tag{7-158}$$

The spread of the conductance values is $2/0.1 = 20$, due to the high value of the pole Q ($Q_p = 10$). These values are still quite practical.

Consider now the sensitivity of this circuit at $s = j\omega = j1$, as we did for the Kerwin circuit. If G_B changes by $+1$ percent, then, by (7-149),

$$A_V = \frac{-1 + 2 + 0.02}{-1 + j(0.1 + 0.001) + 1 + 0.01} = \frac{1.02}{0.01 + j0.101} \approx 0.9902 - j10.001$$

and

$$|A_V| \approx 10.05$$

Since the exact values are $A_V = -j10$ and $|A_V| = 10$, we note that a 1 percent error in G_B results in a less than 1 percent change in $|A_V|$ at $s = j1$. The change due to a 1 percent change of C_A or other element values is comparably small. This is in contrast to the Kerwin filter, where a 1 percent change in the critical parameters K or C gave a 50 percent change in $|A_V|$. The reason for the big difference in behavior is that in the transfer function (7-129) of the Kerwin filter the small value of b_1, that is, the large pole Q, is attained by the near cancellation of terms. For the GIC filter, on the other hand, a large Q_p is obtained simply by choosing the ratio of some of the element values large. This can lead to inconvenient element values for very high pole Q's but no accuracy problems. The price paid for the accuracy improvement is, of course, the extra op-amp needed for the GIC filter circuit.

A check of Eqs. (7-140) to (7-152) reveals that the above argument holds for all GIC filters described, and hence they all represent low-sensitivity filter sections.

Other biquadratic filter sections exist which obtain low sensitivity by using extra op-amps. The best known of these is the *state-variable filter*.[20]† The block diagram of the filter is shown in Fig. 7-52a. Clearly, the relations

$$V = V_1 - a_1 \frac{V}{sT} - a_0 \frac{V}{s^2 T^2}$$

$$V = \frac{V_1}{1 + a_1/sT + a_0/s^2 T^2} = \frac{V_1 s^2}{s^2 + (a_1/T)s + a_0/T^2} \tag{7-159}$$

hold. Hence, if V is the output voltage, i.e., the output terminal is connected to node A, the high-pass transfer function

$$A_V = \frac{V}{V_1} = \frac{s^2}{s^2 + (a_1/T)s + a_0/T^2} \tag{7-160}$$

results. If the output terminal is at B, the transfer function is the bandpass response

$$A_V = \frac{-V/sT}{V_1} = \frac{-(1/T)s}{s^2 + (a_1/T)s + a_0/T^2} \tag{7-161}$$

Finally, if C is used as the output terminal, the low-pass transfer function

$$A_V = \frac{V/s^2 T^2}{V_1} = \frac{1/T^2}{s^2 + (a_1/T)s + a_0/T^2} \tag{7-162}$$

is obtained.

The realization of the adder-subtractor needed in Fig. 7-52a is illustrated in Fig. 7-52b. From the figure

$$I_1 = \frac{V_1 - V_3}{R_1 + R_3} \qquad V^+ = V_1 - I_1 R_1 = \frac{R_3 V_1 + R_1 V_3}{R_1 + R_3}$$

$$I_2 = \frac{V_2 - V}{R_2 + R_4} \qquad V^- = V_2 - I_2 R_2 = \frac{R_4 V_2 + R_2 V}{R_4 + R_2} \tag{7-163}$$

Due to the op-amp action, $V^+ = V^-$:

$$\frac{R_2 V + R_4 V_2}{R_2 + R_4} = \frac{R_3 V_1 + R_1 V_3}{R_1 + R_3} \tag{7-164}$$

which gives

$$V = \frac{1 + R_4/R_2}{1 + R_1/R_3} V_1 + \frac{1 + R_4/R_2}{1 + R_3/R_1} V_3 - \frac{R_4}{R_2} V_2 \tag{7-165}$$

---

† So named because the circuit acts as an analog computer, with its voltages modeling the state variables.

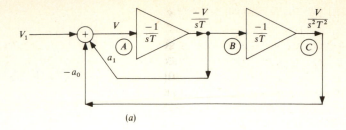

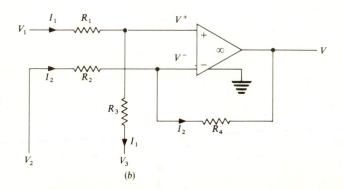

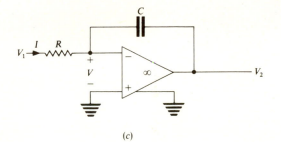

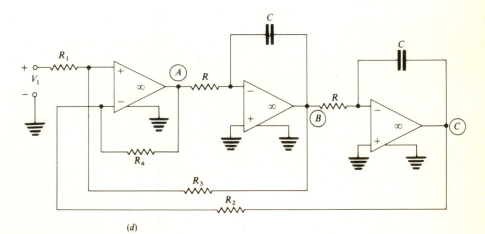

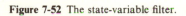

Figure 7-52 The state-variable filter.

Hence, choosing

$$\frac{R_4}{R_2} = a_0 \qquad \frac{1 + R_4/R_2}{1 + R_3/R_1} = \frac{1 + a_0}{1 + R_3/R_1} = a_1 \qquad \frac{R_3}{R_1} = \frac{1 + a_0}{a_1} - 1 \quad (7\text{-}166)$$

we see that the circuit of Fig. 7-52b will operate as the adder needed in Fig. 7-52a. (It will also introduce a scaling of V_1 by

$$\frac{1 + R_4/R_2}{1 + R_1/R_3} = 1 + a_0 - a_1 \qquad (7\text{-}167)$$

which can usually be ignored since it merely represents a constant factor in A_V.)

The integrator needed in the circuit can be realized by the circuit of Fig. 7-52c. Since $V \approx 0$,

$$I \approx \frac{V_1}{R} \approx \frac{-V_2}{1/sC} \qquad V_2 \approx -\frac{V_1}{sRC} \qquad (7\text{-}168)$$

Hence, for $RC = T$, the required transfer function $-1/sT$ results.

Combining the building blocks of Fig. 7-52b and c into the system of Fig. 7-52a gives the circuit shown in Fig. 7-52d. Its possible transfer functions can be obtained using (7-159) to (7-168). For example, choosing node B as output terminal, by (7-161) and (7-166) to (7-168) we obtain the bandpass response

$$A_V = (1 + a_0 - a_1) \frac{-(1/T)s}{s^2 + (a_1/T)s + a_0/T^2}$$

$$= \frac{1 + R_4/R_2}{1 + R_1/R_3} \frac{-(1/RC)s}{s^2 + (1 + R_4/R_2)/[(1 + R_3/R_1)RC]s + R_4/[R_2(RC)^2]}$$

$$(7\text{-}169)$$

If node A or node C is used as the output terminal, low-pass or high-pass response can be obtained.

Example 7-9 Design a state-variable filter to realize the transfer function

$$A_V = \frac{s}{s^2 + 0.1s + 1} \qquad (7\text{-}170)$$

Except for a difference in sign and a constant scale factor, the transfer function of (7-169) is suitable. Hence, choosing $R = 1$ and $C = 1$, we get the relations

$$\frac{R_4}{R_2} = 1 \qquad \frac{1 + R_4/R_2}{1 + R_3/R_1} = \frac{2}{1 + R_3/R_1} = 0.1 \qquad \frac{R_3}{R_1} = 19$$

They can be satisfied by choosing $R_2 = R_4 = 1$ and $R_1 = 0.24$, $R_3 = 4.56$, for example. Clearly, as for the GIC filter, a high pole Q is achieved not by nearly canceling terms but by choosing widely spread element values. Hence, the sensitivity of the response to element-value variations is low.

Finite nonzero $j\omega$-axis transmission zeros can be obtained by summing (at the cost of a fourth op-amp) the signals at nodes A and C. As Eqs. (7-159) and (7-162) illustrate, this introduces a numerator $s^2 + \omega_{0z}^2$ into $A_V(s)$. It is also possible to modify the circuit of Fig. 7-52d by adding resistors to introduce $j\omega$-axis zeros. One such circuit[5,21,22] is analyzed in Prob. 7-26.

The invention of the state-variable filter preceded that of the GIC circuit; it was the first biquadratic filter section suitable for high pole Q applications. However, since it has no significant advantage over the more recently invented GIC filter and requires more op-amps and passive components, it is being replaced by the GIC filter in more and more applications.

As we have seen, the filter sections available for cascade realization range from relatively simple and sensitive circuits to more complicated and insensitive ones. However, it must be recognized that a cascade realization built even from the least sensitive biquadratic circuits is more vulnerable to element-value variations in its passband than a simulated-inductance or simulated-ladder circuit. This is caused by the absence of the inherently low-sensitivity property of doubly terminated LC two-ports, described in Sec. 6-1. The partitioning of the circuit into buffered sections shatters this property. Consider the bandpass gain response a shown in Fig. 7-53. In a typical cascade realization, this response is obtained as the sum† of the gain responses 20 log $|A_{V_1}|$ and 20 log $|A_{V_2}|$, where A_{V_1} and A_{V_2} are biquadratic transfer functions. These are shown as curves b and c in Fig. 7-53. The flatness of the passband at, say, the band center ω_0 is clearly the result of the cancellation of the slopes of the component curves b and c. Especially for high pole Q's, this balance is a precarious one, easily upset by even minor changes in A_{V_1} or A_{V_2}. Clearly, this effect represents an inherent limitation on the slopes and thus on the pole Q's which are practical for the cascade realization, regardless how insensitive the individual sections may be.‡

To give a quantitative measure, Fig. 7-54 shows the standard deviation of the gain of active bandpass filters realized using some of the design techniques described in this chapter[5,23] when the passive components have a uniformly

---

† Note that we are discussing the *logarithmic* gain 20 log $|V_{out}/V_{in}|$. The component gains should therefore be added.

‡ The above argument is due to H. J. Orchard.

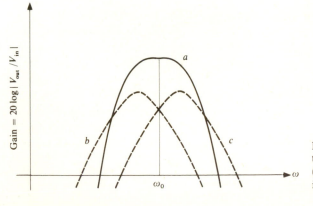

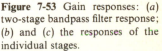

Figure 7-53 Gain responses: (*a*) two-stage bandpass filter response; (*b*) and (*c*) the responses of the individual stages.

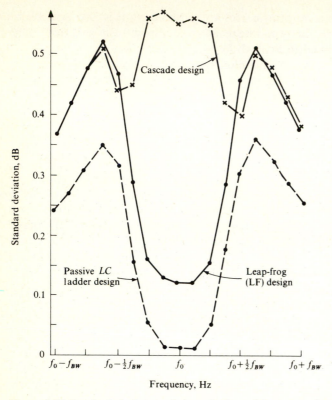

Figure 7-54 Standard deviation of the gain for ± 0.2 percent element tolerance for sixth-degree bandpass filter (from Ref. 5).

distributed tolerance with a 0.1 percent standard deviation. In the passband (near ω_0), the sensitivity of the passive circuit is near zero, and that of the leapfrog filter is also very low, while the response of the cascade realization is distorted by about 0.6 dB, an intolerably large amount. In the stopbands, all active realizations show about the same maximum amount of distortion, 0.5 dB, which is acceptable in most applications.

We close this chapter by listing some of the topics which space limitations prevented us from discussing. These include Linvill's historically important NIC-filter realization (Probs. 7-29 and 7-30), no longer used because of its high sensitivity, Horowitz's method (Probs. 7-31 and 7-32), and many other design techniques fallen into misuse because of one drawback or another; Ref. 3 contains a thorough discussion of many of them.

We have also ignored the many sensitivity and performance measures applicable to biquadratic filter sections. They tend to be confusing and must be interpreted carefully to draw any meaningful conclusions.† The interested reader

---

† According to the tongue-in-cheek hypothesis advanced in Ref. 3, "it is always possible to find a measure of sensitivity by which any given filter is less sensitive than any other."

will find a discussion of these measures and their applications in Refs. 1, 3, and 7.

Similarly, we have omitted any discussion of such important practical topics as stability, dynamic range, noise, nonideal amplifier behavior, tuning, etc. Again, other sources such as Refs. 3 and 5 should be consulted on these subjects.

7-8 SUMMARY

The topics discussed in this chapter were the following:
1. The physical reasons why it is difficult to build high-quality miniature inductors (Sec. 7-1).
2. The necessary conditions for the impedance matrix of an RC two-port and the conclusion that the transfer function of an RC two-port can only have simple negative real poles (although unrestricted zeros) and hence is inefficient for such applications as filtering or delay correction (Sec. 7-1).
3. The properties and limitations of the operational amplifier (op-amp): the vanishing of the input voltage and current and also of the output impedance in a stable feedback configuration (Sec. 7-2).
4. The definition and realization of some active building blocks: VCVS, CCVS, VCCS, CCCS, generalized impedance converter (GIC), negative impedance converter (NIC), generalized impedance inverter (GII), gyrator and frequency-dependent negative resistor (FDNR), using op-amps (Sec. 7-3).
5. Active-circuit design using inductance simulation: the capacitor-terminated gyrator as an inductor, the effects of nonideal gyrators, the simulation of a floating inductor (Sec. 7-4).
6. Active-circuit design using impedance transformation: the invariance of the voltage ratio to the scaling of all impedances by $1/s$; the resulting transformation of a RLC circuit into a CR-FDNR one; the realization of resistive terminations (Sec. 7-5).
7. Active-circuit design using the leapfrog (active-ladder) circuit; modeling of the Kirchhoff equations, realization using op-amps (Sec. 7-5).
8. Cascade realizations: Sallen-Key filter, Kerwin's circuit for the realization of $j\omega$-axis transmission zeros, the calculation of sensitivities (Sec. 7-6).
9. Low-sensitivity biquadratic sections using the GIC (Sec. 7-7).
10. The state-variable circuit as a low-sensitivity biquadratic section; its realization using op-amps (Sec. 7-7).

This chapter contains a large number of topics and circuits in what we hope is a fairly logical order of exposition. A look at the references listed at the end of the chapter and used in assembling this material will convince the reader that this subject is still very much in the center of attention of research workers all over the world. Hence, in contrast to the material in earlier chapters, which has reached maturity and thus some measure of finality, it is possible that major new development will occur in the theory and design of linear active circuits after this book goes into press.

Up to this point, all circuits discussed in this book contained exclusively lumped elements. In the next chapter, this restriction will be dropped, and distributed elements will be permitted in the circuit.

PROBLEMS

7-1 Assume the op-amp model used in Fig. 7-10, as well as $R = R_1 = R_2 = 1$ kΩ, for the circuits shown in Fig. 7-12a to c. Carry out the exact analysis of all circuits and compare your results with the values obtainable when the op-amp is considered to be ideal [Eqs. (7-37) to (7-42)].

7-2 Analyze the circuits of Fig. 7-13 assuming $R = R_1 = R_2 = 1$ kΩ, and (a) ideal op-amps; (b) the op-amp model of Fig. 7-10. Compare your results.

7-3 Show that if the primary port of a GIC is terminated in Z_{L_1} (Fig. 7-14), the impedance seen at the secondary port is $Z_2 = Z_{L_1}/f(s)$. Hence, the GIC operates in both directions, although with different conversion factors.

7-4 Show that [except for the trivial case $f(s) \equiv n^2$, where n is a real constant] a GIC cannot be a reciprocal two-port. What are the circuits corresponding to $f(s) \equiv n^2$?

7-5 Show that the circuits of Fig. 7-55a and b operate as a CNIC and a VNIC, respectively.

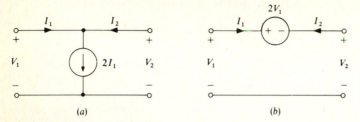

(a)   (b)

Figure 7-55 Negative-impedance converter circuits: (a) CNIC; (b) VNIC.

7-6 Show that if the primary port of a GII is terminated in Z_{L_1} (Fig. 7-16), the impedance at the secondary port is $Z_2 = g'(s)/Z_{L_1}$. What is $g'(s)$?

7-7 Show that the circuits of Fig. 7-56 function as GIIs. What is $g(s)$ for these networks?

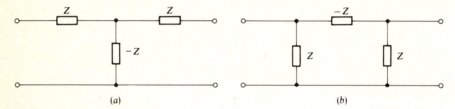

(a)   (b)

Figure 7-56 Generalized impedance inverter circuits.

7-8 Under what conditions is (a) a GIC and (b) a GII a passive two-port? Under what conditions is it lossless?

7-9 Find the chain parameters A and D for a lossless GIC with a conversion factor $f(s)$. What is the circuit if $f(s) = n^2$?

7-10 Analyze the effect of capacitor leakage (represented, as in Fig. 7-2, by a parallel conductance) on the Q of the simulated inductor in the circuit of Fig. 7-21a.

7-11 Analyze the effect of nonzero gyrator admittances

$$y_{11} = y_{22} = \frac{\varepsilon}{R}$$

on the Q of the simulated inductor (Fig. 7-21a).

7-12 Extend the equivalence of Fig. 7-24 to the case of active (rather than lossless) gyrators. Show that if the first gyrator has gyration resistors R_1 and R_2 while the second has R_3 and R_4, the floating-inductor simulation requires $R_1 = R_4$ and $R_2 = R_4$. What is L_{eq}?

7-13 Show what effect a mismatch of the gyration resistances will have on the simulated two-port of Fig. 7-24b. Assume that the first gyrator has gyration resistance R while the second has $(1 + \varepsilon)R$, where $|\varepsilon| \ll 1$.

7-14 Analyze the two-port shown in Fig. 7-25. Compare its chain matrix with that shown in Fig. 7-24b. What are your conclusions? *Warning*: Note that for the output GIC the input and output ports have been interchanged.

7-15 (a) What are all the choices of $Z_1, Z_2, \ldots, Z_5$ in the circuit of Fig. 7-19a which result in an FDNR?

(b) What choice gives $Z = Ks^{-3}$?

7-16 What should be the relations between the element values of the dual circuits shown in Fig. 7-23a and Fig. 7-28b if the transfer functions are to be the same (except for a constant factor)?

7-17 Derive the transfer functions of the circuits shown in Fig. 7-41. Express in terms of C_i, G_k:
    (a) The minimum K value for which the natural modes are complex.
    (b) The maximum value of K for which the circuit is stable.

7-18 Can the circuits of Fig. 7-41 function for negative values of K? Why or why not?

7-19 Design a Sallen-Key filter for $A_V(s) = Ks/(s^2 + 2s + 1)^2$.

7-20 What is the transfer function of the circuit of Fig. 7-45 if $Y \equiv 0$? What is the relation between the moduli of the poles and zeros of $A_V(s)$?

7-21 Analyze the circuit of Fig. 7-45 for $Y = G$ to verify Eq. (7-128) and for $Y = saC$ to derive Eq. (7-129).

7-22 What is the transfer function of the GIC filter (Fig. 7-47c) if $Y_0 = sC_0 + G_0$, $Y_1 = G_1$, $Y_2 = G_2$, $Y_3 = G_3$, $Y_4 = sC_4$, $Y_5 = 0$, $Y_A = G_A$, $Y_B = 0$, and $G_1 G_3 = G_0 G_2$?

7-23 What is the transfer function of the GIC filter of Fig. 7-47c for $Y_0 = sC_0$, $Y_1 = sC_1 + G_1$, $Y_2 = G_2$, $Y_3 = G_3$, $Y_4 = sC_4$, $Y_5 = 0$, $Y_A = G_A$, $Y_B = 0$, and $G_1 G_3 = G_0 G_2$?

7-24 For the GIC filter of Fig. 7-47c, let $Y_0 = G_0$, $Y_1 = G_1$, $Y_2 = sC_2$, $Y_3 = G_3$, $Y_4 = G_4$, $Y_5 = 0$, $Y_A = G_A$, and $Y_B = sC_B$. Under what condition will the section be an allpass circuit?

7-25 (a) Assign the immittances of the circuit of Fig. 7-47c such that the transfer function has one real positive zero and two conjugate complex poles. Repeat for one negative real zero.

    (b) What is the output voltage of the lower op-amp for the circuits of Figs. 7-48 to 7-51? What filters can be obtained by selecting this voltage as V_{out}?

7-26 Analyze the circuit of Fig. 7-57, assuming that node C is the output terminal. Show that A_V has a finite nonzero transmission zero on the $j\omega$ axis. Find the element values so as to realize the transfer function given in Eq. (7-154). Analyze the change in $A_V(j1)$ if R_6 and C (the element in parallel with R_6) changes by 1 percent.

7-27 Analyze the circuit of Fig. 7-57 if the output is (a) at node B; (b) at node C.

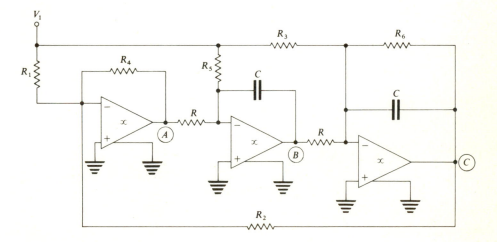

Figure 7-57 Modified state-variable filter with $j\omega$-axis transmission zero.

7-28 Analyze the op-amp filter section shown in Fig. 7-58. Find the element values so as to realize

$$A_V = \frac{1}{s^2 + 0.1s + 1}$$

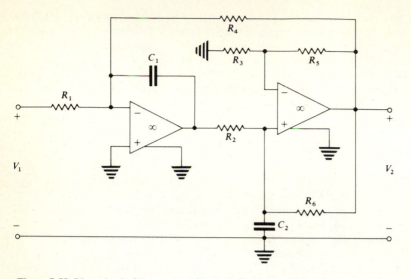

Figure 7-58 Biquadratic filter section. (*P. R. Geffe, RC-Amplifier Resonators for Active Filters, IEEE Trans. Circuit Theory, December 1968, pp. 415–419: fig. 10, p. 419.*)

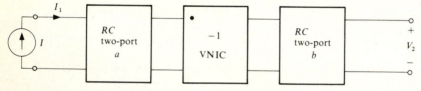

Figure 7-59 *RC*-NIC cascade (Linvill's circuit).

7-29 Show that the basic limitation of *RC* two-ports, i.e., the restriction of their natural modes to the $-\sigma$ axis, can be overcome by including an NIC in the circuit, as shown in Fig. 7-59. *Hint:* Show that the transfer impedance of the circuit is

$$Z_T \triangleq \frac{V_2}{I_1} = \frac{z_{12a}z_{12b}}{z_{22a} - z_{11b}}$$

This is the famous *Linvill circuit*, one of the first practical active filters.

7-30 Realize the transfer function

$$Z_T = \frac{1}{s^2 + 0.1s + 1}$$

using the Linvill circuit shown in Fig. 7-60. How much will $Z_T(j1)$ vary if C_1, or C_2, or the inversion factor K changes by 1 percent? What do you conclude from your results?

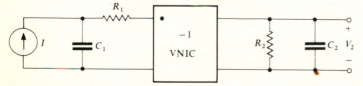

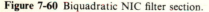

Figure 7-60 Biquadratic NIC filter section.

7-31 Show that the circuit of Fig. 7-61 may have natural modes anywhere in the complex LHP. Find its transfer function $A_V(s)$. (This structure is due to I. M. Horowitz.)

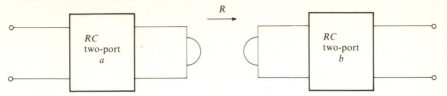

Figure 7-61 RC-gyrator cascade (Horowitz's circuit).

7-32 Design the circuit of Fig. 7-62 from its voltage transfer ratio

$$A_V = \frac{3}{s^2 + 3s + 3}$$

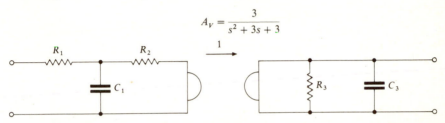

Figure 7-62 Biquadratic RC-gyrator filter section.

REFERENCES

1. Mitra, S. K.: "Analysis and Synthesis of Linear Active Networks," Wiley, New York, 1969.
2. Rupprecht, W.: "Netzwerksynthese," Springer, Berlin, 1972.
3. Heinlein, W. E., and W. H. Holmes: "Active Filters for Integrated Circuits," Prentice-Hall, London, 1974.
4. Orchard, H. J.: Gyrator Circuits, in L. P. Huelsman (ed.), "Active Filters," McGraw-Hill, New York, 1970.
5. A Review of Active Filter Terminology, Technology and Design, report by the Subcommittee on Active Filters of the IEEE Circuits and Systems Society, March 1975.
6. Antoniou, A.: Novel RC-active Network Synthesis Using Generalized Immittance Converters, *IEEE Trans. Circuit Theory*, May 1970, pp. 212–217.
7. Mitra, S. K.: Active Filters with Lumped RC Networks, in G. C. Temes and S. K. Mitra (eds.), "Modern Filter Theory and Design," Wiley, New York, 1973.
8. Riordan, R. H. S.: Simulated Inductors Using Differential Amplifiers, *Electron. Lett.*, February 1967, pp. 50–51.
9. Orchard, H. J.: Inductorless Filters, *Electron. Lett.*, June 1966, pp. 224–225.
10. Holt, A. G. J., and J. Taylor: Method of Replacing Ungrounded Inductors by Grounded Gyrators, *Electron. Lett.*, June 1965, p. 105.
11. Desoer, C. A., and E. S. Kuh: "Basic Circuit Theory," McGraw-Hill, New York, 1969.
12. Deboo, G. J.: Application of a Gyrator-Type Circuit to Realize Ungrounded Inductors, *IEEE Trans. Circuit Theory*, March 1967, pp. 101–102.
13. Bruton, L. T.: Network Transfer Functions Using the Concept of Frequency-dependent Negative Resistance, *IEEE Trans. Circuit Theory*, August 1969, pp. 406–408.
14. Girling, F. E. J., and E. F. Good: Active Filters, pt. 12–16, *Wireless World*, July–December, 1970.
15. Szentirmai, G.: Synthesis of Multi-Feedback Active Filters, *Bell Syst. Tech. J.*, April 1973, pp. 527–555.
16. Cheung, H. W-C.: Sensitivities in the Leapfrog Filter, M.S. thesis, University of California, Los Angeles, 1973.

17. Kerwin, W. J.: Active *RC* Network Synthesis Using Voltage Amplifiers, in L. P. Huelsman (ed.), "Active Filters," McGraw-Hill, New York, 1970.
18. Fliege, N.: "A New Class of Second-Order *RC*-active Filters with Two Operational Amplifiers, *Nachrichtentech. Z.*, June 1973, pp. 279–282.
19. Bhattacharyya, B. B., W. B. Mikhael, and A. Antoniou: Design of *RC*-active Networks Using Generalized Immittance Converters, *J. Franklin Inst.*, January 1974, pp. 45–58.
20. Wait, J. V., L. P. Huelsman, and G. A. Korn: "Introduction to Operational Amplifier Theory and Applications," McGraw-Hill, New York, 1975.
21. Geffe, P. R.: *RC*-Amplifier Resonators for Active Filters, *IEEE Trans. Circuit Theory*, December 1968, pp. 415–419.
22. Muir, A. J. L., and A. E. Robinson: Design of Active *RC* Filters Using Operational Amplifiers, *Sys. Technol.*, April 1968, pp. 18–30.
23. Tow, J.: Design and Evaluation of Shifted Companion Form (Follow the Leader) Active Filters, *Proc. 1974 IEEE Int. Symp. Circuits Syst.*, April 1974, pp. 656–660.
24. Bruton, L. T.: Nonideal Performance of Two-Amplifier Positive-Impedance Converters, *IEEE Trans. Circuit Theory*, November 1970, pp. 541–549.

CHAPTER EIGHT

THE SYNTHESIS OF DISTRIBUTED NETWORKS

8-1 DISTRIBUTED CIRCUITS; THE UNIFORM TRANSMISSION LINE

As defined in Sec. 1-1, a distributed network contains elements whose physical dimensions are comparable to the wavelength of the signal inside the element. In such a network the variations of the voltages and currents in the elements can no longer be approximated by the idealized distribution depicted in Fig. 1-7; instead, their actual dependence on the spatial coordinates, due to the propagation of the electromagnetic waves in the elements and wiring, must be taken into account. As will be seen, even with very restrictive simplifying assumptions, this condition results in more complicated mathematical procedures in both the analysis and the synthesis of these circuits. Why do we then bother to use such networks? There are two basic situations in which distributed circuits are used.

For *high-frequency circuits*, the wavelength λ of the signal (even in free space or air) can be so small that it becomes impossible to construct high-quality components with physical dimensions which are negligible† compared with λ; also, the variation of signal along the wires and leads connecting the components, the radiation of signal energy into the surrounding space, etc., can no longer be ignored. In this situation, therefore, the circuit *must* be regarded and designed as a

---

† Recall the derivations of Sec. 7-1 showing the difficulty of building a high-Q inductor of very small dimensions.

distributed network. To be specific, consider the well-known formula giving the wavelength λ of a propagating sine-wave signal[1]

$$\lambda = \frac{v}{f} \qquad (8\text{-}1)$$

where v is the propagation velocity of the signal and f its frequency. (This formula will be derived below shortly.) Let $f = 10$ GHz $= 10^{10}$ Hz. For such a high frequency the wavelength will be very small; assuming $v = 3 \times 10^8$ m/s (the speed of light), we get only

$$\lambda = \frac{3 \times 10^8 \text{ m/s}}{10^{10} \text{ Hz}} = 0.03 \text{ m}$$

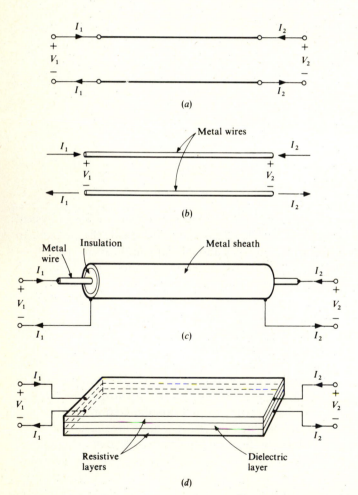

Figure 8-1 (a) Symbol for a transmission line; (b) parallel-wire line; (c) coaxial line; (d) three-layer thin-film line.

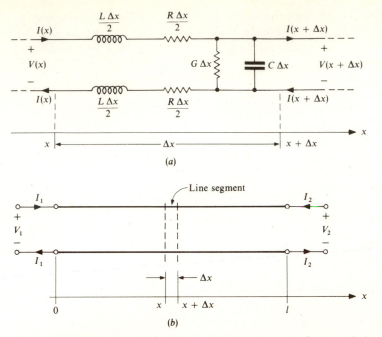

Figure 8-2 (*a*) Equivalent circuit of a Δx-meter-long segment of a transmission line, (*b*) the location of the line segment along the line.

It is not practical to make circuits with lumped discrete components which have dimensions negligibly small compared with 3 cm. In practice, for frequencies above 1 GHz (the 1 to 10^3 GHz frequency band is often called the *microwave band*) we almost invariably design all circuits as distributed networks.

The second situation where distributed circuits are used involves integrated thin-film resistor-capacitor lines. These circuit elements can be manufactured inexpensively and in very small sizes. Furthermore, in some applications, a single such line can replace up to five discrete components. Hence, they may be used as replacements for lumped elements to save cost and/or space.

For these reasons, many circuit designers are sooner or later confronted with a synthesis problem involving distributed circuits. It is therefore important to include at least the fundamentals of the relevant design techniques in a text devoted to circuit synthesis.

The basic building block in the distributed circuits to be discussed in this chapter is the *transmission line*. This is a two-port device; its symbol is illustrated in Fig. 8-1*a*. Some physical realizations of a transmission line are illustrated in Fig. 8-1*b* to *d*. A common feature of these devices is that they contain two current-carrying conductors separated by an insulating dielectric material (which may be air). Hence, an elementary section of such a line, Δx meters long, can be represented by the equivalent circuit† shown in Fig. 8-2*a*. In the circuit, $L/2$ and $R/2$

---

† Note that this circuit already implies a number of simplifying postulates about the line,[1] which, however, are valid for a wide range of geometries.

represent respectively the inductance and resistance of each conductor of the line per unit length, i.e., per meter. Similarly, C and G represent the capacitance and the leakage conductance, respectively, of the insulation between the conductors, also per unit length of the line. We assume that the parameters L, R, G, and C do not depend on the location x of the segment along the line (Fig. 8-2b), i.e., that the device is a *uniform transmission line*.

If the length of the segment Δx is differentially small, then $\Delta x \ll \lambda$ and the Kirchhoff laws can be applied.

In the s domain, therefore,

$$V(x + \Delta x) = V(x) - I(x)(sL + R)\,\Delta x$$

$$I(x + \Delta x) = I(x) - V(x)(sC + G)\,\Delta x \tag{8-2}$$

If $\Delta x \to 0$, then from (8-2)

$$-(sL + R)I(x) = \frac{V(x + \Delta x) - V(x)}{\Delta x} \to \frac{\partial V(x)}{\partial x}$$

$$-(sC + G)V(x) = \frac{I(x + \Delta x) - I(x)}{\Delta x} \to \frac{\partial I(x)}{\partial x} \tag{8-3}$$

Next, we assume a solution of (8-3) in the form

$$V(x) = V_f e^{-\gamma x} + V_r e^{\gamma x} \qquad I(x) = I_f e^{-\gamma x} + I_r e^{\gamma x} \tag{8-4}$$

for the voltage and current. Here, γ, V_f, V_r, I_f, and I_r are functions of s but not of x. Substituting into (8-3) gives

$$-(sL + R)(I_f e^{-\gamma x} + I_r e^{\gamma x}) = -\gamma V_f e^{-\gamma x} + \gamma V_r e^{\gamma x}$$

$$-(sC + G)(V_f e^{-\gamma x} + V_r e^{\gamma x}) = -\gamma I_f e^{-\gamma x} + \gamma I_r e^{\gamma x} \tag{8-5}$$

Since (8-5) must hold for all values of x, the coefficients of $e^{-\gamma x}$ must match on the two sides of both equations. Hence

$$(sL + R)I_f - \gamma V_f = 0 \qquad (sC + G)V_f - \gamma I_f = 0 \tag{8-6}$$

must be valid. Ignoring the trivial case $V_f = I_f = 0$, we therefore have

$$\frac{V_f}{I_f} = \frac{sL + R}{\gamma} = \frac{\gamma}{sC + G} \tag{8-7}$$

and hence

$$\gamma = \sqrt{(sL + R)(sC + G)} \tag{8-8}$$

γ is called the *propagation constant* of the line. Matching the coefficients of $e^{\gamma x}$ in (8-5) gives in an entirely similar way

$$\frac{V_r}{I_r} = -\frac{sL + R}{\gamma} = -\frac{\gamma}{sC + G} \tag{8-9}$$

which also leads to (8-8).

Now let $s = j\omega$. Then V and I can be regarded as phasors of the sinusoid signals

$$v(x, t) = \text{Re } V(x, \omega)e^{j\omega t} \qquad i(x, t) = \text{Re } I(x, \omega)e^{j\omega t} \tag{8-10}$$

as explained in Sec. 2-1. Denoting also

$$\gamma(j\omega) = \alpha(\omega) + j\beta(\omega) \tag{8-11}$$

we obtain from (8-4) and (8-10)

$$v(x, t) = \text{Re } (V_f e^{-(\alpha + j\beta)x + j\omega t} + V_r e^{(\alpha + j\beta)x + j\omega t})$$
$$= |V_f|e^{-\alpha x} \cos (\omega t - \beta x + \varphi_f) + |V_r|e^{\alpha x} \cos (\omega t + \beta x + \varphi_r) \tag{8-12}$$

where φ_f and φ_r are the phase angles of V_f and V_r, respectively. Consider the first term of $v(x, t)$. It represents a cosine function of x and t, with an amplitude which diminishes exponentially toward increasing values of x (Fig. 8-3a). The maxima of the cosine function at a fixed time t occur at those values of x which satisfy

$$\omega t - \beta x + \varphi_f = 2k\pi \qquad k = 0, \pm 1, \pm 2, \ldots \tag{8-13}$$

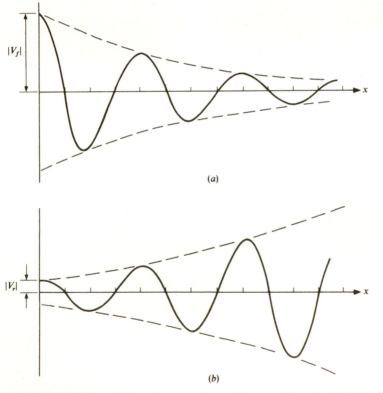

(a)

(b)

Figure 8-3 Variations of the voltage at time t along a uniform transmission line: (a) a forward wave, (b) reflected wave.

At time $t + \Delta t$, these same maxima are at the locations $x + \Delta x$ satisfying

$$\omega(t + \Delta t) - \beta(x + \Delta x) + \varphi_f = 2k\pi \qquad k = 0, \pm 1, \pm 2, \ldots \qquad (8\text{-}14)$$

Substracting (8-13) from (8-14) gives

$$\omega \, \Delta t - \beta \, \Delta x = 0 \qquad (8\text{-}15)$$

Hence, in the time interval Δt the maxima shifted by $\Delta x = (\omega/\beta) \, \Delta t$. Thus, the velocity of the signal on the line is

$$v = \frac{\omega}{\beta} \qquad (8\text{-}16)$$

Here v is the *phase velocity* of the signal. Since $v > 0$, this signal travels toward increasing x, that is, left to right.

For a given time instant t, the wavelength λ of the signal on the line can be found as the distance between two adjacent maxima. From (8-13), this gives

$$-[\omega t - \beta(x + \lambda) + \varphi_f] + (\omega t - \beta x + \varphi_f) = 2\pi \qquad \beta\lambda = 2\pi \qquad \lambda = \frac{2\pi}{\beta} \qquad (8\text{-}17)$$

Combining (8-16) and (8-17), we have

$$\lambda = \frac{2\pi}{\beta} = \frac{2\pi}{\omega/v} = \frac{v}{f} \qquad (8\text{-}18)$$

as stated earlier in (8-1).

Thus, the first term of $v(x, t)$ as given in (8-12) describes a *forward* wave of wavelength $\lambda = 2\pi/\beta$ traveling with a velocity $v = \omega/\beta$ left to right on the line and getting exponentially attenuated due to the $e^{-\alpha x}$ factor (Fig. 8-3a).

An entirely similar examination of the second term in $v(x, t)$ reveals that it corresponds to a *reflected* wave, propagating right to left with a velocity $-\omega/\beta$, having a wavelength $2\pi/\beta$, and also decreasing exponentially in amplitude at the same rate as the forward wave as it travels along the line (Fig. 8-3b).

Identical considerations, again based on (8-4) and (8-10), show that the current $i(x, t)$ can also be separated into forward and reflected waves of the same properties as the corresponding voltage waves. The corresponding phasors are I_f and I_r.

The ratio of the forward-wave phasors V_f and I_f can be found from (8-7) and (8-8):

$$\frac{V_f}{I_f} = \frac{sL + R}{\sqrt{(sL + R)(sC + G)}} = \sqrt{\frac{sL + R}{sC + G}} \qquad (8\text{-}19)$$

The quantity

$$Z_0 = \sqrt{\frac{sL + R}{sC + G}} \qquad (8\text{-}20)$$

is called the *characteristic impedance* of the line. By (8-9), the reflected-wave phasors are also related by Z_0:

$$\frac{V_r}{I_r} = \frac{-(sL + R)}{\sqrt{(sL + R)(sC + G)}} = -Z_0 \tag{8-21}$$

Referring now to Fig. 8-2b, from (8-4) we have

$$I_1 \triangleq I(0) = I_f + I_r \qquad I_2 \triangleq -I(l) = -I_f e^{-\gamma l} - I_r e^{\gamma l} \tag{8-22}$$

Solving for I_f and I_r gives

$$I_f = \frac{I_1 e^{\gamma l} + I_2}{e^{\gamma l} - e^{-\gamma l}} \qquad I_r = \frac{-I_1 e^{-\gamma l} - I_2}{e^{\gamma l} - e^{-\gamma l}} \tag{8-23}$$

Similarly, from (8-4),

$$V_1 \triangleq V(0) = V_f + V_r \qquad V_2 \triangleq V(l) = V_f e^{-\gamma l} + V_r e^{\gamma l} \tag{8-24}$$

Using (8-19) to (8-21), as well as (8-23), gives directly

$$V_1 = Z_0 I_f - Z_0 I_r = \frac{Z_0}{e^{\gamma l} - e^{-\gamma l}} [I_1(e^{\gamma l} + e^{-\gamma l}) + 2I_2]$$

$$V_2 = Z_0 e^{-\gamma l} I_f - Z_0 e^{\gamma l} I_r = \frac{Z_0}{e^{\gamma l} - e^{-\gamma l}} [2I_1 + I_2(e^{\gamma l} + e^{-\gamma l})] \tag{8-25}$$

or, in terms of hyperbolic functions,

$$V_1 = Z_0 \coth (\gamma l) I_1 + Z_0 \operatorname{csch} (\gamma l) I_2$$

$$V_2 = Z_0 \operatorname{csch} (\gamma l) I_1 + Z_0 \coth (\gamma l) I_2 \tag{8-26}$$

A comparison with Eq. (5-4) of Sec. 5-1 shows that the open-circuit impedance matrix of the uniform transmission line is given by

$$\mathbf{Z} = Z_0 \begin{bmatrix} \coth \gamma l & \operatorname{csch} \gamma l \\ \operatorname{csch} \gamma l & \coth \gamma l \end{bmatrix} \tag{8-27}$$

From (5-18), the short-circuit admittance matrix of the line can also be found.
It is

$$\mathbf{Y} = \frac{1}{\Delta_z} \begin{bmatrix} z_{22} & -z_{12} \\ -z_{12} & z_{11} \end{bmatrix} = \frac{1}{Z_0} \begin{bmatrix} \coth \gamma l & -\operatorname{csch} \gamma l \\ -\operatorname{csch} \gamma l & \coth \gamma l \end{bmatrix} \tag{8-28}$$

since

$$\Delta_z \triangleq \det \mathbf{Z} = Z_0^2 \coth^2 \gamma l - Z_0^2 \operatorname{csch}^2 \gamma l = Z_0^2$$

Finally, it is easy to show (see Prob. 8-1) that the chain matrix of the uniform
transmission line is

$$\mathbf{T} = \begin{bmatrix} A & B \\ C & D \end{bmatrix} = \begin{bmatrix} \cosh \gamma l & Z_0 \sinh \gamma l \\ \dfrac{\sinh \gamma l}{Z_0} & \cosh \gamma l \end{bmatrix} \tag{8-29}$$

8-2 THE SYNTHESIS OF LOSSLESS-TRANSMISSION-LINE NETWORKS

Assume that the line parameters satisfy

$$R \ll \omega L \qquad G \ll \omega C \tag{8-30}$$

at all frequencies of interest. Then the approximation $R = 0$, $G = 0$ can be made. A line for which R and G are both zero is called a *lossless transmission line*. Its propagation constant is, by (8-8), purely imaginary for $s = j\omega$:

$$\gamma = j\beta = \sqrt{(j\omega L)(j\omega C)} = j\omega\sqrt{LC} \tag{8-31}$$

and its characteristic impedance, by (8-20), purely real and frequency-independent:

$$Z_0 = \sqrt{\frac{j\omega L}{j\omega C}} = \sqrt{\frac{L}{C}} \tag{8-32}$$

The two-port-parameter matrices then become, from (8-26) to (8-29),

$$\mathbf{Z} = -jZ_0 \begin{bmatrix} \cot\theta & \csc\theta \\ \csc\theta & \cot\theta \end{bmatrix} \tag{8-33}$$

$$\mathbf{Y} = -\frac{j}{Z_0} \begin{bmatrix} \cot\theta & -\csc\theta \\ -\csc\theta & \cot\theta \end{bmatrix} \tag{8-34}$$

and

$$\mathbf{T} = \begin{bmatrix} \cos\theta & jZ_0\sin\theta \\ \dfrac{j\sin\theta}{Z_0} & \cos\theta \end{bmatrix} \tag{8-35}$$

where we introduced the *electric length* θ defined by

$$\theta \triangleq \beta l = \omega\sqrt{LC}\, l \tag{8-36}$$

and used the relations $\sinh j\theta = j\sin\theta$, $\cosh j\theta = \cos\theta$. Using (8-16), we can also express θ as

$$\theta = \beta l = \frac{\omega}{v} l = \omega \frac{l}{v} = \omega T \tag{8-37}$$

Here $T \triangleq l/v = \sqrt{LC}\, l$ is the time needed for the signal to propagate from one end of the line to the other. It is called the *delay time* of the line.

We shall now restrict the discussion to circuits constructed only from resistors and from lines with the same delay time T. Such circuits will be called *commensurate transmission-line networks*.† Assume first that all lines used in the circuit are open- or short-circuited at their output ports. From (8-33), the input impedance of an open-circuited line is

$$Z_{\text{op}} = z_{11} = -jZ_0 \cot\theta \tag{8-38}$$

---

† Note that by cascading lines of delay time T we can obtain lines of delay times $2T$, $3T$, ... in a commensurate-line circuit.

while from (8-34), the short-circuit input impedance is

$$Z_{sh} = \frac{1}{y_{11}} = jZ_0 \tan \theta \qquad (8\text{-}39)$$

Let us introduce now a new complex frequency variable

$$S = \Sigma + j\Omega \triangleq \tanh sT \qquad (8\text{-}40)$$

(The use of S was originally proposed by Richards.[2] It is often called *Richards'*
variable.)

For $s = j\omega$, we have

$$S = \tanh sT = \tanh j\omega T = j \tan \omega T = j \tan \theta \qquad (8\text{-}41)$$

Hence, by (8-38) and (8-39),

$$Z_{op} = \frac{Z_0}{j \tan \theta} = \frac{Z_0}{S} \qquad (8\text{-}42)$$

and

$$Z_{sh} = Z_0 j \tan \theta = Z_0 S \qquad (8\text{-}43)$$

*Thus, an open-circuited line behaves like a capacitor of value $C = 1/Z_0$, and a
short-circuited line like an inductor of value $L = Z_0$ in terms of S. A resistor, being
frequency-independent, remains a resistor whether we use s or S as our frequency
variable.*

The main features of the mapping $s \to S$ given by (8-40) are shown in Fig. 8-4.
The most striking property, evident from Fig. 8-4a, is the *periodicity* of the map-
ping along the $j\omega$ axis, due to the relation $\tanh (sT - j\pi) = \tanh sT$. This causes
the S plane to be mapped into infinitely many strips, each π/T high, in the s plane.
Also of interest is the fact that quadrants II and III, which together represent the
LHP of the S plane, are mapped into the LHP of the s plane.

For $s = j\omega$, (8-40) becomes $S = \Sigma + j\Omega = \tanh j\omega T = j \tan \omega T$ so $\Sigma = 0$ and

$$\Omega = \tan \omega T \qquad (8\text{-}44)$$

The $\Omega \to \omega$ relation is illustrated in Fig. 8-4b. The periodic behavior along the ω
axis is again evident.

Consider now the properties of a commensurate-line network containing resistors and open- and short-circuited transmission lines of equal lengths. By (8-42)
and (8-43), in terms of Richards' variable S, the three element types have impedances R, Z_0/S, and $Z_0 S$, where R and Z_0 are real positive numbers. By
comparison, a passive RLC circuit contains impedances R, $1/sC$, and sL, with R,
C, and L positive real. *It follows that the realizability conditions on any network
function of a commensurate-line network expressed in terms of S are exactly the same
as those on a lumped RLC circuit in terms of s. Similarly, the design procedures of the
lumped RLC circuit can be directly employed in the synthesis of the commensurate-
line network in terms of the transformed variable S. Thus, for example, the im-*

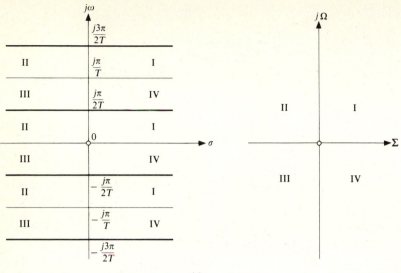

(a)

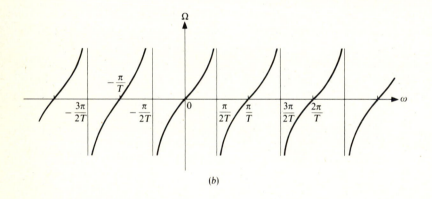

(b)

Figure 8-4 The mapping of (a) the s plane into the S plane and (b) the $j\omega$ axis into the Ω axis.

pedance of a one-port built only from open- and short-circuited commensurate lines is an odd rational function of S, with interlaced simple $j\Omega$-axis zeros and poles positive real residues, and a simple pole or zero at both $S = 0$ and $S \to \infty$. Similarly, the impedance of a one-port containing also resistors in addition to the open- or short-circuited commensurate lines is a rational PR function of S, etc. The realizability conditions on two-port functions can similarly be applied.

Next, the design in the S domain will be illustrated with some examples.

Example 8-1 Examine the realizability of the impedance function

$$Z(s) = 100 \frac{\tanh^3 (10^{-6}s) + 4 \tanh (10^{-6}s)}{\tanh^2 (10^{-6}s) + 1}$$

and realize $Z(s)$.

With the substitution of (8-40), for $T = 1$ µs,

$$Z(S) = 100 \frac{S^3 + 4S}{S^2 + 1} = 100S + \frac{300S}{S^2 + 1}$$

Since, as the results of Chap. 3 show, $Z(S)$ is a reactance function in terms of S, it follows that it can be realized as a commensurate-line network containing only open- and short-circuited lines and no resistors. A hypothetical lumped-reactance circuit which realizes $Z(S)$ in terms of the Richards' variable S is shown in Fig. 8-5a. Since an inductor of value L in the S domain becomes a short-circuited transmission line with characteristic impedance $Z_0 = L$ in the s variable, and a capacitor of value C an open-circuited line with $Z_0 = 1/C$, the S-variable circuit of Fig. 8-5a becomes that of Fig. 8-5b in the s domain. It is instructive

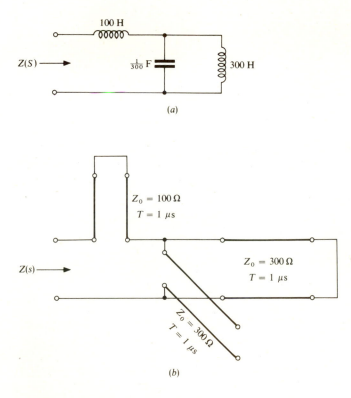

(a)

(b)

Figure 8-5 (a) Lumped S-plane model of a commensurate-line circuit; (b) the commensurate-line network. (*Continued on p. 332.*)

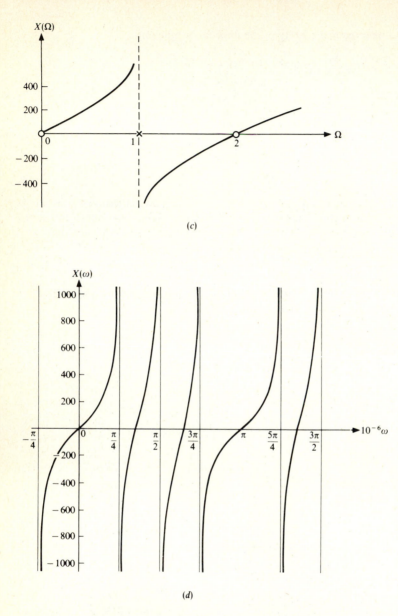

Figure 8-5 (c) Reactance response in the Ω domain; (d) reactance response in the ω domain.

to compare the reactance functions of the two circuits of Fig. 8-5 in their respective frequency domains. They are illustrated in Fig. 8-5c and d. As the curves demonstrate, the physical (ω-axis) response is obtained from the Ω-axis response of the lumped-element model by periodically repeating the (slightly distorted) Ω-axis curve. The exact relation is dictated by Eq. (8-44).

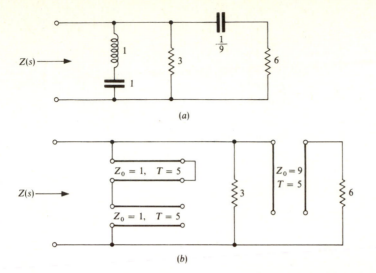

Figure 8-6 (*a*) Lumped *S*-plane model of a commensurate-line circuit; (*b*) commensurate-line network.

Example 8-2 Realize the impedance

$$Z(S) = \frac{2S^3 + 3S^2 + 2S + 3}{S^3 + 3S^2 + 4S + 1}$$

where $S \triangleq \tanh 5s$.

An investigation of this impedance function in terms of the variable s was performed in Sec. 4-2, where it was shown that it is a nonminimum PR impedance and that it can be realized using only pole removals and constant removals. The resulting lumped one-port was shown earlier in Fig. 4-5; it is reproduced in Fig. 8-6*a*. Replacing the reactance elements as before by commensurate lines gives the circuit† shown in Fig. 8-6*b*.

Since, as we have seen, all commensurate-line network functions can be expressed as functions of the mapped variable $S = j\Omega = j \tan \omega T$, all these functions are periodic functions of ω, with a period π/T. Hence, use of commensurate-line networks is restricted to situations where such periodicity of the frequency response is acceptable.

For the important circuit type of *filters*, the periodicity of the loss response means that all commensurate-line filters are multiband filters. Thus, the same filter can be used, in various frequency ranges, as a low-pass, a high-pass, or a bandpass filter (Fig. 8-7*a*). (It must be assured by some additional circuits, of course, that all undesirable signals in the periodically reoccurring passbands are eliminated.) The design of such a commensurate-line filter can then be performed in the following steps:

---

† In this circuit, all element values are normalized. Note that distributed networks, just like lumped ones, are usually designed using impedance and frequency (or time) normalization.

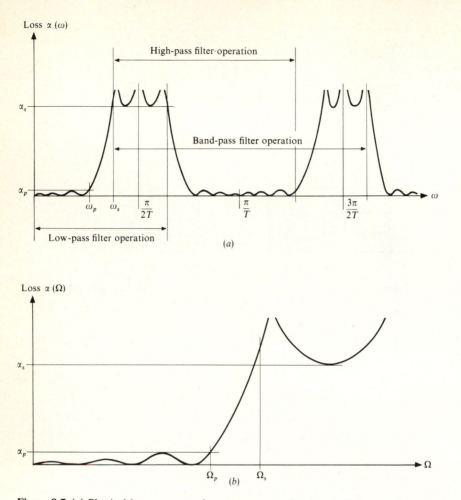

Figure 8-7 (a) Physical loss response of a commensurate-line filter; (b) loss response of a lumped low-pass model filter.

1. A desirable periodic loss response, e.g., that shown in Fig. 8-7a, is specified. The specification usually includes the desired frequency-band limits ω_p and ω_s and the loss limits α_p and α_s (Fig. 8-7a).
2. Using (8-44), the desired features of the response $\alpha(\omega)$ are transformed into the Ω domain (Fig. 8-7b); that is, Ω_p and Ω_s are found. The loss limits α_p and α_s remain unchanged.
3. A lumped prototype circuit is designed which satisfies the Ω-domain loss specifications.
4. The reactance elements of the lumped circuit are transformed, one by one, into open- or short-circuited transmission lines with the aid of Eqs. (8-42) and (8-43).

Example 8-3 The low-pass prototype filter shown in Fig. 8-8a has the response shown in Fig. 8-8b. Calculate the element values and plot the loss response of the commensurate-line distributed filter obtainable from the prototype via the mapping of (8-44). Use $T = 10^{-10}$ s and assume resistive terminations equal to 50 Ω.

From Eq. (8-41), the capacitors c_1 and c_3 of the lumped prototype filter of Fig. 8-8a are transformed into open-circuited transmission lines with characteristic impedances

$$Z_{0_1} = \frac{50}{c_1} = Z_{0_3} = \frac{50}{c_3} \approx 48.6145 \ \Omega$$

Similarly, from Eq. (8-43), the prototype inductance l_2 is transformed into an open-circuited line with characteristic impedance

$$Z_{0_2} = 50 l_2 = (50)(1.1468) = 57.34 \ \Omega$$

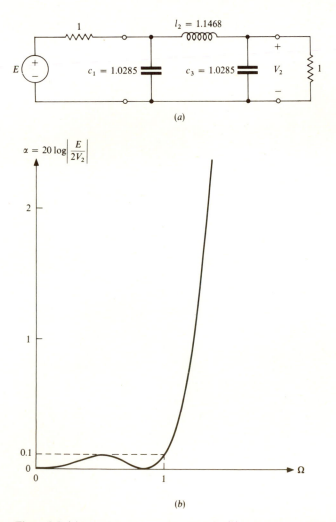

(a)

(b)

Figure 8-8 (a) Lumped low-pass filter circuit; (b) loss response of lumped filter. (*Continued on p. 336.*)

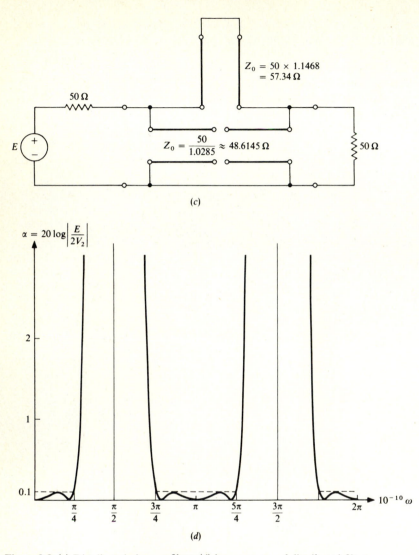

Figure 8-8 (c) Distributed element filter; (d) loss response of distributed filter.

(The factor 50 is due to the change in the terminating resistors from 1 to 50 Ω.) The new circuit is shown in Fig. 8-8c.

The loss response of the transmission-line circuit can be found replacing S by

$$j\Omega = j \tan \omega T = j \tan 10^{-10} \omega$$

in the response of the lumped prototype circuit. The resulting loss response (which has a frequency period of $\pi \times 10^{10}$) is shown in Fig. 8-8d. The first passband limit frequency ω_p is at the transformed image of $\Omega = 1$:

$$\tan 10^{-10} \omega_p = 1 \qquad \omega_p = 10^{10} \tan^{-1} 1 = \frac{\pi}{4} 10^{10}$$

8-3 NETWORK SYNTHESIS USING UNIT ELEMENTS; RICHARDS' THEOREM AND KURODA'S IDENTITIES

The circuits designed using the methods described in the preceding section are often impractical to construct. For example, the network of Fig. 8-8c, constructed using coaxial lines (which very often represent the preferred realization of transmission lines) appears as shown in Fig. 8-9. This circuit has several disadvantages. At microwave frequencies (where these circuits usually operate) the junction points of the lines cause stray capacitances and inductances to appear. The connecting wires behave like lead inductances or as short transmission lines. Also, the sheath of the short-circuited line is not grounded, and hence its capacitance to ground (which is substantial) and the current flowing in it can strongly affect the response. For these reasons, we need additional coupling elements suitable for operation at microwave frequencies. Such an element is a commensurate-length line, used as a two-port rather than as an immittance. It is called a *unit element*. From Eqs. (8-33) to (8-37), for such a lossless line the open-circuit impedance matrix is

$$\mathbf{Z} = -jZ_0 \begin{bmatrix} \cot \omega T & \csc \omega T \\ \csc \omega T & \cot \omega T \end{bmatrix} = Z_0 \begin{bmatrix} \dfrac{1}{S} & \sqrt{S^{-2} - 1} \\ \sqrt{S^{-2} - 1} & \dfrac{1}{S} \end{bmatrix} \tag{8-45}$$

where the trigonometric relation $\csc \theta = (1 + \tan^{-2} \theta)^{1/2}$ was used. Similarly, the other two-port matrices can be found:

$$\mathbf{Y} = \frac{1}{Z_0} \begin{bmatrix} \dfrac{1}{S} & -\sqrt{S^{-2} - 1} \\ -\sqrt{S^{-2} - 1} & \dfrac{1}{S} \end{bmatrix} \qquad \mathbf{T} = \frac{1}{\sqrt{S^{-2} - 1}} \begin{bmatrix} \dfrac{1}{S} & Z_0 \\ \dfrac{1}{Z_0} & \dfrac{1}{S} \end{bmatrix}$$

$$\tag{8-46}$$

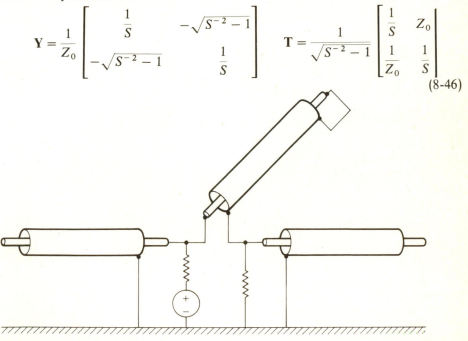

Figure 8-9 Coaxial-filter realization.

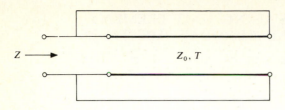

$Z \longrightarrow$ Z_0, T

Figure 8-10 Impedance with irrational dependence on S.

Note that some of the matrix elements are irrational in S. This can cause the driving-point immittance of a circuit containing a unit element also to be an irrational function of S. For example, the impedance of the one-port shown in Fig. 8-10 is given by

$$Z = \frac{Z_0/2}{1/S - \sqrt{S^{-2} - 1}} \tag{8-47}$$

It is possible, however, to design circuits containing unit elements in which all driving-point immittances are rational in S. Such networks (sometimes called *normal networks*) can then again be analyzed and designed using the analogy between the variables s and S. In this chapter, only such normal networks will be discussed.

Consider now the circuit shown in Fig. 8-11. Using Eq. (6-51) [where now R_L is replaced by $Z_1(S)$ and Z_1 by $Z(S)$], we get

$$Z(S) = z_{11}(S) - \frac{z_{12}^2(S)}{z_{22}(S) + Z_1(S)} \tag{8-48}$$

From (8-45), therefore,

$$Z(S) = Z_0\left(\frac{1}{S} - \frac{S^{-2} - 1}{1/S + Z_1/Z_0}\right) = Z_0 \frac{Z_1(S) + SZ_0}{SZ_1(S) + Z_0} \tag{8-49}$$

From (8-49), for $S = 1$,

$$Z(1) = Z_0 \frac{Z_1(1) + Z_0}{Z_1(1) + Z_0} = Z_0 \tag{8-50}$$

It follows from the PR character of $Z(S)$ that $Z_0 = Z(1)$ has a positive real value and the unit element is thus realizable.

Solving (8-49) for $Z_0(S)$ and using (8-50) gives

$$Z_1(S) = Z_0 \frac{Z(S) - SZ_0}{Z_0 - SZ(S)} = Z(1) \frac{Z(S) - SZ(1)}{Z(1) - SZ(S)} \tag{8-51}$$

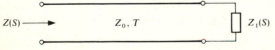

$Z(S) \longrightarrow$ Z_0, T $Z_1(S)$ **Figure 8-11** Unit element with terminating impedance $Z_1(S)$.

$Z_1(S)$ satisfies the following conditions:

Richards' theorem[2] Let $Z(S)$ be a real rational PR function. Then $Z_1(S)$ is also a real rational PR function of S; the degree of $Z_1(S)$ is, in general, no higher than that of $Z(S)$. Finally, if $Z(-1) = -Z(1)$, the degree of $Z_1(S)$ is 1 lower than that of $Z(S)$.

PROOF The real rational character of $Z_1(S)$ follows directly from that of $Z(S)$, by (8-51). The degree of $Z_0(S)$ would, from (8-51), appear to be 1 degree higher than that of $Z(S)$ due to the additional factors S in both numerator and denominator. However, clearly both the numerator and denominator of $Z_1(S)$ contain a factor $S - 1$, since both are rational and both are zero for $S = 1$. The cancellation of this common factor lowers the degree of $Z_1(S)$ by 1, back to what the degree of $Z(S)$ is. Furthermore, if

$$Z(-1) = -Z(1) \tag{8-52}$$

holds, then both the numerator and denominator of $Z_1(S)$ are zero also for $S = -1$, enabling us to cancel a factor $S + 1$ as well. Therefore, in this case the degree of $Z_1(S)$ is 1 degree lower than that of $Z(S)$.

To show the PR character of $Z_1(S)$, the PR conditions a'', b'', and c'' of Sec. 4-1 can be effectively used. Condition a'' requires the real rational character of $Z_1(S)$, which follows directly from that of $Z(S)$ and from (8-51). To show that b'' and c'' also hold, we use impedance normalization so that $Z_0 = Z(1) = 1$. Then, denoting

$$Z(j\Omega) = R(\Omega) + jX(\Omega) \tag{8-53}$$

from (8-51) we have

$$Z_1(j\Omega) = \frac{R + jX - j\Omega}{1 - j\Omega(R + jX)} \tag{8-54}$$

Hence $\ \ \operatorname{Re} Z_1(j\Omega) = \dfrac{R(1 + \Omega X) - (X - \Omega)\Omega R}{(1 + \Omega X)^2 + \Omega^2 R^2} = \dfrac{R(1 + \Omega^2)}{(1 + \Omega X)^2 + \Omega^2 R^2} \geq 0$

$$\tag{8-55}$$

since $R(\Omega) \geq 0$ for a PR function $Z(S)$. Thus, condition b'' is satisfied.

Test c'' restricts the zeros of

$$Z_1(S) + 1 = \frac{Z(S) - S + 1 - SZ(S)}{1 - SZ(S)} = \frac{(1 - S)[1 + Z(S)]}{1 - SZ(S)} \tag{8-56}$$

to the inside of the LHP. This condition is apparently violated by the $1 - S$ factor in the numerator. Recall, however, that this factor cancels since the denominator is zero for $S = 1$. The remaining zeros are those of $1 + Z(S)$, which are inside the LHP since $Z(S)$ is PR. Hence, c'' holds as well, and the PR character of $Z_1(S)$ follows.

This completes the proof of Richards' theorem. A different proof is sketched in Prob. 8-12. The reader is urged to follow through that derivation (which is the one often used to prove Richards' theorem) as well.

Let now $Z(S)$ be a reactance. Then the numerator of $Z_1(S)$, as given in (8-51), is an odd function of S, and its denominator an even function of S. Hence, $Z_1(S)$ is an odd PR function, i.e., another reactance. Furthermore, the odd character of $Z(S)$ guarantees that (8-52) holds; hence, $Z_1(S)$ has a degree 1 lower than that of $Z(S)$. It follows that by repeating the extraction of cascaded unit elements the degree of any reactance function $Z(S)$ can be reduced to zero and the circuit completely synthesized. The last unit element will have a short-circuited (open-circuited) output port if $Z(S) \to 0$ $[Z(S) \to \infty]$ for $S \to 0$.

Example 8-4 The reactance function

$$Z(S) = 100 \frac{S^3 + 4S}{S^2 + 1}$$

was synthesized, using only open- and short-circuited lines, in Sec. 8-2. The resulting circuit, shown in Fig. 8-5b, suffers from the disadvantages discussed in connection with Fig. 8-9 earlier in this section. Hence, a new realization will be obtained utilizing cascaded unit elements. Since each unit-element extraction lowers the degree of $Z(S)$ by 1, the circuit of Fig. 8-12a is anticipated. The characteristic impedances Z_{0_i} can be found by using Eqs. (8-50) and (8-51) repeatedly. Thus, by (8-50),

$$Z_{0_1} = Z(1) = (100)(\tfrac{5}{2}) = 250 \ \Omega$$

and the remainder impedance is, by (8-51),

$$Z_1(S) = 250 \frac{100(S^3 + 4S)/(S^2 + 1) - S250}{250 - S100(S^3 + 4S)/(S^2 + 1)} = 250 \frac{-1.5S^3 + 1.5S}{-S^4 - 1.5S^2 + 2.5}$$

According to our previous discussions, $Z_1(S)$ must contain a common factor $(S + 1) \times (S - 1) = S^2 - 1$ in its numerator and denominator.

Indeed,

$$Z_1(S) = 250 \frac{-1.5(S^2 - 1)S}{-(S^2 - 1)(S^2 + 2.5)} = \frac{375S}{S^2 + 2.5}$$

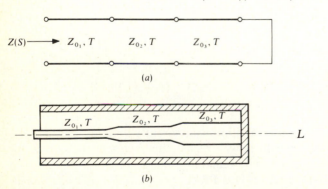

(a)

(b)

Figure 8-12 Unit-element realization of a reactance: (a) circuit; (b) coaxial arrangement.

Repeating these steps, we obtain the characteristic impedance

$$Z_{0_2} = Z_1(1) = \frac{375}{3.5} \approx 107.143 \ \Omega$$

and the remainder impedance

$$Z_2(S) = Z_{0_2}\frac{Z_1(S) - SZ_{0_2}}{Z_{0_2} - SZ_1(S)} = \frac{375}{3.5}\frac{-S^3 + S}{-2.5S^2 + 2.5} = \frac{375}{(3.5)(2.5)}S$$

Finally

$$Z_{0_3} = Z_2(1) = \frac{375}{(3.5)(2.5)} \approx 42.857 \ \Omega$$

$$Z_3(S) = Z_{0_3}\frac{Z_{0_3}S - Z_{0_3}S}{Z_{0_3} - Z_{0_3}S^2} \equiv 0$$

This completes the synthesis. A schematic cross-sectional illustration of the coaxial-line realization of this circuit is given in Fig. 8-12b. Refs. 1 and 3 contain useful information on the dimensioning of coaxial and other transmission lines from specified Z_0 and T.

Example 8-5 Next, the nonminimum impedance function

$$Z(S) = \frac{24S^2 + 2S + 3}{8S^2 + 1}$$

will be synthesized. Since $Z(1) = \frac{29}{9}$, while $Z(-1) = \frac{25}{9}$, Eq. (8-52) is not satisfied and a direct removal of a unit element in cascade does not reduce the degree of $Z(S)$. Hence, we remove instead a series resistance R_1 such that the remainder $Z_1(S) = Z(S) - R_1$ satisfies (8-52). Thus, we require the condition

$$Z(1) - R_1 = -Z(-1) + R_1$$

to hold. This gives

$$R_1 = \tfrac{1}{2}[Z(1) + Z(-1)] = 3$$

The remainder impedance is

$$Z_1(S) = \frac{24S^2 + 2S + 3}{8S^2 + 1} - 3 = \frac{2S}{8S^2 + 1}$$

which is a reactance. Hence, two cascaded unit elements realize the rest of the circuit. The complete network is shown in Fig. 8-13.

It should be pointed out that the design process used in Example 8-4 does not necessarily lead to a realizable structure. As Prob. 8-13 illustrates, the technique may lead to difficulties even for impedance functions which can be realized using open- and short-circuited line stubs.

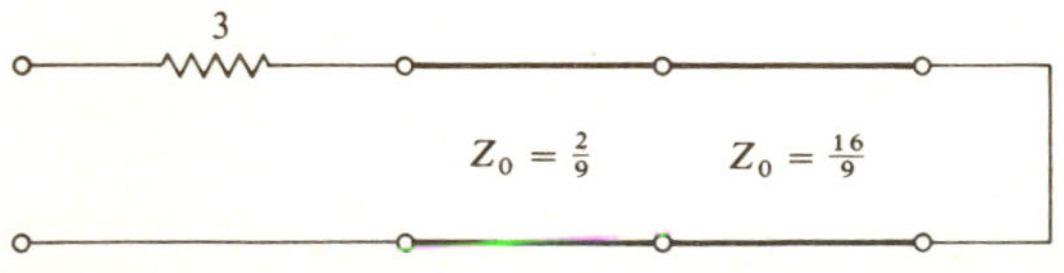

Figure 8-13 Unit-element realization of an impedance.

The application of unit elements in the design of distributed-element reactance two-ports is greatly facilitated by the so-called *Kuroda identities*.[3,4] The general form of these identities is illustrated in Fig. 8-14. The two two-ports shown are equivalent if their chain matrices are the same. Using (8-46), this requires $\mathbf{T} = \mathbf{T}'$ where

$$
\mathbf{T} = \frac{1}{\sqrt{S^{-2}-1}} \begin{bmatrix} A & B \\ C & D \end{bmatrix} \begin{bmatrix} \dfrac{1}{S} & Z_0 \\ \dfrac{1}{Z_0} & \dfrac{1}{S} \end{bmatrix} = \frac{1}{\sqrt{S^{-2}-1}} \begin{bmatrix} \dfrac{A}{S}+\dfrac{B}{Z_0} & AZ_0+\dfrac{B}{S} \\ \dfrac{C}{S}+\dfrac{D}{Z_0} & CZ_0+\dfrac{D}{S} \end{bmatrix}
$$

$$
\mathbf{T}' = \frac{1}{\sqrt{S^{-2}-1}} \begin{bmatrix} \dfrac{1}{S} & Z_0' \\ \dfrac{1}{Z_0'} & \dfrac{1}{S} \end{bmatrix} \begin{bmatrix} A' & B' \\ C' & D' \end{bmatrix} = \frac{1}{\sqrt{S^{-2}-1}} \begin{bmatrix} \dfrac{A'}{S}+C'Z_0' & \dfrac{B'}{S}+D'Z_0' \\ \dfrac{A'}{Z_0'}+\dfrac{C'}{S} & \dfrac{B'}{Z_0'}+\dfrac{D'}{S} \end{bmatrix}
$$

$$(8\text{-}57)$$

Consider now the primary open-circuit parameter z_{11} and z_{11}' of the two two-ports (Fig. 8-14). From Eqs. (5-28) and (8-57),

$$
z_{11} = \frac{A/S + B/Z_0}{C/S + D/Z_0} \tag{8-58}
$$

and

$$
z_{11}' = \frac{A'/S + C'Z_0'}{A'/Z_0' + C'/S} \tag{8-59}
$$

For $S = 1$, equating z_{11}' and z_{11} gives

$$
z_{11}'(1) = \frac{A' + C'Z_0'}{A'/Z_0' + C'} = Z_0' = z_{11}(1) = \frac{A(1) + B(1)/Z_0}{C(1) + D(1)/Z_0} \tag{8-60}
$$

Hence, the characteristic impedance of the transmission line in Fig. 8-14*b* is

$$
Z_0' = \frac{A(1)Z_0 + B(1)}{C(1)Z_0 + D(1)} \tag{8-61}
$$

Next, equating the matrix elements in the first columns of $\mathbf{T}$ and $\mathbf{T}'$ gives

$$
\frac{A'}{S} + C'Z_0' = \frac{A}{S} + \frac{B}{Z_0} \qquad \frac{A'}{Z_0'} + \frac{C'}{S} = \frac{C}{S} + \frac{D}{Z_0} \tag{8-62}
$$

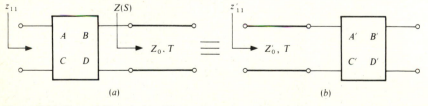

(a) (b)

Figure 8-14 General Kuroda identity.

leading to

$$A' = \frac{A + (S/Z_0)B - (SZ_0')C - S^2(Z_0'/Z_0)D}{1 - S^2}$$

$$C' = \frac{C + (S/Z_0)D - (S/Z_0')A - [S^2/(Z_0 Z_0')]B}{1 - S^2}$$

(8-63)

Similarly, equating elements in the second columns gives

$$B' = \frac{B + (SZ_0)A - (SZ_0')D - (S^2 Z_0 Z_0')C}{1 - S^2}$$

$$D' = \frac{D + (SZ_0)C - (S/Z_0')B - (S^2 Z_0/Z_0')A}{1 - S^2}$$

(8-64)

Hence, the circuit of Fig. 8-14b can be found uniquely from that of Fig. 8-14a.

Example 8-6 Let the two-port of Fig. 8-14a contain a single series-connected short-circuited stub (Fig. 8-15a). Since the impedance of the stub is, by (8-42), equal to $Z_{0_1}S$, the chain matrix can be found with the aid of (5-27):

$$\begin{bmatrix} A & B \\ C & D \end{bmatrix} = \begin{bmatrix} 1 & Z_{0_1}S \\ 0 & 1 \end{bmatrix}$$

(8-65)

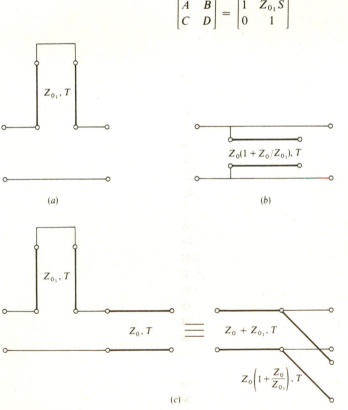

Figure 8-15 Example of the Kuroda identity.

Hence, by (8-61),

$$Z'_0 = \frac{Z_0 + Z_{0_1}}{0 + 1} = Z_0 + Z_{0_1} \tag{8-66}$$

and, by (8-63) and (8-64),

$$A' = \frac{1 + S^2 Z_{0_1}/Z_0 - S^2(Z_0 + Z_{0_1})/Z_0}{1 - S^2} = 1$$

$$B' = \frac{Z_{0_1} + S Z_0 - S(Z_0 + Z_{0_1})}{1 - S^2} = 0$$

$$\tag{8-67}$$

$$C' = \frac{S/Z_0 - S/(Z_0 + Z_{0_1}) - S^3 Z_{0_1}/[Z_0(Z_0 + Z_{0_1})]}{1 - S^2} = \frac{Z_{0_1}}{Z_0(Z_0 + Z_{0_1})}$$

$$D' = \frac{1 - [S/(Z_0 + Z_{0_1})]Z_{0_1}S - S^2 Z_0/(Z_0 + Z_{0_1})}{1 - S^2} = 1$$

It is easy to see that these new chain parameters describe the two-port of Fig. 8-15b. Hence, we obtain the network equivalence shown in Fig. 8-15c.

Proceeding this way, we can obtain the equivalences of Fig. 8-16. In the figure, all lines are commensurate; i.e., they all have the same delay T. Additional identities can be found in Ref. 3; using relations (8-61), (8-63), and (8-64), the reader can derive any number of further ones. It should be noted that some of the identities, e.g., those in Fig. 8-16c and d, introduce ideal transformers; they can be eliminated by moving all parts of the circuit to either the primary or the secondary side of the transformer, as discussed in Sec. 6-5. Also, some identities, e.g., those referred to in Probs. 8-18 and 8-19, introduce negative impedance values, which must be eliminated by combining lines with positive and negative characteristic impedances for physical realizability.

Finally, a trivial but important comment: all Kuroda identities remain valid if both two-ports involved are turned around by interchanging their input and output ports. Thus, the equivalence of Fig. 8-16a gives that of Fig. 8-17, etc.

Next, we introduce two other network equivalences, which are helpful in the utilization of the Kuroda identities. The equivalence shown in Fig. 8-18a can be proved from the relations

$$V_1 = E - Z_0 I_1 = z_{11} I_1 - z_{12} I \qquad V = z_{12} I_1 - z_{22} I \tag{8-68}$$

where the z_{ij} are the impedance parameters of the line. Solving the first equation for I_1 and substituting it into the second relation gives

$$V = z_{12} \frac{E + z_{12} I}{z_{11} + Z_0} - z_{22} I = E \frac{z_{12}}{z_{11} + Z_0} + I \left(\frac{z_{12}^2}{z_{11} + Z_0} - z_{22} \right) \tag{8-69}$$

As (8-35) and (8-37) show,

$$\frac{z_{12}}{z_{11} + Z_0} = \frac{-jZ_0 \csc \theta}{-jZ_0 \cot \theta + Z_0} = e^{-j\theta} = e^{-j\omega T} \tag{8-70}$$

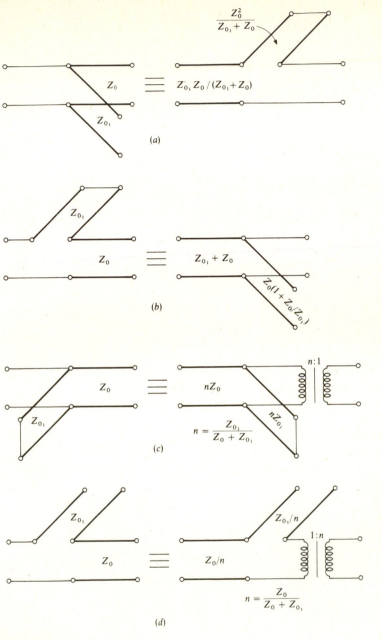

Figure 8-16 Kuroda identities.

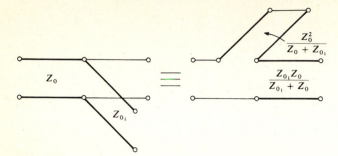

Figure 8-17 Kuroda identity.

and
$$\frac{z_{12}^2}{z_{11} + Z_0} - z_{22} = \frac{-Z_0^2 \csc^2 \theta}{-jZ_0 \cot \theta + Z_0} + jZ_0 \cot \theta = -Z_0 \qquad (8\text{-}71)$$

Hence
$$V = Ee^{-j\omega T} - Z_0 I \qquad (8\text{-}72)$$

which is the voltage-current relation of the simpler generator circuit shown on the right-hand side of Fig. 8-18a. Repeating the process results in the equivalence of Fig. 8-18b.

Consider now the load circuit of Fig. 8-19a. A simple analysis similar to that performed for the circuit of Fig. 8-18a shows that

$$Z = Z_0 \qquad (8\text{-}73)$$

and
$$V_1 = Ve^{-j\omega T} \qquad I_1 = Ie^{-j\omega T} \qquad (8\text{-}74)$$

Repeated application of (8-73) and (8-74) for a cascade of k transmission lines (Fig. 8-19b) gives

$$Z = Z_0 \qquad (8\text{-}75)$$

and
$$V_k = Ve^{-jk\omega T} \qquad I_k = Ie^{-jk\omega T} \qquad (8\text{-}76)$$

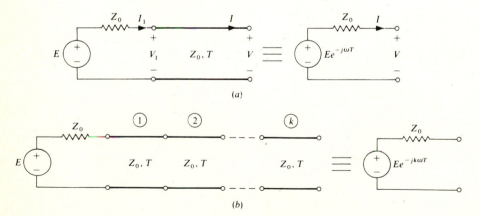

Figure 8-18 A Thevenin equivalence for transmission-line circuits.

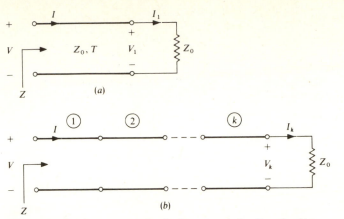

Figure 8-19 Equivalent load circuits for transmission-line circuits.

We conclude from Figs. 8-18 and 8-19, as well as (8-71) to (8-76), that the only effect of inserting unit elements in cascade with the generator (or load) of a doubly terminated two-port is to introduce a linear phase, i.e., flat delay, into the response; however, this conclusion holds only if the line impedances equal the generator (or load) resistance. Since the loss response remains unchanged and the delay is only modified by an added constant, this augmentation of the circuit is almost always allowable.

Example 8-7 To illustrate the use of the Kuroda identities in conjunction with the equivalences of Figs. 8-18 and 8-19, consider a redesign of the circuit of Fig. 8-8a. Inserting two unit elements in series with the generator gives the circuit of Fig. 8-20a. Next, using the Kuroda identity of Fig. 8-17 to two-port A gives the circuit of Fig. 8-20b. Finally, using the Kuroda identity of Fig. 8-16b, with interchanged primary and secondary ports, for two-ports B and C gives the realization of Fig. 8-20c. For coaxial realization, a schematic cross-sectional view is shown in Fig. 8-20d. Note the added lines at both the input and output. As discussed before, these merely introduce additional delays into the response.

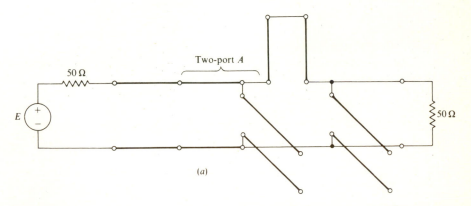

Figure 8-20 The use of the Kuroda identities in the realization of commensurate-line two-ports.

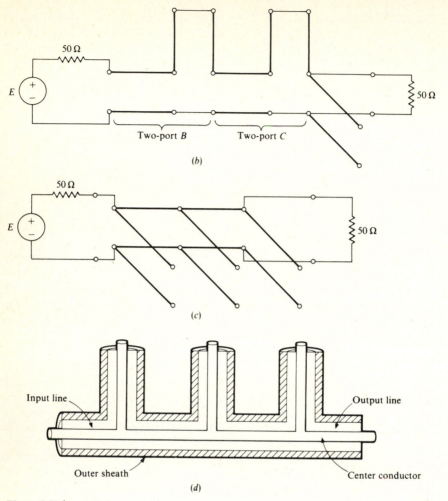

Figure 8-20 (*continued*)

The reader is referred to Probs. 8-15 to 8-23 for additional exercises involving the Kuroda identities and their applications to two-port design. They are, of course, only a small sample of the possible equivalences. Additional material can be found in Refs. 3 and 4.

8-4 ONE-PORT SYNTHESIS USING UNIFORM DISTRIBUTED RC STRUCTURES

Consider now the device illustrated in Fig. 8-1d. If the dimensions and material constants are chosen such that in the equivalent circuit of Fig. 8-2a the conditions

$$R \gg \omega L \qquad \omega C \gg G \tag{8-77}$$

hold, then the device can be regarded as a distributed RC structure. Thus, setting $L = 0$ and $G = 0$ and assuming also that R and C do not vary with x so that the device is uniform, from Eq. (8-20) we have for this case the characteristic impedance

$$Z_0 = \sqrt{\frac{R}{sC}} \tag{8-78}$$

and from Eq. (8-8) the propagation constant

$$\gamma = \sqrt{sRC} \tag{8-79}$$

Note that now both Z_0 and γ are complex for $s = j\omega$.

The two-port matrices of the structure are, by (8-26) to (8-29),

$$\mathbf{Z} = \sqrt{\frac{R}{sC}} \begin{bmatrix} \coth (a\sqrt{s}) & \operatorname{csch} (a\sqrt{s}) \\ \operatorname{csch} (a\sqrt{s}) & \coth (a\sqrt{s}) \end{bmatrix} \tag{8-80}$$

$$\mathbf{Y} = \sqrt{\frac{sC}{R}} \begin{bmatrix} \coth (a\sqrt{s}) & -\operatorname{csch} (a\sqrt{s}) \\ -\operatorname{csch} (a\sqrt{s}) & \coth (a\sqrt{s}) \end{bmatrix} \tag{8-81}$$

and

$$\mathbf{T} = \begin{bmatrix} \cosh (a\sqrt{s}) & \sqrt{\frac{R}{(sC)}} \sinh (a\sqrt{s}) \\ \sqrt{\frac{sC}{R}} \sinh (a\sqrt{s}) & \cosh (a\sqrt{s}) \end{bmatrix} \tag{8-82}$$

Here, we introduced for convenience the constant

$$a \triangleq l\sqrt{RC} = \sqrt{R_t C_t} \tag{8-83}$$

where $R_t \triangleq lR$ is the total resistance and $C_t \triangleq lC$ the total capacitance of the element.

The device described by Eqs. (8-78) to (8-83) is called a *uniform distributed RC structure*, abbreviated $\overline{URC}$ *structure*. The symbol commonly used for it is shown in Fig. 8-21a. A possible implementation which also realizes the short circuit between the lower terminals of the ports is shown in Fig. 8-21b.

From (8-80), the input impedance of a $\overline{URC}$ structure when its output port is open-circuited is given by

$$Z_{op} \equiv z_{11} = \frac{1}{\sqrt{s}} \sqrt{\frac{R}{C}} \coth (a\sqrt{s}) \tag{8-84}$$

Similarly, the input impedance of the device with a short-circuited output port is, by (8-81),

$$Z_{sh} \equiv \frac{1}{y_{11}} = \frac{1}{\sqrt{s}} \sqrt{\frac{R}{C}} \tanh (a\sqrt{s}) \tag{8-85}$$

Both Z_{op} and Z_{sh} are complex for $s = j\omega$. As (8-84) and (8-85) show, if we introduce (in close analogy to the Richards' variable S used for lossless lines) the new variable[7]

$$\hat{S} = \hat{\Sigma} + j\hat{\Omega} \triangleq \tanh (a\sqrt{s}) \tag{8-86}$$

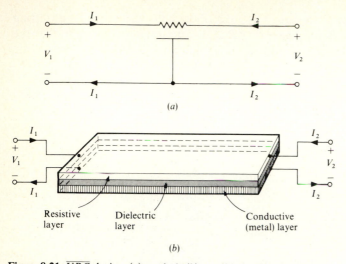

Figure 8-21 $\overline{URC}$ device: (a) symbol; (b) possible implementation.

the open- and short-circuit impedances satisfy

$$\sqrt{s}\,Z_{\text{op}} = \sqrt{\frac{R}{C}\frac{1}{\hat{S}}} = \frac{1}{\sqrt{C_t/R_t}\,\hat{S}} \qquad \sqrt{s}\,Z_{\text{sh}} = \sqrt{\frac{R}{C}}\,\hat{S} = \sqrt{\frac{R_t}{C_t}}\,\hat{S} \qquad (8\text{-}87)$$

Consider now a one-port circuit built exclusively from $\overline{URC}$ devices. To avoid horrible mathematical difficulties, assume that all these $\overline{URC}$ devices are *commensurate*, i.e., that they all have the same constant $a^2 \triangleq R_t C_t$. Assume also that all devices are open- or short-circuited at their output ports. Then all internal component impedances of the one-port are given by one or the other of the relations in (8-87). Multiplying every impedance in the circuit by $\sqrt{s}$ means that the port impedance $Z(s)$ is also multiplied by $\sqrt{s}$; this is instinctively obvious but can also be seen from Eq. (2-26) of Sec. 2-1. Hence, if the circuit of Fig. 8-22a has the port impedance $Z(s)$, then that of Fig. 8-22b is $\sqrt{s}\,Z(s)$. Using (8-87), we conclude that the scaled impedance function $\sqrt{s}\,Z(s)$ can be regarded as the port impedance of a one-port containing element impedances of the forms $(\sqrt{C_t/R_t}\,\hat{S})^{-1}$ and $\sqrt{R_t/C_t}\,\hat{S}$ only (Fig. 8-22c). In terms of $\hat{S}$, these correspond to capacitive and inductive impedances, respectively. Hence, *the necessary and sufficient condition for $Z(s)$ to be the port impedance of a commensurate $\overline{URC}$ one-port is that $\sqrt{s}\,Z(s)$ be a reactance function in terms of $\hat{S}$.* For such a realizable $Z(s)$, the synthesis can therefore by Fig. 8-22 proceed in the following steps:

1. $\sqrt{s}\,Z(s)$ is formed, and $\tanh\,(a\sqrt{s})$ is replaced by $\hat{S}$. This results in a reactance function in terms of $\hat{S}$:

$$Z(\hat{S}) \triangleq \sqrt{s}\,Z(s)\Big|_{\tanh\,(a\sqrt{s})\,=\,\hat{S}} \qquad (8\text{-}88)$$

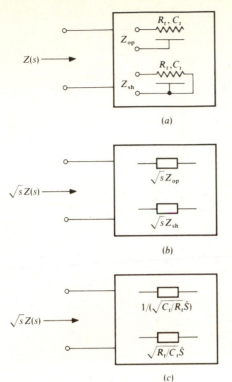

(a)

(b)

(c)

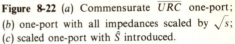

Figure 8-22 (a) Commensurate $\overline{URC}$ one-port;
(b) one-port with all impedances scaled by $\sqrt{s}$;
(c) scaled one-port with $\hat{S}$ introduced.

2. $Z(\hat{S})$ is synthesized, using the methods of Chap. 3, by a circuit containing hypothetical inductors with impedances $\hat{L}\hat{S}$ and hypothetical capacitors with impedances $1/\hat{C}\hat{S}$.

3. Each inductor is replaced by a short-circuited $\overline{URC}$ element. The constants of this element can be derived from (8-83) and (8-87), which give now

$$\sqrt{R_t C_t} = a \qquad \sqrt{\frac{R_t}{C_t}} = \hat{L} \qquad (8\text{-}89)$$

so that

$$R_t = lR = a\hat{L} \qquad C_t = lC = \frac{a}{\hat{L}} \qquad (8\text{-}90)$$

where R_t and C_t are the *total* resistance and capacitance, respectively, of the $\overline{URC}$ element.

4. Similarly, each capacitor is replaced by an open-circuited $\overline{URC}$ element. Again using (8-83) and (8-87), the parameters of the $\overline{URC}$ device are obtainable from

$$\sqrt{R_t C_t} = a \qquad \sqrt{\frac{C_t}{R_t}} = \hat{C} \qquad (8\text{-}91)$$

Equation (8-91) gives the device parameters

$$R_t \triangleq lR = \frac{a}{\hat{C}} \qquad C_t \triangleq lC = a\hat{C} \tag{8-92}$$

The process will be illustrated by an example.

Example 8-8 Synthesize the normalized impedance function

$$Z(s) = \frac{1}{\sqrt{s}} \frac{\tanh^4 (5\sqrt{s}) + 10 \tanh^2 (5\sqrt{s}) + 9}{\tanh^3 (5\sqrt{s}) + 2 \tanh (5\sqrt{s})}$$

Clearly, here $a = 5$ and

$$\sqrt{s}\, Z(s) = Z(\hat{S}) = \frac{\hat{S}^4 + 10\hat{S}^2 + 9}{\hat{S}^3 + 2\hat{S}}$$

If $Z(\hat{S})$ is synthesized using, say, Cauer 1 expansion, the ladder network of Fig. 8-23a is obtained. Next, $\hat{L}_1$ is transformed into a short-circuited $\overline{URC}$ element; its parameter values are, by (8-90),

$$R_{t_1} = a\hat{L}_1 = 5 \qquad C_{t_1} = \frac{a}{\hat{L}_1} = 5$$

Similarly, $\hat{C}_2$ becomes an open-circuited $\overline{URC}$ device with parameters

$$R_{t_2} = \frac{a}{\hat{C}_2} = 40 \qquad C_{t_2} = a\hat{C}_2 = 0.625$$

Proceeding element by element this way, we end up with the circuit of Fig. 8-23b.

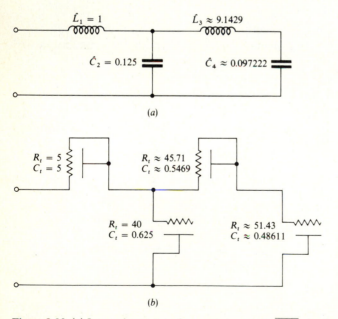

(a)

(b)

Figure 8-23 (a) Lumped prototype for a commensurate $\overline{URC}$ circuit; (b) actual circuit.

Like lossless transmission-line circuits, the networks obtainable by this simple method are unsuitable for actual construction since they require many interconnections and the metal electrodes of some devices are not grounded. More practical configurations can be obtained by using the $\overline{URC}$ device as a two-port, i.e., as a unit element. Consider the circuit of Fig. 8-24a. Using (8-48) and (8-80), we obtain

$$
\begin{aligned}
Z(s) &= z_{11}(s) - \frac{z_{12}^2(s)}{z_{22}(s) + Z_1(s)} \\
&= \sqrt{\frac{R_t}{sC_t}} \coth(a\sqrt{s}) - \frac{(R_t/sC_t)\operatorname{csch}^2(a\sqrt{s})}{\sqrt{R_t/sC_t}\coth(a\sqrt{s}) + Z_1(s)} \\
&= \sqrt{\frac{R_t}{sC_t}} \frac{\sqrt{R_t/sC_t} + \coth(a\sqrt{s})Z_1(s)}{\sqrt{R_t/sC_t}\coth(a\sqrt{s}) + Z_1(s)}
\end{aligned}
\tag{8-93}
$$

(Here, the relation $\coth^2\theta - \operatorname{csch}^2\theta = 1$ was utilized.) Introducing $\hat{S}$ via (8-86) and using the notation of Eq. (8-88) to define $Z(\hat{S})$ and $Z_1(\hat{S})$, we obtain

$$
\begin{aligned}
Z(\hat{S}) = \sqrt{s}\, Z(s) &= \sqrt{\frac{R_t}{C_t}} \frac{\sqrt{R_t/C_t} + (1/\hat{S})\sqrt{s}\, Z_1(s)}{\sqrt{R_t/C_t}/\hat{S} + \sqrt{s}\, Z_1(s)} \\
&= \sqrt{\frac{R_t}{C_t}} \frac{Z_1(\hat{S}) + \hat{S}\sqrt{R_t/C_t}}{\hat{S}Z_1(\hat{S}) + \sqrt{R_t/C_t}}
\end{aligned}
\tag{8-94}
$$

Furthermore, for $\hat{S} = 1$,

$$
Z(1) = \sqrt{\frac{R_t}{C_t}} \frac{Z_1(1) + \sqrt{R_t/C_t}}{Z_1(1) + \sqrt{R_t/C_t}} = \sqrt{\frac{R_t}{C_t}}
\tag{8-95}
$$

A comparison of (8-94) and (8-95) on the one hand and (8-49) and (8-50) on the other reveals that these equations are in exactly the same form. The only difference is that Z_0 is here replaced by $\sqrt{R_t/C_t}$. It follows, therefore, that Rich-

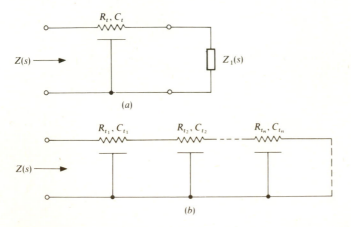

(a)

(b)

Figure 8-24 (a) Unit element in $\overline{URC}$ circuit; (b) unit-element realization of a commensurate $\overline{URC}$ impedance.

ards' theorem holds for $Z(\hat{S})$ and $Z_1(\hat{S})$. Consequently, if $Z(\hat{S})$ is a reactance function, and if the parameters R_t and C_t of the $\overline{URC}$ device are chosen such that

$$\sqrt{\frac{R_t}{C_t}} = \sqrt{\frac{R}{C}} = Z(1) \tag{8-96}$$

holds for the unit element,† then the remainder impedance

$$Z_1(\hat{S}) = Z(1)\frac{Z(\hat{S}) - \hat{S}Z(1)}{Z(1) - \hat{S}Z(\hat{S})} \tag{8-97}$$

is also a reactance function, of a degree 1 lower than that of $Z(\hat{S})$. This shows that any realizable $Z(s)$ can be synthesized in the form shown in Fig. 8-24b. The output port of the last (nth) element will be short-circuited if $Z(\hat{S}) = 0$ for $\hat{S} = 0$; it will be open-circuited if $Z(\hat{S}) \to \infty$ for $\hat{S} \to 0$.

The design process will be illustrated by an example.

Example 8-9 Consider again the impedance

$$Z(s) = \frac{1}{\sqrt{s}}\frac{\tanh^4 (5\sqrt{s}) + 10 \tanh^2 (5\sqrt{s}) + 9}{\tanh^3 (5\sqrt{s}) + 2 \tanh (5\sqrt{s})}$$

synthesized earlier in the form shown in Fig. 8-23. To find a unit-element realization, we start, as before, from

$$Z(\hat{S}) = \frac{\hat{S}^4 + 10\hat{S}^2 + 9}{\hat{S}^3 + 2\hat{S}}$$

For the first unit element, by (8-96),

$$\sqrt{\frac{R_{t_1}}{C_{t_1}}} = Z(1) = \tfrac{20}{3}$$

holds.

Also, $a = \sqrt{R_{t_1} C_{t_1}} = 5$, and hence

$$R_{t_1} = aZ(1) = \tfrac{100}{3} \qquad C_{t_1} = \frac{a}{Z(1)} = \frac{3}{4}$$

The remainder $Z_1(\hat{S})$ (Fig. 8-25a) is, by (8-97), obtained as

$$Z_1(\hat{S}) = \frac{20}{3}\frac{Z(\hat{S}) - \tfrac{20}{3}\hat{S}}{\tfrac{20}{3} - \hat{S}Z(\hat{S})} = \frac{20}{3}\frac{-17\hat{S}^4 - 10\hat{S}^2 + 27}{-3\hat{S}^5 - 10\hat{S}^3 + 13\hat{S}} = \frac{20}{3}\frac{17\hat{S}^2 + 27}{3\hat{S}^3 + 13\hat{S}}$$

When we repeat the process, the parameters of the second unit element satisfy

$$\sqrt{\frac{R_{t_2}}{C_{t_2}}} = Z_1(1) = \frac{(20)(44)}{(3)(16)} = \frac{55}{3}$$

Hence, $R_{t_2} = aZ_1(1) = \tfrac{275}{3}$ and $C_{t_2} = a/Z_1(1) = \tfrac{3}{11}$. The remainder impedance is

$$Z_2(\hat{S}) = \frac{55}{3}\frac{Z_1(\hat{S}) - \tfrac{55}{3}\hat{S}}{\tfrac{55}{3} - \hat{S}Z_1(\hat{S})} = \frac{55}{3}\frac{-33\hat{S}^4 - 75\hat{S}^2 + 108}{-35\hat{S}^3 + 35\hat{S}} = \frac{11}{7}\frac{11\hat{S}^2 + 36}{\hat{S}}$$

Repeating the unit-element extraction twice more gives the complete circuit of Fig. 8-25a. A possible implementation of the circuit is shown in Fig. 8-25b.

† It follows from the PR character of $Z(\hat{S})$ that $\sqrt{R_t/C_t} = Z(1)$ is a positive real number.

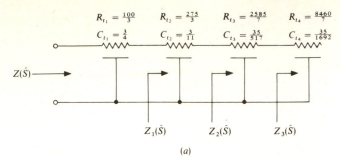

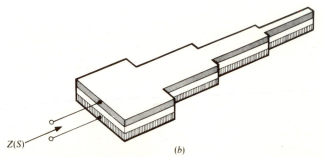

Figure 8-25 $\overline{URC}$ impedance; (*a*) circuit; (*b*) realization.

Note that when the above-described design process involving the $\hat{S}$ variable is used for the synthesis of commensurate $\overline{URC}$ circuits, only $\overline{URC}$ elements and *no* lumped resistors are permitted in the circuit. This is in contrast to the use of the S-variable design for lossless transmission-line circuits discussed earlier, where lumped resistors were allowed. The reason for this difference is that the introduction of the $\hat{S}$ variable (Fig. 8-22) involved the multiplication of all impedances in the circuit by $\sqrt{s}$. This causes a resistor R to have an impedance

$$Z_R = R\sqrt{s} = \frac{R}{a} \tanh^{-1} \hat{S} \tag{8-98}$$

a transcendental function of $\hat{S}$.

The network identities illustrated in Figs. 8-18 and 8-19 do not hold for $\overline{URC}$ circuits, and Kuroda's identities are not useful either for $\overline{URC}$ circuits.

Finally, it should be noted that an altogether different approach, based on the transformed variable

$$P \triangleq \cosh\left(a\sqrt{s}\right) \tag{8-99}$$

is also available for the design of $\overline{URC}$ one-ports. This technique, due to O'Shea,[8] is described in Refs. 5 and 6. The resulting circuit, however, suffers from the same shortcomings as mentioned for the one-port of Fig. 8-23. Hence, it is not as useful for practical realization as the configuration of Fig. 8-25 obtainable using the $\hat{S}$ variable.

8-5 TWO-PORT DESIGN USING $\overline{URC}$, R, C, AND ACTIVE COMPONENTS

Since $\overline{URC}$ elements are basically RC circuits, two-ports built exclusively from $\overline{URC}$ and R, C elements have all their natural modes on the negative real s axis ($-\sigma$ axis). As explained in Sec. 7-1, this makes such circuits badly suited for filtering or phase-correcting applications. Hence, even though design techniques exist which use either the $\hat{S}$ or P variable[5-8] for the design of commensurate $\overline{URC}$ two-ports, these circuits are of limited usefulness and will not be discussed in this book.

Some useful two-port networks can be obtained, by contrast, if we combine $\overline{URC}$ with lumped passive and/or active elements. Even though such hybrid circuits can usually be analyzed only by approximations or graphical procedures, they are economical because the $\overline{URC}$ element can replace several lumped components. A few such circuits will be briefly discussed. Additional circuits and more detailed treatment can be found in Refs. 9 to 11.

Consider, for example, the circuit of Fig. 8-26a. We can regard the two-port as a series combination of two elementary two-ports (Fig. 8-26b), one with impedance matrix $\mathbf{Z}_1$ given by (8-80) and the other, containing R, with impedance matrix

$$\mathbf{Z}_2 = \begin{bmatrix} R & R \\ R & R \end{bmatrix} \tag{8-100}$$

Hence, for the complete two-port,

$$\mathbf{Z} = \mathbf{Z}_1 + \mathbf{Z}_2 = \begin{bmatrix} R + \sqrt{\dfrac{R_t}{sC_t}}\coth(a\sqrt{s}) & R + \sqrt{\dfrac{R_t}{sC_t}}\operatorname{csch}(a\sqrt{s}) \\ R + \sqrt{\dfrac{R_t}{sC_t}}\operatorname{csch}(a\sqrt{s}) & R + \sqrt{\dfrac{R_t}{sC_t}}\coth(a\sqrt{s}) \end{bmatrix} \tag{8-101}$$

Since $I_2 = 0$,

$$A_v \triangleq \frac{V_2}{V_1} = \frac{I_1 z_{12} + 0}{I_1 z_{11} + 0} = \frac{z_{12}}{z_{11}} = \frac{R + \sqrt{R_t/sC_t}\operatorname{csch}(a\sqrt{s})}{R + \sqrt{R_t/sC_t}\coth(a\sqrt{s})}$$

$$= \frac{a\sqrt{s}\sinh(a\sqrt{s}) + R_t/R}{a\sqrt{s}\sinh(a\sqrt{s}) + (R_t/R)\cosh(a\sqrt{s})} \tag{8-102}$$

There are transmission zeros for those values of s which satisfy

$$a\sqrt{s}\sinh(a\sqrt{s}) = -\frac{R_t}{R} \tag{8-103}$$

For some special values of R_t/R a transmission zero will lie on the $j\omega$ axis. Setting $s = j\omega$ and using the formula

$$\sinh(u + jv) = \sinh u \cos v + j \cosh u \sin v \tag{8-104}$$

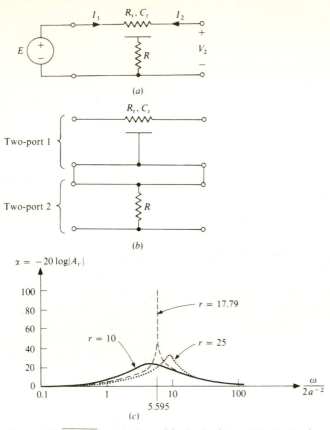

Figure 8-26 $\overline{\text{URC-R}}$ notch filter: (a) circuit. (*From W. M. Kaufman, Proc. IRE, September 1960, pp. 1540–1545.*) (b) Equivalent representation. (c) Response for various R_t/R ratios.

(8-103) gives

$$a\sqrt{\frac{\omega}{2}}(1+j)\left|\sinh\left(a\sqrt{\frac{\omega}{2}}\right)\cos\left(a\sqrt{\frac{\omega}{2}}\right)+j\cosh\left(a\sqrt{\frac{\omega}{2}}\right)\sin\left(a\sqrt{\frac{\omega}{2}}\right)\right|=-\frac{R_t}{R}$$

(8-105)

The imaginary part of the left-hand side must be zero; this gives for $x \triangleq a\sqrt{\omega/2}$

$$x(\sinh x \cos x + \cosh x \sin x) = 0$$

$$\tan x = -\tanh x$$

(8-106)

Figure 8-27 illustrates the possible solutions of the last equation. The first (lowest-frequency) zero occurs at $x_A \approx 2.365$. The corresponding value of $r \triangleq R_t/R$ is then given by the real part of (8-105):

$$r_A = -x_A(\sinh x_A \cos x_A - \cosh x_A \sin x_A) = -2x_A \sinh x_A \cos x_A \approx 17.79$$

(8-107)

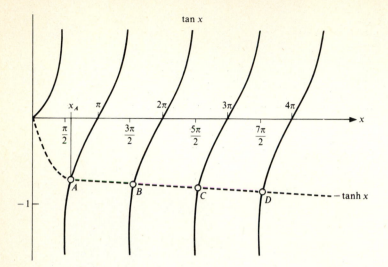

Figure 8-27 Graphical solution of Eq. (8-106).

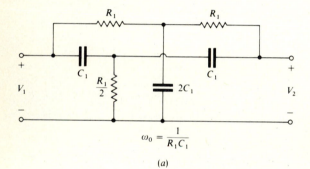

$$\omega_0 = \frac{1}{R_1 C_1}$$

(a)

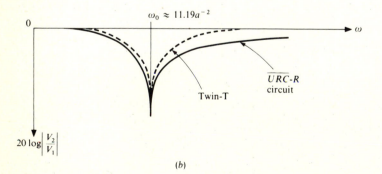

(b)

Figure 8-28 (a) Twin-T circuit; (b) comparison of the gain responses of a twin-T and a $\overline{URC}$-R circuit for $R_t/R = 17.79$.

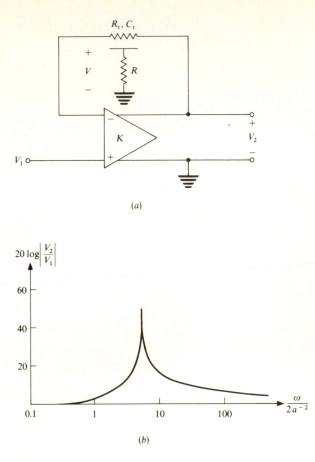

(a)

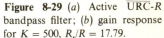

(b)

Figure 8-29 (a) Active $\overline{URC}$-R bandpass filter; (b) gain response for $K = 500$, $R_t/R = 17.79$.

Hence, for $R_t/R = 17.79$, the circuit of Fig. 8-26a should provide a $j\omega$-axis transmission zero around

$$\omega_A = 2\left(\frac{x_A}{a}\right)^2 \approx \frac{11.19}{a^2} \tag{8-108}$$

This can be confirmed by evaluating the voltage ratio of (8-102) for various values of $s = j\omega$ and also $r \triangleq R_t/R$. Figure 8-26c shows the loss response of the circuit for $r = 10$, 17.79, and 25. As predicted, the $r = 17.79$ response has a transmission zero (loss pole) for $\omega_A \approx (5.595)(2a^{-2}) = 11.19a^{-2}$.

The response of the notch circuit for $r = 17.79$ is similar to that of a twin-T circuit (Fig. 8-28a). Since the latter requires six elements while the $\overline{URC}$-R circuit requires only two, the $\overline{URC}$ element is, in effect, replacing five lumped elements in this configuration.

It is well known that if a twin-T two-port is placed in the feedback path of an amplifier, a bandpass filter results. Hence, we can immediately conjecture that the circuit of Fig. 8-29a will also act as a bandpass filter. In fact, for $K = 500$ and

$r = 17.79$, the gain response of Fig. 8-29b results. The analytical expression for the voltage gain is readily obtained from the relations

$$V_2 = K(V_1 - V) \qquad V = A_v V_2 \tag{8-109}$$

where A_v is given by (8-102). This is left as an exercise (see Prob. 8-37) for the reader.

A simple active $\overline{URC}$ lowpass filter is shown in Fig. 8-30a. The voltage relations are

$$V_1 = V_a + V_2 \qquad V = V_b + V_2 \qquad V_2 = KV \qquad V_b = A_v^0 V_a \tag{8-110}$$

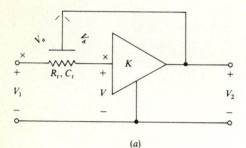

(a)

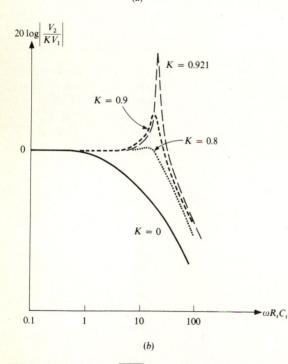

(b)

Figure 8-30 (a) Active $\overline{URC}$ lowpass filter. (b) Gain responses. (c) Design chart. (*From W. J. Kerwin, in L. P. Huelsman (ed.), "Active Filters," McGraw-Hill, New York, 1970, by permission.*) (d) Approximation error.

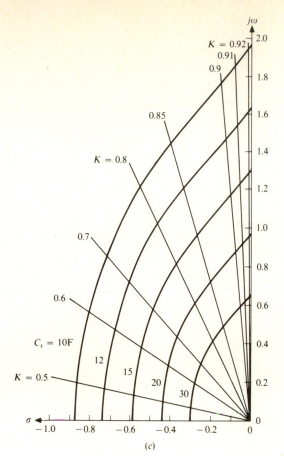

$C_t = 10F$

(c)

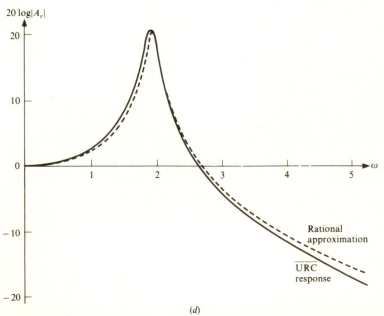

Rational approximation

$\overline{URC}$ response

(d)

361

Here, A_v^0 is the voltage ratio of the open-circuited $\overline{URC}$ element; it is, by (8-82),

$$A_v^0 = \frac{1}{A} = \frac{1}{\cosh(a\sqrt{s})} \qquad (8\text{-}111)$$

From (8-110) and (8-111),

$$V_2 = KV = K(V_b + V_2) = K(A_v^0 V_a + V_2) = K[A_v^0(V_1 - V_2) + V_2]$$
$$= KA_v^0 V_1 + K(1 - A_v^0)V_2 \qquad (8\text{-}112a)$$

$$V_2 = \frac{KA_v^0 V_1}{1 - K(1 - A_v^0)} = \frac{KV_1}{K + (1 - K)\cosh(a\sqrt{s})} \qquad (8\text{-}112b)$$

Hence, the voltage ratio defined by†

$$A_v \triangleq \frac{V_2}{KV_1} = \frac{1}{K + (1 - K)\cosh(a\sqrt{s})} \qquad (8\text{-}113)$$

gives the logarithmic gain responses shown in Fig. 8-30b. For $K \approx 0.8$, the circuit behaves like a lowpass filter. For $K \approx 0.921$, the voltage gain becomes infinite at the frequency $\omega_0 = 2\pi^2/R_t C_t$, and the circuit becomes an oscillator (see Prob. 8-36).

Calculations show that for an appropriate choice of the parameters R_t, C_t, and K the circuit of Fig. 8-30a has a gain response which is very close to that obtainable with the Sallen-Key filter of Fig. 7-41a. Hence, the $\overline{URC}$ element replaces here four lumped RC components.

The voltage ratio of the Sallen-Key circuit is in the form

$$A_v = \frac{|s_1|^2}{(s - s_1)(s - s_2)} \qquad (8\text{-}114)$$

where $s_2 = s_1^*$. In order to use the active $\overline{URC}$ circuit as a replacement of a Sallen-Key filter, therefore, it should be possible to calculate the parameters a and K from given s_1, s_2 such that

$$|A_v| = \frac{1}{|K + (1 - K)\cosh(a\sqrt{j\omega})|} \approx \frac{|s_1|^2}{|(j\omega - s_1)(j\omega - s_1^*)|} \qquad (8\text{-}115)$$

and, vice versa, it greatly facilitates the analysis of the active $\overline{URC}$ circuit if the location of the *equivalent poles* s_1, s_2 can be directly deduced from a and K.

These tasks are made possible by the availability of design charts[9,11] showing the constant-K and constant-a contours in the s_1, s_2 plane. Such a chart is shown in Fig. 8-30c, with the normalization $R_t = 1\ \Omega$.

Example 8-10 Find C_t and K such that the circuit of Fig. 8-30a realizes approximately a rational transfer function of the form of (8-114), with a pole pair at $s_{1,2} = -0.09 \pm j1.9$.

---

† This definition keeps $A_v(0) = 1$, independent of K.

From the chart the appropriate parameter values are about $K \approx 0.91$ and $C_t \approx 10$. Figure 8-30d compares the gain responses of the $\overline{URC}$-active circuit thus obtained and the two-pole function

$$-20 \log \left| \frac{(j\omega - s_1)(j\omega - s_2)}{s_1 s_2} \right|$$

being approximated. As the figure illustrates, the approximation is very good over the $0 \leq \omega \leq 4$ range. For high frequencies, the exponential decrease of the $|A_v|$ of the $\overline{URC}$ circuit is faster than that of any rational function of s.

Another useful lumped RC-active filter, Kerwin's circuit, was introduced in Fig. 7-45. Since the circuit utilizes a twin-T circuit, which (as shown in Fig. 8-28) is nearly equivalent to the $\overline{URC}$-R circuit of Fig. 8-26a, it seems plausible that for appropriate values of the parameters R, R_t, C_t, K, and Y, the circuit of Fig. 8-31a becomes nearly equivalent to that of Fig. 7-45. Since the equivalence of the $\overline{URC}$-R two-port to the twin T requires $r = R_t/R \approx 17.79$, we can choose the normalized resistor values $R = 1$ and $R_t \approx 17.79$. The transmission zero of the overall active two-port is at the same frequency as where the transmission zero of the $\overline{URC}$-R two-port is (compare Prob. 8-38). As derived in connection with Fig. 8-26a, the latter occurs at

$$\omega_A \approx (5.595)(2a^{-2}) = \frac{11.19}{R_t C_t} \tag{8-116}$$

Hence, we can obtain a loss pole at the normalized frequency $\omega_A = 1$ by choosing

$$C_t \approx \frac{11.19}{R_t} \approx \frac{11.19}{17.79} \approx 0.629 \tag{8-117}$$

This leaves only the value of K and the nature and value of Y undetermined. It can be anticipated that choosing Y the same way as for the lumped circuit of Fig. 7-45 will be the correct procedure. Hence, if the circuit is to provide (approximately) the response

$$A_v \propto \frac{s^2 + \omega_A^2}{(s - s_1)(s - s_1^*)} \tag{8-118}$$

then the choice

$$Y \equiv \begin{cases} G_1 & \text{if } |s_1| > \omega_A \\ sC_1 & \text{if } |s_1| < \omega_A \end{cases} \tag{8-119}$$

will be made. ($Y \equiv 0$ for $|s_1| = \omega_A$.)

The values of K and G_1 (or C_1) can again be obtained from design charts. *Assuming* $\omega_A = 1$ *and* $|s_1| > \omega_A$ (high-pass response), the normalized circuit of Fig. 8-31b results; the missing element values K and $R_1 = 1/G_1$ can then be read from the chart of Fig. 8-31c in terms of the coordinates of $s_1 = \sigma + j\omega$.

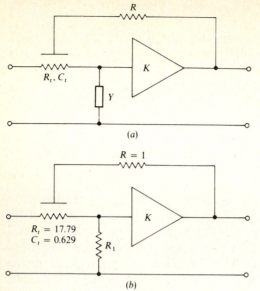

(a)

(b)

Figure 8-31 (a) Active $\overline{URC}$-RC lowpass or highpass filter (distributed Kerwin circuit). (b) Schematic for highpass circuit. (c) Design chart for finding K and R_1. (b and c from W. J. Kerwin, in L. P. Huelsman (ed.), "Active Filters," McGraw-Hill, New York, 1970, by permission.)

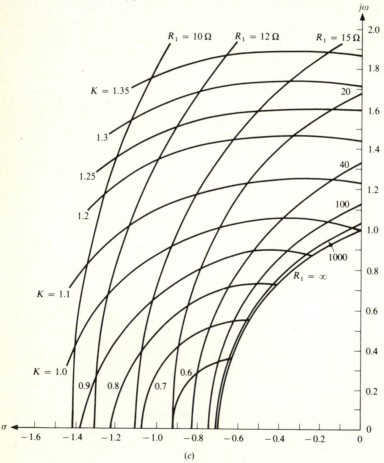

(c)

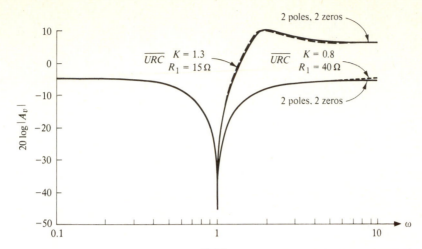

Figure 8-32 A comparison of active $\overline{URC}$ network responses and responses obtained from biquadratic rational functions. (*From W. J. Kerwin, in L. P. Huelsman (ed.), "Active Filters," McGraw-Hill, New York, 1970, by permission.*)

Example 8-11 Let the normalized value of the desired natural modes be $s_{1,2} = -0.37 \pm j1.75$. Then Fig. 8-31c gives $K \approx 1.3$ and $R_1 = 15 \, \Omega$. The resulting response of the active $\overline{URC}$ circuit of Fig. 8-31b is compared with that of the rational function[†]

$$A_v = \frac{C(s^2 + 1)}{(s - s_1)(s - s_1^*)} = \frac{C(s^2 + 1)}{s^2 + 0.74s + 3.2}$$

in Fig. 8-32. The figure also shows a comparison of the behavior of the active $\overline{URC}$ circuit for $K = 0.8$, $R_1 = 40 \, \Omega$ with that of a rational function. The equivalent poles are, from Fig. 8-31c, at $s_{1,2} \approx -0.65 \pm 70.72$.

If $|s_1| < \omega_A = 1$ (low-pass response), $Y \equiv sC_1$ and the normalized circuit is that shown in Fig. 8-33a. The unknown element values K and C_1 can again be determined such that the response approximates the rational function given in (8-118). Using the chart shown in Fig. 8-33b for specified $s_1 = \sigma + j\omega$ the appropriate values of K and C_1 can be read.

Example 8-12 Realize the active $\overline{URC}$-RC circuit of Fig. 8-33a such that it approximates the rational A_v function of Eq. (8-118) with $\omega_A = 1$ and $s_{1,2} = -0.4 \pm j0.2$.

From Fig. 8-33b, the corresponding values are $K = 0.6$ and $C_1 = 0.1$. The rational-function response is compared with the actual one in Fig. 8-33c, which also shows the behavior of the $K = 1.2$, $C_1 = 0.1$ circuit in comparison with the corresponding rational response.

---

† Here, C is a constant chosen to optimize the match between the two responses.

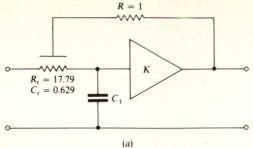

(a)

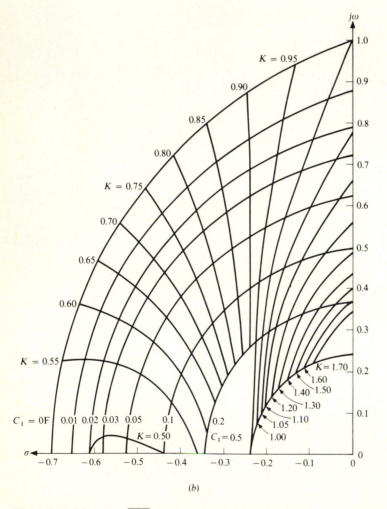

(b)

Figure 8-33 (a) Active $\overline{URC}$-RC low-pass circuit, (b) design chart for finding K and C_1, (c) A comparison of active $\overline{URC}$ responses with rational-function responses. (*From W. J. Kerwin, in L. P. Huelsman (ed.), "Active Filters," McGraw-Hill, New York, 1970, by permission.*)

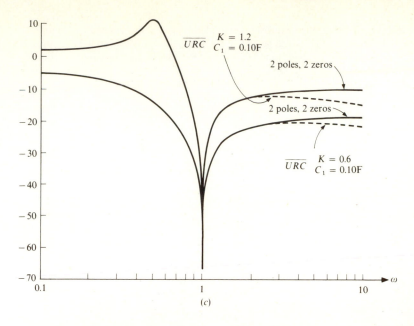

(c)

8-6 SUMMARY

In Chap. 8, the design of circuits (both one-ports and two-ports) containing distributed elements was discussed. The main topics were:

1. The uniform transmission line: its differential equations; the definitions of the propagation constant, phase velocity, forward and reflected waves, and characteristic impedance; the immittance and chain matrices of a transmission line (Sec. 8-1).
2. The uniform *lossless* transmission line: its immittance and chain matrices, electrical length, and delay time (Sec. 8-2).
3. Commensurate transmission-line networks: Richards' variable and the corresponding transformation of open- and short-circuited lines into LC elements; the realizability conditions and synthesis of commensurate-line networks; the periodic nature of their frequency response (Sec. 8-2).
4. The unit element; Richards' theorem; reactance synthesis using cascaded unit elements; Kuroda's identities and their application to obtain practical circuit configurations (Sec. 8-3).
5. The uniform distributed RC ($\overline{URC}$) structure; commensurate $\overline{URC}$ circuits; impedance design using a modified Richards' variable (Sec. 8-4).
6. Two-ports containing $\overline{URC}$, R, C and active elements: the $\overline{URC}$-R notch circuit, the simulation of lumped active two-ports (twin-T feedback circuit, Sallen-Key circuit, and Kerwin's circuit) using design charts (Sec. 8-5).

Some of the circuits which were discussed in this chapter were of very general character; however, the more general they became, the less powerful the techniques available for their design. In fact, for the most general circuits containing $\overline{URC}$, R, C and active components, we could only use approximate methods based on analogies with lumped filters and (horror of horrors) design charts. In fact, only a few types of the circuits described (mostly commensurate lossless-line networks, used as microwave filters, etc.) are used in practical design, mainly because of a lack of suitable design apparatus for the other circuits.

Another disadvantage of distributed networks, not evident at this point, concerns the finding of the transfer function (approximation). As will be seen in Chap. 12, there exists a veritable treasury of methods for finding suitable (real rational) transfer functions for lumped circuits. Since the transfer function of a distributed circuit is a much more complicated function of s, few comparable methods and results exist for it. In fact, the only simple available technique is to use a transformation of the frequency variable, such as given in Eq. (8-44), to utilize transfer functions derived originally for lumped networks in the design of commensurate-line circuits.

Our discussions up to now concerned the synthesis of circuits. In the next two chapters, we shall explore a different but important related subject: the calculation of sensitivities for circuits already designed. As we have shown for active circuits in Chap. 7, high sensitivities can doom an otherwise useful circuit. Hence, it is important to learn efficient methods for finding all sensitivities of networks. Such methods will be described in Chaps. 9 and 10.

PROBLEMS

8-1 Find the chain matrix **T** of a uniform transmission line. *Hint:* Use Eq. (5-26).

8-2 Express the real and imaginary parts of γ and Z_0 for $s = j\omega$ in terms of L, R, C, G, and ω.

8-3 Using the results of Prob. 8-2, find γ and Z_0 for the special values of line parameters satisfying $R/L = G/C$. What special properties do α, β, and Z_0 have now? *Hint:* A line with such parameters is often called a *distortionless line.* Why?

8-4 Find expressions for the elements of the equivalent **T** and Π networks of Fig. 8-34 such that they have the same two-port parameters as a uniform transmission line. *Hint:* Use Eqs. (8-26) to (8-27).

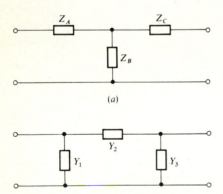

(a)

(b)

Figure 8-34 (a) T equivalent; (b) Π equivalent of a transmission line.

8-5 Show that the input impedance Z of the terminated uniform transmission line (Fig. 8-35) is given by

$$Z = Z_0 \frac{Z_1/Z_0 + \tanh \gamma l}{1 + (Z_1/Z_0) \tanh \gamma l}$$

Analyze the expression for $Z_1 = 0$, $Z_1 \to \infty$, and $Z_1 = Z_0$.

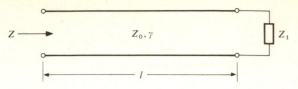

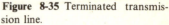

Figure 8-35 Terminated transmission line.

8-6 Derive the expressions given in (8-46) for the admittance matrix **Y** and chain matrix **T** of a lossless unit element in terms of Richards' variable S.

8-7 Derive Eq. (8-47) for the impedance shown in Fig. 8-10.

8-8 Show that the cascade of two unit elements has two-port matrices which are rational in S.

8-9 Realize the reactance functions

(a) $\dfrac{2S}{S^2 + 1}$ (b) $\dfrac{S^2 + 2}{S(S^2 + 4)}$

(c) $\dfrac{S(S^2 + 3)}{(S^2 + 1)(S^2 + 5)}$ (d) $\dfrac{1}{S} + \dfrac{3S}{S^2 + 4} + 5S$

with a commensurate-line network using open- and short-circuited line stubs.

8-10 Repeat the synthesis calculations of Prob. 8-9 using only cascaded unit elements.

8-11 Prove that the impedance

$$Z_1(s) = \frac{kZ(s) - sZ(k)}{kZ(k) - sZ(s)}$$

is PR if $Z(s)$ is PR and k is a positive real number. *Hint:* Use the technique used to prove that the $Z_1(S)$ given in Eq. (8-51) was PR. This relation is the basis of the Bolt-Duffin *RLC* impedance synthesis, briefly mentioned in Sec. 4-3.

8-12 A proof of Richards' theorem, different from that given in the text, can be obtained in the following steps:

(a) It is shown that since $Z(S)$ is PR, the hypothetical reflection factor defined by $\rho(S) \triangleq [Z(S) - Z(1)]/[Z(S) + Z(1)]$ satisfies, for $S = j\Omega$, $|\rho(j\Omega)| \leq 1$.

(b) It is shown that a second reflection factor, defined by $\rho_1(S) = [Z_1(S) - Z(1)]/[Z_1(S) + Z(1)]$ is related to the first one by $\rho_1(S) = [(1 + S)/(1 - S)]\rho(S)$.

(c) From the results of steps (a) and (b), it is shown that $\rho_1(S)$ is a real rational function of S, that $\rho_1(S)$ is analytic for Re $S \geq 0$ (in spite of the apparent pole at $S = 1$), and that $|\rho_1(j\Omega)| \leq 1$.

(d) From the above, it is deduced that $\rho_1(S)$ is real for real values of S and that $|\rho_1(S)| \leq 1$ for Re $S \geq 0$. The latter conclusion can be reached through the maximum-modulus theorem.

(e) Finally, from the result of step (d) and the definition of $\rho_1(S)$, it is easily seen that $Z_1(S)$ is PR. Carry out the detailed proof.

8-13 Attempt the synthesis of the impedance

$$Z(S) = \frac{2S^3 + 3S^2 + 2S + 3}{S^3 + 3S^2 + 4S + 1}$$

using the design techniques described in connection with Fig. 8-13. Explain your difficulties.

8-14 Prove the network equivalences shown in Fig. 8-36 for commensurate-line networks. What should the values of Z_{0_3}, Z_{0_4}, Z_{0_5}, and Z_{0_6} be?

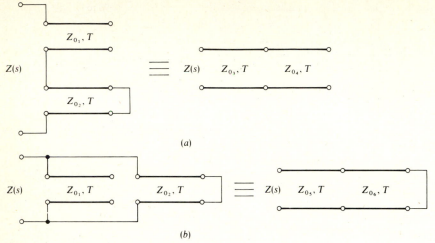

(a)

(b)

Figure 8-36 Commensurate-line network equivalences.

8-15 Prove the Kuroda identity shown in Fig. 8-16a.

8-16 Prove the Kuroda identity shown in Fig. 8-16c.

8-17 Prove the Kuroda identity of Fig. 8-16d.

8-18 Find Z_0' and the two-port T in the equivalent circuit of Fig. 8-37b.

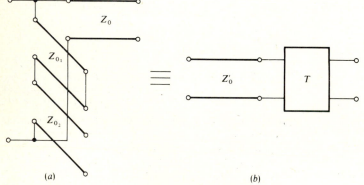

(a)   (b)

Figure 8-37 A commensurate-line network equivalence.

8-19 Find Z_0' and the two-port T in the equivalent circuit of Fig. 8-38b.

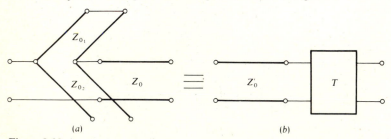

(a)   (b)

Figure 8-38 A commensurate-line network equivalence.

8-20 Prove relations (8-73) and (8-74) for the circuit of Fig. 8-19a.

8-21 Prove relations (8-75) and (8-76) for the circuit of Fig. 8-19b.

8-22 Using the design technique of Fig. 8-20, transform the circuit of Fig. 8-39*a* to that of Fig. 8-39*b*. Find all element values.

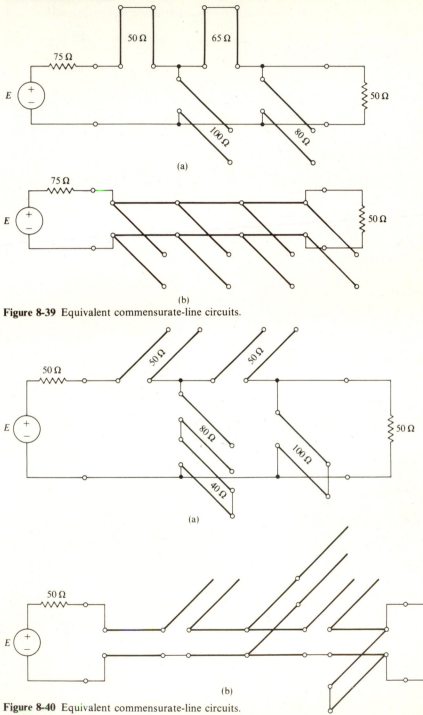

(a)

(b)

Figure 8-39 Equivalent commensurate-line circuits.

(a)

(b)

Figure 8-40 Equivalent commensurate-line circuits.

8-23 (*a*) Use the Kuroda identities of Fig. 8-16*d* and Fig. 8-37 as well as the transformation of Fig. 8-36*a* to transform the high-pass filter of Fig. 8-40*a* into that of Fig. 8-40*b*. First work out the sequence of partial transformations schematically. Is the transformed circuit always realizable? Next, work out the given problem.

(*b*) The circuit of Fig. 8-40*b* has nearly twice as many elements (lines) as that of Fig. 8-40*a*. Why do you think it can be more useful in spite of that?

8-24 What are the conditions on the length, width, sheet thicknesses, resistivity, and dielectric constant of the structure of Fig. 8-1*d* for the inequalities (8-77)?

8-25 Obtain the Cauer 2, Foster 1, and Foster 2 forms of the lumped LC prototype circuit of Fig. 8-23*a*. Derive the corresponding $\overline{URC}$ networks.

8-26 Find the zeros and poles of Z_{op} and Z_{sh}, defined in Eqs. (8-84) and (8-85), for $\overline{URC}$ devices. *Hint*: Use the product expansions

$$\sinh \theta = \theta \prod_{i=1}^{\infty} \left(1 + \frac{\theta^2}{i^2 \pi^2}\right) \qquad \cosh \theta = \prod_{i=1}^{\infty} \left[1 + \frac{(2\theta)^2}{(2i-1)^2 \pi^2}\right]$$

8-27 Find the port impedance $Z(s)$ for the $\overline{URC}$ circuits shown in Fig. 8-41.

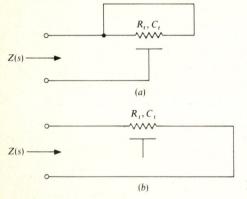

(*a*)

(*b*)

Figure 8-41 Simple $\overline{URC}$ circuits.

8-28 Synthesize the $\overline{URC}$ impedance

$$Z(s) = \frac{1}{\sqrt{s}} \frac{\tanh^3 \sqrt{s} + 2 \tanh \sqrt{s}}{(\tanh^2 \sqrt{s} + 1)(\tanh^2 \sqrt{s} + 3)}$$

(*a*) In the Foster forms.
(*b*) In the Cauer forms.
(*c*) Using cascade synthesis.

Can you predict $\sum_{i=1}^{4} R_{t_i}$?

8-29 Carry out the cascade synthesis of the $\overline{URC}$ impedance

$$Z(s) = \frac{1}{\sqrt{s}} \frac{\tanh^2 (2\sqrt{s}) + 1}{\tanh^3 (2\sqrt{s}) + 4 \tanh (2\sqrt{s})}$$

8-30 Calculate the open-circuit voltage ratio V_2/V_1 of a $\overline{URC}$ line (Fig. 8-21*a*). Assume $R_t = 10 \text{ k}\Omega$ and $C_t = 1000 \text{ pF}$. At what frequency will $|V_2/V_1|$ reach half of its zero-frequency value?

8-31 Calculate $|A_v|$ from (8-102) for the $\overline{URC}$-R notch circuit of Fig. 8-26*a*, for $\omega/\omega_1 = 1, 5, 10$, and 20. Use $r = 17.8$, and $\omega_1 = 2/R_t C_t$.

8-32 Calculate the real and imaginary parts of Z_0, γ, Z, Y, T, Z_{op}, Z_{sh}, and $\hat{S}$, as defined for $\overline{URC}$ circuits in Sec. 8-4, for $s = j\omega$.

8-33 Analyze the properties of the mapping $\hat{S} \triangleq \tanh(a\sqrt{s})$ used for $\overline{URC}$ circuits.

8-34 Calculate the voltage ratio $A_v = V_2/V_1$ for the $\overline{URC}$-C circuit of Fig. 8-42. Compare the function with the voltage ratio of the $\overline{URC}$-R circuit of Fig. 8-26a. Under what condition are the two circuits equivalent?

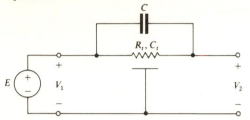

Figure 8-42 A $\overline{URC}$-C two-port.

8-35 Calculate $r = R_t/R$ if the solution B of (8-106) is selected in Fig. 8-27. Is the circuit realizable? Repeat for solutions C and D. *Hint:* Assume $\tanh x \approx 1$ for $x > \pi$.

8-36 At what frequency does the circuit of Fig. 8-30a become oscillatory for $K \approx 0.921$? What is the exact value of K? *Hint:* Use the relation $\cosh(u + jv) = \cosh u \cos v + j \sinh u \sin v$.

8-37 (a) Calculate the voltage ratio A_v^k for the circuit of Fig. 8-29a. *Hint:* Use Eqs. (8-109) and (8-102).
(b) Verify a few points on the gain curve of Fig. 8-29 for $K = 500$ and $r = 17.79$.

8-38 Prove that the transmission zero of the active-$\overline{URC}$-R circuit of Fig. 8-31a is at the same frequency ω_A as the transmission zero of the $\overline{URC}$-R two-port (Fig. 8-26a) alone. *Hint:*† Ground the input of the active element. Then its output is also at zero ac potential. Assume that $\omega = \omega_A$ and remove the ground from the input of the amplifier. What will be the new output voltage?

REFERENCES

1. Chipman, R. A.: "Theory and Problems of Transmission Lines," McGraw-Hill, New York, 1968.
2. Richards, P. I.: Resistor–Transmission-Line Circuits, *Proc. IRE*, February 1948, pp. 217–220.
3. Matsumoto, A. (ed.): "Microwave Filters and Circuits," Academic, New York, 1970.
4. Crystal, E. G.: Microwave Filters, in G. C. Temes and S. K. Mitra (eds.), "Modern Filter Theory and Design," Wiley, New York, 1973.
5. Chirlian, P. M.: "Integrated and Active Network Analysis and Synthesis," Prentice-Hall, Englewood Cliffs, N.J., 1967.
6. Ghausi, M. S., and J. J. Kelly: "Introduction to Distributed Parameter Networks," Holt, New York, 1968.
7. Wyndrum, R. W., Jr.: The Exact Synthesis of Distributed RC Networks, *New York Univ. Lab. Electrosci.* TR-400-76, May 1963.
8. O'Shea, R. P.: Synthesis Using Distributed RC Networks, *IEEE Trans. Circuit Theory*, December 1965, pp. 546–554.
9. Wyndrum, R. W., Jr.: Active Distributed RC Networks, in G. C. Temes and S. K. Mitra (eds.), "Modern Filter Theory and Design," Wiley, New York, 1973.
10. Kaufman, W. M.: Theory of a Monolithic Null Device and Some Novel Circuit Applications, *Proc. IRE*, September 1960, pp. 1540–1545.
11. Kerwin, W. J.: Active RC Network Synthesis Using Voltage Amplifiers, in L. P. Huelsman (ed.), "Active Filters," McGraw-Hill, 1970.

† This simple nonmathematical argument was advanced by H. J. Orchard.

CHAPTER NINE

SENSITIVITY ANALYSIS OF
MEMORYLESS CIRCUITS

9-1 TOLERANCE ANALYSIS; CIRCUIT SENSITIVITIES

In practice, no matter how accurately and carefully the engineer designs his circuit, the final product will contain imperfect elements which will cause the circuit performance to deviate from the anticipated response. One imperfection will be the inaccuracy of the element values. These values for the actual circuit can lie anywhere in a range, called the *tolerance range*, which the designer declares acceptable for the purpose. The assignment of tolerances is, in fact, one of the most important parts of the circuit designer's task.

In addition to the unavoidable element-value deviations, or *tolerances*, other effects, such as aging, temperature and humidity variations, etc., affect the circuit's performance. Also, parasitic elements, e.g., stray capacitances, lead inductances, and winding resistances in inductors, can have appreciable and usually detrimental effects on the circuit response.

In this chapter, some methods will be described which enable the designer to predict the effects of such imperfections. Since the basic question to be answered is how *sensitive* the circuit is to these perturbations, the process is usually called *sensitivity analysis*.

Example 9-1 As an illustration, consider the simple circuit shown in Fig. 9-1a. Assume that both resistors have a tolerance range of ± 5 percent; that is, the inequalities $47.5 \leq R_1 \leq 52.5$ and $95 \leq R_2 \leq 105$ limit the possible values of the elements. How accurate will v_0 be?

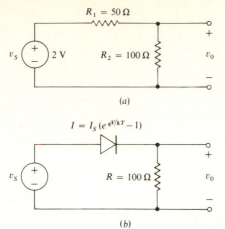

(a)

(b)

Figure 9-1 Simple circuits illustrating sensitivity analysis.

If the deviations $\Delta R_1 \triangleq R_{1_{act}} - 50$ and $\Delta R_2 \triangleq R_{2_{act}} - 100$ are small compared with R_1 and R_2, respectively, i.e., if

$$\left| \frac{\Delta R_i}{R_i} \right| \ll 1 \qquad i = 1, 2 \tag{9-1}$$

holds, we can reasonably expect that the first few terms of the Taylor-series expansion

$$v_0(R_1 + \Delta R_1, R_2 + \Delta R_2) = v_0(R_1, R_2) + \frac{\partial v_0}{\partial R_1} \Delta R_1 + \frac{\partial v_0}{\partial R_2} \Delta R_2 + \frac{1}{2} \frac{\partial^2 v_0}{\partial R_1^2} \Delta R_1^2$$

$$+ \frac{\partial^2 v_0}{\partial R_1\, \partial R_2} \Delta R_1\, \Delta R_2 + \frac{1}{2} \frac{\partial^2 v_0}{\partial R_2^2} \Delta R_2^2 + \cdots \tag{9-2}$$

of $v_0(R_1 + \Delta R_1, R_2 + \Delta R_2)$ around the nominal values R_1, R_2 will give an acceptable estimate.

To simplify the analysis, we shall neglect second- and higher-order terms and use the approximate relation

$$v_0(R_1 + \Delta R_1, R_2 + \Delta R_2) \approx v_0(R_1, R_2) + \frac{\partial v_0}{\partial R_1} \Delta R_1 + \frac{\partial v_0}{\partial R_2} \Delta R_2 \tag{9-3}$$

[In (9-2) and (9-3), all derivatives should, of course, be evaluated using the nominal values R_1, R_2.] The *deviation* of the output voltage from its nominal value is then

$$\Delta v_0 \triangleq v_0(R_1 + \Delta R_1, R_2 + \Delta R_2) - v_0(R_1, R_2) \approx \frac{\partial v_0}{\partial R_1} \Delta R_1 + \frac{\partial v_0}{\partial R_2} \Delta R_2 \tag{9-4}$$

For our circuit, the values are

$$v_0 = \frac{R_2}{R_1 + R_2} v_s = \tfrac{4}{3} \text{ V}$$

$$\frac{\partial v_0}{\partial R_1} = -\frac{R_2}{(R_1 + R_2)^2} v_s \approx -0.0089 \text{ V}/\Omega \tag{9-5}$$

and

$$\frac{\partial v_0}{\partial R_2} = \frac{R_1}{(R_1 + R_2)^2} v_s \approx 0.00444 \text{ V}/\Omega$$

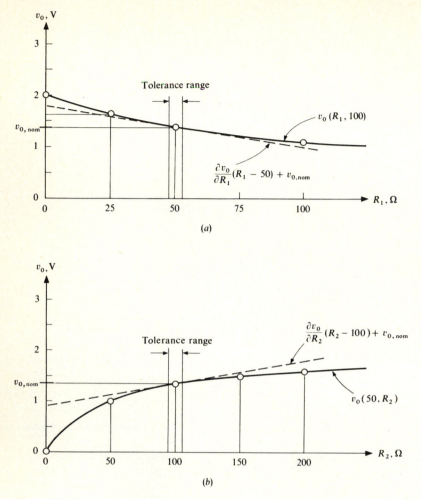

Figure 9-2 Linear approximation.

Some measure of the accuracy of (9-4) can be gained by plotting (Fig. 9-2) v_0 as a function of R_1 with R_2 fixed at 100 Ω and as a function of R_2 with R_1 fixed at 50 Ω. The relation (9-3) is also plotted as a broken straight line in the figures. The diagrams indicate that the straight-line approximation is valid in the tolerance range. [In fact, the curves give only a rough indication of the accuracy.† The correct, but more involved, process is to compare the surface $v_0 = f(R_1, R_2)$ given by (9-5) on the one hand and the tangent plane given by (9-3) on the other in the tolerance range. The excellent match indicated in Fig. 9-2 makes it more than likely, however, that the accuracy of the linear approximation will be acceptable.]

---

† They show that the constant and linear terms dominate the terms containing $\partial^i v_0/\partial R_1^i$, $\partial^i v_0/\partial R_2^i$, $i = 2, 3, \ldots$, in Eq. (9-2). However, they neglect the effect of *mixed* partial derivatives $\partial^{j+k} v_0/(\partial R_1^j \, \partial R_2^k)$.

Having obtained a linear approximation to $v_0(R_1, R_2)$, it is easy to predict the toler-ance effects. The *worst case* occurs when the two terms are of the same sign and ΔR_1 and ΔR_2 are as large as possible. Hence

$$|\Delta v_0|_{max} = \left| \frac{\partial v_0}{\partial R_1} \Delta R_{1\,max} \right| + \left| \frac{\partial v_0}{\partial R_2} \Delta R_{2\,max} \right|$$

$$\approx (8.9 \times 10^{-3})(2.5) + (4.44 \times 10^{-3})(5) \approx 0.0444 \text{ V} \tag{9-6}$$

Thus, v_0 satisfies $1.2889 \leq v_0 \leq 1.3777$ for all permissible values of R_1 and R_2.

For *statistical analysis*,[1] the linear approximation also simplifies the calculations. If there is no correlation between the random values assumed by R_1 and R_2, and if the variances† of their value distributions are $\sigma^2_{R_1}$ and $\sigma^2_{R_2}$, then it can be shown (Ref. 1, chap. 7) that the variance $\sigma^2_{v_0}$ of v_0 is

$$\sigma^2_{v_0} = \left(\frac{\partial v_0}{\partial R_1} \right)^2 \sigma^2_{R_1} + \left(\frac{\partial v_0}{\partial R_2} \right)^2 \sigma^2_{R_2} \tag{9-7}$$

We can therefore conclude that once an approximate relation of the form (9-3) is obtained, the calculation of tolerance effects becomes an easy task. Hence, the calculation of the partial derivatives $\partial v_0 / \partial R_1$ and $\partial v_0 / \partial R_2$ (often called *sensiti-vities*) is of crucial importance in tolerance analysis.

It will be shown in Chap. 11 that the *optimization* (automated design) of circuits also requires the circuit sensitivities $\partial v_0 / \partial p_i$, where the p_i's are the variable ("designable") parameters of the circuit. These derivatives give an indication of how the p_i should be changed to improve the performance of the circuit.

For very simple circuits, like those shown in Fig. 9-1a and b, it is possible to calculate the sensitivities analytically. For even moderately large circuits, however, the analytic method becomes very tedious. Since computers are badly suited for analytic differentiation, numerical techniques must be found to accom-plish the computation of the sensitivities.

Conceptually, the easiest numerical procedure is to use difference quotients as approximations of the derivatives. For example, one can use the *forward-difference quotient*

$$\frac{\partial v_0}{\partial R_1} \approx \frac{v_0(R_1 + \Delta R_1, R_2) - v_0(R_1, R_2)}{\Delta R_1} \tag{9-8}$$

or the somewhat more accurate *central-difference quotient*

$$\frac{\partial v_0}{\partial R_1} \approx \frac{v_0(R_1 + \Delta R_1, R_2) - v_0(R_1 - \Delta R_1, R_2)}{2 \, \Delta R_1} \tag{9-9}$$

These approximations are illustrated in Fig. 9-3.

These simple methods have two important disadvantages. The *first* one in-volves the accuracy of the approximation. If the change in the parameter, say ΔR_1, is too small, $v_0(R_1 + \Delta R_1, R_2) - v_0(R_1, R_2)$ will be the difference of nearly equal quantities and hence inaccurate. If ΔR_1 is too large, the difference quotient will be

---

† The variance σ^2_x of a quantity x is defined as the average value of $(\Delta x)^2$ (see Ref. 1, p. 176).

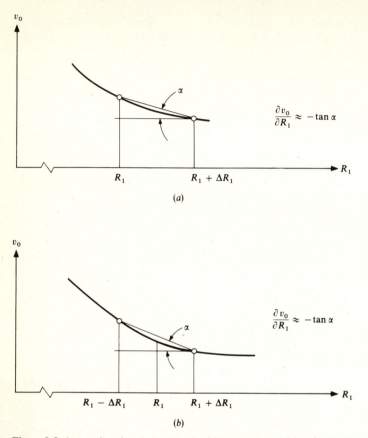

Figure 9-3 Approximating derivatives by (a) forward- and (b) central-difference quotient.

a poor approximation of the derivative. By judicious choice of ΔR_1 (say, using $|\Delta R_1/R_1| = 1$ percent) this problem can be overcome. More important is the *second* disadvantage, which is the large amount of calculation needed to evaluate (9-8) or (9-9). Specifically, if there are n variable parameters p_i in the circuit, the calculation of all $\partial v_0/\partial p_i$ requires $n + 1$ analyses of the circuit if forward differences are used. This is because the augmented values of v_0: $v_0(p_1 + \Delta p_1, p_2, \ldots, p_n)$, $v_0(p_1, p_2 + \Delta p_2, \ldots, p_n)$, $\ldots$, $v_0(p_1, p_2, \ldots, p_n + \Delta p_n)$ as well as its nominal value $v_0(p_1, p_2, \ldots, p_n)$ must all be found. For central differences, the situation is even worse: $2n$ analyses must be performed. If n is large (for integrated circuits, it may be of the order of 100 or more), the computational effort and cost may become overwhelming.

In this chapter and the next, a numerical method will be described for the calculation of circuit sensitivities. The method is exact yet well suited for implementation on a digital computer. Most important, it is very economical in terms of computational effort: the calculations required are at most equivalent to *two* (rather than $n + 1$) circuit analyses. Finally, the process can handle all practical circuits, including those containing nonlinear and distributed elements. It is based on the concept of the *adjoint network*, to be described in the next section. This concept, derived by S. W. Director and R. A. Rohrer[2,3] in 1969, proved to be a powerful aid for the visualization and application of some abstract results and techniques of mathematical system theory.

9-2 THE ADJOINT NETWORK

In Chap. 2, we derived the following forms of Tellegen's theorem:

$$\sum_{k=1}^{N} v_k j'_k = \mathbf{v}^T \mathbf{j}' = 0 \qquad \sum_{k=1}^{N} v'_k j_k = \mathbf{v}'^T \mathbf{j} = 0 \qquad (9\text{-}10)$$

[See Eqs. (2-36) and (2-37).] Here, the v_k and j_k are the branch voltages and currents of a network N, with associated reference directions being used (Fig. 2-4). Similarly, v'_k and j'_k are the branch variables of a second network N', also with associated directions. Since N and N' have the same graph, i.e., the same topological configuration, they have the same number of nodes and branches; corresponding branches are incident on corresponding nodes. Furthermore, the node and branch numbering and the reference directions are the same for the two circuits (see, for example, the networks in Figs. 2-4 and 2-5).

Consider now the three circuits shown in Fig. 9-4. In the figure, N is the *physical network* whose sensitivities we have to calculate. N_Δ is the *incremented network* in which all variable elements p_i have been replaced by their incremented values $p_i + \Delta p_i$. Finally, $\hat{N}$ is the *adjoint network*, to be discussed. All three circuits satisfy pairwise the conditions described above for N and N'. Thus, all three must have the same topology, same node and branch numbering, reference directions, etc. Furthermore, we postulate that the generators (independent sources) of $\hat{N}$ are located in the same branches as those of N and N_Δ. The types of these sources (voltage or current) must also be the same; however, their values may be different. Applying Tellegen's theorem to N and $\hat{N}$ gives

$$\mathbf{v}^T \hat{\mathbf{j}} = 0 \qquad \hat{\mathbf{v}}^T \mathbf{j} = 0 \qquad (9\text{-}11)$$

and so also $\qquad \qquad \mathbf{v}^T \hat{\mathbf{j}} - \hat{\mathbf{v}}^T \mathbf{j} = 0 \qquad (9\text{-}12)$

Applying Tellegen's theorem to N_Δ and $\hat{N}$ gives

$$(\mathbf{v} + \Delta \mathbf{v})^T \hat{\mathbf{j}} = 0 \qquad \hat{\mathbf{v}}^T (\mathbf{j} + \Delta \mathbf{j}) = 0 \qquad (9\text{-}13)$$

where $\Delta \mathbf{j}$ and $\Delta \mathbf{v}$ indicate that the currents and voltages of N_Δ will differ from those of N, due to the changes in the p_i. From (9-13),

$$(\mathbf{v} + \Delta \mathbf{v})^T \hat{\mathbf{j}} - \hat{\mathbf{v}}^T (\mathbf{j} + \Delta \mathbf{j}) = 0 \qquad (9\text{-}14)$$

Subtracting (9-12) from (9-14) gives

$$\Delta \mathbf{v}^T \hat{\mathbf{j}} - \hat{\mathbf{v}}^T \Delta \mathbf{j} = 0 \qquad (9\text{-}15)$$

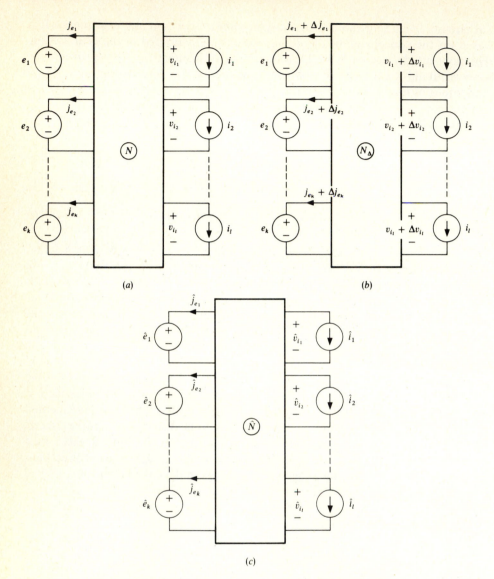

Figure 9-4 (a) Physical network; (b) incremental network; (c) adjoint network.

sometimes called the *differential Tellegen's theorem*. In scalar form, Eq. (9-15) becomes

$$\sum_{\text{all branches}} (\hat{j}_k \, \Delta v_k - \hat{v}_k \, \Delta j_k) = 0 \tag{9-16}$$

Here, the Δv_k and Δj_k are the results of the changes in the variable circuit parameters from p_i to $p_i + \Delta p_i$, $i = 1, 2, \ldots, n$.

We shall next show how to determine the sensitivities of the simple circuit of

Fig. 9-1a from (9-16). The circuit is redrawn in Fig. 9-5a with the notation of Fig. 9-4. Note, in particular, that the open-circuit output "branch" has been replaced by an equivalent 0-A current source. This does not represent any change, but it makes the following discussions somewhat easier. The adjoint network is shown in Fig. 9-5b. We only know the configuration and the location of its sources; hence, the other branches are represented by black boxes.

Applying the differential Tellegen's theorem to the two circuits of Fig. 9-5a and b gives

$$(\hat{j}_{e_1} \Delta e_1 - \hat{e}_1 \Delta j_{e_1}) + (\hat{j}_1 \Delta v_1 - \hat{v}_1 \Delta j_1) + (\hat{j}_2 \Delta v_2 - \hat{v}_2 \Delta j_2)$$
$$+ (\hat{i}_1 \Delta v_{i_1} - \hat{v}_{i_1} \Delta i_1) = 0 \quad (9\text{-}17)$$

Our purpose is to bring (9-17) into the form of (9-4). Since $v_0 \equiv v_{i_1}$, we have to (1) introduce ΔR_1 and ΔR_2 into (9-17) and (2) suppress all terms except those containing ΔR_1, ΔR_2, and Δv_{i_1}. These objectives will be achieved by choosing the structure and element values of $\hat{N}$, hitherto unspecified, in an appropriate way.

Consider the first two terms on the left-hand side of (9-17). Since e_1 is a

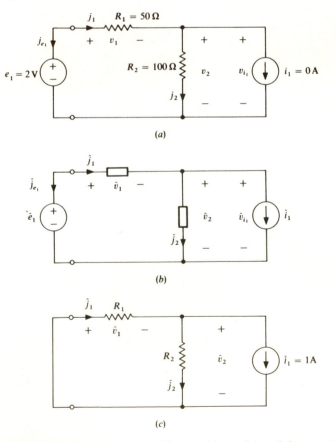

(a)

(b)

(c)

Figure 9-5 (a) Physical network; (b) general form of the adjoint network $\hat{N}$; (c) the actual $\hat{N}$.

generator which we assume to be unaffected by tolerances,† $\Delta e_1 \equiv 0$. Since the second term does not contain ΔR_1, ΔR_2, or Δv_{i_1}, we suppress it by choosing $\hat{e}_1 \equiv 0$ in $\hat{N}$ (Fig. 9-5c).

In the second two terms, ΔR_1 can be introduced through the branch relations of the original network N:

$$v_1 = R_1 j_1 \tag{9-18}$$

and of the incremented network N_Δ:

$$v_1 + \Delta v_1 = (R_1 + \Delta R_1)(j_1 + \Delta j_1) \tag{9-19}$$

Subtracting (9-18) from (9-19) gives

$$\Delta v_1 = R_1 \, \Delta j_1 + \Delta R_1 j_1 + \Delta R_1 \, \Delta j_1 \tag{9-20}$$

If all increments are small, $\Delta R_1 \, \Delta j_1$ is a second-order small term and can be neglected. Hence

$$\hat{j}_1 \, \Delta v_1 - \hat{v}_1 \, \Delta j_1 \approx \hat{j}_1(R_1 \, \Delta j_1 + \Delta R_1 j_1) - \hat{v}_1 \, \Delta j_1 = j_1 \hat{j}_1 \, \Delta R_1 + (\hat{j}_1 R_1 - \hat{v}_1) \, \Delta j_1 \tag{9-21}$$

The first term on the right-hand side of (9-21) contains ΔR_1 and should hence be retained. The second term can be suppressed by making

$$\hat{v}_1 = R_1 \hat{j}_1 \tag{9-22}$$

which implies that *this branch of $\hat{N}$ should contain a resistor R_1* (Fig. 9-5c). Then the right-hand side of (9-21) becomes $j_1 \hat{j}_1 \, \Delta R_1$.

An analog derivation shows that

$$\hat{j}_2 \, \Delta v_2 - \hat{v}_2 \, \Delta j_2 = j_2 \hat{j}_2 \, \Delta R_2 \tag{9-23}$$

provided that

$$\hat{v}_2 = R_2 \hat{j}_2 \tag{9-24}$$

i.e., provided that *this branch of $\hat{N}$ contains the resistor R_2* (Fig. 9-5c).

Finally, in the last two terms of (9-17), $\Delta i_1 \equiv 0$. The term $\hat{i}_1 \, \Delta v_{i_1}$ must be kept. If, for convenience, we choose $\hat{i}_1 = 1$ A (Fig. 9-5c), then (9-17) becomes finally

$$j_1 \hat{j}_1 \, \Delta R_1 + j_2 \hat{j}_2 \, \Delta R_2 + \Delta v_{i_1} = 0 \tag{9-25}$$

or

$$\Delta v_0 \equiv \Delta v_{i_1} = -j_1 \hat{j}_1 \, \Delta R_1 - j_2 \hat{j}_2 \, \Delta R_2 \tag{9-26}$$

A comparison of (9-26) and (9-4) shows immediately that the desired sensitivities are

$$\frac{\partial v_0}{\partial R_1} \equiv \frac{\partial v_{i_1}}{\partial R_1} = -j_1 \hat{j}_1 \qquad \frac{\partial v_0}{\partial R_2} \equiv \frac{\partial v_{i_1}}{\partial R_2} = -j_2 \hat{j}_2 \tag{9-27}$$

Here, j_1 and j_2 are the currents through R_1 and R_2 in N (Fig. 9-5a). By inspection,

$$j_1 = j_2 = \frac{e_1}{R_1 + R_2} \approx 0.01333 \text{ A}$$

---

† The calculation of the effects of source variations will be discussed in Sec. 9-4.

The currents $\hat{j}_1$ and $\hat{j}_2$ are the corresponding currents in the adjoint network $\hat{N}$. From Fig. 9-5c, also by inspection,

$$\hat{v}_2 = -\hat{i}_1 \frac{R_1 R_2}{R_1 + R_2} \approx -33.33 \text{ V}$$

and
$$\hat{j}_1 = -\frac{\hat{v}_2}{R_1} \approx 0.6667 \text{ A} \qquad \hat{j}_2 = \frac{\hat{v}_2}{R_2} \approx -0.3333 \text{ A}$$

Hence, by (9-27), we obtain the sensitivity values

$$\frac{\partial v_0}{\partial R_1} = -j_1 \hat{j}_1 \approx -(0.01333)(0.6667) \approx -0.0089 \text{ V}/\Omega$$

and
$$\frac{\partial v_0}{\partial R_2} = -j_2 \hat{j}_2 \approx (0.01333)(0.3333) \approx 0.00444 \text{ V}/\Omega$$

which agree with the results obtained by differentiation in Sec. 9-1.

The process followed in this example will be generalized for an arbitrary linear resistive circuit. The circuit may also contain current-controlled voltage sources (CCVS). The branch relations of the resistors are

$$v_k = R_k j_k \tag{9-28}$$

and the controlled branch voltages of the CCVS satisfy

$$v_l = R_{lm} j_m \tag{9-29}$$

These branch relations can now be merged into the vector relation

$$\mathbf{v}_B = \mathbf{R} \mathbf{j}_B \tag{9-30}$$

Here, $\mathbf{R}$ is the *branch resistance matrix*. The vectors $\mathbf{v}_B$ and $\mathbf{j}_B$ contain the variables for all *internal* branches of N, that is, for those branches which do *not* contain independent sources. (These are contained inside the box N in Fig. 9-4a.) Similarly, let the source voltages and currents of N be collected in the vectors $\mathbf{e}$ and $\mathbf{i}$, respectively. A zero-valued current source is to be placed at an open-circuited output port (Fig. 9-6a); this source must be included in $\mathbf{i}$. Similarly, if the sensitivities $\partial i_0/\partial R_k$ of an output current i_0 are required, a zero-valued voltage source should be inserted in the output branch (Fig. 9-6b); it must be included in $\mathbf{e}$.

The branch variables of $\hat{N}$ will similarly be collected in $\hat{\mathbf{v}}_B$ and $\hat{\mathbf{j}}_B$, the sources of $\hat{N}$ in $\hat{\mathbf{e}}$ and $\hat{\mathbf{i}}$. Then (9-16) can be written in the form

$$\left(\hat{\mathbf{j}}_e^T \Delta \mathbf{e} - \hat{\mathbf{e}}^T \Delta \mathbf{j}_e\right) + \left(\hat{\mathbf{j}}_B^T \Delta \mathbf{v}_B - \hat{\mathbf{v}}_B^T \Delta \mathbf{j}_B\right) + \left(\hat{\mathbf{i}}^T \Delta \mathbf{v}_i - \hat{\mathbf{v}}_i^T \Delta \mathbf{i}\right) = 0 \tag{9-31}$$

Since we assume constant sources, $\Delta \mathbf{e} = \mathbf{0}$ and $\Delta \mathbf{i} = \mathbf{0}$. Also, from (9-30),

$$\Delta \mathbf{v}_B = \Delta \mathbf{R} \, \mathbf{j}_B + \mathbf{R} \, \Delta \mathbf{j}_B \tag{9-32}$$

where we neglected the second-order small term $\Delta \mathbf{R} \, \Delta \mathbf{j}_B$ just as we did in (9-20). Then

$$-\hat{\mathbf{e}}^T \Delta \mathbf{j}_e + \hat{\mathbf{i}}^T \Delta \mathbf{v}_i = -\hat{\mathbf{j}}_B^T (\Delta \mathbf{R} \, \mathbf{j}_B + \mathbf{R} \, \Delta \mathbf{j}_B) + \hat{\mathbf{v}}_B^T \Delta \mathbf{j}_B = -\hat{\mathbf{j}}_B^T \Delta \mathbf{R} \, \mathbf{j}_B + (\hat{\mathbf{v}}_B^T - \hat{\mathbf{j}}_B^T \mathbf{R}) \Delta \mathbf{j}_B \tag{9-33}$$

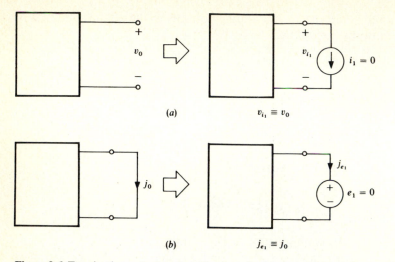

(a)

$v_{i_1} \equiv v_0$

(b)

$j_{e_1} \equiv j_0$

Figure 9-6 Terminating output ports with zero-valued independent sources.

The terms ΔR_k, ΔR_{lm} are contained in $\Delta \mathbf{R}$; to eliminate all other terms except $-\hat{\mathbf{j}}_B^T \Delta \mathbf{R} \mathbf{j}_B$ on the right-hand side we set

$$\hat{\mathbf{v}}_B^T - \hat{\mathbf{j}}_B^T \mathbf{R} = \mathbf{0} \tag{9-34}$$

or

$$\hat{\mathbf{v}}_B = \mathbf{R}^T \hat{\mathbf{j}}_B \tag{9-35}$$

Equation (9-35) gives the internal branch relations of $\hat{N}$. A comparison with (9-30) shows that the branch relations for N and $\hat{N}$ are *different* in that $\mathbf{R}$ has been replaced by $\hat{\mathbf{R}} \triangleq \mathbf{R}^T$ for $\hat{N}$. However, if N contains *only* resistors and *no* CCVS, all branch relations are of the form of (9-28) and hence $\mathbf{R}$ is diagonal. Then $\mathbf{R}^T = \mathbf{R}$, and the branch relations for N and $\hat{N}$ are the same.

Example 9-2 For the simple circuit of Fig. 9-5 the branch relations were

$$v_1 = R_1 j_1 \qquad v_2 = R_2 j_2 \tag{9-36}$$

so that in vector form

$$\begin{bmatrix} v_1 \\ v_2 \end{bmatrix} = \begin{bmatrix} R_1 & 0 \\ 0 & R_2 \end{bmatrix} \begin{bmatrix} j_1 \\ j_2 \end{bmatrix} \tag{9-37}$$

with $\mathbf{R}$ a diagonal matrix. For this network therefore, $\mathbf{R}^T = \mathbf{R}$. This implies that the internal branches of N and $\hat{N}$ are identical, as shown in Fig. 9-5.

Example 9-3 Consider, by contrast, the circuit of Fig. 9-7a. The internal branch relations are

$$v_1 = R_1 j_1 \qquad v_2 = 0 \qquad v_3 = R_{32} j_2 \qquad v_4 = R_4 j_4 \tag{9-38}$$

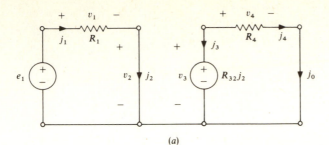

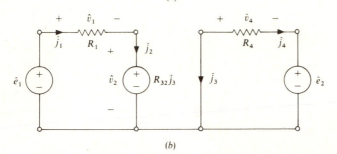

Figure 9-7 (*a*) Circuit containing a CCVS; (*b*) its adjoint network. The element values are $e_1 = 5$ V, $R_1 = 10$ kΩ, $R_4 = 50$ kΩ, and $R_{32} = 20$ kΩ.

Thus, the branch resistance matrix of N is

$$\mathbf{R} = \begin{bmatrix} R_1 & 0 & 0 & 0 \\ 0 & 0 & 0 & 0 \\ 0 & R_{32} & 0 & 0 \\ 0 & 0 & 0 & R_4 \end{bmatrix} \tag{9-39}$$

while that of $\hat{N}$ should be chosen as

$$\hat{\mathbf{R}} = \mathbf{R}^T = \begin{bmatrix} R_1 & 0 & 0 & 0 \\ 0 & 0 & R_{32} & 0 \\ 0 & 0 & 0 & 0 \\ 0 & 0 & 0 & R_4 \end{bmatrix} \tag{9-40}$$

$\hat{\mathbf{R}}$ corresponds to the branch relations

$$\hat{v}_1 = R_1 \hat{j}_1 \qquad \hat{v}_2 = R_{32} \hat{j}_3 \qquad \hat{v}_3 = 0 \qquad \hat{v}_4 = R_4 \hat{j}_4$$

Hence, $\hat{N}$ has the structure shown in Fig. 9-7*b*. Its internal branches are *different* from those of N. Specifically, the controlling and controlled branches of the CCVS are interchanged in $\hat{N}$.

This result is easily generalized. Let the mth branch current control the lth branch voltage in a general circuit (Fig. 9-8*a*). Then the branch relations of the CCVS are

$$v_l = R_{lm} j_m \qquad v_m = 0 \tag{9-41}$$

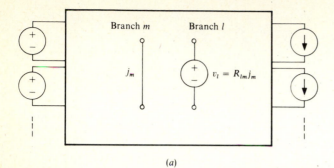

(a)

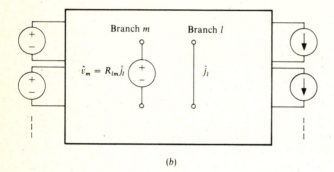

(b)

Figure 9-8 (a) General circuit with CCVS; (b) its adjoint network.

so that the branch resistance matrix $\mathbf{R}$ of N is

$$
\begin{array}{c} \text{row } l \\ \\ \text{row } m \end{array}
\begin{bmatrix}
0 & 0 & \cdots & 0 & \cdots & R_{lm} & \cdots & 0 \\
\\
0 & 0 & \cdots & 0 & \cdots & 0 & \cdots & 0
\end{bmatrix}
\qquad (9\text{-}42)
$$

$$
\qquad\qquad\qquad\qquad\quad \text{col. } l \qquad \text{col. } m
$$

The branch resistance matrix $\hat{\mathbf{R}}$ of $\hat{N}$ is then the transpose of $\mathbf{R}$, or

$$
\begin{array}{c} \text{row } l \\ \\ \text{row } m \end{array}
\begin{bmatrix}
0 & 0 & \cdots & 0 & \cdots & 0 & \cdots & 0 \\
\\
0 & 0 & \cdots & R_{lm} & \cdots & 0 & \cdots & 0
\end{bmatrix}
\qquad (9\text{-}43)
$$

$$
\qquad\qquad\qquad\qquad\quad \text{col. } l \qquad \text{col. } m
$$

Evidently now the current of branch l controls the voltage of branch m; that is, the controlling and controlled branches have changed places† (Fig. 9-8b).

---

† Note that the controlling branch of the CCVS in the above derivations contains no circuit elements; it is a short circuit. This definition of the branch is for convenience; it simplifies the discussions somewhat.

Let us return now to Eq. (9-33). If (9-34) holds, then the right-hand side of (9-33) becomes

$$-\hat{\mathbf{j}}_B^T \Delta \mathbf{R} \, \mathbf{j}_B = -\sum_{l, \, m} \hat{j}_l j_m \, \Delta R_{lm}$$

$$= -\sum_{\substack{\text{all} \\ \text{resistors}}} \hat{j}_l j_l \, \Delta R_l - \sum_{\substack{\text{all CCVS} \\ l \neq m}} \hat{j}_l j_m \, \Delta R_{lm} \qquad (9\text{-}44)$$

If the output is the voltage v_{i_k}, choosing all $\hat{e}_l = 0$ and all $\hat{i}_l = 0$ except for $\hat{i}_k = 1$ A causes the left-hand side of (9-33) to be equal to Δv_{i_k}. If the output is the current j_{e_k}, then choosing all $\hat{i}_l = 0$ and all $\hat{e}_l = 0$ except for $\hat{e}_k = -1$ V makes the left-hand side equal to Δj_{e_k}. Hence, (9-33) becomes

$$\Delta v_{i_k} \text{ or } \Delta j_{e_k} = -\sum_{\text{all } Rl} \hat{j}_l j_l \, \Delta R_l - \sum_{\substack{\text{all CCVS} \\ l \neq m}} \hat{j}_l j_m \, \Delta R_{lm} \qquad (9\text{-}45)$$

Thus the sensitivity of the output to a resistance R_l is $-\hat{j}_l j_l$; to the gain R_{lm} of a CCVS (Fig. 9-8a) it is $-\hat{j}_l j_m$. Note that the *controlling* currents $\hat{j}_l$ and j_m enter the sensitivity expression for the CCVS.

We can now recapitulate the steps leading to the sensitivities of a resistive circuit containing CCVSs:

1. Construct the internal branches of $\hat{N}$ from those of N. Specifically, branches containing resistors in N are simply duplicated in $\hat{N}$. Branches containing the controlling and controlled variables of a CCVS in N are interchanged in $\hat{N}$.
2. Set all independent sources to zero in $\hat{N}$ except the one at the output port. If the output quantity is a voltage, excite the output port with a 1-A current source; if it is a current, excite the output port with a -1-V voltage source.
3. Analyze N and $\hat{N}$ to find all internal branch currents j_k and $\hat{j}_l$.
4. The sensitivities of the output to resistances are given by

$$\frac{\partial v_{i_k}}{\partial R_l} \text{ or } \frac{\partial j_{e_k}}{\partial R_l} = -\hat{j}_l j_l \qquad (9\text{-}46)$$

and to CCVS gains by

$$\frac{\partial v_{i_k}}{\partial R_{lm}} \text{ or } \frac{\partial j_{e_k}}{\partial R_{lm}} = -\hat{j}_l j_m \qquad (9\text{-}47)$$

Example 9-4 For the circuit of Fig. 9-7a, by inspection $j_1 = j_2 = e_1/R_1 = 0.5$ mA, $j_3 = -j_4 = -R_{32} j_2/R_4 = -0.2$ mA. In the adjoint network (Fig. 9-7b), we set $\hat{e}_1 = 0$ and $\hat{e}_2 = -1$ V. Then $\hat{j}_3 = -\hat{j}_4 = \hat{e}_2/R_4 = -0.02$ mA, and $\hat{j}_1 = \hat{j}_2 = -R_{32}\hat{j}_3/R_1 = 0.04$ mA. Hence, the sensitivities are

$$\frac{\partial j_0}{\partial R_1} = -\hat{j}_1 j_1 = -0.02 \times 10^{-6} \text{ A}/\Omega = -0.02 \text{ mA/k}\Omega$$

$$\frac{\partial j_0}{\partial R_4} = -\hat{j}_4 j_4 = -0.004 \times 10^{-6} \text{ A}/\Omega = -0.004 \text{ mA/k}\Omega$$

$$\frac{\partial j_0}{\partial R_{32}} = -\hat{j}_3 j_2 = 0.01 \times 10^{-6} \text{ A}/\Omega = 0.01 \text{ mA/k}\Omega$$

As a test, we can perform a mathematical analysis of the circuit. This gives for the output current

$$j_o = \frac{e_1}{R_1} \frac{R_{32}}{R_4}$$

and hence the sensitivities

$$\frac{\partial j_o}{\partial R_1} = -\frac{e_1 R_{32}}{R_1^2 R_4} = -0.02 \text{ mA/k}\Omega$$

$$\frac{\partial j_o}{\partial R_4} = -\frac{e_1 R_{32}}{R_1 R_4^2} = -0.004 \text{ mA/k}\Omega$$

$$\frac{\partial j_o}{\partial R_{32}} = \frac{e_1}{R_1 R_4} = 0.01 \text{ mA/k}\Omega$$

which agree with the results obtained using the adjoint-network approach.

Consider now a circuit containing in its internal branches only resistors and *voltage-controlled current sources* (VCCS). Such circuits will have the internal branch relations $j_k = (1/R_k)v_k = G_k v_k$ for the resistors and $j_l = G_{lm} v_m$ for the VCCS.

These relations can be assembled into the vector relation

$$\mathbf{j}_B = \mathbf{G} \mathbf{v}_B \tag{9-48}$$

where $\mathbf{G}$ is the *branch conductance matrix*.

An argument dual to that proving (9-45) now shows that if $\mathbf{G}^T$ is chosen as the branch conductance matrix $\hat{\mathbf{G}}$ of $\hat{N}$, then (9-31) becomes

$$-\hat{\mathbf{e}}^T \Delta \mathbf{j}_e + \hat{\mathbf{i}}^T \Delta \mathbf{v}_i = \hat{\mathbf{v}}_B^T \Delta \mathbf{G} \mathbf{v}_B = \sum_{\substack{\text{all} \\ \text{branches}}} \hat{v}_l v_m \Delta G_{lm}$$

$$= \sum_{\substack{\text{all} \\ \text{resistors}}} \hat{v}_l v_l \Delta G_l + \sum_{\substack{\text{all VCCS} \\ l \neq m}} \hat{v}_l v_m \Delta G_{lm} \tag{9-49}$$

The relation $\hat{\mathbf{G}} = \mathbf{G}^T$ can again be shown to result in an interchange of the controlling and controlled branches of all VCCS (Fig. 9-9). If the $\hat{\mathbf{e}}$ and $\hat{\mathbf{i}}$ are chosen the same way as for the CCVS circuits, then (9-49) gives

$$\Delta v_{i_k} \text{ or } \Delta j_{e_k} = \sum_{\text{all } G_l} \hat{v}_l v_l \Delta G_l + \sum_{\substack{\text{all VCCS} \\ l \neq m}} \hat{v}_l v_m \Delta G_{lm} \tag{9-50}$$

Hence, the sensitivities of the output to the resistances are†

$$\frac{\partial v_{i_k}}{\partial G_l} \text{ or } \frac{\partial j_{e_k}}{\partial G_l} = \hat{v}_l v_l \tag{9-51}$$

† Note that (9-51) is equivalent to (9-46) since, say

$$\frac{\partial v_{i_k}}{\partial G_l} = \frac{\partial v_{i_k}}{\partial (1/R_l)} = \frac{\partial v_{i_k}/\partial R_l}{\partial (1/R_l)/\partial R_l} = \frac{-\hat{j}_l j_l}{-1/R_l^2} = \hat{v}_l v_l$$

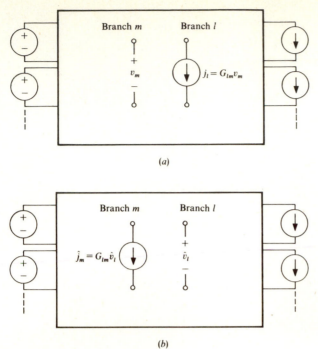

Figure 9-9 (*a*) General circuit with VCCS; (*b*) its adjoint network.

and to VCCS gains

$$\frac{\partial v_{ik}}{\partial G_{lm}} \quad \text{or} \quad \frac{\partial j_{ek}}{\partial G_{lm}} = \hat{v}_l v_m \tag{9-52}$$

Note that again only the *controlling* voltages enter the sensitivities.

Example 9-5 Calculate the sensitivities of v_0 in the circuit of Fig. 9-10a.

The adjoint network $\hat{N}$ is found by turning the VCCS around and choosing the generators $\hat{i}_1 = 0$ and $\hat{i}_2 = 1$ A (Fig. 9-10b). Both N and $\hat{N}$ can again be analyzed by inspection: from Fig. 9-10a, $v_1 = v_2 = -i_1/G_1 = -5$ V, $v_3 = v_4 = -G_{32}v_2/G_4 = 0.625$ V. Similarly from Fig. 9-10b, with $\hat{i}_1 = 0$ and $\hat{i}_2 = 1$ A, $\hat{v}_3 = \hat{v}_4 = -\hat{i}_2/G_4 = -250$ V and $\hat{v}_1 = \hat{v}_2 = -G_{32}\hat{v}_3/G_1 = 125$ V.†

Hence, by (9-51) and (9-52), the sensitivities are

$$\frac{\partial v_0}{\partial G_1} = \hat{v}_1 v_1 = -625 \text{ V/℧} = -0.625 \text{ V/m℧}$$

$$\frac{\partial v_0}{\partial G_4} = \hat{v}_4 v_4 = -156.25 \text{ V/℧} \approx -0.156 \text{ V/m℧}$$

$$\frac{\partial v_0}{\partial G_{32}} = \hat{v}_3 v_2 = 1250 \text{ V/℧} = 1.25 \text{ V/m℧}$$

† The large $\hat{v}$ values are due to the choice of $\hat{i}_2 = 1$ A, a convenient but uncommonly large value.

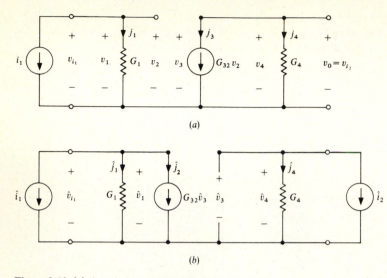

Figure 9-10 (*a*) A resistive circuit with a VCCS; (*b*) its adjoint network. The elements values are $i_1 = 5$ mA, $G_1 = 1$ m℧, $G_4 = 4$ m℧, and $G_{32} = 0.5$ m℧.

Mathematical analysis, performed as a test, gives the output voltage

$$v_0 = \frac{i_1 G_{32}}{G_1 G_4}$$

and hence

$$\frac{\partial v_0}{\partial G_1} = -\frac{i_1 G_{32}}{G_1^2 G_4} = -0.625 \text{ V/m℧}$$

$$\frac{\partial v_0}{\partial G_4} = -\frac{i_1 G_{32}}{G_1 G_4^2} \approx -0.156 \text{ V/m℧}$$

$$\frac{\partial v_0}{\partial G_{23}} = \frac{i_1}{G_1 G_4} = 1.25 \text{ V/m℧}$$

as before.

Example 9-6 The previous examples were chosen to be simple, to permit easy understanding of the method and also easy checking. The following example† demonstrates the power of the method under practical circumstances. Consider the differential amplifier shown in Fig. 9-11*a*. We want to analyze the sensitivities of the small-signal voltage gain $A_V = v_0/v_{\rm in}$. If we assume the small-signal model shown in Fig. 9-11*b* for all transistors and utilize the symmetry of the circuit about ground (Fig. 9-11*a*) to bisect it, we get the half-circuit model of Fig. 9-11*c*. Computer analysis gives the branch currents indicated (in milliamperes) in this circuit. This analysis is routine, and many computer programs are available to carry it out.

Next, using the simple rules established earlier, the adjoint network of the circuit of Fig. 9-11*c* is drawn. It is shown in Fig. 9-11*d*. The branch current values (in amperes) are also indicated and are also obtained routinely on the computer.

† Paraphrased by permission from D. A. Calahan, "Computer-aided Network Design," rev. ed., pp. 116–118, McGraw-Hill, New York, 1972.

From the voltages and currents shown in Fig. 9-11c and d, *all* sensitivities of $v_0/2$ can immediately be found using (9-51) [or its equivalent, (9-46)] and (9-52). For example, the sensitivity of $v_0/2$ to the 100-Ω resistor is

$$\frac{\partial(v_0/2)}{\partial R_2} = -\hat{j}_{R_2}j_{R_2} = (-9.4)(4.9 \times 10^{-3}) \approx -0.046 \text{ V}/\Omega$$

while the sensitivity to the transconductance of the first transistor is

$$\frac{\partial(v_0/2)}{\partial G_{m_1}} = \hat{v}_{c_1}v_{b_1} = (306)(6.46 \times 10^{-3}) \approx 1.975 \text{ V}/\mho \approx 2 \text{ mV/m}\mho$$

It should be evident that for this circuit the analytic approach, consisting of finding an explicit mathematical expression for $v_0/2 = f(R_1, R_2, \ldots, G_{m_1}, G_{m_2})$

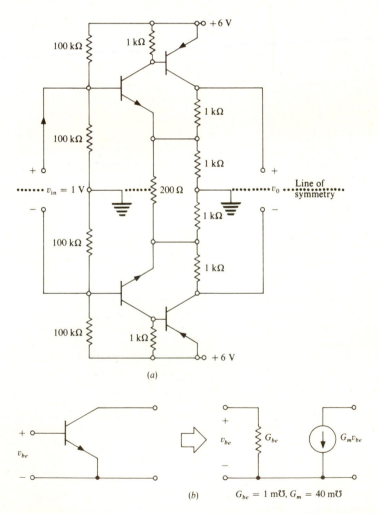

Figure 9-11 Sensitivity analysis of a differential amplifier: (*a*) full circuit; (*b*) transistor model. (*Continued on p. 392.*)

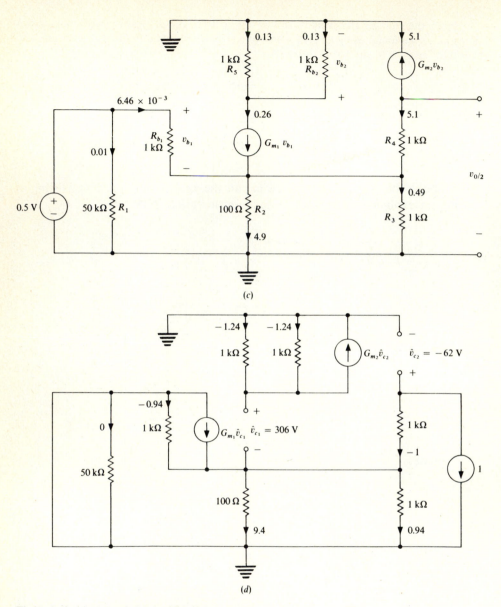

Figure 9-11 (c) Bisected circuit; (d) adjoint network.

and differentiating it formally, is impractical. Yet this is a very small circuit compared with those used routinely in, say, integrated op-amps. For such large circuits the computer analysis of the branch voltages and currents is still practical and hence so is the adjoint network technique of sensitivity analysis. Other approaches (mathematical differentiation, difference quotients) are impossible or at least prohibitingly expensive.

9-3 HYBRID REPRESENTATION FOR MEMORYLESS CIRCUITS

For some circuits, neither the branch resistance matrix $\mathbf{R}$ nor the branch conductance matrix $\mathbf{G}$ can be used to describe the branch relations.

Example 9-7 Consider the feedback amplifier circuit of Fig. 9-12a. Using the model of Fig. 9-12b for the amplifier gives the circuit shown in Fig. 9-12c. The branch relations are in the form

$$j_1 = G_1 v_1 \qquad j_2 = G_2 v_2 \qquad j_3 = 0 \qquad v_4 = -\mu v_3 \qquad (9\text{-}53)$$

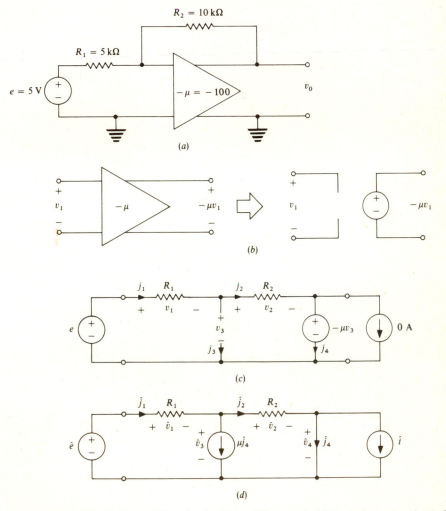

(a)

(b)

(c)

(d)

Figure 9-12 (a) Feedback amplifier circuit; (b) amplifier model, (c) equivalent network N; (d) the adjoint network $\hat{N}$.

These relations cannot in any way be grouped into either of the forms $\mathbf{v}_B = \mathbf{R}\mathbf{j}_B$ or $\mathbf{j}_B = \mathbf{G}\mathbf{v}_B$, but they can be written in the hybrid matrix equation

$$\mathbf{j}_{B_1} \begin{bmatrix} j_1 \\ j_2 \\ j_3 \\ --- \\ v_4 \end{bmatrix} \mathbf{v}_{B_2} = \begin{array}{c} \overset{\mathbf{G}}{} \quad\quad\quad \overset{\mathbf{A}}{} \\ \begin{bmatrix} G_1 & 0 & 0 & 0 \\ 0 & G_2 & 0 & 0 \\ 0 & 0 & 0 & 0 \\ ---&---&---&--- \\ 0 & 0 & -\mu & 0 \end{bmatrix} \\ \underset{\mathbf{M}}{} \quad\quad\quad \underset{\mathbf{R}}{} \end{array} \begin{bmatrix} v_1 \\ v_2 \\ v_3 \\ --- \\ j_4 \end{bmatrix} \begin{array}{l} \\ \mathbf{v}_{B_1} \\ \\ \\ \mathbf{j}_{B_2} \end{array} \qquad (9\text{-}54)$$

The square matrix on the right-hand side is called a *hybrid branch matrix*. Its general form is

$$\mathbf{H} \triangleq \begin{bmatrix} \mathbf{G} & \mathbf{A} \\ \mathbf{M} & \mathbf{R} \end{bmatrix} \qquad (9\text{-}55)$$

where the elements of the submatrix $\mathbf{G}$ are in mhos, the elements of $\mathbf{R}$ in ohms, the element of $\mathbf{A}$ in amperes per ampere, and those of $\mathbf{M}$ in volts per volt. Equation (9-54) shows how the $\mathbf{H}$ of the example can be partitioned into $\mathbf{G}$, $\mathbf{A}$, $\mathbf{M}$, and $\mathbf{R}$.

In terms of the hybrid branch matrix, the general form of the branch relations is

$$\begin{bmatrix} \mathbf{j}_{B_1} \\ \mathbf{v}_{B_2} \end{bmatrix} = \mathbf{H} \begin{bmatrix} \mathbf{v}_{B_1} \\ \mathbf{j}_{B_2} \end{bmatrix} = \begin{bmatrix} \mathbf{G} & \mathbf{A} \\ \mathbf{M} & \mathbf{R} \end{bmatrix} \begin{bmatrix} \mathbf{v}_{B_1} \\ \mathbf{j}_{B_2} \end{bmatrix} \qquad (9\text{-}56)$$

or, defining the hybrid vectors

$$\mathbf{x} \triangleq \begin{bmatrix} \mathbf{v}_{B_1} \\ \mathbf{j}_{B_2} \end{bmatrix} \qquad \mathbf{y} \triangleq \begin{bmatrix} \mathbf{j}_{B_1} \\ \mathbf{v}_{B_2} \end{bmatrix} \qquad (9\text{-}57)$$

we have

$$\mathbf{y} = \mathbf{H}\mathbf{x} \qquad (9\text{-}58)$$

For example, for the circuit of Fig. 9-12c, $\mathbf{v}_{B_1} = [v_1 \quad v_2 \quad v_3]^T$, $\mathbf{j}_{B_2} = j_4$, $\mathbf{j}_{B_1} = [j_1 \quad j_2 \quad j_3]^T$ and $\mathbf{v}_{B_2} = v_4$. Hence, $\mathbf{x} = [v_1 \quad v_2 \quad v_3 \quad j_4]^T$ and $\mathbf{y} = [j_1 \quad j_2 \quad j_3 \quad v_4]^T$.

We shall now duplicate the derivations performed for the branch-impedance-matrix description in Eqs. (9-28) to (9-35). From (9-58) and from the hybrid branch relations of the incremented network N_Δ

$$\mathbf{y} + \Delta\mathbf{y} = (\mathbf{H} + \Delta\mathbf{H})(\mathbf{x} + \Delta\mathbf{x}) \qquad (9\text{-}59)$$

by subtraction

$$\Delta\mathbf{y} = \mathbf{H}\,\Delta\mathbf{x} + \Delta\mathbf{H}\,\mathbf{x} \qquad (9\text{-}60)$$

Here, the second-order term $\Delta\mathbf{H}\,\Delta\mathbf{x}$ has been neglected. For the adjoint network, we use slightly different definitions. Let the hybrid branch relations be

$$\begin{bmatrix} \hat{\mathbf{j}}_{B_1} \\ \hat{\mathbf{v}}_{B_2} \end{bmatrix} = \begin{bmatrix} \hat{\mathbf{G}} & \hat{\mathbf{A}} \\ \hat{\mathbf{M}} & \hat{\mathbf{R}} \end{bmatrix} \begin{bmatrix} \hat{\mathbf{v}}_{B_1} \\ \hat{\mathbf{j}}_{B_2} \end{bmatrix} \qquad (9\text{-}61)$$

which can also be rewritten as

$$\begin{bmatrix} \hat{\mathbf{j}}_{B_1} \\ -\hat{\mathbf{v}}_{B_2} \end{bmatrix} = \begin{bmatrix} \hat{\mathbf{G}} & -\hat{\mathbf{A}} \\ -\hat{\mathbf{M}} & \hat{\mathbf{R}} \end{bmatrix} \begin{bmatrix} \hat{\mathbf{v}}_{B_1} \\ -\hat{\mathbf{j}}_{B_2} \end{bmatrix} \tag{9-62}$$

Next, we make the definitions†

$$\hat{\mathbf{x}} \triangleq \begin{bmatrix} \hat{\mathbf{v}}_{B_1} \\ -\hat{\mathbf{j}}_{B_2} \end{bmatrix} \qquad \hat{\mathbf{y}} \triangleq \begin{bmatrix} \hat{\mathbf{j}}_{B_1} \\ -\hat{\mathbf{v}}_{B_2} \end{bmatrix} \qquad \hat{\mathbf{H}} \triangleq \begin{bmatrix} \hat{\mathbf{G}} & -\hat{\mathbf{A}} \\ -\hat{\mathbf{M}} & \hat{\mathbf{R}} \end{bmatrix} \tag{9-63}$$

so that
$$\hat{\mathbf{y}} = \hat{\mathbf{H}}\hat{\mathbf{x}} \tag{9-64}$$

Next, we substitute into the internal-branch terms in the differential Tellegen's theorem (9-31) the quantities $\hat{\mathbf{j}}_B^T = [\hat{\mathbf{j}}_{B_1}^T \quad \hat{\mathbf{j}}_{B_2}^T]$, etc.:

$$\hat{\mathbf{j}}_B^T \, \Delta\mathbf{v}_B - \hat{\mathbf{v}}_B^T \, \Delta\mathbf{j}_B = \hat{\mathbf{j}}_{B_1}^T \, \Delta\mathbf{v}_{B_1} + \hat{\mathbf{j}}_{B_2}^T \, \Delta\mathbf{v}_{B_2} - \hat{\mathbf{v}}_{B_1}^T \, \Delta\mathbf{j}_{B_1} - \hat{\mathbf{v}}_{B_2}^T \, \Delta\mathbf{j}_{B_2} \tag{9-65}$$

The right-hand side can also be written

$$[\hat{\mathbf{j}}_{B_1}^T \quad -\hat{\mathbf{v}}_{B_2}^T]\begin{bmatrix} \Delta\mathbf{v}_{B_1} \\ \Delta\mathbf{j}_{B_2} \end{bmatrix} - [\hat{\mathbf{v}}_{B_1}^T \quad -\hat{\mathbf{j}}_{B_2}^T]\begin{bmatrix} \Delta\mathbf{j}_{B_1} \\ \Delta\mathbf{v}_{B_2} \end{bmatrix}$$

$$= \hat{\mathbf{y}}^T \, \Delta\mathbf{x} - \hat{\mathbf{x}}^T \, \Delta\mathbf{y} = (\hat{\mathbf{y}}^T - \hat{\mathbf{x}}^T\mathbf{H}) \, \Delta\mathbf{x} - \hat{\mathbf{x}}^T \, \Delta\mathbf{H}\,\mathbf{x} \tag{9-66}$$

Since the increments of the element values are contained in $\Delta\mathbf{H}$ while $\Delta\mathbf{x}$ is unknown, we want to eliminate the first term on the right-hand side. This requires

$$\hat{\mathbf{y}} = \mathbf{H}^T\hat{\mathbf{x}} \tag{9-67}$$

Now (9-63), (9-64), and (9-55) show that the hybrid matrices must be related by

$$\hat{\mathbf{H}} \triangleq \begin{bmatrix} \hat{\mathbf{G}} & -\hat{\mathbf{A}} \\ -\hat{\mathbf{M}} & \hat{\mathbf{R}} \end{bmatrix} = \mathbf{H}^T = \begin{bmatrix} \mathbf{G}^T & \mathbf{M}^T \\ \mathbf{A}^T & \mathbf{R}^T \end{bmatrix} \tag{9-68}$$

so that the submatrices of the adjoint hybrid matrix are given by

$$\hat{\mathbf{G}} = \mathbf{G}^T \qquad \hat{\mathbf{A}} = -\mathbf{M}^T \qquad \hat{\mathbf{M}} = -\mathbf{A}^T \qquad \hat{\mathbf{R}} = \mathbf{R}^T \tag{9-69}$$

Note especially the negative signs in the relations giving $\hat{\mathbf{A}}$ and $\hat{\mathbf{M}}$.

In conclusion, if the branches of N are described by (9-56), the branch relations of $\hat{N}$ are

$$\begin{bmatrix} \hat{\mathbf{j}}_{B_1} \\ \hat{\mathbf{v}}_{B_2} \end{bmatrix} = \begin{bmatrix} \mathbf{G}^T & -\mathbf{M}^T \\ -\mathbf{A}^T & \mathbf{R}^T \end{bmatrix} \begin{bmatrix} \hat{\mathbf{v}}_{B_1} \\ \hat{\mathbf{j}}_{B_2} \end{bmatrix} \tag{9-70}$$

and from (9-65) to (9-67)

$$\sum_{\substack{\text{all internal} \\ \text{branches}}} (\hat{j}_k \, \Delta v_k - \hat{v}_k \, \Delta j_k) = -\hat{\mathbf{x}}^T \, \Delta\mathbf{H}\,\mathbf{x}$$

$$= -[\hat{\mathbf{v}}_{B_1}^T \quad -\hat{\mathbf{j}}_{B_2}^T]\begin{bmatrix} \Delta\mathbf{G} & \Delta\mathbf{A} \\ \Delta\mathbf{M} & \Delta\mathbf{R} \end{bmatrix}\begin{bmatrix} \mathbf{v}_{B_1} \\ \mathbf{j}_{B_2} \end{bmatrix}$$

$$= -\hat{\mathbf{v}}_{B_1}^T \, \Delta\mathbf{G}\,\mathbf{v}_{B_1} - \hat{\mathbf{v}}_{B_1}^T \, \Delta\mathbf{A}\,\mathbf{j}_{B_2}$$

$$+ \hat{\mathbf{j}}_{B_2}^T \, \Delta\mathbf{M}\,\mathbf{v}_{B_1} + \hat{\mathbf{j}}_{B_2}^T \, \Delta\mathbf{R}\,\mathbf{j}_{B_2} \tag{9-71}$$

† The introduction of the negative signs into $\hat{\mathbf{x}}$, $\hat{\mathbf{y}}$, and $\hat{\mathbf{H}}$ simplifies subsequent derivations.

From (9-31) and (9-71), the differential Tellegen's theorem now becomes

$$-\hat{\mathbf{e}}^T \, \Delta \mathbf{j}_e + \hat{\mathbf{i}}^T \, \Delta \mathbf{v}_i = \hat{\mathbf{x}}^T \, \Delta \mathbf{H} \, \mathbf{x} \qquad (9\text{-}72)$$

Hence, choosing $\hat{\mathbf{e}}$ and $\hat{\mathbf{i}}$ the same manner as described for the R-CCVS and R-VCCS circuits, the sensitivity relation is

$$\Delta v_{i_m} \text{ or } \Delta j_{e_m} = \underbrace{\sum \hat{v}_k v_l \, \Delta H_{kl}}_{\substack{\text{all } H_{kl} \\ \text{in } \mathbf{G}}} + \underbrace{\sum \hat{v}_k j_l \, \Delta H_{kl}}_{\substack{\text{all } H_{kl} \\ \text{in } \mathbf{A}}} - \underbrace{\sum \hat{j}_k v_l \, \Delta H_{kl}}_{\substack{\text{all } H_{kl} \\ \text{in } \mathbf{M}}} - \underbrace{\sum \hat{j}_k j_l \, \Delta H_{kl}}_{\substack{\text{all } H_{kl} \\ \text{in } \mathbf{R}}}$$

$$(9\text{-}73)$$

Therefore, the sensitivities are

$$\frac{\partial v_{i_m}}{\partial H_{kl}} \text{ or } \frac{\partial j_{e_m}}{\partial H_{kl}} = \begin{cases} \hat{v}_k v_l & \text{if } H_{kl} \text{ is in } \mathbf{G} \\ \hat{v}_k j_l & \text{if } H_{kl} \text{ is in } \mathbf{A} \\ -\hat{j}_k v_l & \text{if } H_{kl} \text{ is in } \mathbf{M} \\ -\hat{j}_k j_l & \text{if } H_{kl} \text{ is in } \mathbf{R} \end{cases} \qquad (9\text{-}74)$$

The derivations leading to the results (9-70) to (9-74) were quite tedious. Fortunately, however, the application of these results to practical problems is straightforward, as some examples will demonstrate.

Example 9-8 For the circuit of Fig. 9-12c, the branch relations were given in (9-54). Using (9-70), we obtain by inspection for $\tilde{N}$ the branch relations

$$\hat{\mathbf{j}}_{B_1} \begin{bmatrix} \hat{j}_1 \\ \hat{j}_2 \\ \hat{j}_3 \\ --- \\ \hat{v}_4 \end{bmatrix} = \begin{matrix} \overset{\mathbf{G}^T \quad -\mathbf{M}^T}{\begin{bmatrix} G_1 & 0 & 0 & 0 \\ 0 & G_2 & 0 & 0 \\ 0 & 0 & 0 & \mu \\ ---&---&---&--- \\ 0 & 0 & 0 & 0 \end{bmatrix}} \\ \underset{-\mathbf{A}^T \qquad \mathbf{R}^T}{} \end{matrix} \begin{bmatrix} \hat{v}_1 \\ \hat{v}_2 \\ \hat{v}_3 \\ --- \\ \hat{j}_4 \end{bmatrix} \begin{matrix} \hat{\mathbf{v}}_{B_1} \\ \\ \\ \hat{\mathbf{j}}_{B_2} \end{matrix} \qquad (9\text{-}75)$$

The circuit $\hat{N}$ described by these relations is shown in Fig. 9-12d. Note that the controlling and controlled branches of the controlled source have been interchanged; also, the VCVS of gain $-\mu$ contained in N has been replaced in $\hat{N}$ by a CCCS of gain $+\mu$.

From (9-74), the sensitivities are given by the formulas

$$\frac{\partial v_0}{\partial G_1} = \hat{v}_1 v_1 \qquad \frac{\partial v_0}{\partial G_2} = \hat{v}_2 v_2 \qquad \frac{\partial v_0}{\partial(-\mu)} = -\frac{\partial v_0}{\partial \mu} = -\hat{j}_4 v_3$$

where $\hat{N}$ is excited by $\hat{e} = 0$ V and $\hat{i} = 1$ A.

Analysis of the circuits N and $\hat{N}$ shows that

$$v_1 \approx 4.9029 \text{ V} \qquad \hat{v}_1 \approx -9708.7 \text{ V}$$

$$v_2 \approx 9.8058 \text{ V} \qquad \hat{v}_2 \approx 9708.7 \text{ V}$$

$$v_3 \approx 0.0971 \text{ V} \qquad \hat{j}_4 \approx -0.0291 \text{ A}$$

and hence
$$\frac{\partial v_0}{\partial G_1} \approx -47{,}601 \text{ V/}\mho \qquad \frac{\partial v_0}{\partial G_2} \approx 95{,}202 \text{ V/}\mho$$

$$\frac{\partial v_0}{\partial \mu} \approx -0.002828 \text{ V/unit gain}$$

Mathematical analysis gives the symbolic relation

$$v_0 = -\frac{\mu R_2 e}{(1 + \mu)R_1 + R_2} = -\frac{\mu G_1 e}{(1 + \mu)G_2 + G_1}$$

and so differentiation gives the sensitivities

$$\frac{\partial v_0}{\partial G_1} = -\frac{\mu(1 + \mu)G_2 e}{[(1 + \mu)G_2 + G_1]^2} \approx -47{,}601 \text{ V/}\mho$$

$$\frac{\partial v_0}{\partial G_2} = +\frac{\mu(1 + \mu)G_1 e}{[(1 + \mu)G_2 + G_1]^2} \approx 95{,}202 \text{ V/}\mho$$

and
$$\frac{\partial v_0}{\partial \mu} = -\frac{G_1(G_1 + G_2)e}{[(1 + \mu)G_2 + G_1]^2} = -0.0028278 \text{ V/unit gain}$$

All results agree with those obtained using the adjoint-network approach.

It is easy to show that the transformation of a VCVS in N into a CCCS in $\hat{N}$ is generally valid. Consider an arbitrary circuit containing a VCVS (Fig. 9-13a). The VCVS branch relations enter into rows m and l of $\mathbf{H}$ as follows:

$$
\begin{array}{c}
\\
\text{row } m \\
\\
\\
\text{row } l \\
\\
\end{array}
\begin{bmatrix}
\cdots \\
j_m \\
\cdots \\
-- \\
v_l \\
\cdots
\end{bmatrix}
\overset{\overset{\displaystyle G \qquad\qquad A}{}}{
\begin{bmatrix}
0 & 0 & \cdots & \cdots & 0 & 0 \\
& & & & & \\
0 & \cdots & M_{lm} & \cdots & 0 & 0 \\
\end{bmatrix}}
\underset{\displaystyle \mathbf{M} \qquad\quad \mathbf{R}}{}
\begin{bmatrix}
\cdots \\
v_m \\
\cdots \\
-- \\
j_l \\
\cdots
\end{bmatrix}
\tag{9-76}
$$

In the adjoint network, $\hat{\mathbf{A}} = -\mathbf{M}^T$, and, vice versa, $\hat{\mathbf{M}} = -\mathbf{A}^T$. Hence, the all-zero elements in the lth column† of $\mathbf{H}$ form the lth row of the hybrid matrix of $\hat{N}$. Also, the mth column of $\mathbf{H}$, which contains zeros except for the element M_{lm}, becomes the mth row of the hybrid matrix of $\hat{N}$ with M_{lm} changed to $-M_{lm}$. Thus, for $\hat{N}$ the branch relations are given by the matrix equation

$$
\begin{bmatrix}
\cdots \\
\hat{j}_m \\
\cdots \\
-- \\
\hat{v}_l \\
\cdots
\end{bmatrix}
=
\overset{\overset{\displaystyle \hat{G} \qquad\qquad\qquad \hat{A}}{}}{
\begin{bmatrix}
0 & 0 & \cdots & \cdots & \cdots & \cdots & -M_{lm} & \cdots \\
& & & & & & & \\
0 & \cdots & \cdots & 0 & \cdots & \cdots & 0 & 0 \\
\end{bmatrix}}
\underset{\displaystyle \hat{\mathbf{M}} \qquad\qquad\quad \hat{\mathbf{R}}}{}
\begin{bmatrix}
\cdots \\
\hat{v}_m \\
\cdots \\
-- \\
\hat{j}_l \\
\cdots
\end{bmatrix}
\tag{9-77}
$$

† These elements are all zeros since the lth branch of N is the *controlled* branch of a VCVS. Hence, it has no resistor in it, nor is it the controlling branch of any controlled source in N.

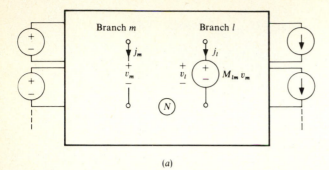

(a)

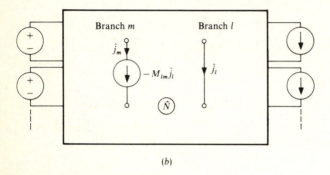

(b)

Figure 9-13 (a) General circuit N with VCVS; (b) its adjoint network $\hat{N}$.

The relations represented by rows m and l of (9-77) are

$$\hat{j}_m = -M_{lm}\hat{j}_l \qquad \hat{v}_l = 0 \tag{9-78}$$

which correspond to a CCCS with gain $-M_{lm}$, as illustrated in Fig. 9-13b.

An entirely analogous derivation shows that the reciprocal of this transformation is also valid (Fig. 9-14). Instead of carrying out the argument, however, it is simpler to observe from the transformation

$$\begin{bmatrix} \mathbf{G} & \mathbf{A} \\ \mathbf{M} & \mathbf{R} \end{bmatrix} \Rightarrow \begin{bmatrix} \mathbf{G}^T & -\mathbf{M}^T \\ -\mathbf{A}^T & \mathbf{R}^T \end{bmatrix} \tag{9-79}$$

generating the hybrid matrix of $\hat{N}$ from that of N that repeating the operation gives

$$\begin{bmatrix} \mathbf{G}^T & -\mathbf{M}^T \\ -\mathbf{A}^T & \mathbf{R}^T \end{bmatrix} \Rightarrow \begin{bmatrix} (\mathbf{G}^T)^T & -(-\mathbf{A}^T)^T \\ -(-\mathbf{M}^T)^T & (\mathbf{R}^T)^T \end{bmatrix} = \begin{bmatrix} \mathbf{G} & \mathbf{A} \\ \mathbf{M} & \mathbf{R} \end{bmatrix} \tag{9-80}$$

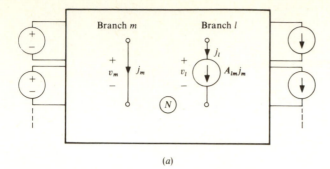

(a)

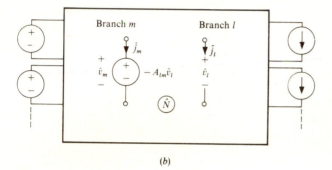

(b)

Figure 9-14 (a) General circuit N with CCCS; (b) its adjoint network $\hat{N}$.

Therefore, *the adjoint network of the adjoint network $\hat{N}$ is again the physical network N; or, formally, $\hat{\hat{N}} \equiv N$.* Hence, all correspondences between N and $\hat{N}$, illustrated in Figs. 9-8, 9-9, and 9-13, can be interpreted both ways, reading down or up. Interchanging the roles of N and $\hat{N}$ in Fig. 9-13 gives the result shown in Fig. 9-14 directly.

At this point, we can compile a collection of circuit elements in N, their counterparts in the adjoint network $\hat{N}$, and the sensitivities of the output of N to their values. This tabulation is given in Table 9-1.

We can now state that *the calculation of all sensitivities of a linear memoryless circuit can be carried out at the cost of two circuit analyses: one is needed to find all internal branch variables (v_k, j_k) of the physical circuit and one to find the internal branch variables $(\hat{v}_k, \hat{j}_k)$ of the adjoint network $\hat{N}$.* The internal structure of $\hat{N}$ can be constructed using Table 9-1; it has only one source, of value 1 A or -1 V, located at the port which corresponds to the output port of N. All sensitivities can simply be found as products of the branch variables of N and $\hat{N}$.

The two analyses needed for the sensitivity calculation using the adjoint method should be contrasted with the $n + 1$ analyses needed to derive the sensitivities using difference quotients, as in (9-8).

Table 9-1 Circuit elements together with their adjoint counterparts and sensitivities

Element in N			Element in $\hat{N}$			Sensitivity
Element type	Branch relations	Symbol	Element type	Branch relations	Symbol	$S_p \triangleq \dfrac{\partial v_0}{\partial p}$ or $\dfrac{\partial j_0}{\partial p}$
R_l	$v_l = R_l j_l$		R_l	$\hat{v}_l = R_l \hat{j}_l$		$S_{R_l} = -\hat{j}_l j_l$
G_l	$j_l = G_l v_l$		G_l	$\hat{j}_l = G_l \hat{v}_l$		$S_{G_l} = \hat{v}_l v_l$
CCVS	$v_m = 0$ $v_l = R_{lm} j_m$		CCVS	$\hat{v}_m = R_{lm} \hat{j}_l$ $\hat{v}_l = 0$		$S_{R_{lm}} = -\hat{j}_l j_m$

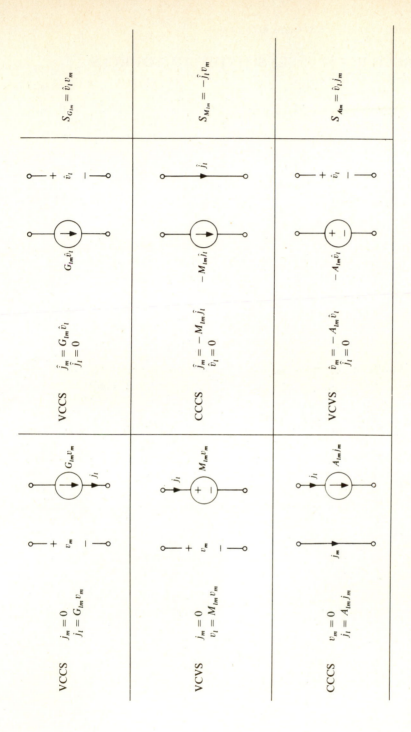

$$S_{G_{lm}} = \hat{v}_l v_m$$

$$S_{M_{lm}} = -\hat{j}_l v_m$$

$$S_{A_{lm}} = \hat{v}_l j_m$$

VCCS
$$\hat{j}_m = G_{lm}\hat{v}_l$$
$$\hat{j}_l = 0$$

CCCS
$$\hat{j}_m = -M_{lm}\hat{j}_l$$
$$\hat{v}_l = 0$$

VCVS
$$\hat{v}_m = -A_{lm}\hat{v}_l$$
$$\hat{j}_l = 0$$

VCCS
$$j_m = 0$$
$$j_l = G_{lm}v_m$$

VCVS
$$j_m = 0$$
$$v_l = M_{lm}v_m$$

CCCS
$$v_m = 0$$
$$j_l = A_{lm}j_m$$

9-4 INTERRECIPROCITY; SENSITIVITY TO SOURCES; HIGHER DERIVATIVES

Consider the network N and its adjoint $\hat{N}$ shown in Fig. 9-15. By Tellegen's theorem as given in Eqs. (2-36) and (2-37), the circuits satisfy

$$\sum_{\substack{\text{all} \\ \text{branches}}} (\hat{v}_k j_k - \hat{j}_k v_k) = 0 \tag{9-81}$$

so that

$$\sum_{\substack{\text{source} \\ \text{branches}}} (\hat{v}_k j_k - \hat{j}_k v_k) = - \sum_{\substack{\text{internal} \\ \text{branches}}} (\hat{v}_k j_k - \hat{j}_k v_k) \tag{9-82}$$

Let the hybrid internal branch relations of N be given by (9-56); then those of $\hat{N}$ will be given by (9-70). The right-hand side of (9-82) can be written as

$$\hat{\mathbf{j}}_B^T \mathbf{v}_B - \hat{\mathbf{v}}_B^T \mathbf{j}_B = \hat{\mathbf{j}}_{B_1}^T \mathbf{v}_{B_1} + \hat{\mathbf{j}}_{B_2}^T \mathbf{v}_{B_2} - \hat{\mathbf{v}}_{B_1}^T \mathbf{j}_{B_1} - \hat{\mathbf{v}}_{B_2}^T \mathbf{j}_{B_2} \tag{9-83}$$

When we use Eqs. (9-57), (9-63) and (9-58), (9-64), the right-hand side becomes

$$\hat{\mathbf{y}}^T \mathbf{x} - \hat{\mathbf{x}}^T \mathbf{y} = \hat{\mathbf{x}}^T \hat{\mathbf{H}}^T \mathbf{x} - \hat{\mathbf{x}}^T \mathbf{H} \mathbf{x} \tag{9-84}$$

The right-hand side of (9-84) is, by (9-68), identically zero. Hence for the special case of a network and its adjoint (9-82) can be more specific:

$$\sum_{\substack{\text{source} \\ \text{branches}}} (\hat{v}_k j_k - \hat{j}_k v_k) = - \sum_{\substack{\text{internal} \\ \text{branches}}} (\hat{v}_k j_k - \hat{j}_k v_k) = 0 \tag{9-85}$$

Two circuits which have this property are called *interreciprocal*.[4] Hence any network N and its adjoint network $\hat{N}$ form an interreciprocal pair.

Next, the interreciprocity relation (9-85) will be used to calculate the hitherto neglected effect of source variations on the output. Consider the special choice of sources made for $\hat{N}$ in Fig. 9-15b. From (9-85),

$$\sum_{\substack{l=1 \\ \text{voltage} \\ \text{sources}}}^{k} (\hat{v}_l j_l - \hat{j}_l e_l) + \sum_{\substack{l=k+1 \\ \text{current} \\ \text{sources}}}^{n} (\hat{v}_l i_l - \hat{j}_l v_l) + (\hat{v}_0 i_0 - \hat{i}_0 v_0) = - \sum_{l=1}^{k} \hat{j}_l e_l + \sum_{l=k+1}^{n} \hat{v}_l i_l - v_0$$

$$= 0 \tag{9-86}$$

Hence

$$v_0 = - \sum_{l=1}^{k} \hat{j}_l e_l + \sum_{l=k+1}^{n} \hat{v}_l i_l \tag{9-87}$$

and the desired sensitivities of v_0 to the source values are simply

$$\frac{\partial v_0}{\partial e_l} = -\hat{j}_l \qquad l = 1, 2, \ldots, k$$

$$\frac{\partial v_0}{\partial i_l} = \hat{v}_l \qquad l = k+1, k+2, \ldots, n \tag{9-88}$$

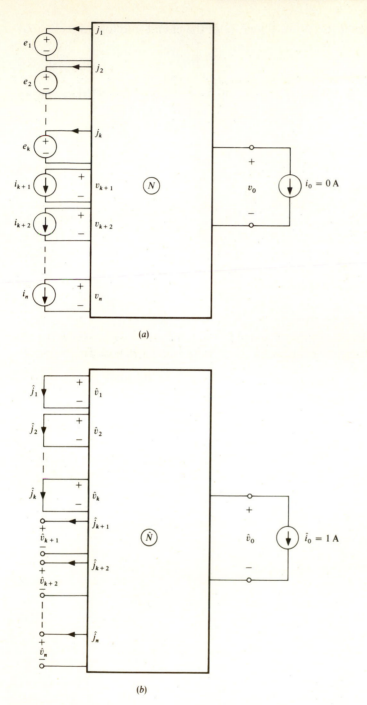

Figure 9-15 (*a*) Multisource circuit; (*b*) its adjoint network.

It should be noted that from the linearity of the circuit it follows directly that

$$v_0 = \sum_{l=1}^{k} A_l e_l + \sum_{l=k+1}^{n} B_l i_l \tag{9-89}$$

where the constant coefficients A_l, B_l are the sensitivities. Clearly, they can be obtained from the formulas:

$$
A_m = v_0, e_l = \begin{cases} 1 & \text{if } l = m \\ 0 & \text{if } l \neq m \end{cases} \qquad i_l = 0 \text{ for all } l
$$

$$
B_m = v_0, i_l = \begin{cases} 1 & \text{if } l = m \\ 0 & \text{if } l \neq m \end{cases} \qquad e_l = 0 \text{ for all } l
\tag{9-90}
$$

Thus, these sensitivities can be found at the cost of n single-source circuit analyses without recourse to the adjoint network $\hat{N}$. Since the adjoint network approach requires only *one* analysis of the single-source circuit $\hat{N}$, it is more economical even for a two-source circuit; it becomes imperative for circuits with many ($n > 10$) independent sources.

Example 9-9 Calculate the sensitivities of the output voltage v_0 in the circuit shown in Fig. 9-16a to variations in the values of the independent sources e and i.

Drawing the adjoint network $\hat{N}$ with the aid of Table 9-1 and Fig. 9-15 gives the circuit of Fig. 9-16b. This circuit can be analyzed by inspection to get

$$\hat{v}_2 = \frac{\partial v_0}{\partial i} = 100 \text{ V/A} = 0.1 \text{ V/mA} \qquad -\hat{j}_1 = \frac{\partial v_0}{\partial e} = -0.5 \text{ V/V}$$

It will be shown in Chap. 11 that optimization (automated design) may require the calculation of the second partial derivatives of the output with respect to the circuit parameters. These derivatives can also be found efficiently with the aid of the adjoint-network

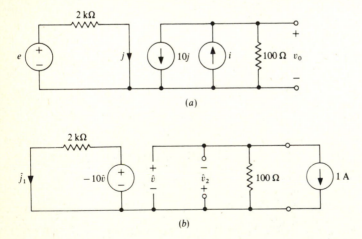

Figure 9-16 (a) Circuit with two independent sources; (b) its adjoint network.

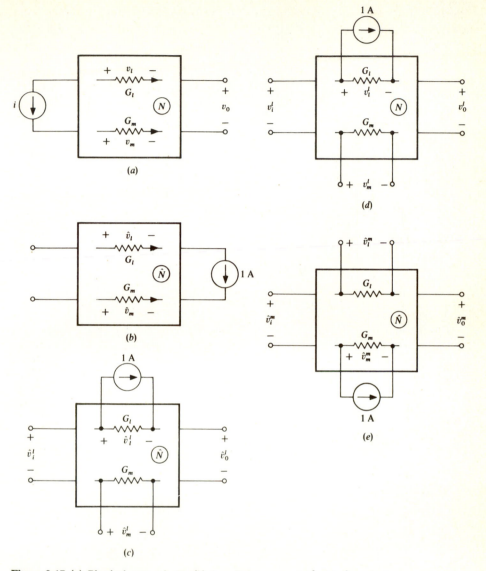

Figure 9-17 (a) Physical network N; (b) its adjoint network $\hat{N}$; (c) $\hat{N}$ excited across G_l; (d) N excited across G_l; (e) $\hat{N}$ excited across G_m.

concept. Consider, for example, the calculation of $\partial^2 v_0/(\partial G_l\,\partial G_m)$ in a conductance circuit (Fig. 9-17a). Using (9-51), we have

$$\frac{\partial^2 v_0}{\partial G_l\,\partial G_m} = \frac{\partial}{\partial G_m}(\hat{v}_l v_l) = \frac{\partial \hat{v}_l}{\partial G_m} v_l + \frac{\partial v_l}{\partial G_m} \hat{v}_l \tag{9-91}$$

where $\hat{v}_l$ is the voltage across G_l in the adjoint network $\hat{N}$ when $\hat{N}$ is excited by a 1-A source at its output (Fig. 9-17b).

Next, consider $\partial v_l / \partial G_m$, the sensitivity of the voltage v_l across G_l (Fig. 9-17a) to the variations of G_m. Since *we can regard v_l as the output voltage*, Eq. (9-51) can be used again to obtain

$$\frac{\partial v_l}{\partial G_m} = \hat{v}_m^l v_m \tag{9-92}$$

where $\hat{v}_m^l$ is the voltage across G_m in the adjoint network *when a 1-A source is placed across G_l* (Fig. 9-17c). Note that the ports corresponding to the original input and output ports are now both open-circuited in $\hat{N}$.

Similarly, $\partial \hat{v}_l / \partial G_m$ can be found from

$$\frac{\partial \hat{v}_l}{\partial G_m} = \hat{\hat{v}}_m^l \hat{v}_m \tag{9-93}$$

where $\hat{v}_m$ is the voltage across G_m in $\hat{N}$ and $\hat{\hat{v}}_m^l$ is the voltage across G_m in *the adjoint network of $\hat{N}$* when a 1-A source is placed across G_l. But, as we have seen from Eq. (9-80), *the adjoint network of $\hat{N}$ is N*. Hence, $\hat{\hat{v}}_m^l \equiv v_m^l$. The calculation of v_m^l is from the circuit shown in Fig. 9-17d.

At this stage, (9-91) takes the form

$$\frac{\partial^2 v_0}{\partial G_l \, \partial G_m} = v_m^l \hat{v}_m v_l + \hat{v}_m^l v_m \hat{v}_l \tag{9-94}$$

The calculation of all v_l, v_m and $\hat{v}_l$, $\hat{v}_m$ requires the two circuit analyses indicated in Fig. 9-17a and b. The calculation of all v_m^l requires n analyses of N with the source connected sequentially across all n conductances in the circuit (of Fig. 9-17d). Finally, the $\hat{v}_m^l$ are found through n analyses of $\hat{N}$ with the source shifted from element to element (Fig. 9-17c). Thus, it appears that $2n + 2$ analyses are needed to find all second derivatives. In fact, this number can be reduced. Using the interreciprocity property (9-85) of the two circuits shown in Fig. 9-17d and e gives

$$[(\hat{v}_i^m)(0) - (0)(v_i^l)] + [(\hat{v}_0^m)(0) - (0)(v_0^l)]$$

$$+ [(\hat{v}_l^m)(1) - (0)(v_l^l)] + [(\hat{v}_m^m)(0) - (1)(v_m^l)] = 0 \tag{9-95}$$

and hence

$$\hat{v}_l^m = v_m^l \tag{9-96}$$

Equation (9-96) (which illustrates well the meaning of the term interreciprocity) enables us to rewrite (9-94) in its final form

$$\frac{\partial^2 v_0}{\partial G_l \, \partial G_m} = \hat{v}_l^m \hat{v}_m v_l + \hat{v}_m^l \hat{v}_l v_m \tag{9-97}$$

Equation (9-97) shows that only $\hat{N}$ needs to be analyzed with its source moved from element to element.† Hence, the total number of analyses needed to find $\partial^2 v_0 /(\partial G_l \, \partial G_m)$ for all l and m is $n + 2$. This contrasts with the $n(n + 1)/2$ analyses needed to calculate the second derivatives using difference quotients from the first derivatives.

The above results were derived for derivatives of the form $\partial^2 v_0 /(\partial G_l \, \partial G_m)$; however, they can readily be extended to the calculation of second partial deriva-

---

† It is of course possible to eliminate $\hat{v}_m^l$ rather than v_m^l from (9-94). The resulting formula requires n shifted-source analyses of N.

tives with respect to the gain parameters R_{lm}, G_{lm}, M_{lm}, and A_{lm} of controlled sources, as will be shown next, in the framework of a simple example.

Example 9-10 Find the second derivatives with respect to G_1, G_4, and G_{32} for the circuit of Fig. 9-10a, used as an example in Sec. 9-2.

Applying (9-97), we get

$$\frac{\partial^2 v_0}{\partial G_1^2} = \hat{v}_1^1 \hat{v}_1 v_1 + \hat{v}_1^1 \hat{v}_1 v_1 = 2\hat{v}_1^1 \hat{v}_1 v_1$$

Here, by inspection of Fig. 9-10a and b, $v_1 = -5$ V and $\hat{v}_1 = 125$ V. Also $\hat{v}_1^1$ is the voltage across G_1 when a 1-A source is placed across it. Hence, $\hat{v}_1^1 = -1/G_1 = -10^3$ V, and

$$\frac{\partial^2 v_0}{\partial G_1^2} = (2)(-10^3)(125)(-5) = 1.25 \times 10^6 \ \text{V}/\mho^2 = 1.25 \ \text{V}/m\mho^2$$

Next, using (9-52), (9-92), and (9-93) gives

$$\frac{\partial^2 v_0}{\partial G_1 \, \partial G_{32}} = \frac{\partial}{\partial G_1}\left(\frac{\partial v_0}{\partial G_{32}}\right) = \frac{\partial}{\partial G_1}(\hat{v}_3 v_2)$$

$$= \frac{\partial \hat{v}_3}{\partial G_1} v_2 + \hat{v}_3 \frac{\partial v_2}{\partial G_1} = v_1^3 \hat{v}_1 v_2 + \hat{v}_3 \hat{v}_1^2 v_1$$

Here, v_1^3 is the voltage across G_1 when a 1-A source is connected in parallel with branch 3 (containing the controlled source). Clearly, $v_1^3 \equiv 0$. Also, $\hat{v}_1^2 = -1/G_1 = -10^3$ V, $v_1 = -5$ V, and $\hat{v}_3 = -250$ V. Hence,

$$\frac{\partial^2 v_0}{\partial G_1 \, \partial G_{32}} = (-250)(-10^3)(-5) = -1.25 \times 10^6 \ \text{V}/\mho^2$$

$$= -1.25 \times \ \text{V}/m\mho^2$$

Next, using (9-97), we obtain

$$\frac{\partial^2 v_0}{\partial G_1 \, \partial G_4} = \hat{v}_1^4 \hat{v}_4 v_1 + \hat{v}_4^1 \hat{v}_1 v_4$$

Here, from Fig. 9-10b, $\hat{v}_1^4 = \hat{v}_1 = 125$ V, $\hat{v}_4^1 \equiv 0$, and $\hat{v}_4 = -250$ V. From Fig. 9-10a, $v_1 = -5$ V, and $v_4 = 0.625$ V. Hence,

$$\frac{\partial^2 v_0}{\partial G_1 \, \partial G_4} = (125)(-250)(-5) = 0.15625 \times 10^6 \ \text{V}/m\mho^2 = 0.15625 \ \text{V}/m\mho^2$$

This completes the calculation of the first row of the second-derivative matrix:

$$\begin{bmatrix} \dfrac{\partial^2 v_0}{\partial G_1^2} & \dfrac{\partial^2 v_0}{\partial G_1 \, \partial G_{32}} & \dfrac{\partial^2 v_0}{\partial G_1 \, \partial G_4} \\[2em] \dfrac{\partial^2 v_0}{\partial G_{32} \, \partial G_1} & \dfrac{\partial^2 v_0}{\partial G_{32}^2} & \dfrac{\partial^2 v_0}{\partial G_{32} \, \partial G_4} \\[2em] \dfrac{\partial^2 v_0}{\partial G_4 \, \partial G_1} & \dfrac{\partial^2 v_0}{\partial G_4 \, \partial G_{32}} & \dfrac{\partial^2 v_0}{\partial G_4^2} \end{bmatrix}$$

Since the matrix is symmetrical, only three more elements need to be calculated. This is left as an exercise for the reader.

Our calculations can be checked by using the analytic expressions

$$\frac{\partial^2 v_0}{\partial G_1^2} = \frac{2 i_1 G_{32}}{G_1^3 G_4} = \tfrac{5}{4} \times 10^6 \text{ V}/\mho^2$$

$$\frac{\partial^2 v_0}{\partial G_1 \, \partial G_{32}} = -\frac{i_1}{G_1^2 G_4} = -\tfrac{5}{4} \times 10^6 \text{ V}/\mho^2$$

and

$$\frac{\partial^2 v_0}{\partial G_1 \, \partial G_4} = \frac{i_1 G_{32}}{G_1^2 G_4^2} = \frac{2.5 \times 10^6}{16} \text{ V}/\mho^2$$

These agree with our previous results. It should be kept in mind that for a more practical circuit, e.g., that shown in Fig. 9-11a, the analytic differentiation is no longer feasible although the adjoint-network technique remains practical.

Finally, two important remarks should be made. First, the reader is at this stage left with the impression that the analyses of N and $\hat{N}$ constitute two separate tasks. In fact, due to the close relation between the structures of N and $\hat{N}$, the data obtained during the analysis of N can be utilized in finding the branch variables of $\hat{N}$. Hence, the analysis of $\hat{N}$ requires only a very small amount of additional calculations once N has been analyzed.† The reader should consult to Refs. 6 and 7 for the details of this procedure.

The second remark concerns *nonlinear* memoryless circuits. It turns out, amazingly, that with a minimum amount of additional effort the derivations presented in this chapter can be applied to this important circuit class. Hence, for example, the stability of the bias conditions of a transistor circuit can be efficiently checked using the adjoint-network method. This topic is beyond the scope of this book; the interested reader will find a discussion of it and additional references in Refs. 1 and 2.

9-5 SUMMARY

In this chapter we attempted to give a motivation for studying sensitivity analysis as a tool of tolerance analysis. For small variations of the parameter values, the corresponding variation of the response can be predicted from the circuit sensitivities. Hence, it is important to learn efficient techniques for finding these sensitivities. Such techniques were described for resistive networks in this chapter.

The main subjects discussed were:

1. The use of Taylor-series expansion to reduce the calculation of tolerance effects to sensitivity analysis (Sec. 9-1).

---

† Specifically, the LU factors of the circuit matrix of $\hat{N}$ ($\hat{\mathbf{Y}}$, $\hat{\mathbf{Z}}$, or $\hat{\mathbf{H}}$) can readily be found from those of N without any need to know what $\hat{N}$ looks like. Hence the circuit equations of $\hat{N}$ need not be solved.

2. The calculation of sensitivities using difference quotients and the disadvantages of this technique (Sec. 9-1).
3. The differential Tellegen's theorem, the adjoint network, and its application to the calculation of sensitivities for resistor-CCVS circuits (Sec. 9-2).
4. The calculation of sensitivities for resistor-VCCS circuits (Sec. 9-2).
5. The calculation of sensitivities for resistor-VCVS-VCCS-CCVS-CCCS circuits using the adjoint-network approach (Sec. 9-3).
6. Interreciprocity; the calculation of response sensitivities to the variations of independent source values (Sec. 9-4).
7. The calculation of second- and higher-order derivatives (Sec. 9-4).

In Chap. 9, all sensitivity calculations were restricted to *resistive* circuits. The extension of the concepts and methods derived here will be extended to the frequency-domain sensitivity analysis of *dynamic* circuits in Chap. 10.

PROBLEMS

9-1 Find analytically the derivatives $\partial v_0/\partial T$, $\partial v_0/\partial I_S$, and $\partial v_0/\partial R$ for the circuit of Fig. 9-1b. The numerical values are $T = 300$ K, $q/kT \approx 40$ V^{-1}, $I_S = 10^{-9}$ A, $v_S = 2$ V.

9-2 Let the tolerances of R_1 and R_4 of the circuit of Fig. 9-7a be 5 percent; let that of R_{32} be 10 percent. What are the limits on the value of j_0?

9-3 Find all sensitivities of the circuit of Fig. 9-11c using the numerical values of voltages and currents given in Fig. 9-11c and d.

9-4 Develop the entries in Table 9-1 for the gyrator described by the branch relations

$$v_l = \alpha j_m \qquad v_m = -\alpha j_l$$

and for the ideal transformer given by the branch relations

$$v_l = n v_m \qquad i_m = -n i_l$$

9-5 Find the sensitivities of the output current j_0 (Fig. 9-18) to changes in the resistances R_1, R_4 and the gyration ratio α. The branch relations for the gyrator are $v_2 = \alpha j_3$, $v_3 = -\alpha j_2$. In what range will j_0 be if R_1, R_4, and α all have 10 percent tolerances?

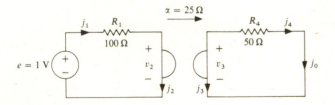

Figure 9-18 Circuit containing a gyrator.

9-6 Analyze the sensitivities of the small-signal output v_0 (Fig. 9-19a) to variations of R_1, R_2, R_3, r, and β using the transistor model of Fig. 9-19b. What is the possible range of v_0 if all tolerances are 5 percent?

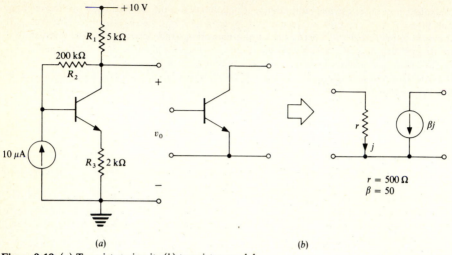

(a) (b)

Figure 9-19 (a) Transistor circuit; (b) transistor model.

9-7 Find the sensitivities of the dc collector current i_c in the amplifier shown in Fig. 9-20a. Use the dc transistor model of Fig. 9-20b. *Hint:* Use a computer to analyze N and $\hat{N}$.

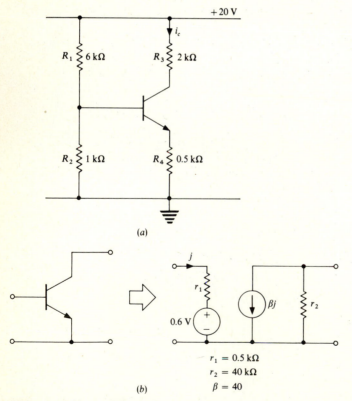

(a)

(b)

Figure 9-20 (a) Amplifier circuit; (b) transistor model.

9-8 Check, using Eqs. (9-89) and (9-90), the source sensitivities derived for the circuit of Fig. 9-16a. Calculate the sensitivities of v_0 to the resistances and to the CCCS gain using the adjoint network $\hat{N}$; assume $e = 10$ V and $i = 25$ mA. Check your calculations using symbolic analysis and differentiation. In what range will v_0 be if the sources are accurate to ± 1 percent and the resistances and the CCCS gain to ± 5 percent?

9-9 How many circuit analyses are required to find all first and second derivatives $\partial v_0/\partial p_k$ and $\partial^2 v_0/(\partial p_k \, \partial p_l)$, where the p_k are the circuit parameters $R_l, G_l, G_{lm}, \ldots$, if only difference quotients are used? Assume that the total number of the parameters p_i is n.

9-10 Find all second derivatives $\partial^2 v_0/(\partial p_k \, \partial p_l)$ in the circuits of Figs. 9-7a, 9-10a, 9-12c, and 9-16a.

9-11 (a) Show that the output resistance

$$\hat{R}_0 \triangleq \frac{-\hat{v}_0}{\hat{i}_0} = -\hat{v}_0$$

of the adjoint network $\hat{N}$ shown in Fig. 9-15b is the same as that of N under similar excitations. Hence, $R_0 = -\hat{v}_0$. *Hint:* Use Eq. (9-85).

(b) Using your result and Eq. (9-87), show that a *single analysis of $\hat{N}$* is sufficient to construct a Thevenin or Norton model of N.†

9-12 Using the results of Prob. 9-11, construct the Thevenin equivalents of the circuits shown in Figs. 9-7a, 9-10a, 9-12c, and 9-16a.

9-13 Show that by repeating the reduction process of Eqs. (9-91) to (9-97), *all* partial derivatives $\partial^n v_0/(\partial p_i^k \, \partial p_r^m \cdots)$ can be found from $n + 2$ analyses.‡

9-14 Show how to calculate the sensitivities of a dc driving-point resistance or conductance using the adjoint-network method. *Hint:* For a voltage source e_{in} at the input, $Y_{in} = j_{in}/e_{in}$, so that $\partial Y_{in}/\partial G_k = (1/e_{in}) \, \partial j_{in}/\partial G_k$. Similarly, for current-source input, $\partial Z_{in}/\partial G_k = (1/i_{in}) \, \partial v_{in}/\partial G_k$.

9-15 Using the results of Prob. 9-14, find $\partial Z_{in}/\partial G_1$, $\partial Z_{in}/\partial G_4$, and $\partial Z_{in}/\partial G_{32}$ for the circuit of Fig. 9-10a.

REFERENCE

1. Calahan, D. A.: "Computer-aided Network Design," rev. ed., McGraw-Hill, New York, 1972.
2. Director, S. W., and R. A. Rohrer: The Generalized Adjoint Network and Network Sensitivities, *IEEE Trans. Circuit Theory*, August 1969, pp. 318–323.
3. Director, S. W., and R. A. Rohrer: Automated Network Design: The Frequency Domain Case, *IEEE Trans. Circuit Theory*, August 1969, pp. 330–337.
4. Bordewijk, J. L.: Inter-Reciprocity Applied to Electric Networks, *Appl. Sci. Res., Sec. B.,* 1956, pp. 1–74.
5. Director, S. W., and D. A. Wayne: Computational Efficiency in the Determination of Thevenin and Norton Equivalents, *IEEE Trans. Circuit Theory*, January 1972, pp. 96–97.
6. Director, S. W.: LU Factorization in Network Sensitivity Computations, *IEEE Trans. Circuit Theory*, January 1971, pp. 184–185.
7. Gadenz, R. N., and G. C. Temes: Efficient Hybrid and State-Space Analysis of the Adjoint Network, *IEEE Trans. Circuit Theory*, September 1972, pp. 520–521.

† This was originally proved in Ref. 4.
‡ S. Director, personal communication.

CHAPTER TEN

SENSITIVITY ANALYSIS OF DYNAMIC CIRCUITS IN THE FREQUENCY DOMAIN

10-1 THE ADJOINT-NETWORK CONCEPT IN THE FREQUENCY DOMAIN

In Chap. 9, the problems of sensitivity and tolerance analysis were introduced in the time domain for linear memoryless circuits. An efficient technique, utilizing the adjoint-network concept, was described for carrying out the sensitivity calculations. It was also mentioned that the method can readily be extended to the sensitivity analysis of nonlinear memoryless circuits.

It will be shown next that all the results of Chap. 9 are applicable, with trivial modifications, to the important problem of frequency-domain sensitivity analysis of dynamic circuits. As already mentioned in Sec. 2-2, Tellegen's theorem, which is based on the KCL given in (2-28) and on the KVL given in (2-32), remains valid for *any* vector **j** satisfying (2-28) whether it contains physical currents or not and for any vector **v** having a representation given in (2-32) whether it contains physical voltages or not. Since the *complex phasors* associated with the steady-state sine-wave branch currents and voltages of a network certainly satisfy Kirchhoff's laws, the derivations of Sec. 2-2 remain valid. Hence, we have, as in (2-36) and (2-37),

$$\mathbf{V}^T \hat{\mathbf{J}} = \hat{\mathbf{V}}^T \mathbf{J} = 0 \qquad (10\text{-}1)$$

Here, the vectors $\mathbf{V}$ and $\mathbf{J}$ contain the complex phasors of the branch voltages and branch currents, respectively, of the network N. $\hat{\mathbf{V}}$ and $\hat{\mathbf{J}}$ contain the corresponding phasors for another network $\hat{N}$ (the adjoint network) which has the same incidence matrix $\mathbf{A}$ as N.

From (10-1), we can follow the same steps as in Sec. 9-2 to arrive at the differential Tellegen's theorem valid for the complex phasors of branch voltages and currents:

$$\hat{\mathbf{J}}^T \, \Delta \mathbf{V} - \hat{\mathbf{V}}^T \, \Delta \mathbf{J} = \sum_{\text{all branches}} (\hat{J}_k \, \Delta V_k - \hat{V}_k \, \Delta J_k) = 0 \qquad (10\text{-}2)$$

which is closely analogous to (9-16).

Next, exactly as for the memoryless circuits, we separate the branches of N into branches containing independent voltage sources (for which $\Delta V_k = 0$), branches containing independent current sources (with $\Delta J_k = 0$), and internal branches. The output branch is again treated as a 0-A current source if the output is a voltage and a 0-V voltage source if the output is a current. We choose, as before, all independent sources equal to zero in $\hat{N}$ except for the one which corresponds to the zero-valued source at the output of N. This source is set equal to 1 A if the output of N is a voltage and to -1 V if the output is a current.

With these choices for the generator branches of $\hat{N}$, (10-2) can be rewritten in the form

$$\Delta V_0 \text{ or } \Delta J_0 = - \sum_{\substack{\text{all internal} \\ \text{branches}}} (\hat{J}_k \, \Delta V_k - \hat{V}_k \, \Delta J_k) \qquad (10\text{-}3)$$

where V_0 or J_0 is the output of N. Assume now that the internal branch phasors V_k and J_k of N can be related through a *branch impedance matrix* $\mathbf{Z}$

$$\mathbf{V}_B = \mathbf{Z} \mathbf{J}_B \qquad (10\text{-}4)$$

Here $\mathbf{V}_B$ and $\mathbf{J}_B$ contain the complex phasors V_k and J_k, respectively, for all internal branches of N; $\mathbf{Z}$ contains the complex branch impedances and, for CCVS, the gains Z_{lm}. Then a derivation, exactly duplicating that which proved (9-35), shows that the branch variables of $\hat{N}$ should be chosen so as to satisfy

$$\hat{\mathbf{V}}_B = \mathbf{Z}^T \hat{\mathbf{J}}_B \qquad (10\text{-}5)$$

Hence, the branch impedance matrix of $\hat{N}$ is $\hat{\mathbf{Z}} = \mathbf{Z}^T$. For this choice of $\hat{\mathbf{Z}}$, the right-hand side of (10-3) becomes

$$-\hat{\mathbf{J}}_B^T \, \Delta \mathbf{Z} \, \mathbf{J}_B = - \sum_{\substack{\text{all internal} \\ \text{branches}}} \hat{J}_l J_m \, \Delta Z_{lm}$$

$$= - \sum_{\substack{\text{all} \\ \text{impedances}}} \hat{J}_l J_l \, \Delta Z_l - \sum_{\substack{\text{all CCVS} \\ l \neq m}} \hat{J}_l J_m \, \Delta Z_{lm} \qquad (10\text{-}6)$$

Unlike the case in a memoryless circuit, here the ΔZ_l are not simply identical to the element tolerances; they are, however, closely related. From Ohm's law valid for phasors, for an inductive branch

$$\Delta Z_l = j\omega \, \Delta L_l \qquad (10\text{-}7)$$

and for a capacitive branch

$$\Delta Z_l = \Delta \frac{1}{j\omega C_l} = - \frac{\Delta C_l}{j\omega C_l^2} \tag{10-8}$$

where the element values L_l and C_l are treated as toleranced elements. For a resistive branch or for the two branches of a resistive CCVS, $\Delta Z_l \equiv \Delta R_l$ and $\Delta Z_{lm} \equiv \Delta R_{lm}$, as before. Combining these relations with (10-3) and (10-6) gives the sensitivities of the output voltage V_0:

$$
\begin{aligned}
\frac{\partial V_0}{\partial L_l} &= -\hat{J}_l J_l j\omega & \frac{\partial V_0}{\partial C_l} &= \frac{\hat{J}_l J_l}{j\omega C_l^2} \\[2mm]
\frac{\partial V_0}{\partial R_l} &= -\hat{J}_l J_l & \frac{\partial V_0}{\partial R_{lm}} &= -\hat{J}_l J_m
\end{aligned}
\tag{10-9}
$$

The analogy between (10-9) and Eqs. (9-46) and (9-47) is obvious. For a current output, the output current J_0 replaces V_0.

Example 10-1 Find the sensitivities of the output voltage V_0 to the element values of the simple circuit shown in Fig. 10-1a.

Since the internal branches of the circuit contain only impedances, $\hat{Z} \equiv Z^T = Z$ and hence the internal branches of N and $\hat{N}$ are the same. The only nonzero source of $\hat{N}$ is a 1-A current source corresponding to V_0. The resulting circuit is shown in Fig. 10-1b.

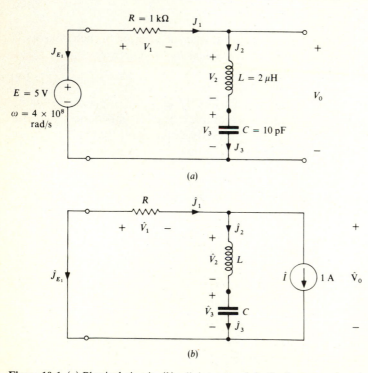

(a)

(b)

Figure 10-1 (a) Physical circuit; (b) adjoint network in the frequency domain.

From N, by inspection,

$$J_1 = J_2 = J_3 = \frac{E}{R + j\omega L + 1/j\omega C} \approx 3.84 - j2.11 \text{ mA}$$

For $\hat{N}$, from Fig. 10-1b,

$$\hat{V}_0 = -1 \frac{R(j\omega L + 1/j\omega C)}{R + j\omega L + 1/j\omega C} \approx -232 - j422 \text{ V}$$

and hence

$$\hat{J}_1 = -\frac{\hat{V}_0}{R} \approx +0.232 + j0.422 \text{ A}$$

and

$$\hat{J}_2 = \hat{J}_3 = \frac{\hat{V}_0}{j\omega L + 1/j\omega C} \approx -0.767 + j0.422 \text{ A}$$

(As a check, we note that $\hat{J}_1 - \hat{J}_2 = 1$ A.) Therefore, using (10-9), we have

$$\frac{\partial V_0}{\partial L} = -\hat{J}_2 J_2 \, j\omega \approx (12.95 + j8.22) \times 10^5 \text{ V/H}$$

$$= 1.295 + j0.822 \text{ V/}\mu\text{H}$$

$$\frac{\partial V_0}{\partial C} = +\frac{\hat{J}_3 J_3}{j\omega C^2} \approx (0.81 + j0.514) \times 10^{11} \text{ V/F}$$

$$= 0.081 + j0.0514 \text{ V/pF}$$

and finally

$$\frac{\partial V_0}{\partial R} = -\hat{J}_1 J_1 \approx (-1.78 - j1.131) \times 10^{-3} \text{ V/}\Omega$$

$$= -1.78 - j1.131 \text{ V/k}\Omega$$

The meaning of the last relation, for example, is of course that a 1-Ω change in R results in a change of $-1.78 - j1.131$ mV in the complex output phasor V_0. (This statement is valid to a first-order approximation only, as discussed in Sec. 9-1.) The other two sensitivities can be interpreted similarly.

From the formula

$$V_0 = E \frac{j\omega L + 1/j\omega C}{R + j\omega L + 1/j\omega C}$$

one can find the sensitivities analytically. For example,

$$\frac{\partial V_0}{\partial L} = E \frac{j\omega R}{(R + j\omega L + 1/j\omega C)^2} \approx (1.295 + j0.822) \times 10^6 \text{ V/H}$$

confirming our previously calculated result. The calculation of the other two sensitivities is left to the reader.

It should now be obvious how all other results obtained in Chap. 9 can be extended to frequency-domain sensitivity analysis. If the internal branch relations of N can be described by the *branch admittance matrix* **Y**

$$\mathbf{J}_B = \mathbf{Y}\mathbf{V}_B \tag{10-10}$$

then the adjoint network $\hat{N}$ has the branch admittance matrix $\hat{Y} = Y^T$ and the sensitivities of an output voltage V_0 are given by

$$\frac{\partial V_0}{\partial L_l} = -\frac{\hat{V}_l V_l}{j\omega L_l^2} \qquad \frac{\partial V_0}{\partial C_l} = +\hat{V}_l V_l j\omega$$

$$\frac{\partial V_0}{\partial G_l} = +\hat{V}_l V_l \qquad \frac{\partial V_0}{\partial G_{lm}} = +\hat{V}_l V_m$$

(10-11)

where the last relation gives the sensitivity to the gain of a VCCS.

For the calculation of the sensitivities of an output *current*, J_0 in all relations J_0 replaces V_0.

The analogy between (10-11) on the one hand and Eqs. (9-51) and (9-52) on the other is manifest.

Often only a hybrid formulation of the internal branch relations is possible, i.e., often

$$\begin{bmatrix} J_{B_1} \\ V_{B_2} \end{bmatrix} = \begin{bmatrix} Y & A \\ M & Z \end{bmatrix} \begin{bmatrix} V_{B_1} \\ J_{B_2} \end{bmatrix}$$

(10-12)

represents the only feasible description of the equations connecting the branch vectors

$$J_B = \begin{bmatrix} J_{B_1} \\ J_{B_2} \end{bmatrix} \qquad \text{and} \qquad V_B = \begin{bmatrix} V_{B_1} \\ V_{B_2} \end{bmatrix}$$

(10-13)

Then a derivation essentially identical to that leading to (9-70) to (9-74) shows that the internal structure of $\hat{N}$ must be such that

$$\begin{bmatrix} \hat{J}_{B_1} \\ \hat{V}_{B_2} \end{bmatrix} = \begin{bmatrix} Y^T & -M^T \\ -A^T & Z^T \end{bmatrix} \begin{bmatrix} \hat{V}_{B_1} \\ \hat{J}_{B_2} \end{bmatrix}$$

(10-14)

holds. Then the sensitivities of the output voltage V_0 are given by the relations

$$\frac{\partial V_0}{\partial H_{lm}} = \begin{cases} \hat{V}_l V_m & \text{if } H_{lm} \text{ is in } Y \\ \hat{V}_l J_m & \text{if } H_{lm} \text{ is in } A \\ -\hat{J}_l V_m & \text{if } H_{lm} \text{ is in } M \\ -\hat{J}_l J_m & \text{if } H_{lm} \text{ is in } Z \end{cases}$$

(10-15)

V_0 is to be replaced by J_0 if the output is the current J_0.

Note that the elements of A are the CCCS gains and hence are usually real. Similarly, the elements of M are the (normally real) VCVS gains. The diagonal elements of Y are the branch admittances and are therefore in the form $G_l, j\omega C_l$, or $1/j\omega L_l$; the off-diagonal elements are the (usually real) gains of the VCCSs. Dually, the diagonal elements of Z are branch impedances ($R_l, j\omega L_l$, or $1/j\omega C_l$), while its off-diagonal elements are the (normally real) CCVS gains. For example, if $Z_l = j\omega L_l$ is the H_{ll} element which is contained in Z, then, by (10-15), $\partial V_0/\partial L_l = (\partial V_0/\partial H_{ll})(\partial H_{ll}/\partial L_l) = (-\hat{J}_l J_l)(j\omega)$. Other sensitivities can be found similarly.

Example 10-2 Calculate the gain sensitivities of the active LC circuit shown in Fig. 10-2a using the amplifier model shown in Fig. 10-2b.

Incorporating the model into the circuit leads to the network N of Fig. 10-2c. The internal branch relations of N are

$$J_1 = \frac{1}{j\omega L} V_1 \qquad J_2 = \frac{1}{R} V_2 \qquad J_3 = 0 \qquad J_4 = j\omega C V_4 \qquad V_5 = -\mu V_3$$

or

$$
\mathbf{J}_{B_1} \begin{bmatrix} J_1 \\ J_2 \\ J_3 \\ J_4 \\ --- \\ V_5 \end{bmatrix} \mathbf{V}_{B_2}
=
\begin{bmatrix}
\frac{1}{j\omega L} & 0 & 0 & 0 & \vdots & 0 \\
0 & \frac{1}{R} & 0 & 0 & \vdots & 0 \\
0 & 0 & 0 & 0 & \vdots & 0 \\
0 & 0 & 0 & j\omega C & \vdots & 0 \\
\hdashline
0 & 0 & -\mu & 0 & \vdots & 0
\end{bmatrix}
\begin{bmatrix} V_1 \\ V_2 \\ V_3 \\ V_4 \\ --- \\ J_5 \end{bmatrix}
\begin{matrix} \\ \\ \mathbf{V}_{B_1} \\ \\ \\ \mathbf{J}_{B_2} \end{matrix}
$$

where the top braces are labeled $\mathbf{Y}$, $\mathbf{A}$ and the bottom braces are labeled $\mathbf{M}$, $\mathbf{Z}$.

where a possible partitioning is also shown using dashed lines.†

The branch relations of the adjoint network $\hat{N}$ can now be written by inspection using (10-14):

$$
\hat{\mathbf{J}}_{B_1} \begin{bmatrix} \hat{J}_1 \\ \hat{J}_2 \\ \hat{J}_3 \\ \hat{J}_4 \\ --- \\ \hat{V}_5 \end{bmatrix} \hat{\mathbf{V}}_{B_2}
=
\begin{bmatrix}
\frac{1}{j\omega L} & 0 & 0 & 0 & \vdots & 0 \\
0 & \frac{1}{R} & 0 & 0 & \vdots & 0 \\
0 & 0 & 0 & 0 & \vdots & \mu \\
0 & 0 & 0 & j\omega C & \vdots & 0 \\
\hdashline
0 & 0 & 0 & 0 & \vdots & 0
\end{bmatrix}
\begin{bmatrix} \hat{V}_1 \\ \hat{V}_2 \\ \hat{V}_3 \\ \hat{V}_4 \\ --- \\ \hat{J}_5 \end{bmatrix}
\begin{matrix} \\ \\ \hat{\mathbf{V}}_{B_1} \\ \\ \\ \hat{\mathbf{J}}_{B_2} \end{matrix}
$$

where the top braces are labeled $\mathbf{Y}^T$, $-\mathbf{M}^T$ and the bottom braces are labeled $-\mathbf{A}^T$, $\mathbf{Z}^T$.

This corresponds to the circuit $\hat{N}$ shown in Fig. 10-2d. ($\hat{N}$ could also have been obtained using Table 9-1.)

The desired sensitivities are, by (10-15),

$$\frac{\partial V_0}{\partial L} = \frac{\partial H_{11}}{\partial L} \frac{\partial V_0}{\partial H_{11}} = -\frac{1}{j\omega L^2} \hat{V}_1 V_1$$

$$\frac{\partial V_0}{\partial R} = \frac{\partial H_{22}}{\partial R} \frac{\partial V_0}{\partial H_{22}} = -\frac{1}{R^2} \hat{V}_2 V_2$$

$$\frac{\partial V_0}{\partial C} = \frac{\partial H_{44}}{\partial C} \frac{\partial V_0}{\partial H_{44}} = j\omega \hat{V}_4 V_4$$

† Resistive, capacitive, and inductive branches may be included in either $\mathbf{Y}$ or $\mathbf{Z}$. Controlled-source gains must enter the appropriate unique submatrix, however.

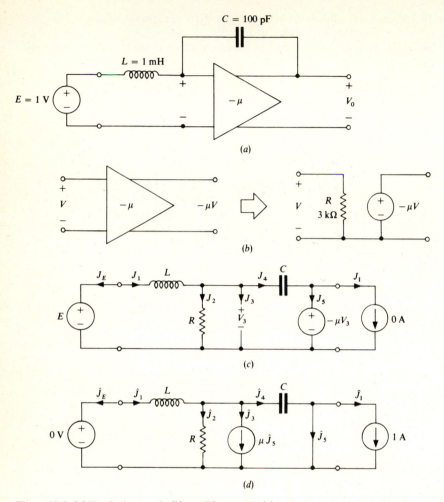

Figure 10-2 (a) Physical network; (b) amplifier model; (c) equivalent network N; (d) adjoint network N. The radian frequency is $\omega = 2\pi 10^6$ rad/s; the gain is $\mu = 100$.

and

$$\frac{\partial V_0}{\partial \mu} = \frac{\partial H_{53}}{\partial \mu} \frac{\partial V_0}{\partial H_{53}} = (-1)(-\hat{J}_s V_3)$$

Analysis of the circuit of Fig. 10-2c gives

$$V_1 = \frac{Ej\omega L[1/R + j\omega C(1 + \mu)]}{D(\omega)} = 1.0025 + j1.324 \times 10^{-5} \text{ V}$$

where $D(\omega) \triangleq 1 - \omega^2 LC(1 + \mu) + j\omega L/R$. Also,

$$V_2 = V_3 = \frac{E}{D(\omega)} = -2.5142 \times 10^{-3} - j1.324 \times 10^{-5} \text{ V}$$

$$V_4 = \frac{(1 + \mu)E}{D(\omega)} = -0.25393 - j1.3372 \times 10^{-3} \text{ V}$$

Similarly, analysis of the adjoint network (Fig. 10-2d) gives

$$-\hat{V}_1 = \hat{V}_2 = \hat{V}_3 = \hat{V}_4 = \frac{\mu j \omega L}{D(\omega)} = 8.3185 - j1.5797 \times 10^3 \text{ V}$$

$$J_5 = -\frac{1 + j\omega L/R - \omega^2 LC}{D(\omega)} = -7.4391 \times 10^{-3} + j5.2267 \times 10^{-3} \text{ A}$$

Hence, the sensitivities of V_0 are

$$\frac{\partial V_0}{\partial L} = -252.05 - j1.3306 \text{ V/H}$$

$$= -2.5205 \times 10^{-4} - j1.3306 \times 10^{-6} \text{ V/}\mu\text{H}$$

$$\frac{\partial V_0}{\partial R} = 4.6476 \times 10^{-9} - j4.4129 \times 10^{-7} \text{ V/}\Omega$$

$$= 4.6476 \times 10^{-6} - j4.4129 \times 10^{-4} \text{ V/k}\Omega$$

$$\frac{\partial V_0}{\partial C} = -2.5204 \times 10^9 - j2.6544 \times 10^7 \text{ V/F}$$

$$= -2.5204 \times 10^{-3} - j2.6544 \times 10^{-5} \text{ V/pF}$$

$$\frac{\partial V_0}{\partial \mu} = 1.8773 \times 10^{-5} - j1.3042 \times 10^{-5} \text{ V/unit gain change}$$

The reader is again reminded that in a practical situation *the analyses of both N and $\hat{N}$ would be routinely carried out on a digital computer*, using one of the many circuit-analysis programs available. Hence, these analyses are readily and painlessly performed even for large circuits.

As a test, we shall perform the differentiations analytically as well. Since

$$V_0 = -\frac{\mu E}{D(\omega)} = -\frac{\mu E}{1 - \omega^2 LC(1 + \mu) + j\omega L/R}$$

we obtain

$$\frac{\partial V_0}{\partial L} = \frac{\mu E[j\omega/R - \omega^2 C(1 + \mu)]}{[D(\omega)]^2} = -252.05 - j1.3306 \text{ V/H}$$

$$\frac{\partial V_0}{\partial R} = -\frac{\mu E j\omega L}{[RD(\omega)]^2} = 4.6476 \times 10^{-9} - j4.4129 \times 10^{-7} \text{ V/}\Omega$$

$$\frac{\partial V_0}{\partial C} = -\frac{\mu E \omega^2 L(1 + \mu)}{[D(\omega)]^2} = -2.5204 \times 10^9 - j2.6544 \times 10^7 \text{ V/F}$$

$$\frac{\partial V_0}{\partial \mu} = -\frac{E(1 + j\omega L/R - \omega^2 LC)}{[D(\omega)]^2}$$

$$= 1.8773 \times 10^{-5} - j1.3042 \times 10^{-5} \text{ V/unit gain change}$$

which agrees with our results obtained using the adjoint-network concept but which could *not* have been obtained without human analytic work.

10-2 SENSITIVITY ANALYSIS OF CIRCUITS CONTAINING TRANSMISSION LINES

As described in Chap. 8, a uniform transmission line can be characterized in terms of its propagation constant

$$\gamma = \sqrt{(R + j\omega L)(G + j\omega C)} \tag{10-16}$$

and its characteristic impedance

$$Z_0 = \sqrt{\frac{R + j\omega L}{G + j\omega C}} \tag{10-17}$$

where R, L, G, and C are the series resistance and inductance and shunt conductance and capacitance, respectively, per unit length. Then the terminal voltages V_{l_1} and V_{l_2} and currents J_{l_1} and J_{l_2} (Fig. 10-3a) are related by

$$V_{l_1} = Z_0 \coth (\gamma l) J_{l_1} + Z_0 \operatorname{csch} (\gamma l) J_{l_2}$$
$$V_{l_2} = Z_0 \operatorname{csch} (\gamma l) J_{l_1} + Z_0 \coth (\gamma l) J_{l_2} \tag{10-18}$$

where l is the length of the transmission line. In matrix form

$$\mathbf{V}_l = \mathbf{Z}_l \, \mathbf{J}_l \tag{10-19}$$

where
$$\mathbf{V}_l \triangleq \begin{bmatrix} V_{l_1} \\ V_{l_2} \end{bmatrix} \qquad \mathbf{J}_l \triangleq \begin{bmatrix} J_{l_1} \\ J_{l_2} \end{bmatrix}$$

$$\mathbf{Z}_l = \begin{bmatrix} Z_{l_{11}} & Z_{l_{12}} \\ Z_{l_{21}} & Z_{l_{22}} \end{bmatrix} \triangleq Z_0 \begin{bmatrix} \coth \gamma l & \operatorname{csch} \gamma l \\ \operatorname{csch} \gamma l & \coth \gamma l \end{bmatrix} \tag{10-20}$$

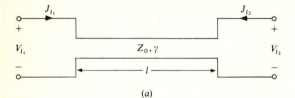

(a)

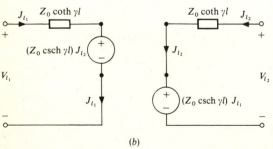

(b)

Figure 10-3 (a) Uniform transmission line; (b) equivalent lumped two-port.

It is easy to see that (10-18) and (10-20) are also valid for the lumped network of Fig. 10-3b, which is therefore completely equivalent to the uniform line shown in Fig. 10-3a as far as its terminal behavior is concerned. The circuit of Fig. 10-3b, which contains two impedances and two CCVSs, is of course suitable for adjoint-network calculations. Since all branches contain impedances and CCVSs, they can be used in a branch-impedance-matrix description such as given in (10-4). The branches corresponding to the transmission line in the adjoint network can be obtained from the rule $\hat{\mathbf{Z}} = \mathbf{Z}^T$. Since $\mathbf{Z}_l$ enters $\mathbf{Z}$ along a diagonal, as shown below,

$$
\begin{bmatrix} \vdots \\ \hline V_{l_1} \\ V_{l_2} \\ \hline \vdots \end{bmatrix}
=
\begin{bmatrix} \ddots & & & \\ \hline & Z_{l_{11}} & Z_{l_{12}} & \\ & Z_{l_{21}} & Z_{l_{22}} & \\ \hline & & & \ddots \end{bmatrix}
\begin{bmatrix} \vdots \\ \hline J_{l_1} \\ J_{l_2} \\ \hline \vdots \end{bmatrix}
\tag{10-21}
$$

and since $Z_{l_{12}} = Z_{l_{21}}$, the corresponding elements entering $\hat{\mathbf{Z}}$ will be the same as for $\mathbf{Z}$. Hence, the branches corresponding to the transmission line in $\hat{N}$ will have the same voltage-current relations of the transmission line in N. (This same conclusion can be reached by using Table 9-1.) *Hence, $\hat{N}$ will contain the same transmission line in the same location.* The contribution of the changes in Z_0, γ, and l can be derived from (10-6):

$$
-(\hat{J}_{l_1} J_{l_1} \, \Delta Z_{l_{11}} + \hat{J}_{l_2} J_{l_2} \, \Delta Z_{l_{22}})
$$

$$
-(\hat{J}_{l_1} J_{l_2} \, \Delta Z_{l_{12}} + \hat{J}_{l_2} J_{l_1} \, \Delta Z_{l_{21}})
$$

$$
= -(\hat{J}_{l_1} J_{l_1} + \hat{J}_{l_2} J_{l_2}) \, \Delta(Z_0 \coth \gamma l)
$$

$$
-(\hat{J}_{l_1} J_{l_2} + \hat{J}_{l_2} J_{l_1}) \, \Delta(Z_0 \operatorname{csch} \gamma l)
\tag{10-22}
$$

Here, by the chain rule of differentiation,

$$
\Delta(Z_0 \coth \gamma l) = \coth(\gamma l) \, \Delta Z_0 - Z_0 l \operatorname{csch}^2 (\gamma l) \, \Delta \gamma - Z_0 \gamma \operatorname{csch}^2 (\gamma l) \, \Delta l
\tag{10-23}
$$

and

$$
\Delta(Z_0 \operatorname{csch} \gamma l) = \operatorname{csch}(\gamma l) \, \Delta Z_0 - Z_0 l \coth \gamma l \operatorname{csch}(\gamma l) \, \Delta \gamma
$$

$$
- Z_0 \gamma \coth \gamma l \operatorname{csch}(\gamma l) \, \Delta l
\tag{10-24}
$$

Hence, we obtain as the sensitivity contribution of ΔZ_0

$$
-[(\hat{J}_{l_1} J_{l_1} + \hat{J}_{l_2} J_{l_2}) \coth \gamma l + (\hat{J}_{l_1} J_{l_2} + \hat{J}_{l_2} J_{l_1}) \operatorname{csch} \gamma l] \, \Delta Z_0
\tag{10-25}
$$

Using (10-20), we can rewrite this in the form

$$
-\hat{\mathbf{J}}_l^T \mathbf{Z}_l \, \mathbf{J}_l \, \frac{\Delta Z_0}{Z_0}
\tag{10-26}
$$

The equality of the two expressions can be shown simply by carrying out the indicated operations in (10-26).

Similarly, the effect of a change in γ is given by

$$Z_0 \, l\,[(\hat{J}_{l_1}J_{l_1} + \hat{J}_{l_2}J_{l_2})\operatorname{csch}^2 \gamma l + (\hat{J}_{l_1}J_{l_2} + \hat{J}_{l_2}J_{l_1})\coth \gamma l \operatorname{csch} \gamma l] \, \Delta\gamma \quad (10\text{-}27)$$

which can concisely be written as

$$\hat{\mathbf{J}}_l^T \mathbf{Z}_l \begin{bmatrix} J_{l_2} \\ J_{l_1} \end{bmatrix} l \operatorname{csch} \gamma l \, \Delta\gamma \quad (10\text{-}28)$$

Similarly, the contribution of a change in l is given by

$$\hat{\mathbf{J}}_l^T \mathbf{Z}_l \begin{bmatrix} J_{l_2} \\ J_{l_1} \end{bmatrix} \gamma \operatorname{csch} \gamma l \, \Delta l \quad (10\text{-}29)$$

We can now extend the design information contained in Table 9-1 by including the circuit elements needed in frequency-domain sensitivity analysis. This extension is given in Table 10-1.† Of course, Table 9-1 remains usable in the frequency domain if the voltages and currents are replaced by their complex phasors.

Equation (10-18) can be solved for J_{l_1} and J_{l_2} to give

$$
\begin{aligned}
J_{l_1} &= \frac{1}{Z_0}\coth(\gamma l)V_{l_1} - \frac{1}{Z_0}\operatorname{csch}(\gamma l)V_{l_2} \\[2mm]
J_{l_2} &= -\frac{1}{Z_0}\operatorname{csch}(\gamma l)V_{l_1} + \frac{1}{Z_0}\coth(\gamma l)V_{l_2}
\end{aligned}
\quad (10\text{-}30)
$$

which is in the form $\mathbf{J}_l = \mathbf{Y}_l \mathbf{V}_l$. Hence, the transmission-line relations are also compatible with branch-admittance-matrix formulation of (10-10). In a hybrid analysis, the transmission-line relations may be included either in $\mathbf{Z}$ or in $\mathbf{Y}$.

The sensitivities to the line constants R, L, G, and C can also be be readily found by the chain rule of derivatives. For example, in the case of an RC distributed line we have $L = 0$ and $G = 0$, and hence

$$
\begin{aligned}
\Delta Z_0 &= \Delta\sqrt{\frac{R}{j\omega C}} = \frac{\Delta R}{2\sqrt{j\omega RC}} - \frac{1}{2}\sqrt{\frac{R}{j\omega C^3}}\,\Delta C \\[2mm]
\Delta\gamma &= \Delta\sqrt{j\omega RC} = \frac{1}{2}\sqrt{\frac{j\omega C}{R}}\,\Delta R + \frac{1}{2}\sqrt{\frac{j\omega R}{C}}\,\Delta C
\end{aligned}
\quad (10\text{-}31)
$$

Substituting (10-31) into (10-26) and (10-28) and adding the results gives $S_R \triangleq \partial V_0/\partial R$ and $S_C \triangleq \partial V_0/\partial C$ directly.

Example 10-3 Calculate the sensitivities of the output current J_0 to variations of Z_0, l, R, and C for the circuit shown in Fig. 10-4a.

Using Table 10-1, we see that the adjoint network becomes that shown in Fig. 10-4b. Here a branch impedance formulation is possible for both N and $\hat{N}$:

$$\mathbf{V}_B = \mathbf{Z}\mathbf{J}_B \qquad \hat{\mathbf{V}}_B = \hat{\mathbf{Z}}\hat{\mathbf{J}}_B$$

---

† For completeness, Table 10-1 contains also the information on transformers and gyrators. The reader should work out these relations as an exercise.

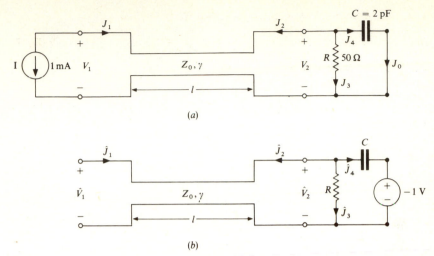

Figure 10-4 (a) Circuit containing a transmission line, $\omega = 2\pi \times 10^9$ rad/s, $Z_0 = 50\,\Omega$, $\gamma = j10\pi$ rad/m, $l = 0.07$ m; (b) the corresponding adjoint network.

where

$$\mathbf{Z} = \hat{\mathbf{Z}} = \begin{bmatrix} Z_{l_{11}} & Z_{l_{12}} & 0 & 0 \\ Z_{l_{21}} & Z_{l_{22}} & 0 & 0 \\ 0 & 0 & R & 0 \\ 0 & 0 & 0 & \dfrac{1}{j\omega C} \end{bmatrix}$$

The sensitivity contribution of Z_0 is then given by (10-25) and that of l by (10-29). Using (10-6), we also have $S_R = -\hat{J}_3 J_3$ and $S_c = \hat{J}_4 J_4 / j\omega C^2$. Hence, the branch currents of N and $\hat{N}$ must be found.

From Fig. 10-4a, denoting $Y \triangleq 1/R + j\omega C$, we have

$$J_1 = -I \qquad J_2 = I\frac{Z_{l_{21}} Y}{Z_{l_{22}} Y + 1} \qquad J_3 = -I\frac{Z_{l_{21}}/R}{Z_{l_{22}} Y + 1} \qquad J_4 = -I\frac{Z_{l_{21}} j\omega C}{Z_{l_{22}} Y + 1}$$

From Fig. 10-4b, similarly,

$$\hat{J}_1 = 0 \qquad \hat{J}_2 = -\frac{j\omega C}{Z_{l_{22}} Y + 1} \qquad \hat{J}_3 = -\frac{Z_{l_{22}} j\omega C/R}{Z_{l_{22}} Y + 1} \qquad \hat{J}_4 = \frac{j\omega C(Z_{l_{22}}/R + 1)}{Z_{l_{22}} Y + 1}$$

Since here

$$Z_{l_{11}} = Z_{l_{22}} = Z_0 \coth \gamma l = -jZ_0 \cot \frac{\gamma l}{j} = -j50 \cot 0.7\pi \approx j36.327\,\Omega$$

$$Z_{l_{12}} = Z_{l_{21}} = Z_0 \operatorname{csch} \gamma l = -jZ_0 \csc \frac{\gamma l}{j} = -j50 \csc 0.7\pi \approx -j61.803\,\Omega$$

we find

$$J_1 = -10^{-3}\,\text{A} \qquad\qquad J_2 \approx (-0.5781 - j1.501) \times 10^{-3}\,\text{A}$$

$$J_3 \approx (1.0908 + j0.816) \times 10^{-3}\,\text{A} \qquad J_4 \approx (-0.5127 + j0.6854) \times 10^{-3}\,\text{A}$$

Table 10-1 Circuit elements with their adjoint counterparts and sensitivities in the frequency domain

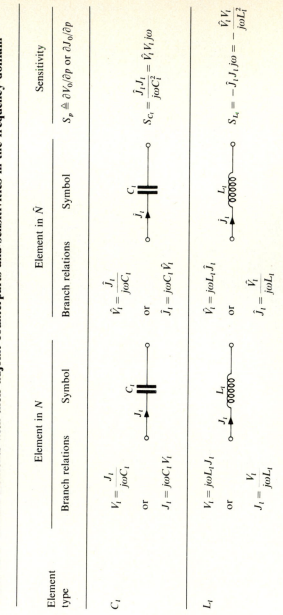

Element type	Element in N		Element in $\hat{N}$		Sensitivity
	Branch relations	Symbol	Branch relations	Symbol	$S_p \triangleq \partial V_0/\partial p$ or $\partial J_0/\partial p$
C_l	$V_l = \dfrac{J_l}{j\omega C_l}$ or $J_l = j\omega C_l V_l$		$\hat{V}_l = \dfrac{\hat{J}_l}{j\omega C_l}$ or $\hat{J}_l = j\omega C_l \hat{V}_l$		$S_{C_l} = \dfrac{\hat{J}_l J_l}{j\omega C_l^2} = \hat{V}_l V_l j\omega$
L_l	$V_l = j\omega L_l J_l$ or $J_l = \dfrac{V_l}{j\omega L_l}$		$\hat{V}_l = j\omega L_l \hat{J}_l$ or $\hat{J}_l = \dfrac{\hat{V}_l}{j\omega L_l}$		$S_{L_l} = -\hat{J}_l J_l j\omega = -\dfrac{\hat{V}_l V_l}{j\omega L_l^2}$

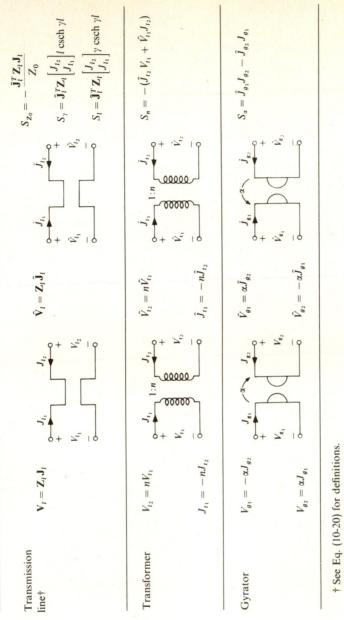

Transmission line†

$$\mathbf{V}_l = \mathbf{Z}_l \mathbf{J}_l \qquad \hat{\mathbf{V}}_l = \mathbf{Z}_l \hat{\mathbf{J}}_l$$

$$S_{Z_0} = -\frac{\hat{\mathbf{J}}_l^T \mathbf{Z}_l \mathbf{J}_l}{Z_0}$$

$$S_\gamma = \hat{\mathbf{J}}_l^T \mathbf{Z}_l \begin{bmatrix} J_{l_2} \\ J_{l_1} \end{bmatrix} l \operatorname{csch} \gamma l$$

$$S_l = \hat{\mathbf{J}}_l^T \mathbf{Z}_l \begin{bmatrix} J_{l_2} \\ J_{l_1} \end{bmatrix} \gamma \operatorname{csch} \gamma l$$

Transformer

$$V_{l_2} = n V_{l_1} \qquad \hat{V}_{l_2} = n \hat{V}_{l_1}$$

$$J_{l_1} = -n J_{l_2} \qquad \hat{J}_{l_1} = -n \hat{J}_{l_2}$$

$$S_n = -(\hat{J}_{l_2} V_{l_1} + \hat{V}_{l_1} J_{l_2})$$

Gyrator

$$V_{g_1} = -\alpha J_{g_2} \qquad \hat{V}_{g_1} = \alpha \hat{J}_{g_2}$$

$$V_{g_2} = \alpha J_{g_1} \qquad \hat{V}_{g_2} = -\alpha \hat{J}_{g_1}$$

$$S_\alpha = \hat{J}_{g_1} J_{g_2} - \hat{J}_{g_2} J_{g_1}$$

† See Eq. (10-20) for definitions.

425

As a check, we note that $J_2 + J_3 + J_4 \approx 0$. Also,

$$\hat{J}_1 = 0 \qquad\qquad \hat{J}_2 = (-11.09 - j8.2962) \times 10^{-3} \text{ A}$$

$$\hat{J}_3 = (6.025 - j8.0573) \times 10^{-3} \text{ A} \qquad \hat{J}_4 = (5.065 + j16.354) \times 10^{-3} \text{ A}$$

and $\hat{J}_2 + \hat{J}_3 + \hat{J}_4 \approx 0$. Hence, we have

$$S_R = -\hat{J}_3 J_3 \approx (-13.147 + j3.8725) \times 10^{-6} \text{ A/}\Omega \approx -0.01315 + j0.00387 \text{ mA/}\Omega$$

$$S_C = \frac{\hat{J}_4 J_4}{j\omega C^2} \approx (-5.493 - j1.955) \times 10^8 \text{ A/F} \approx -0.549 - j0.196 \text{ mA/pF}$$

Also, by (10-26)

$$S_{Z_0} = -(\hat{J}_1 J_1 + \hat{J}_2 J_2)\frac{Z_{l_{11}}}{Z_0} - (\hat{J}_1 J_2 + \hat{J}_2 J_1)\frac{Z_{l_{12}}}{Z_0}$$

$$= -\hat{J}_2 \frac{J_2 Z_{l_{11}} + J_1 Z_{l_{12}}}{Z_0} \approx (5.324 + j18.1) \times 10^{-6} \text{ A/}\Omega$$

$$\approx 0.0053 + j0.018 \text{ mA/}\Omega$$

and by (10-29)

$$S_l = [\hat{J}_1 \hat{J}_2]\begin{bmatrix} Z_{l_{11}} J_2 + Z_{l_{12}} J_1 \\ Z_{l_{12}} J_2 + Z_{l_{11}} J_1 \end{bmatrix} \gamma \text{ csch } \gamma l$$

$$= \hat{J}_2 (J_1 Z_{l_{11}} + J_2 Z_{l_{12}}) \gamma \frac{Z_{l_{12}}}{Z_0} = (-63.16 + j1.144) \times 10^{-3} \text{ A/m}$$

$$= -0.6316 + j0.01144 \text{ mA/cm}$$

Again, we emphasize that for larger circuits all $\hat{J}_i$ and J_k can conveniently be computed using commercially available circuit-analysis programs. The only additional effort required consists of the few complex multiplications and additions needed to construct all the sensitivities from these branch currents. Hence, the adjoint-network method is useful and practical for the analysis of large circuits containing distributed elements as well as lumped ones.

10-3 OTHER FREQUENCY-DOMAIN APPLICATIONS

Consider the linear active two-port shown in Fig. 10-5.† In terms of the voltage ratio V_0/E, we can obtain the loss $\alpha(\omega)$ and the phase $\beta(\omega)$ from

$$\alpha(\omega) + j\beta(\omega) = -\ln \frac{V_0}{E} \tag{10-32}$$

where α is measured in nepers and β in radians. Hence, if the parameters p_k inside

---

† This circuit is used as an illustration only. The results obtained below can be readily extended to other input-output arrangements.

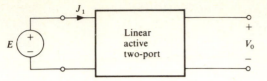

Figure 10-5 Linear active two-port with voltage input and output.

the two-port change by some small amounts Δp_k, then $\alpha(\omega)$ and $\beta(\omega)$ also change to a first order by

$$\Delta\alpha(\omega) \approx \frac{\partial\alpha}{\partial V_0}\,\Delta V_0 \approx -\mathrm{Re}\left(\frac{1}{V_0}\sum_k \frac{\partial V_0}{\partial p_k}\,\Delta p_k\right)$$

$$\Delta\beta(\omega) \approx \frac{\partial\beta}{\partial V_0}\,\Delta V_0 \approx -\mathrm{Im}\left(\frac{1}{V_0}\sum_k \frac{\partial V_0}{\partial p_k}\,\Delta p_k\right)$$

(10-33)

Since we usually calculate the sensitivities with respect to *real* parameter changes Δp_k, for example, ΔR, ΔC, ΔZ_0, Δl, the sensitivities of α and β with respect to a parameter p_k are then given by

$$S_{p_k}^{\alpha} \triangleq \frac{\partial\alpha}{\partial p_k} = -\mathrm{Re}\left(\frac{1}{V_0}\frac{\partial V_0}{\partial p_k}\right)$$

$$S_{p_k}^{\beta} \triangleq \frac{\partial\beta}{\partial p_k} = -\mathrm{Im}\left(\frac{1}{V_0}\frac{\partial V_0}{\partial p_k}\right)$$

(10-34)

Here $\partial V_0/\partial p_k$ can be computed exactly using the adjoint-network approach; a small amount of additional calculation gives $S_{p_k}^{\alpha}$ and $S_{p_k}^{\beta}$.

An interesting application[5] of (10-34) is the computation of the *group delay*, defined by

$$T_G \triangleq \frac{\partial\beta(\omega)}{\partial\omega} = S_{\omega}^{\beta}$$

(10-35)

Since ω and its change $\Delta\omega$ are certainly real, by (10-34),

$$T_G = S_{\omega}^{\beta} = -\mathrm{Im}\left(\frac{1}{V_0}\frac{\partial V_0}{\partial\omega}\right)$$

(10-36)

Since the only frequency-dependent elements in the circuit are usually the capacitors, inductors, and transmission lines, we have

$$\frac{\partial V_0}{\partial\omega} = \sum_{\substack{\text{all} \\ \text{capacitors}}} \frac{\partial V_0}{\partial Y_c}\frac{\partial Y_c}{\partial\omega} + \sum_{\substack{\text{all} \\ \text{inductors}}} \frac{\partial V_0}{\partial Z_L}\frac{\partial Z_L}{\partial\omega}$$

$$+ \sum_{\substack{\text{all trans-} \\ \text{mission lines}}}\left(\frac{\partial V_0}{\partial Z_0}\frac{\partial Z_0}{\partial\omega} + \frac{\partial V_0}{\partial\gamma}\frac{\partial\gamma}{\partial\omega}\right)$$

(10-37)

where $Y_c = j\omega C$ and $Z_L = j\omega L$.

Consider now the term $(\partial V_0/\partial Y_c)(\partial Y_c/\partial\omega)$. By (10-15),

$$\frac{\partial V_0}{\partial Y_c} = \hat{V}_c V_c \tag{10-38}$$

where V_c is the voltage across C in N and $\hat{V}_c$ the voltage across C in $\hat{N}$. Since also $\partial Y_c/\partial\omega = jC$, the contribution of C to the group delay is

$$-\mathrm{Im}\left(\frac{1}{V_0}\frac{\partial V_0}{\partial Y_c}\frac{\partial Y_c}{\partial\omega}\right) = -\mathrm{Re}\,\frac{C\hat{V}_c V_c}{V_0} \tag{10-39}$$

Repeating the process for the inductors, we find that in the absence of transmission lines, i.e., for a lumped circuit,

$$T_G = -\mathrm{Re}\left(\frac{1}{V_0}\sum_{\substack{\text{all capa-}\\\text{citors}}} C\hat{V}_c V_c\right) + \mathrm{Re}\left(\frac{1}{V_0}\sum_{\substack{\text{all induc-}\\\text{tors}}} L\hat{J}_L J_L\right) \tag{10-40}$$

Using Eqs. (10-16), (10-17), (10-25), and (10-28), we can also readily find the contribution of the transmission lines. This is left as an exercise for the reader.

Example 10-4 Consider the active integrator circuit shown in Fig. 10-6a. When the op-amp is treated simply as a VCVS, the equivalent circuit N of Fig. 10-6b is obtained. The adjoint network $\hat{N}$ shown in Fig. 10-6c is easily found, either from Tables 9-1 and 10-1 or from the hybrid branch matrix of N.

Since the circuit has only one element C which contributes to the group delay, by (10-40) here

$$T_G = -\mathrm{Re}\,\frac{C\hat{V}_3 V_3}{V_4}$$

Straightforward circuit analysis gives for N from Fig. 10-6b

$$V_3 = \frac{(1+\mu)E}{1+(1+\mu)j\omega RC} \approx 5\times10^{-6} - j0.05\ \mathrm{V}$$

and

$$V_4 = \frac{-\mu E}{1+(1+\mu)j\omega RC} = -\frac{\mu}{\mu+1}V_3 \approx -V_3$$

For $\hat{N}$, from Fig. 10-6c

$$\hat{V}_3 = \frac{\mu R\hat{I}}{1+(1+\mu)j\omega RC} = 0.02 - j200\ \mathrm{V}$$

Hence $T_G = -\mathrm{Re}\left(C\dfrac{\hat{V}_3 V_3}{V_4}\right) = -\mathrm{Re}\left(C\dfrac{\mu+1}{-\mu}\hat{V}_3\right) = C\left(1+\dfrac{1}{\mu}\right)\mathrm{Re}\,\hat{V}_3 \approx 10^{-10}\ \mathrm{s}$

Note that $T_G \ll RC = 10^{-5}$ s. This is a measure of the proper operation of the integrator: ideally, $\beta = (\omega/|\omega|)(\pi/2)$ so that $T_G = 0$ holds everywhere except for a singularity at $\omega = 0$.

The *sensitivities* of T_G are in the form

$$S_{p_i}^{T_G} \triangleq \frac{\partial T_G}{\partial p_i} = \frac{\partial^2\beta(\omega, p_i)}{\partial\omega\,\partial p_i} \tag{10-41}$$

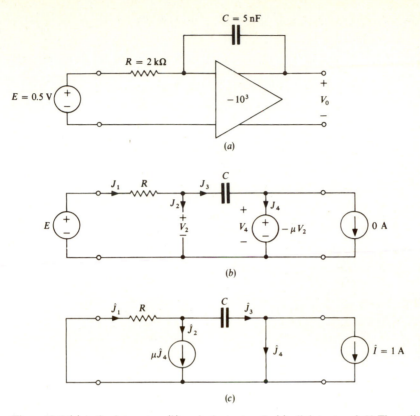

Figure 10-6 (a) Active integrator; (b) equivalent network; (c) adjoint network N. The radian frequency is $\omega = 10^6$ rad/s; the gain $\mu = 1000$.

If the parameter p_i is real (as usually is the case), then by (10-40)

$$\frac{\partial T_G}{\partial p_i} = \text{Re} \left[\frac{1}{V_0^2} \frac{\partial V_0}{\partial p_i} \left(\sum_{\text{all } C} C \hat{V}_c V_c - \sum_{\text{all } L} L \hat{J}_L J_L \right) \right]$$

$$- \text{Re} \left[\frac{1}{V_0} \frac{\partial}{\partial p_i} \left(\sum_{\text{all } C} C \hat{V}_c V_c - \sum_{\text{all } L} L \hat{J}_L J_L \right) \right] \tag{10-42}$$

The techniques described for the computation of second derivatives in Sec. 9-4 (which of course remain valid if all voltages and currents are replaced by their complex phasors) can then be used to calculate $\partial V_c/\partial p_i$, $\partial \hat{V}_c/\partial p_i$, $\partial J_L/\partial p_i$, and $\partial \hat{J}_L/\partial p_i$.

The above derivations can easily be extended to the calculation of the *loss slope* $\partial \alpha/\partial \omega$ or the *gain slope* $-\partial \alpha/\partial \omega$, which are of interest in stability studies.

Another application of the adjoint-network concept concerns the calculation of pole and zero sensitivities.[6] These are of interest for feedback amplifiers, active filters, etc., where the movement of some critical natural mode due to element-value variations can lead to oscillation. Following established usage,[7] we define the root sensitivity by

$$S_{p_k}^{r_i} \triangleq \frac{\partial r_i}{\partial p_k / p_k} = \frac{\partial r_i}{\partial (\ln p_k)} \tag{10-43}$$

where r_i is a root of either the numerator or the denominator polynomial of the transfer function and p_k is a circuit parameter, e.g., element value. To be specific, let us again consider the circuit of Fig. 10-5, and let the transfer function be

$$\frac{V_0(s)}{E} = \frac{N(s)}{D(s)} \tag{10-44}$$

Assume first that $r_i \equiv z_i$, a *zero* of $N(s)$ and hence of V_0/E. Then, by definition,

$$V_0(z_i) = 0 \tag{10-45}$$

Now let all elements p_k change by various small amounts Δp_k. To a first-order approximation, the change in $V_0(s)$ is then

$$\Delta V_0(s) = \sum_k \frac{\partial V_0(s)}{\partial p_k} \Delta p_k \tag{10-46}$$

and the new output voltage $V_0(s) + \Delta V_0(s)$ will have its zero at a new location $z_i + \Delta z_i$, rather than at z_i. Hence,

$$V_0(z_i + \Delta z_i) + \Delta V_0(z_i + \Delta z_i) = 0 \tag{10-47}$$

Since $V_0(z_i + \Delta z_i) \cong [\partial V_0(s)/\partial s]_{s=z_i} \Delta z_i$ and

$$\Delta V_0(z_i + \Delta z_i) \cong \Delta V_0(z_i) + (\partial \Delta V_0/\partial s)_{s=z_i} \Delta z_i$$

to a first-order approximation we have

$$\left[\frac{\partial V_0(s)}{\partial s} \right]_{s=z_i} \Delta z_i + \Delta V_0(z_i) \approx 0 \tag{10-48}$$

Here, $\Delta V_0(s)$ is given by (10-46), where the $\partial V_0/\partial p_k$ can be calculated routinely from the internal branch voltages and currents of N and $\hat{N}$. Also, $\partial V_0(s)/\partial s$ is easily obtainable from these same quantities since for a lumped circuit s affects $V_0(s)$ through the inductive impedances $Z_L = sL$ and capacitive admittances $Y_c = sC$ only and therefore

$$\frac{\partial V_0}{\partial s} = \sum_{\text{all } L} \frac{\partial V_0}{\partial Z_L} \frac{\partial Z_L}{\partial s} + \sum_{\text{all } C} \frac{\partial V_0}{\partial Y_c} \frac{\partial Y_c}{\partial s} \tag{10-49}$$

Using (10-15) therefore gives

$$\frac{\partial V_0}{\partial s} = \sum_{\text{all } L} (-\hat{J}_L J_L) L + \sum_{\text{all } C} \hat{V}_c V_c C \tag{10-50}$$

Substituting (10-46) and (10-50) into (10-48), results in

$$\left(\sum_{\text{all }C} C\hat{V}_c V_c - \sum_{\text{all }L} L\hat{J}_L J_L\right)_{s=z_i} \Delta z_i = -\sum_k \left(\frac{\partial V_0}{\partial p_k}\right)_{s=z_i} \Delta p_k \qquad (10\text{-}51)$$

From (10-51), it follows that

$$S_{p_k}^{z_i} \triangleq \frac{\partial z_i}{\partial p_k} p_k = \left(\frac{p_k \, \partial V_0/\partial p_k}{\sum_{\text{all }L} L\hat{J}_L J_L - \sum_{\text{all }C} C\hat{V}_c V_c}\right)_{s=z_i} \qquad (10\text{-}52)$$

This formula looks forbidding, but in fact it is easy to apply, as will be shown next in a simple example.

Example 10-5 Calculate the zero sensitivities of the circuit of Fig. 10-1a,

By inspection, the zeros of $V_0(s)$ are at $z_{1,2} = \pm j/\sqrt{LC}$. This circuit has been analyzed in Sec. 10-1 at $\omega = 4 \times 10^8$ rad/s. Here, we need the currents in N and $\hat{N}$ at $s = z_1$ and $s = z_2$. Now

$$z_{1,2} = \frac{\pm j}{\sqrt{2 \times 10^{-6} \times 10^{-11}}} \approx \pm j2.2361 \times 10^8 \text{ rad/s}$$

and we have at z_1 and z_2

$$J_1 = J_2 = J_3 = \frac{E}{R + z_{1,2}L + 1/z_{1,2}C} = \frac{E}{R} = 5 \text{ mA}$$

$$\hat{J}_2 = \hat{J}_3 = -\hat{I} = -1 \text{ A} \qquad \hat{J}_1 = 0 \text{ A}$$

$$V_3 = \frac{J_3}{z_{1,2}C} = \frac{5 \times 10^{-3}}{\pm j2.2361 \times 10^8 \times 10^{-11}} = \mp j2.2361 \text{ V}$$

$$\hat{V}_3 = \frac{\hat{J}_3}{z_{1,2}C} = \frac{-1}{\pm j2.2361 \times 10^8 \times 10^{-11}} = \pm j447.21 \text{ V}$$

where the upper sign holds for z_1 and the lower one for z_2. Hence, by (10-52) and (10-9)

$$S_L^{z_i} = \frac{L \, \partial V_0/\partial L}{LJ_2\hat{J}_2 - CV_3\hat{V}_3} = \frac{-Lz_{1,2}\hat{J}_2 J_2}{LJ_2\hat{J}_2 - CV_3\hat{V}_3}$$

$$= \frac{(-2 \times 10^{-6})(\pm j2.2361 \times 10^8)(-1)(5 \times 10^{-3})}{(2 \times 10^{-6})(5 \times 10^{-3})(-1) - (10^{-11})(2.2361)(447.221)}$$

$$= \mp j1.118034 \times 10^8 \text{ rad/s}$$

which turns out to be $-z_{1,2}/2$. This means that a change of, say, $+1$ percent in L results in a change of $-j1.118 \times 10^6$ rad/s (or -0.5 percent) in the value of z_1. Similarly,

$$S_C^{z_i} = \frac{C \, \partial V_0/\partial C}{LJ_2\hat{J}_2 - CV_3\hat{V}_3} = \frac{Cz_{1,2}\hat{V}_3 V_3}{LJ_2\hat{J}_2 - CV_3\hat{V}_3} = S_L^{z_i}$$

and, as expected,

$$S_R^{z_i} = \frac{-R\hat{J}_1 J_1}{LJ_2\hat{J}_2 - CV_3\hat{V}_3} = 0 \text{ rad/s}$$

since $\hat{J}_1 = 0$ A. These results can be confirmed from $z_{1,2} = \pm j/\sqrt{LC}$ by differentiation.

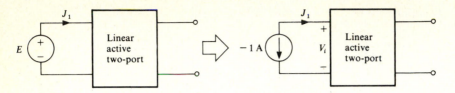

Figure 10-7 Changed excitation for calculating the pole sensitivities of a linear active two-port.

We next consider the calculation of *pole*, i.e., *natural-mode*, sensitivities. The natural modes s_l of the circuit of Fig. 10-5 are defined[9] as the nonzero roots of the circuit determinant $\Delta(s)$. The circuit determinant is here the determinant formed from the right-hand side coefficients Z_{ij} of the loop equations

$$Z_{11}J_1 + Z_{12}J_2 + \cdots + Z_{1n}J_n = E$$
$$Z_{21}J_1 + Z_{22}J_2 + \cdots + Z_{2n}J_n = 0 \qquad (10\text{-}53)$$
$$\cdots\cdots\cdots\cdots\cdots\cdots\cdots\cdots\cdots\cdots$$
$$Z_{n1}J_1 + Z_{n2}J_2 + \cdots + Z_{nn}J_n = 0$$

Since the input impedance†

$$Z(s) \triangleq \frac{E}{J_1} = \frac{\Delta(s)}{\Delta_{11}(s)} \qquad (10\text{-}54)$$

is zero at the s_l, we can write

$$V_i(s_l) = 0 \qquad (10\text{-}55)$$

where $V_i(s) = [Z(s)](1)$ *is the input voltage of the two-port when excited by a* -1-A *current source at the input* (Fig. 10-7).

Since (10-55) is in the same form as (10-45), we can repeat the derivation leading to Eq. (10-52) to obtain

$$S_{p_k}^{s_l} \triangleq \frac{\partial s_l}{\partial p_k} p_k = \left(\frac{p_k\, \partial V_i / \partial p_k}{\sum\limits_{\text{all } L} L\hat{J}_L J_L - \sum\limits_{\text{all } C} C\hat{V}_c V_c} \right)_{s=s_l} \qquad (10\text{-}56)$$

Note that (10-56) is formally identical to (10-52), but since $V_i \neq V_0$ and $s_l \neq z_i$ and since the excitations are different in the two cases, the constants entering the two equations are unrelated.

Example 10-6 Calculate the pole sensitivities of the circuit of Fig. 10-1.

By inspection, the poles are the nonzero roots of the input impedance

$$Z = R + sL + \frac{1}{sC} = \frac{s^2 LC + sRC + 1}{sC}$$

---

† Note the duality between this discussion and that given in Sec. 2-1. Note also that we are ignoring the possible cancellation of common factors in $\Delta(s)$ and $\Delta_{11}(s)$.

which are

$$s_{1,2} = -\frac{R}{2L} \pm \left(\frac{R^2}{4L^2} - \frac{1}{LC}\right)^{1/2} = -2.5 \times 10^8 \pm 1.118 \times 10^8 \text{ rad/s}$$

The analysis of N and $\hat{N}$ when they are excited by a -1-A current source at the input is trivial: all J_i and $\hat{J}_i$ are equal to 1 A. Hence, by (10-56) and (10-9)

$$S_L^{s_1} = \left(\frac{L \, \partial V_i / \partial L}{LJ_2 \hat{J}_2 - CV_3 \hat{V}_3}\right)_{s=s_1} = \left(\frac{-L\hat{J}_2 J_2 s}{LJ_2 \hat{J}_2 - J_3 \hat{J}_3 / s^2 C}\right)_{s=s_1}$$

$$= \frac{-Ls_1}{L - 1/s_1^2 C} \approx -8.541 \times 10^7 \text{ rad/s}$$

$$S_C^{s_1} = \left(\frac{C \, \partial V_i / \partial C}{LJ_2 \hat{J}_2 - J_3 \hat{J}_3 / s^2 C}\right)_{s=s_1} = \left(\frac{\hat{J}_3 J_3 / sC}{LJ_2 \hat{J}_2 - J_3 \hat{J}_3 / s^2 C}\right)_{s=s_1}$$

$$= \frac{1/s_1 C}{L - 1/s_1^2 C} \approx 2.236 \times 10^8 \text{ rad/s}$$

$$S_R^{s_1} = \left(\frac{R \, \partial V_i / \partial R}{LJ_2 \hat{J}_2 - J_3 \hat{J}_3 / s^2 C}\right)_{s=s_1} = \left(\frac{-R\hat{J}_1 J_1}{LJ_2 \hat{J}_2 - J_3 \hat{J}_3 / s^2 C}\right)_{s=s_1}$$

$$= \frac{-R}{L - 1/s_1^2 C} \approx 3.09 \times 10^8 \text{ rad/s}$$

The above results can be verified by analytic differentiation using the formula

$$s_1 = -\frac{R}{2L} + \left(\frac{R^2}{4L^2} - \frac{1}{LC}\right)^{1/2}$$

This calculation, along with the computation of $S_L^{s_2}$, $S_C^{s_2}$, and $S_R^{s_2}$, is left for the reader as an exercise.

To demonstrate the use of the adjoint-network method for a practical circuit, a somewhat more complex example is worked out next.

Example 10-7 Calculate the pole sensitivities of the active resonator circuit shown in Fig. 10-8a. Use the amplifier model of Fig. 10-8b.

Replacing the amplifier by its model gives the circuit N shown in Fig. 10-8c, where we also replaced the voltage source by a -1-A current source. The corresponding adjoint network $\hat{N}$ is shown in Fig. 10-8d.

Analysis of N shows that the natural modes are the zeros of the polynomial†

$$s^2(\mu + 1)R_1 C_2 R_3' C_4 + s(R_1 C_2 + R_1 C_4 + R_3' C_2) + 1$$

They are

$$s_{1,2} \approx -148.515 \pm j983.891 \text{ rad/s}$$

† Computer programs are available which compute the natural modes of a given circuit numerically, along with its voltages and currents.

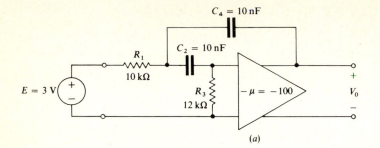

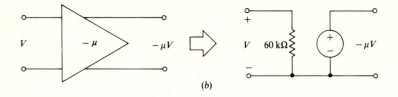

(a)

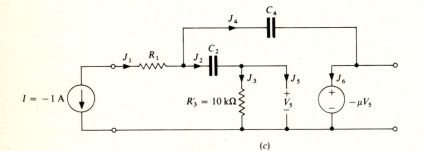

(b)

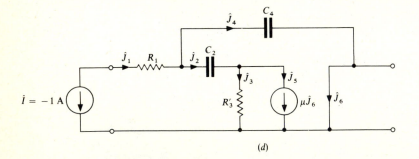

(c)

(d)

Figure 10-8 (a) Physical circuit; (b) amplifier model; (c) equivalent network N; (d) adjoint network $\hat{N}$. The element values are $R_1 = R'_3 = 10 \text{ k}\Omega$, $C_2 = C_4 = 10 \text{ nF}$, $\mu = 100$.

The voltages and currents of N and $\hat{N}$, evaluated for $s = s_i$ and needed for the sensitivity formula, are

$$J_1 = 1 \text{ A} \qquad V_2 = \frac{1}{s_1 P(s_1)} = -1.00505 \times 10^4 + j1.00377 \times 10^3 \text{ V}$$

$$J_3 = \frac{C_2}{P(s_1)} = 5.0505 \times 10^{-3} - j0.100377 \text{ A}$$

$$V_4 = \frac{(\mu + 1)R_3' C_2 + 1/s_1}{P(s_1)} = -4.9495 \times 10^3 - j1.00377 \times 10^5 \text{ V}$$

$$V_5 = \frac{R_3' C_2}{P(s_1)} = 50.505 - j1.00377 \times 10^3 \text{ V}$$

and $\qquad \hat{J}_1 = 1 \text{ A} \qquad \hat{V}_2 = \frac{\mu R_3' C_4 + 1/s_1}{P(s_1)} = -5 \times 10^3 - j9.93731 \times 10^4 \text{ V}$

$$\hat{J}_3 = \frac{C_2 - \mu C_4}{P(s_1)} = -0.5 + j9.9373 \text{ A}$$

$$\hat{V}_4 = \frac{R_3' C_2 + 1/s_1}{P(s_1)} = -1 \times 10^4 + j2.19727 \times 10^{-3} \text{ V}$$

$$\hat{J}_6 = \frac{C_4(s_1 R_3' C_2 + 1)}{P(s_1)} = 1.48515 \times 10^{-2} - j9.83891 \times 10^{-2} \text{ A}$$

where

$$P(s_1) \triangleq s_1(\mu + 1)R_3' C_2 C_4 + (C_2 + C_4) = 5 \times 10^{-9} + j9.9373 \times 10^{-8} \text{ F}$$

The pole sensitivities are then, from (10-56) and (10-15),

$$S_{R_1}^{s_1} = \left(\frac{R_1 \hat{J}_1 J_1}{C_2 \hat{V}_2 V_2 + C_4 \hat{V}_4 V_4} \right)_{s=s_1} = 49.5052 - j495.682 \text{ rad/s}$$

$$S_{R_3'}^{s_1} = \left(\frac{R_3' \hat{J}_3 J_3}{C_2 \hat{V}_2 V_2 + C_4 \hat{V}_4 V_4} \right)_{s=s_1} = 99.0101 - j488.209 \text{ rad/s}$$

$$S_{C_2}^{s_1} = \left(\frac{-sC_2 \hat{V}_2 V_2}{C_2 \hat{V}_2 V_2 + C_4 \hat{V}_4 V_4} \right)_{s=s_1} = 49.505 - j495.682 \text{ rad/s}$$

$$S_{C_4}^{s_1} = \left(\frac{-sC_4 \hat{V}_4 V_4}{C_2 \hat{V}_2 V_2 + C_4 \hat{V}_4 V_4} \right)_{s=s_1} = 99.0101 - j488.209 \text{ rad/s}$$

and $\qquad S_{\mu}^{s_1} = \left(\frac{-\mu \hat{J}_6 V_5}{C_2 \hat{V}_2 V_2 + C_4 \hat{V}_4 V_4} \right)_{s=s_1} = 147.045 - j475.977 \text{ rad/s}$

As the last example illustrates, analytical calculations become tedious for even moderately large circuits; they are impossible for very large ones. Hence, computer analysis must be used. Then the computation of the zero or pole sensitivities proceeds in the following steps:

1. Determination of the zeros and/or poles. These are usually known a priori as parts of the specification, but if necessary they can be determined numerically

from the circuit using such computer programs as LISA, CORNAP, NASAP, SLIC, etc.

2. The s-plane analysis of N and $\hat{N}$, with appropriate excitations, at the zero or pole. This is simple for a zero which lies on the $j\omega$ axis. However, the zeros *often* are complex, and for stable circuits the natural modes are *always* either complex or on the negative real axis. For a complex zero or pole $\sigma_i + j\omega_i$ a standard ac analysis program can still be used[10] with a minor modification, as follows. We add a series resistor of a value $\sigma_i L_l$ ohms to every inductor L_l in the circuit, and we add a shunt conductance $\sigma_i C_l$ mhos to each capacitor C_l. The source exciting the circuit is assumed to be a sine-wave generator with a radian frequency ω_i. Then the branch relation of the augmented inductor will be

$$V_l = j\omega_i L_l J_l + \sigma_i L_l J_l = (\sigma_i + j\omega_i) L_l J_l \qquad (10\text{-}57)$$

as required. Similarly, the relation for the augmented capacitor is

$$J_l = j\omega_i C_l V_l + \sigma_i C_l V_l = (\sigma_i + j\omega_i) C_l V_l \qquad (10\text{-}58)$$

as the analysis at $s_i = \sigma_i + j\omega_i$ dictates. Figure 10-9 illustrates this process for the circuit of Fig. 10-8c, analyzed at its pole $s_1 \approx -148.515 + j983.891$ rad/s. The added resistors are negative, but this causes no difficulties in the analysis. The currents J_l and voltages V_l obtained this way agree of course with those obtained earlier.

3. The actual sensitivities are calculated from the currents and voltages of N and $\hat{N}$, using (10-52) and/or (10-56). This is a trivial operation conveniently performed on a computer or even a small calculator.

In conclusion, the calculation of the group delay and its sensitivities, as well as the zero and pole sensitivities of a circuit, can be conveniently and exactly performed on a computer utilizing the adjoint-network concept.

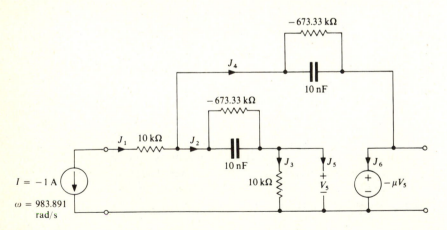

Figure 10-9 Augmenting the circuit of Fig. 10-8 for a complex-pole analysis using an ac-analysis program.

10-4 SENSITIVITY ANALYSIS OF DOUBLY TERMINATED REACTANCE TWO-PORTS

The sensitivity relations of doubly terminated reactance two-ports are of importance for two reasons: (1) this class of circuits is perhaps most commonly used among all passive networks, as filters, delay lines, and delay equalizers, and (2) because of their low-sensitivity properties, these circuits often serve as models for the low-sensitivity realization of *active and digital filters*. It turns out that these circuits can be analyzed very simply using the adjoint-network approach.

Consider the circuit shown in Fig. 10-10. We shall calculate the sensitivities of the output power

$$P_2 = \tfrac{1}{2}|J_2|^2 R_L \tag{10-59}$$

with respect to variations of the internal reactive elements C_k and L_l as well as the terminations R_G and R_L.

We recall from Sec. 6-1 that here the input power

$$P_1 \triangleq \tfrac{1}{2}|J_1|^2 \operatorname{Re} Z_1 = \frac{E^2/2}{|R_G + Z_1|^2} \operatorname{Re} Z_1 \tag{10-60}$$

equals the output power P_2. Hence, denoting $\operatorname{Re} Z_1$ by R_1 and $\operatorname{Im} Z_1$ by X_1, so that $Z_1 = R_1 + jX_1$, we have

$$P_2 = \frac{1}{2} \frac{E^2 R_1}{(R_G + R_1)^2 + X_1^2} \tag{10-61}$$

The sensitivity to variations of C_k is therefore

$$\frac{\partial P_2}{\partial C_k} = \frac{E^2}{2} \frac{(R_G^2 - R_1^2 + X_1^2)\,\partial R_1/\partial C_k - 2R_1 X_1\,\partial X_1/\partial C_k}{[(R_G + R_1)^2 + X_1^2]^2} \tag{10-62}$$

Since, as calculation easily shows,

$$\operatorname{Re}\left(\frac{R_G - Z_1^*}{R_G + Z_1}\frac{\partial Z_1}{\partial C_k}\right) = \operatorname{Re}\left[\frac{R_G - R_1 + jX_1}{R_G + R_1 + jX_1}\left(\frac{\partial R_1}{\partial C_k} + j\frac{\partial X_1}{\partial C_k}\right)\right]$$

$$= \frac{(R_G^2 - R_1^2 + X_1^2)\,\partial R_1/\partial C_k - 2R_1 X_1\,\partial X_1/\partial C_k}{(R_G + R_1)^2 + X_1^2}$$

$$\tag{10-63}$$

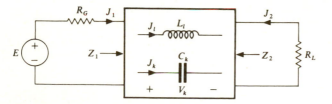

Figure 10-10 Doubly terminated reactance two-port.

relation (10-62) can be rewritten as

$$\frac{\partial P_2}{\partial C_k} = \frac{E^2/2}{|R_G + Z_1|^2} \, \text{Re} \left(\frac{R_G - Z_1^*}{R_G + Z_1} \frac{\partial Z_1}{\partial C_k} \right) \tag{10-64}$$

Also, since R_G is a constant,

$$\frac{\partial Z_1}{\partial C_k} = \frac{\partial}{\partial C_k} (R_G + Z_1) = \frac{\partial}{\partial C_k} \left(\frac{E}{J_1} \right) = -\frac{E}{J_1^2} \frac{\partial J_1}{\partial C_k} \tag{10-65}$$

From (10-9),

$$\frac{\partial J_1}{\partial C_k} = \frac{\hat{J}_k J_k}{j\omega C_k^2} = j\omega \hat{V}_k V_k \tag{10-66}$$

where $\hat{V}_k$ is the voltage across C_k in the adjoint network $\hat{N}$ when the input of $\hat{N}$ is driven by a 1-V voltage source. However, for the *RLCM* circuit discussed, all branches of $\hat{N}$ are the same as those of N. Hence, the voltages and currents of $\hat{N}$ are obtained from those of N simply through multiplication by the scale factor $(1 \text{ V})/(E \text{ volts})$. Therefore $\hat{V}_k = +V_k/E$, and, by (10-66),

$$\frac{\partial J_1}{\partial C_k} = +\frac{j\omega}{E} V_k^2 \tag{10-67}$$

From (10-65)

$$\frac{\partial Z_1}{\partial C_k} = -\frac{j\omega}{J_1^2} V_k^2 = -\frac{(R_G + Z_1)^2}{E^2} j\omega V_k^2 \tag{10-68}$$

and from (10-64), assuming $E = 1$ V,

$$\frac{\partial P_2}{\partial C_k} = \frac{\frac{1}{2}}{|R_G + Z_1|^2} \, \text{Re} \left[-(R_G - Z_1^*)(R_G + Z_1) j\omega V_k^2 \right]$$

$$= +\tfrac{1}{2} \, \text{Im} \left[\frac{(R_G - Z_1^*)(R_G + Z_1)}{(R_G + Z_1^*)(R_G + Z_1)} \omega V_k^2 \right]$$

$$= +\tfrac{1}{2} \, \text{Im} \left(\frac{R_G - Z_1^*}{R_G + Z_1^*} \omega V_k^2 \right) = + \, \text{Im} \, \frac{\rho_1^* \omega V_k^2}{2} \tag{10-69}$$

where $\rho_1 \triangleq (R_G - Z_1)/(R_G + Z_1)$ is the input reflection coefficient of the two-port.

If the transducer loss α is measured in nepers, then

$$\alpha \triangleq \tfrac{1}{2} \ln \frac{P_{\text{max}}}{P_2} \tag{10-70}$$

Hence, the logarithmic sensitivity of α to the variations of C_k is

$$C_k \frac{\partial \alpha}{\partial C_k} = -\frac{C_k}{2P_2} \frac{\partial P_2}{\partial C_k} = -\frac{\omega}{2P_2} \, \text{Im} \, \frac{\rho_1^* C_k V_k^2}{2} \tag{10-71}$$

An entirely analogous derivation gives also

$$L_l \frac{\partial \alpha}{\partial L_l} = -\frac{L_l}{2P_2} \frac{\partial P_2}{\partial L_l} = +\frac{\omega}{2P_2} \, \text{Im} \, \frac{\rho_1^* L_l J_l^2}{2} \tag{10-72}$$

Since $j\omega C_k V_k = J_k$ and $j\omega L_l J_l = V_l$, Eqs. (10-71) and (10-72) can be rewritten in the common form

$$p_k \frac{\partial \alpha}{\partial p_k} = -\frac{p_k}{2P_2} \frac{\partial P_2}{\partial p_k} = \frac{\pm 1}{4P_2} \operatorname{Re} \rho_1^* V_k J_k \qquad (10\text{-}73)$$

Here, the positive (negative) sign holds if p_k is a capacitor (inductor).

It is easy to see that $|\rho_1| \to 0$ and hence $P_2 \to P_{max}$; $p_k \, \partial\alpha/\partial p_k$ goes to zero as $|\rho_1|$ does. *Hence, the smaller the reflection factor, the less sensitive the circuit is.* Also, since the average stored energy in C_k or L_k is given by $|V_k J_k|/4\omega$, the sensitivity is proportional to the energy stored in the element.

It is interesting to note that for this special circuit a *single* analysis of the branch voltages and currents of N gives the sensitivities with respect to all reactive components. The proportionality between $p_k \, \partial\alpha/\partial p_k$ and $|\rho|$ for small $|\rho|$ explains the low sensitivity of doubly terminated reactance two-ports in their "pass-bands," i.e., in their low-loss frequency regions. This property was briefly and heuristically discussed in Sec. 6-1.

Example 10-8 The circuit of Fig. 10-11 is a third-degree low-pass filter which has a small reflection coefficient (10 percent) and hence a low loss (less than 0.05 dB) between $\omega = 0$ and $\omega = 1$ Mrad/s. Find the sensitivities of the loss with respect to L_1, L_2, C_2, and L_3 at $\omega = 0.5$ Mrad/s.

Analysis of the filter gives

$$J_1 = 4.7238 - j0.4145 \text{ mA} \qquad J_2 = 0.58364 + j2.3443 \text{ mA}$$

$$J_3 = -4.1402 + j2.7588 \text{ mA} \qquad V_{C_2} = 0.53892 - j0.13417 \text{ V}$$

From Fig. 10-10, we have

$$J_1 = \frac{E}{R_G + Z_1} \qquad Z_1 = \frac{E}{J_1} - R_G$$

and hence

$$\rho_1 \triangleq \frac{R_G - Z_1}{R_G + Z_1} = \frac{2R_G J_1 - E}{E} = -0.05524 - j0.08290$$

Also
$$P_2 = \tfrac{1}{2}|J_3|^2 R_L = 1.2376 \text{ mW}$$

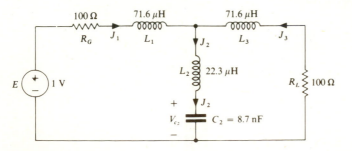

Figure 10-11 Third-degree low-pass filter.

Substituting into (10-72) gives

$$L_1 \frac{\partial \alpha}{\partial L_1} = +1.4839 \times 10^{-2} \text{ Np/unit change}$$

$$L_2 \frac{\partial \alpha}{\partial L_2} = -1.3030 \times 10^{-3} \text{ Np/unit change}$$

$$L_3 \frac{\partial \alpha}{\partial L_3} = +1.4839 \times 10^{-2} \text{ Np/unit change}$$

and from (10-71)

$$C_2 \frac{\partial \alpha}{\partial C_2} = -2.6865 \times 10^{-2} \text{ Np/unit change}$$

Thus, for example, a 5 percent change in C_2 (which is the most critical element) results in a loss change $\Delta \alpha = (2.6865 \times 10^{-2})(0.05) \approx 1.34325 \times 10^{-3} \text{ Np} \approx 0.0117 \text{ dB}$, which is, for all practical purposes, negligible.

In practice, the reactances of such circuits are *not* adjusted or chosen independently. Instead, the capacitors are selected or adjusted to the required accuracy, and then the inductors are tuned to resonate at the correct frequency with an adjacent capacitor. For the circuit of Fig. 10-11 this would mean that L_2 is tuned such that L_2 and C_2 resonate at 2.27 Mrad/s.

Assume now that C_2 has a relative error of $\varepsilon_{C_2} = 0.02$, that is, 2 percent. Since thanks to the tuning of L_2, the $L_2 C_2$ product is accurate, the relative error ε_{L_2} can be found from

$$C_2(1 + 0.02)L_2(1 + \varepsilon_{L_2}) = C_2 L_2 \tag{10-74}$$

which gives $\varepsilon_{L_2} = -0.01961 \approx -\varepsilon_{C_2}$. The combined effect of ε_{C_2} and ε_{L_2} is therefore, by (10-73),

$$\Delta \alpha = \varepsilon_{C_2} C_2 \frac{\partial \alpha}{\partial C_2} + \varepsilon_{L_2} L_2 \frac{\partial \alpha}{\partial L_2} = \frac{\varepsilon_{C_2}}{4P_2} \text{ Re} \left[\rho_1^* J_2 (V_{C_2} + V_{L_2}) \right] \tag{10-75}$$

where

$$V_{C_2} + V_{L_2} = \frac{J_2}{j\omega C_2} + J_2 j\omega L_2 = V_{C_2}(1 - \omega^2 L_2 C_2) \tag{10-76}$$

Substituting in (10-75), we get

$$\Delta \alpha = \frac{\varepsilon_{C_2}}{4P_2} (1 - \omega^2 L_2 C_2) \text{ Re} (\rho_1^* J_2 V_{C_2}) \tag{10-77}$$

A comparison with (10-73) shows that the *combined* error effects of C_2 and L_2 are smaller now (due to the $1 - \omega^2 L_2 C_2$ factor) in the frequency range near $\omega_2 = 1/\sqrt{L_2 C_2}$ than the effect of C_2 alone would be. This illustrates the beneficial effect of tuning.

Example 10-9 The loss pole of the ladder circuit of Fig. 10-11 is at $\omega_2 = 2.27$ Mrad/s. To illustrate the effect of tuning, find the loss change $\Delta \alpha$ for $\varepsilon_{C_2} = |\varepsilon_{L_2}| = 2$ percent at

$\omega = 2.5$ Mrad/s if (a) L_2 and C_2 are tuned to resonate at ω_2, so that $\varepsilon_{L_2} = -\varepsilon_{C_2}$; (b) the worst case occurs with $\varepsilon_{L_2} = \varepsilon_{C_2}$.

Analysis at 2.5 Mrad/s gives

$$J_1 = 2.2010 - j4.1376 \text{ mA} \qquad\qquad J_2 = 2.0235 - j4.0175 \text{ mA}$$

$$J_3 = -0.17759 + j0.12013 \text{ mA} \qquad V_{C_2} = -0.18471 - j0.093032 \text{ V}$$

Hence
$$\rho_1 = \frac{2R_G J_1}{E} - 1 = -0.55979 - j0.82752$$

(note that $|\rho_1| \approx 1$) and

$$P_2 = \frac{|J_3|^2 R_L}{2} = 2.2985 \ \mu\text{W}$$

Hence, for *tuned* L_2 and C_2, by (10-77),

$$\Delta\alpha_2 = \frac{\varepsilon_{C_2}}{4P_2} \left(1 - \frac{\omega^2}{\omega_2^2}\right) \text{Re} \ (\rho_1^* J_2 V_{C_2}) \approx +1.8460 \times 10^{-2} \text{ Np} \approx +0.16034 \text{ dB}$$

which (in the stopband) is negligible, *while in the worst case*

$$\Delta\alpha_2 = \frac{\varepsilon_{C_2}}{4P_2} \left(1 + \frac{\omega^2}{\omega_2^2}\right) \text{Re} \ (\rho_1^* J_2 V_{C_2}) \approx 0.1919 \text{ Np} \approx 1.6665 \text{ dB}$$

since now the two terms add rather than subtract. This is more than 10 times the loss change obtained for the tuned circuit.

The effects of L_1 and L_3 can be obtained from (10-72). Assuming $\varepsilon_{L_1} = \varepsilon_{L_3} = 2$ percent, we have

$$\Delta\alpha_1 = \varepsilon_{L_1} L_1 \frac{\partial\alpha}{\partial L_1} = \frac{\varepsilon_{L_1}\omega}{2P_2} \text{Im} \ \frac{\rho_1^* L_1 J_1^2}{2} = +1.48126 \times 10^{-2} \text{ Np} = +0.12866 \text{ dB}$$

and
$$\Delta\alpha_3 = \varepsilon_{L_3} L_3 \frac{\partial\alpha}{\partial L_3} = \frac{\varepsilon_{L_3}\omega}{2P_2} \text{Im} \ \frac{\rho_1^* L_3 J_3^2}{2} = +1.48126 \times 10^{-2} \text{ Np} = +0.12866 \text{ dB}$$

(Note that $\Delta\alpha_1 = \Delta\alpha_3$. This result follows from the symmetric structure of the circuit and from the reciprocity theorem.)

In general, the loss poles are obtained in a ladder network either through resonance in a shunt branch (Fig. 10-12a) or through antiresonance in a series

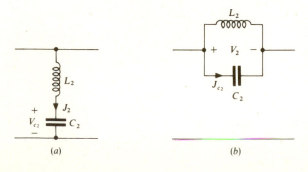

(a)   (b)

Figure 10-12 Loss poles are obtained in ladder circuits through (a) resonance in a shunt branch or (b) antiresonance in a series branch.

branch (Fig. 10-12b). The errors of the elements of these branches have the greatest effect on the loss in the stopbands, i.e., in the high-loss frequency regions. As shown above, this effect can be reduced considerably by tuning the elements of a resonant branch. An entirely dual derivation gives for the antiresonant branch (Fig. 10-12b)

$$\Delta\alpha = \frac{\varepsilon_{C2}}{4P_2}\left(1 - \frac{1}{\omega^2 L_2 C_2}\right) \text{Re}\,(\rho_1^* J_{C_2} V_2) \tag{10-78}$$

which also results in a large reduction of loss sensitivity near the radian frequency $(L_2 C_2)^{-1/2}$.

We conclude that a doubly terminated reactance ladder has a small loss sensitivity $S_{p_k}^\alpha$ in the low-loss regions (passbands) by virtue of the proportionality of $S_{p_k}^\alpha$ and $|\rho|$. It also has a low sensitivity in the high-loss regions (stopbands) because (1) the loss is significantly affected by only those elements which produce the loss-poles, for example, L_2 and C_2 in Fig. 10-12, and (2) the effects of the inaccuracies of those elements can be reduced to very low values by tuning them to their correct resonant or antiresonant frequencies.

These properties make this circuit class very desirable for practical applications. In situations where active or discrete-signal circuits must be used, successful attempts have been made to base the design on modeling such doubly terminated reactance ladders.

Finally, the sensitivities to changes in the terminations R_G and R_L will be derived. From (10-70),

$$\alpha = \tfrac{1}{2}\ln\frac{P_{\max}}{P_2} = \tfrac{1}{2}\ln\frac{E^2/8R_G}{P_2} \tag{10-79}$$

Substituting for P_2 from (10-61), we get

$$\alpha = \tfrac{1}{2}\ln\frac{(R_G + R_1)^2 + X_1^2}{4R_G R_1} \tag{10-80}$$

where R_1 and X_1 are *not* functions of R_G.

Hence, differentiation gives

$$\frac{\partial\alpha}{\partial R_G} = \frac{1}{2R_G}\frac{R_G^2 - R_1^2 - X_1^2}{(R_G + R_1)^2 + X_1^2} \tag{10-81}$$

Since

$$\text{Re}\,\rho_1 = \text{Re}\,\frac{R_G - R_1 - jX_1}{R_G + R_1 + jX_1} = \frac{R_G^2 - R_1^2 - X_1^2}{(R_G + R_1)^2 + X_1^2} \tag{10-82}$$

we have

$$\frac{\partial\alpha}{\partial R_G} = \frac{1}{2R_G}\,\text{Re}\,\rho_1 \qquad R_G\frac{\partial\alpha}{\partial R_G} = \text{Re}\,\frac{\rho_1}{2} \tag{10-83}$$

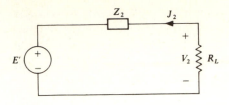

Figure 10-13 Thevenin-equivalent representation of the reactance two-port and the generator.

The effect of changes in R_L can be deduced by replacing the rest of the circuit by a Thevenin equivalent (Fig. 10-13). From the figure

$$P_2 = \tfrac{1}{2}|J_2|^2 R_L = \frac{1}{2} \frac{E'^2 R_L}{|R_L + Z_2|^2} \qquad (10\text{-}84)$$

and hence

$$\alpha = \tfrac{1}{2} \ln\left[\left(\frac{E}{E'}\right)^2 \frac{(R_L + R_2)^2 + X_2^2}{4R_L R_G}\right] \qquad (10\text{-}85)$$

This has the same general form in terms of R_L as (10-80) in terms of R_G. Differentiation gives

$$\frac{\partial \alpha}{\partial R_L} = \frac{1}{2R_L} \frac{R_L^2 - R_2^2 - X_2^2}{(R_L + R_2)^2 + X_1^2} = \frac{1}{2R_L} \operatorname{Re} \rho_2$$
$$(10\text{-}86)$$

$$R_L \frac{\partial \alpha}{\partial R_L} = \operatorname{Re} \frac{\rho_2}{2}$$

We note that once again the sensitivity becomes zero if $|\rho| \to 0$. Hence, the low-sensitivity property of the doubly terminated reactance ladders extends to the variations of the terminating resistances.

Example 10-10 For the filter of Fig. 10-11, $|\rho| \le 10$ percent in the passband, for $0 \le \omega \le 1$ Mrad/s. Hence for a 5 percent change in R_G or R_L

$$|\Delta\alpha| \approx 0.05 \left|\operatorname{Re} \frac{\rho}{2}\right| \le \frac{0.05|\rho|}{2} = 2.5 \times 10^{-3} \text{ Np} \approx 0.02171 \text{ dB}$$

which is negligible. For high frequencies, $\rho_1 \to \rho_2 \to -1$, and so

$$|\Delta\alpha| \approx 0.05 \times 0.5 = 2.5 \times 10^{-2} \text{ Np} \approx 0.2171 \text{ dB}$$

This too is negligible in the stopband, where α is typically 20 dB or more.

10-5 SUMMARY

In Chap. 10, we have extended the sensitivity-analysis techniques derived for resistive circuits in Chap. 9 to the frequency-domain sensitivity analysis of dynamic circuits, lumped or distributed. We also discussed in detail the sensitivity relations of the important class of doubly terminated reactance two-ports. The specific topics discussed were the following:

1. The frequency-domain versions of the differential Tellegen's theorem and the basic relation giving the output change ΔV_0 or ΔJ_0 for lumped active circuits (Sec. 10-1).

2. The extension of these relations to circuits containing uniform transmission lines as well as lumped elements (Sec. 10-2).
3. The calculation of loss and phase sensitivities (Sec. 10-3).
4. The calculation of the group delay and its sensitivities (Sec. 10-3).
5. The calculation of the zero and pole sensitivities of active networks (Sec. 10-3).
6. A derivation of the loss sensitivities of a doubly loaded reactance two-port in terms of its branch voltages and currents and its input reflection coefficient (Sec. 10-4).
7. The beneficial effects of element-value tuning in reducing the loss sensitivities (Sec. 10-4).
8. The sensitivities of the loss to variations in the values of the terminating resistors (Sec. 10-4).

As mentioned earlier, the circuit sensitivities are utilized not only in tolerance analysis but also in the optimization of circuits which cannot be designed using the step-by-step synthesis methods described in Chaps. 3 to 8. This topic will be discussed in the next chapter.

PROBLEMS

10-1 Find analytically all sensitivities of the output voltage V_0 of the circuit of Fig. 10-1a. Compare your results with those obtained in Sec. 10-1 using the adjoint-network concept.

10-2 Use the adjoint-network concept to calculate the sensitivities of the output V_0 of the active integrator circuit shown in Fig. 10-14. Check your results using analytic differentiation.

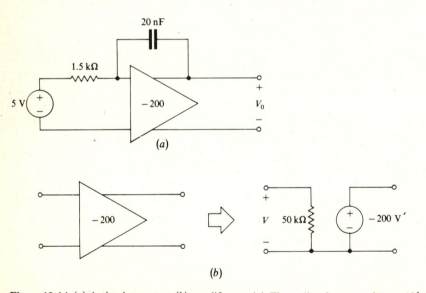

Figure 10-14 (a) Active integrator; (b) amplifier model. The radian frequency is $\omega = 10^4$ rad/s.

10-3 Use the adjoint-network method to find all sensitivities of the output V_0 of the RC-active resonator of Fig. 10-8.

10-4 Find the sensitivities of the input impedance $Z_{in} \triangleq E/J_1$ for the circuits in Figs. 10-1a, 10-2a, 10-8a, and 10-14a.

10-5 Find the sensitivities of a circuit containing a transmission line, to the line parameters R, L, G, and C. *Hint:* Use Eqs. (10-22) to (10-29) as well as (10-16) and (10-17).

10-6 Find the sensitivities of the open-circuit transmission-line impedance to variations in R, L, G, C, and l. Repeat for the short-circuit line impedance. *Hint*: All sensitivities can be expressed in terms of $\hat{J}_{l_1}$ and J_{l_1}.

10-7 Rewrite the sensitivity relations given in line 3 of Table 10-1 for the special cases of:
  (a) Lossless transmission lines ($R = 0$, $G = 0$).
  (b) Distortionless transmission lines ($R/L = G/C$).
  (c) RC transmission lines ($L = 0$, $G = 0$).

10-8 Find the expression for the group delay of a circuit containing transmission lines. *Hint*: Follow the derivation leading to (10-40); use Eqs. (10-16), (10-17), (10-25), and (10-28). Use your formula to find the group delay of the circuit of Fig. 10-4a at $\omega = 2\pi 10^9$ rad/s.

10-9 Calculate all sensitivities of V_0 for the circuit of Fig. 10-6a, using the adjoint-network concept. Check your calculations using analytic differentiation.

10-10 Calculate the group delay of the circuit of Fig. 10-2 at $\omega = 2\pi$ Mrad/s, using the adjoint-network method.

10-11 Calculate the group delay of the circuit of Fig. 10-1 at $\omega = 400$ Mrad/s using the adjoint-network method. Confirm the result by analytic differentiation.

11-12 Find the group delay of the circuit of Fig. 10-14a at $\omega = 10$ krad/s using the adjoint-network concept.

10-13 Compute the group-delay sensitivities $S_R^{T_G}$, $S_C^{T_G}$, and $S_\mu^{T_G}$ of the circuit of Fig. 10-6a. Confirm your results by differentiation.

10-14 Calculate the group-delay sensitivities $S_R^{T_G}$, $S_L^{T_G}$, and $S_C^{T_G}$ of the circuit of Fig. 10-1a. Check the results by analytic differentiation.

10-15 (a) Verify the pole-sensitivity values $S_L^{s_1}$, $S_C^{s_1}$, and $S_R^{s_1}$ obtained in Sec. 10-3 for the circuit of Fig. 10-1 using analytic differentiation.
  (b) Calculate $S_L^{s_2}$, $S_C^{s_2}$, $S_R^{s_2}$ for the same circuit.

10-16 Let s_i be a zero of the polynomial $P(s) = a_n s^n + a_{n-1} s^{n-1} + \cdots + a_1 s + a_0$. Let the coefficients a_i depend on the parameters $p_1, p_2, \ldots, p_k$. Show that the sensitivity of s_i to the parameter p_l is given by

$$S_{p_l}^{s_i} \triangleq p_l \frac{\partial s_i}{\partial p_l} = -p_l \left[\frac{\partial P(s)/\partial p_l}{\partial P(s)/\partial s} \right]_{s = s_i}$$

Hint: Follow the derivation leading to Eq. (10-52).

10-17 Use the equation derived in the preceding problem to verify the values found for the pole sensitivities of the circuits of Figs. 10-1 and 10-8.

10-18 Show that for the circuit of Fig. 10-12b the change $\Delta\alpha$ in the loss is given by (10-77) with the factor $1 - \omega^2 LC_2$ replaced by $1 - 1/\omega^2 L_2 C_2$.

10-19 Calculate the sensitivities of the filter shown in Fig. 10-15 to its reactances and terminations at $\omega_1 = 0.3$ krad/s and $\omega_2 = 10$ krad/s. What are the loss values at ω_1 and ω_2? How do your results relate to the discussions of Sec. 10-4?

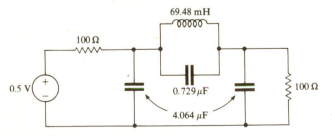

Figure 10-15 Third-degree low-pass filter.

REFERENCES

1. Bandler, J. W.: Computer-aided Circuit Optimization, in G. C. Temes and S. K. Mitra (eds.), "Modern Filter Theory and Design," Wiley, New York, 1973.
2. Bandler, J. W., and R. E. Seviora: Computation of Sensitivities for Noncommensurate Networks, *IEEE Trans. Circuit Theory*, January 1971, pp. 174–178.
3. Calahan, D. A.: "Computer-aided Network Design," rev. ed., McGraw-Hill, New York, 1972.
4. Director, S. W., and R. A. Rohrer: Automated Network Design: The Frequency Domain Case, *IEEE Trans. Circuit Theory*, August 1969, pp. 330–337.
5. Temes, G. C.: Exact Computation of Group Delay and Its Sensitivities Using the Adjoint-Network Concept, *Electron. Lett.*, July 1970, pp. 483–485.
6. Hayes, L. L., and G. C. Temes: An Efficient Numerical Method for the Computation of Zero and Pole Sensitivities, *Int. J. Electron.*, January 1972, pp. 21–31.
7. Mitra, S. K.: Active Filters with Lumped RC Networks, in G. C. Temes and S. K. Mitra (eds.), "Modern Filter Theory and Design," Wiley, New York, 1973.
8. Geffe, P. R.: RC-Amplifier Resonators for Active Filters, *IEEE Trans. Circuit Theory*, December 1968, pp. 415–419.
9. Desoer, C. A., and E. S. Kuh: "Basic Circuit Theory," McGraw-Hill, New York, 1969.
10. Gadenz, R. N., and G. C. Temes: Iterative Compensation Techniques for Lossy or Mismatched Two-Ports, *IEEE Trans. Circuit Theory*, September 1973, pp. 599–603.
11. Temes, G. C., and H. J. Orchard: First-Order Sensitivity and Worst-Case Analysis of Doubly Terminated Reactance Two-Ports, *IEEE Trans. Circuit Theory*, September 1973, pp. 567–571.
12. Blostein, M. L.: Sensitivity Analysis of Parasitic Effects in Resistance Terminated LC Two-Ports, *IEEE Trans. Circuit Theory*, March 1967, pp. 21–25.

CHAPTER ELEVEN

CIRCUIT DESIGN USING OPTIMIZATION

11-1 WHEN IS CIRCUIT OPTIMIZATION NEEDED?

In Chaps. 2 to 8 some powerful techniques were described for step-by-step circuit synthesis. All these methods assume, however, that the following conditions hold:

1. A transfer function can be found or is given which simultaneously satisfies all specifications *and* the realizability conditions of the chosen circuit configuration.
2. Ideal elements are available for building the circuit.
3. Neither the circuit configuration nor the element types or values are restricted by practical or economical considerations.

Often, these assumptions are indeed valid for practical purposes, and we *can* use the methods discussed in Chaps. 2 to 8. However, especially in the design of active and distributed-element circuits, the listed conditions sometimes do not hold and other design techniques must be used.

Example 11-1 Let us resurrect Example 1-2, concerning the design of an amplifier-equalizer which must provide the linearly falling gain vs. frequency response illustrated in

447

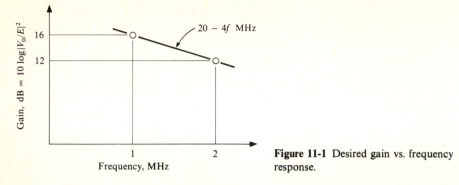

Figure 11-1 Desired gain vs. frequency response.

Fig. 11-1 in the 1- to 2-MHz frequency range. The circuit must be built from the following components:

1. Resistors, with values in the range of 50 Ω to 10 kΩ
2. Capacitors, with values in the range of 10 to 1000 pF
3. Op-amps, with 40-dB voltage gain in the 1- to 2-MHz region and with a 10-kΩ input impedance and 100-Ω output impedance

As explained in Sec. 1-3, an analysis of the circuits shown in Fig. 1-20 leads us to consider the circuit given in Fig. 11-2. When the input and output impedances of the active element are included, the circuit then becomes that shown in Fig. 11-3. The problem now is to find R and C such that the match between the desired gain response shown in Fig. 11-1 and the actual response of the circuit of Fig. 11-3 is the best possible.

Clearly, here conditions 2 and 3 do not hold. In fact, as a survey of approximation techniques (to be discussed in Chap. 12) reveals, neither does condition 1. Hence, straightforward synthesis cannot be used, and optimization must be applied. Since we anticipate that tedious calculations lie ahead, we first use normalization and reasonable approximations to simplify all expressions and thus reduce the complexity of the problem. The prescribed gain response is in the form

$$F(\omega) = 20 - \frac{4}{2\pi} \frac{\omega}{10^6} \tag{11-1}$$

where ω is measured in radians per second. The actual gain response, readily obtained from the circuit of Fig. 11-3, is

$$F_a(\omega, R, C) = 10 \log \frac{(R_2 - RG)^2 + G^2/\omega^2 C^2}{\left[(G + 1)R_0 + (R + R_2)\left(1 + \dfrac{R_0}{R_1}\right) \right]^2 + \dfrac{(1 + R_0/R_1)^2}{\omega^2 C^2}} \tag{11-2}$$

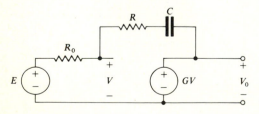

Figure 11-2 Active amplifier-equalizer circuit.

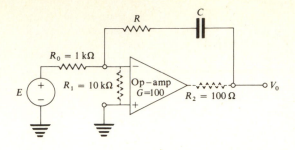

Figure 11-3 Complete circuit of the amplifier-equalizer.

To simplify the calculations which follow, we introduce normalized quantities. Choosing the obviously convenient scale factors

$$\omega_0 = 10^6 \qquad Z_0 = 10^3 \qquad (11\text{-}3)$$

and finding from Eq. (1-21)

$$C_0 = \frac{1}{\omega_0 Z_0} = 10^{-9} \qquad (11\text{-}4)$$

we see that in terms of the normalized frequency $\Omega \triangleq \omega/\omega_0$ Eq. (11-1) becomes

$$F(\Omega) = 20 - \frac{2}{\pi}\Omega \qquad (11\text{-}5)$$

while (11-2) can be rewritten in terms of $r \triangleq R/Z_0$ and $c \triangleq C/C_0$ as

$$F_a(\Omega, r, c) = 10 \log \frac{(100 - 10^5 r)^2 + (10^4 \times 10^6)/\Omega^2 c^2}{\{101 \times 10^3 + [(r + 0.1) \times 10^3](1.1)\}^2 + [(1.1)^2 \times 10^6]/\Omega^2 c^2}$$

$$= 10 \log \frac{(0.1 - 100r)^2 + 10^4/\Omega^2 c^2}{(101.11 + 1.1r)^2 + 1.21/\Omega^2 c^2} \qquad (11\text{-}6)$$

The permissible values for the normalized quantities are $0.05 \le r \le 10$, $2\pi \le \Omega \le 4\pi$, and $0.01 \le c \le 1$. Hence, in the numerator we have $0.1 \ll 100r$ for all permissible values of r, and in the denominator we have $101.11 \gg 1.1r$ and $(101.11)^2 \gg 1.21/\Omega^2 c^2$ for all permissible r, Ω, and c values. Therefore, to a good approximation

$$F_a(\Omega, r, c) \approx 10 \log \frac{10^4(r^2 + 1/\Omega^2 c^2)}{(101.11)^2} \approx 10 \log \left(r^2 + \frac{1}{\Omega^2 c^2}\right) \qquad (11\text{-}7)$$

It is worth noting the great simplifications achieved by normalizing and by examining the possible magnitudes of the various terms in the exact expression and neglecting the unimportant ones. Such a priori investigation is almost always worth carrying out in optimization problems.

We are to obtain a match between the specified $F(\Omega)$ given in (11-5) and the $F_a(\Omega, r, c)$ of (11-7) for $2\pi \le \Omega \le 4\pi$. A simple process to achieve this is to equate the two functions at $\Omega_1 = 2\pi$ and $\Omega_2 = 4\pi$. The resulting relations are

$$10 \log \left(r^2 + \frac{1}{\Omega_1^2 c^2}\right) = 20 - \frac{2}{\pi}\Omega_1$$

$$\qquad (11\text{-}8)$$

$$10 \log \left(r^2 + \frac{1}{\Omega_2^2 c^2}\right) = 20 - \frac{2}{\pi}\Omega_2$$

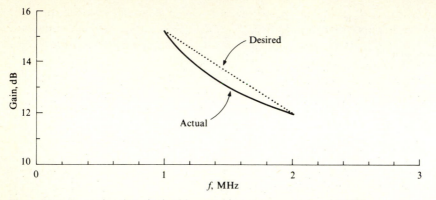

Figure 11-4 Approximation obtained by matching responses at 1 and 2 MHz.

These can readily be solved for r and c, giving (in rounded numbers)

$$r \approx 2.8 \qquad c \approx 0.028$$

or, in denormalized units [compare (11-3) and (11-4)],

$$R \approx 2800 \ \Omega \qquad C \approx 28 \text{ pF}$$

The corresponding response is shown, along with the desired response, in Fig. 11-4. The maximum error is about 0.54 dB, which may be acceptable in some applications. The error can be further reduced by judiciously changing the matching points so that they lie inside the $2\pi < \Omega < 4\pi$ range.

The point-matching technique used above to find R and C is conceptually simple. However, there is a considerable amount of arbitrariness in the choice of the matching points Ω_1 and Ω_2. Furthermore, for a more complicated problem, involving, say, 5 to 20 designable circuit elements,† it necessitates the solution of 5

† This is the usual range in many practical circuit design problems.

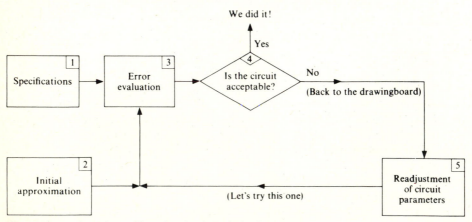

Figure 11-5 A flow chart of iterative circuit optimization.

to 20 simultaneous nonlinear equations. This is a difficult numerical problem, which will be further discussed in Sec. 11-8. If the matching points are improperly chosen, this method may still not give a usable circuit. Hence a more general approach is usually needed.

The flow chart of a general circuit optimization process, given in Fig. 1-23, is reproduced in Fig. 11-5. The steps represented by the various boxes of the flow chart were briefly discussed in Sec. 1-3. In the next section, a somewhat more detailed description will be given.

In all discussions pertaining to the flow chart of Fig. 11-5 it should be remembered that these calculations are invariably performed on a digital computer. The amount of computation involved in the optimization of a large circuit is so great, in fact, that it can strain the capabilities of even a modern fast computer. Hence, it is important to use design techniques which are comparatively efficient and economical.

11-2 THE OPTIMIZATION PROCESS

The steps of the general optimization procedure, illustrated in Fig. 11-5, are the following.

Step 1: Specifications These may include the frequency response $F(\omega)$ of the circuit; they may include the time response (transient response) $f(t)$; they may include the total power P_d dissipated in the circuit, etc. Often, several responses and other quantities such as P_d may be specified simultaneously. In addition, the allowable deviations from the specified values must always be given, so that the designer can tell when a necessarily imperfect design is acceptable from a practical viewpoint.

Step 2: Initial approximation In contrast to step-by-step synthesis, iterative optimization can only refine or improve an *existing* circuit. Hence, to start the process up, the designer must provide an initial circuit with all its element values tentatively assigned. This can be obtained by the point-matching technique illustrated in Sec. 11-1 or from physical considerations or from experience, etc. In any case, it is worth investing a great deal of effort in finding the best possible initial approximation since, as will be seen, the optimization process may fail altogether or may converge to an inferior final circuit if it starts from poor initial values.

Step 3: Error evaluation The error is defined, reasonably enough, as the difference between the specified response [say $F(\omega)$] and the actual one. Since, as the flow chart indicates, optimization is an iterative procedure, the actual circuit parameters p_i are going to change from iteration to iteration. Hence, so will the actual response. In recognition of this, we shall indicate the value of the parameter p_i in the jth iteration by p_i^j, including the superscript (*not* exponent) j in the symbol.

Thus, p_2^4 will denote the value of p_2 in the fourth iteration.† When this same convention is used for the actual response, in the jth iteration the circuit response will be $F^j(\omega, p_1^j, p_2^j, \ldots, p_n^j)$. Then the error in the jth iteration will be given by

$$e^j(\omega, p_1^j, p_2^j, \ldots, p_n^j) = F(\omega) - F^j(\omega, p_1^j, p_2^j, \ldots, p_n^j) \tag{11-9}$$

Since e^j is a function of ω, it is sometimes called the *error function*. We can also include a *weight function* $w(\omega)$ in e^j by defining

$$e^j(\omega, p_1^j, p_2^j, \ldots, p_n^j) \triangleq w(\omega)[F(\omega) - F^j(\omega, p_1^j, p_2^j, \ldots, p_n^j)] \tag{11-10}$$

At frequencies where $w(\omega)$ is small, the error will appear to be small even if $F - F^j$ is large, and vice versa. Hence the weight function enables us to emphasize or deemphasize the error in the various regions of the ω axis.‡

In the actual optimization process, we normally use a single number E^j, called the *error criterion* or *performance index* or *optimality index*, to indicate the quality of the performance achieved by the circuit in the jth iteration. The smaller E^j, the smaller the error.

E^j can be defined various ways. Usually, its definition is such that $E^j \geq 0$ and $E^j = 0$ indicates zero error, i.e., an exact match for all ω values between $F(\omega)$ and $F^j(\omega, p_1^j, p_2^j, \ldots, p_n^j)$. The two most commonly used definitions are the *least-squares* error criterion and the *min-max* error criterion.§ The least-squares error criterion E_2^j is given by

$$E_2^j \triangleq \int_{\omega_l}^{\omega_u} [e^j(\omega, p_1^j, p_2^j, \ldots, p_n^j)]^2 \, d\omega$$

$$= \int_{\omega_l}^{\omega_u} \{w(\omega)[F(\omega) - F^j(\omega, p_1^j, p_2^j, \ldots, p_n^j)]\}^2 \, d\omega \tag{11-11}$$

where $\omega_l \leq \omega \leq \omega_u$ is the frequency region of interest.¶

Since the calculations must be performed numerically, the integration in (11-11) will be replaced by summation. We choose a total of *m sample points* ω_1, $\omega_2, \ldots, \omega_m$ between ω_l and ω_u and rewrite (11-11) in the form

$$E_2^j \triangleq \sum_{i=1}^{m} \{w(\omega_i)[F(\omega_i) - F^j(\omega_i, \mathbf{p}^j)]\}^2 \tag{11-12}$$

Note that in (11-12) we have, for convenience, collected $p_1^j, p_2^j, \ldots, p_n^j$ into the *parameter vector* $p^j \triangleq (p_1^j, p_2^j, \ldots, p_n^j)^T$.

It should be clear that $E_2^j \geq 0$, since all terms of the summation are squared, and also that $E_2^j = 0$ if and only if $F(\omega_i) = F^j(\omega_i, \mathbf{p}^j)$ at all ω_i; that is, the

---

† If the iteration index j is unimportant, or if the p_i can be obtained without iteration, the superscript j will be omitted.

‡ Here and in what follows ω will be used as the independent variable in F, F^j, e^j, etc. It should be kept in mind that often the independent variable is the time t and sometimes the temperature T.

§ Also often called the *Chebyshev* error criterion.

¶ The observant reader will note that E_2^j is closely related to the variance and the rms value of the error function.

specifications are exactly met. In general, the smaller E_2^j, the less $F(\omega)$ and $F^j(\omega, \mathbf{p}^j)$ differ.

For (11-12) to be useful, the sample points ω_i must be chosen densely enough so that neither $F(\omega)$ nor $F^j(\omega, \mathbf{p}^j)$ changes significantly between two adjacent samples. In any case, however, it is usual to choose at least $5n$ sample points, where n is the total number of designable circuit parameters p_i.

Example 11-2 For the amplifier-equalizer discussed in Sec. 11-1, the specified response is given by (11-5) and the simplified actual response by (11-7). The independent variable is Ω; the parameters are $p_1 \triangleq r$ and $p_2 \triangleq c$. If we choose, for simplicity, only five sample points equally spaced in the $2\pi \leq \Omega \leq 4\pi$ range, the ith sample point is given by $\Omega_i = 2\pi + (i-1)2\pi/4 = (i+3)\pi/2$, for $i = 1, 2, 3, 4,$ and 5. Hence, if we do not assign a weight function, i.e., choose $w(\omega) \equiv 1$, then (11-12) takes the form

$$E_2^j = \sum_{i=1}^{5} \left[20 - \frac{2(i+3)\pi}{\pi} \frac{}{2} - 10 \log \left\{ r^2 + \frac{1}{[(i+3)(\pi/2)c]^2} \right\} \right]^2$$

$$= \sum_{i=1}^{5} \left\{ 17 - i - 10 \log \left[r^2 + \frac{4}{(i+3)^2\pi^2c^2} \right] \right\}^2 \qquad (11\text{-}13)$$

The definition of the *min-max error criterion* is simply

$$E_\infty^j \triangleq \max_{\omega_l \leq \omega \leq \omega_u} |e^j(\omega, \mathbf{p}^j)| = \max_{\omega_l \leq \omega \leq \omega_u} |w(\omega)[F(\omega) - F^j(\omega, \mathbf{p}^j)]| \qquad (11\text{-}14)$$

In words, the min-max error criterion is *the largest absolute error in the frequency range of interest*. Again, this can be evaluated by choosing m sample points ω_i between ω_l and ω_u and replacing (11-14) by

$$E_\infty^j \triangleq \max_i |e^j(\omega_i, \mathbf{p}^j)| \qquad (11\text{-}15)$$

Example 11-3 For the amplifier-equalizer, if the sample points Ω_i are chosen as before, the Chebyshev (min-max) error criterion is given by

$$E_\infty^j = \max_{1 \leq i \leq 5} \left| 17 - i - 10 \log \left[r^2 + \frac{4}{(i+3)^2\pi^2c^2} \right] \right| \qquad (11\text{-}16)$$

Example 11-4 Again for the amplifier-equalizer, with the same Ω_i and using the values $r = 2.8$ and $c = 0.028$ obtained in Sec. 11-1, we can calculate† the values shown in Table 11-1.

From (11-12), therefore,

$$E_2^0 = \sum_{i=1}^{5} [e^0(\Omega_i)]^2 \approx 0.616511$$

while from (11-15)

$$E_\infty^0 = \max_i |e^0(\Omega_i)| \approx 0.536553$$

---

† The nonzero error at Ω_1 and Ω_5 is due to the rounding in the values of r and c.

Table 11-1

i	Ω_i	$F(\Omega_i)$	$F^0(\Omega_i)$	$e^0(\Omega_i)$
1	2π	16	16.036753	-0.036753
2	2.5π	15	14.551159	0.448841
3	3π	14	13.463447	0.536553
4	3.5π	13	12.645792	0.354208
5	4π	12	12.018684	-0.018684

Two important extensions of our concepts should be mentioned. First, as mentioned earlier, often *several* responses and additional quantities (power dissipation, direct currents, etc.) need to be optimized simultaneously, and we have to define an error criterion which includes all these. A possibility is using a weighted sum. For example, if the time response $f(t)$ and the frequency response $F(\omega)$ are both prescribed, and if the dissipated power P_d is to be minimized simultaneously, we can define a least-squares error criterion of the form

$$E_2^j = K_\omega \sum_{i=1}^{m_\omega} |w_\omega(\omega_i)[F(\omega_i) - F^j(\omega_i, \mathbf{p}^j)]|^2$$

$$+ K_t \sum_{i=1}^{m_t} |w_t(t_i)[f(t_i) - f^j(t_i, \mathbf{p}^j)]|^2 + K_d P_d^2 \qquad (11\text{-}17)$$

where K_ω, K_t, and K_d = weight factors assigned by designer to
        frequency-domain, time-domain, and power-drain
        specifications
    m_ω, m_t = numbers of sample points in frequency and
        time domains
    w_ω, w_t = corresponding weight functions

These parameters must all be chosen carefully and, if necessary, changed to obtain a good compromise between all (often conflicting) requirements. A similar weighted sum of the maximum frequency and time-domain errors, as well as P_d, gives the corresponding min-max error criterion.

The second generalization concerns the exponent in the least-squares error criterion. We can define the *least-pth error criterion* by the relation

$$E_p^j \triangleq \sum_{i=1}^{m} |e^j(\omega_i, \mathbf{p}^j)|^p = \sum_{i=1}^{m} |w(\omega_i)[F(\omega_i) - F^j(\omega_i, \mathbf{p}^j)]|^p \qquad (11\text{-}18)$$

Here, p is an integer.† Clearly, for $p = 2$, we return to the special case of least-squares error criterion E_2^j.

We conclude that the performance of the actual circuit obtained in the jth iteration can be measured by the size of the nonnegative error criterion E^j. The smaller E^j, the better the response; $E^j = 0$ indicates a perfect match between the desired and the actual responses.

---

† It is hoped that the reader will not confuse p, an integer exponent in E_p^j, and $\mathbf{p}^j$ or p_i^j, the designable circuit parameters.

Step 4: Is the circuit acceptable? In this step, the behavior of the jth circuit is compared with (1) the specifications and (2) the behavior of the $(j-1)$st circuit. If the deviation of the desired response is within the (specified) acceptable limits, the iteration stops. It also stops if the values of the circuit parameters p_i^j and/or the error criterion E^j no longer change significantly from iteration to iteration, i.e., if the change is less than a pre-programmed amount (say 0.01 or 0.1 percent). In addition, it is advisable to incorporate an upper limit on j which would halt the iteration after, say, 50 or 100 iterations, regardless of performance, to limit the computational costs incurred.

If all these tests are negative (we have not yet met the specifications; the parameters still change significantly; the maximum number of iterations has not been reached), then the process goes on to step 5. Otherwise, typically, all values obtained in the last iteration are printed and the computation stops.

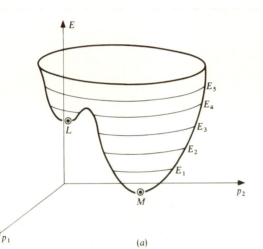

(a)

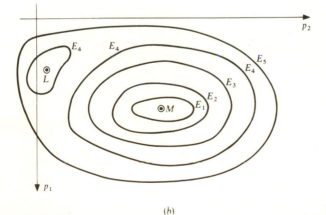

(b)

Figure 11-6 (a) Error surface; (b) its contours.

Step 5: Readjustment of the circuit parameters This step is the crucial part of the optimization procedure. It consists of changing the p_i^j in such a way that the error criterion E^j is decreased. Ideally, of course, the p_i^j should be chosen so that E^j becomes the *smallest* possible; however, except for trivially simple circuits, this cannot be achieved in one iteration. Hence, we limit our goal to decreasing E^j by a significant amount.

In considering the variations E^j with the p_i^j, it is useful to introduce the concept of the *error surface*, illustrated in Fig. 11-6a, and its contours, shown in Fig. 11-6b, for $n = 2$. The ultimate purpose of the optimization process is to reach the minimum M of the surface. However, in each iteration we can only climb down a limited amount, say from E_5 to E_4, or E_4 to E_3, etc. The figure also illustrates that in addition to the optimum at M (often called *global optimum*), there can be *local optima* like that at L.

The error surfaces which arise in circuit optimization are typically very rich in local optima. Since most optimization algorithms simply search for the *nearest* minimum point on the error surface, it is important to start the process near the global optimum or at least near a good local optimum. Hence, the choice of the initial approximation is crucial.

The rest of this chapter will be devoted to describing algorithms for carrying out step 5, the readjustment of the p_i^j, so as to decrease E^j and thus eventually to reach the minimum of the error surface.

11-3 GRADIENT METHODS OF OPTIMIZATION

Assume that in the jth iteration the circuit parameters p_i^j have been obtained.† The corresponding error is E^j. In the $(j + 1)$st iteration the p_i^j will be changed by Δp_i^j to obtain the new parameters

$$p_i^{j+1} = p_i^j + \Delta p_i^j \qquad i = 1, 2, \ldots, n \tag{11-19}$$

or in vector form $\mathbf{p}^{j+1} = \mathbf{p}^j + \Delta \mathbf{p}^j$. If the Δp_i^j are small, i.e., if $\Delta p_i^j \ll p_i^j$ for all i, then a good estimate for the new error E^{j+1} can be obtained from the Taylor-series expansion

$$E^{j+1}(\mathbf{p}^{j+1}) = E^{j+1}(\mathbf{p}^j + \Delta \mathbf{p}_j) = E^j(\mathbf{p}^j) + \sum_{i=1}^{n} \left(\frac{\partial E^j}{\partial p_i^j} \right) \Delta p_i^j$$

$$+ \frac{1}{2} \sum_{k=1}^{n} \sum_{l=1}^{n} \left(\frac{\partial^2 E^j}{\partial p_k^j \, \partial p_l^j} \right) \Delta p_k^j \, \Delta p_l^j + \cdots \tag{11-20}$$

where all derivatives are evaluated for $\mathbf{p} = \mathbf{p}^j$. We shall keep only the terms shown in Eq. (11-20) and neglect the remaining terms, which are cubic or of higher order

---

† For $j = 0$, the p_i^0 are the initial values given by the designer.

in Δp_i. The retained terms can be compactly expressed by introducing two new quantities. The *error gradient* is defined as the gradient vector of $E^j(\mathbf{p}^j)$

$$\nabla E^j(\mathbf{p}^j) \triangleq \begin{bmatrix} \dfrac{\partial E^j}{\partial p_1^j} \\ \vdots \\ \dfrac{\partial E^j}{\partial p_n^j} \end{bmatrix} \tag{11-21}$$

while the *Hessian matrix* is defined as the matrix of the second partial derivatives of E^j:

$$\mathbf{H}^j(\mathbf{p}^j) \triangleq \begin{bmatrix} \dfrac{\partial^2 E^j}{\partial p_1^{j2}} & \dfrac{\partial^2 E^j}{\partial p_1^j \, \partial p_2^j} & \cdots & \dfrac{\partial^2 E^j}{\partial p_1^j \, \partial p_n^j} \\[2mm] \dfrac{\partial^2 E^j}{\partial p_2^j \, \partial p_1^j} & \dfrac{\partial^2 E^j}{\partial p_2^{j2}} & \cdots & \dfrac{\partial^2 E^j}{\partial p_2^j \, \partial p_n^j} \\ \cdots\cdots\cdots\cdots\cdots\cdots\cdots\cdots \\ \dfrac{\partial^2 E^j}{\partial p_n^j \, \partial p_1^j} & \dfrac{\partial^2 E^j}{\partial p_n^j \, \partial p_2^j} & \cdots & \dfrac{\partial^2 E^j}{\partial p_n^{j2}} \end{bmatrix} \tag{11-22}$$

By carrying out the matrix operations, it is easy to show that the terms shown in (11-20) can be rewritten in the form

$$E^{j+1}(\mathbf{p}^{j+1}) \approx E^j(\mathbf{p}^j) + \nabla E^j(\mathbf{p}^j)^T \, \Delta \mathbf{p}^j + \tfrac{1}{2} \, \Delta \mathbf{p}^{j\,T} \mathbf{H}^j(\mathbf{p}^j) \, \Delta \mathbf{p}^j \tag{11-23}$$

where

$$\Delta \mathbf{p}^j \triangleq \begin{bmatrix} \Delta p_1 \\ \vdots \\ \Delta p_n \end{bmatrix} \tag{11-24}$$

is the *element increment vector*.

The *gradient methods* of optimization utilize (11-23) to find suitable values for the Δp_i^j.

Example 11-5 For the amplifier-equalizer, using a least-fourth-power error criterion and five equally spaced sample points, we have

$$E_4^j \triangleq \sum_{i=1}^{5} \left| 17 - i - 10 \log \left[p_1^{j2} + \frac{4}{(i+3)^2 \pi^2 p_2^{j2}} \right] \right|^4 \tag{11-25}$$

Here, we designated $p_1 \triangleq r$ and $p_2 \triangleq c$. Hence, the two elements of the gradient vector are

$$\frac{\partial E_4^j}{\partial p_1^j} = \sum_{i=1}^{5} 4 \left| 17 - i - 10 \log \left[p_1^{j2} + \frac{4}{(i+3)^2 \pi^2 p_2^{j2}} \right] \right|^3$$

$$\times \frac{(-20 \log e) p_1^j}{p_1^{j2} + 4/(i+3)^2 \pi^2 p_2^{j2}} \tag{11-26}$$

and

$$\frac{\partial E_4^j}{\partial p_2^j} = \sum_{i=1}^{5} 4 \left\{ 17 - i - 10 \log \left[p_1^{j2} + \frac{4}{(i+3)^2 \pi^2 p_2^{j2}} \right] \right\}^3$$

$$\times \frac{(20 \log e)[4/(i+3)^2 \pi^2 p_2^{j3}]}{p_1^{j2} + 4/(i+3)^2 \pi^2 p_2^{j2}} \tag{11-27}$$

The analytic calculation of the Hessian matrix is a nightmare which will not be inflicted upon the reader.

We conclude that using analytic differentiation as a way to find $\mathbf{V}E^j$ and $\mathbf{H}^j$ is impractical, even for such a simple circuit. Hence, other methods are needed. As discussed in Sec. 9-1, the derivatives involved can be approximated by difference quotients obtained by incrementing the p_i^j one at a time by, say, 1 percent; however, this is a tedious and inaccurate process. Fortunately, the adjoint-network method discussed at length in Chaps. 9 and 10 is applicable to this problem and provides a convenient and precise algorithm for finding all necessary derivatives.

For the least-pth error criterion, assuming even p and real responses† $F(\omega)$ and $F^j(\omega, \mathbf{p})$,

$$E_p^j = \sum_{i=1}^{m} \{w(\omega_i)[F(\omega_i) - F^j(\omega_i, p_1, p_2, \ldots, p_n)]\}^p \tag{11-28}$$

so that

$$\frac{\partial E_p^j}{\partial p_k} = p \sum_{i=1}^{m} \{w(\omega_i)[F(\omega_i) - F^j(\omega_i, \mathbf{p}^j)]\}^{p-1} \left\{ -w(\omega_i) \frac{\partial F^j(\omega_i, \mathbf{p}^j)}{\partial p_k} \right\}$$

$$= -p \sum_{i=1}^{m} w(\omega_i)[e^j(\omega_i, \mathbf{p}^j)]^{p-1} \frac{\partial F^j(\omega_i, \mathbf{p}^j)}{\partial p_k} \tag{11-29}$$

The only derivatives needed, therefore, are those of the actual circuit response F^j with respect to the element values p_k. These can readily be obtained using the adjoint-network method. Only two circuit analyses, those of N and $\hat{N}$, are needed to find all derivatives.

Example 11-6 For the amplifier-equalizer, the normalized equivalent circuit N is shown in Fig. 11-7a. We select, as before, the initial values $r = 2.8$ and $c = 0.028$. Next, using the process described in Chaps. 9 and 10, we obtain the adjoint network $\hat{N}$ given in Fig. 11-7b.

Straightforward analysis of N and $\hat{N}$ gives the branch currents shown in Table 11-2. (The other branch currents can be obtained trivially from the ones in Table 11-2; however, they are not needed for the gradient calculation.)

From the branch currents, the derivatives of V_0 are easily found. They are given in Table 11-3.

Again, all calculations can be simplified greatly if physical reasoning is used. We know that the input of a high-gain voltage amplifier with negative feedback becomes a virtual

---

† For complex responses we can use the absolute value of $F(\omega) - F^j(\omega, \mathbf{p})$.

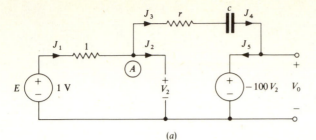

(a)

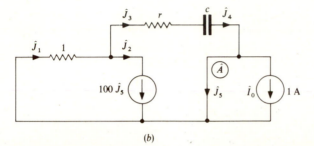

(b)

Figure 11-7 (a) Normalized circuit for the amplifier-equalizer; (b) its adjoint network.

ground, i.e., that the voltage V_2 at point A (Fig. 11-7a) is close to zero. Hence, $J_2 \approx 0$, and $J_3 = J_4 \approx J_1 \approx E/1 = 1$ A; also, $V_0 = -J_3(r + 1/j\Omega c) \approx -r + j/\Omega c$. Similarly, the input current $\hat{J}_5$ of the feedback current amplifier (at point $\hat{A}$ in Fig. 11-7b) is nearly zero; hence, by inspection, $\hat{J}_3 = \hat{J}_4 = 1$ A. Therefore, $\partial V_0/\partial r = -J_3 \hat{J}_3 \approx -1 + j0$, $\partial V_0/\partial c = J_4 \hat{J}_4/j\Omega c^2 \approx 1/[j(0.028)^2(i + 3)\pi/2]$, and the values shown in Table 11-4 are easily obtained.

A comparison with Table 11-3 shows that these values agree reasonably well with the ones obtained using exact analysis.

In the rest of this chapter, we shall use the approximate formula (11-7) and the approximate sensitivities given in Table 11-4.

Table 11-2

i	Ω_i	$J_1 = J_3 = J_4$	$\hat{J}_1$	$\hat{J}_3 = \hat{J}_4$
1	2π	0.970116 $+j5.31237 \times 10^{-2}$	-2.98840 $+j5.31237$	0.960511 $+j5.25977 \times 10^{-2}$
2	2.5π	0.971161 $+j4.25448 \times 10^{-2}$	-2.88388 $+j4.25447$	0.961546 $+j4.21235 \times 10^{-2}$
3	3π	0.971730 $+j3.54747 \times 10^{-2}$	-2.82700 $+j3.54747$	0.962109 $+j3.51235 \times 10^{-2}$
4	3.5π	0.972073 $+j3.04176 \times 10^{-2}$	-2.79268 $+j3.04176$	0.962449 $+j3.01165 \times 10^{-2}$
5	4π	0.972296 $+j2.66215 \times 10^{-2}$	-2.77039 $+j2.66215$	0.962670 $+j2.63580 \times 10^{-2}$

Table 11-3

i	Ω_i	V_0	$\dfrac{\partial V_0}{\partial r} \equiv \dfrac{\partial V_0}{\partial p_1} = -J_3 \hat{J}_3$	$\dfrac{\partial V_0}{\partial c} \equiv \dfrac{\partial V_0}{\partial p_2} = \dfrac{J_4 \hat{J}_4}{j\Omega_i c^2}$
1	2π	-2.98840 $+j5.31237$	-0.929013 $-j0.102052$	20.7169 $-j188.593$
2	2.5π	-2.88388 $+j4.25447$	-0.932024 $-j0.0818174$	13.2874 $-j151.364$
3	3π	-2.82700 $+j3.54747$	-0.933664 $-j0.0682611$	9.23817 $-j126.358$
4	3.5π	-2.79268 $+j3.04176$	-0.934655 $-j0.0585508$	6.79203 $-j108.422$
5	4π	-2.77039 $+j2.66215$	-0.935299 $-j0.0512555$	5.20253 $-j94.9345$

Next, the gradient $\mathbf{V}E^0$ and the Hessian matrix $\mathbf{H}^0$ will be found. From (11-29), using $p = 4$ and $w(\Omega_i) \equiv 1$, the components of the gradient can be obtained:

$$\frac{\partial E_4^0}{\partial p_k} = -4 \sum_{i=1}^{5} [e^0(\Omega_i, \mathbf{p}^0)]^3 \frac{\partial}{\partial p_k} \left(10 \log \left| \frac{V_0}{E} \right|^2 \right) \qquad k = 1, 2 \qquad (11\text{-}30)$$

Here

$$\frac{\partial}{\partial p_k} \left(10 \log \left| \frac{V_0}{E} \right|^2 \right) = \frac{10 \log e}{|V_0|^2} \frac{\partial |V_0|^2}{\partial p_k} \qquad (11\text{-}31)$$

and

$$\frac{1}{|V_0|^2} \frac{\partial |V_0|^2}{\partial p_k} = \frac{\partial (V_0 V_0^*)}{|V_0|^2 \partial p_k} = \frac{V_0^* \, \partial V_0 / \partial p_k + V_0 \, \partial V_0^* / \partial p_k}{|V_0|^2}$$

$$= \frac{2}{|V_0|^2} \text{Re} \left(V_0^* \frac{\partial V_0}{\partial p_k} \right) = 2 \text{Re} \left(\frac{1}{V_0} \frac{\partial V_0}{\partial p_k} \right) \qquad (11\text{-}32)$$

Hence (11-30) becomes

$$\frac{\partial E_4^0}{\partial p_k} = -80 \log e \sum_{i=1}^{5} [e^0(\Omega_i, \mathbf{p}^0)]^3 \text{Re} \left(\frac{1}{V_0} \frac{\partial V_0}{\partial p_k} \right) \qquad k = 1, 2 \qquad (11\text{-}33)$$

Table 11-4

i	Ω_i	$V_0 \approx -r + \dfrac{j}{\Omega_i c}$	$\dfrac{\partial V_0}{\partial r} \equiv \dfrac{\partial V_0}{\partial p_1} = -J_3 \hat{J}_3$	$\dfrac{\partial V_0}{\partial c} \equiv \dfrac{\partial V_0}{\partial p_2} = \dfrac{J_4 \hat{J}_4}{j\Omega c^2}$
1	2π	$-2.8 + j5.68410$	$-1 + j0$	$0 - j203.004$
2	2.5π	$-2.8 + j4.54728$	$-1 + j0$	$0 - j162.403$
3	3π	$-2.8 + j3.78940$	$-1 + j0$	$0 - j135.336$
4	3.5π	$-2.8 + j3.24806$	$-1 + j0$	$0 - j116.001$
5	4π	$-2.8 + j2.84205$	$-1 + j0$	$0 - j101.502$

When the values obtained earlier for e^0, V_0, and $\partial V_0/\partial p_k$ are substituted, the gradient vector is found to be

$$\nabla E_4^0(\mathbf{p}^0) \triangleq \begin{bmatrix} \dfrac{\partial E_4^0}{\partial p_1} \\[2mm] \dfrac{\partial E_4^0}{\partial p_2} \end{bmatrix} = \begin{bmatrix} -1.22033 \\[2mm] 236.924 \end{bmatrix} \tag{11-34}$$

With the technique discussed in Sec. 9-4 for the calculation of second derivatives, the Hessian matrix $\mathbf{H}^0(\mathbf{p}^0)$ can also be calculated by analyzing N and $\hat{N}$ with appropriate excitations. The resulting $\mathbf{H}^0(\mathbf{p}^0)$ is

$$\begin{bmatrix} 8.42072 & -1.63787 \times 10^3 \\ -1.63787 \times 10^3 & 2.96056 \times 10^5 \end{bmatrix}$$

A somewhat different technique can also be used for the calculation of ∇E_p using the adjoint network.[1] Assume, for illustration, that the error criterion chosen is

$$E_p \triangleq \sum_{i=1}^{m} w(\omega_i)\,|F(\omega_i) - V_0(\omega_i)|^p \tag{11-35}$$

where p is an integer and $F(\omega_i)$ is the specified output-voltage response. If the designable parameters p_k change, V_0 and hence E_p also change. The change in E_p is

$$\Delta E_p = \sum_{k=1}^{n} \frac{\partial E_p}{\partial p_k}\,\Delta p_k = \sum_{i=1}^{m} w(\omega_i)\,\Delta\,|F(\omega_i) - V_0(\omega_i)|^p \tag{11-36}$$

Here

$$\Delta\,|F(\omega_i) - V_0(\omega_i)|^p = |F(\omega_i) - V_0(\omega_i) - \Delta V_0(\omega_i)|^p - |F(\omega_i) - V_0(\omega_i)|^p \tag{11-37}$$

For small $|\Delta V_0|$, omitting for convenience the argument ω_i,

$$|F - V_0 - \Delta V_0|^p = (F - V_0 - \Delta V_0)^{p/2}(F^* - V_0^* - \Delta V_0^*)^{p/2}$$

$$= (F - V_0)^{p/2}\left(1 - \frac{\Delta V_0}{F - V_0}\right)^{p/2}(F^* - V_0^*)^{p/2}\left(1 - \frac{\Delta V_0^*}{F^* - V_0^*}\right)^{p/2}$$

$$\approx |F - V_0|^p\left(1 - \frac{p}{2}\frac{\Delta V_0}{F - V_0}\right)\left(1 - \frac{p}{2}\frac{\Delta V_0^*}{F^* - V_0^*}\right)$$

$$\approx |F - V_0|^p - p\,\mathrm{Re}\,[\,|F - V_0|^{p-2}(F^* - V_0^*)\,\Delta V_0] \tag{11-38}$$

Hence $\qquad \Delta|F - V_0|^p = -p\,\mathrm{Re}\,[\,|F - V_0|^{p-2}(F^* - V_0^*)\,\Delta V_0] \tag{11-39}$

and interchanging summation and real-part calculations gives

$$\Delta E_p = \mathrm{Re}\left\{\sum_{i=1}^{m}[-p\,|F - V_0|^{p-2}(F^* - V_0^*)]_{\omega_i}\,\Delta V_0(\omega_i)\right\} \tag{11-40}$$

Denoting

$$\hat{I}_0(\omega_i) \triangleq -p|F(\omega_i) - V_0(\omega_i)|^{p-2}[F^*(\omega_i) - V_0^*(\omega_i)] \qquad (11\text{-}41)$$

we obtain

$$\Delta E_p = \text{Re}\left[\sum_{i=1}^{m} \hat{I}_0(\omega_i)\,\Delta V_0(\omega_i)\right] \qquad (11\text{-}42)$$

Consider now the physical circuit N of Fig. 11-8a and its adjoint network $\hat{N}$ excited as shown in Fig. 11-8b. From the differential Tellegen's theorem given in Eq. (10-2)

$$\hat{I}_0\,\Delta V_0 = -\sum_{\substack{\text{all internal}\\\text{branches}}} (\hat{J}_k\,\Delta V_k - \hat{V}_k\,\Delta J_k) \qquad (11\text{-}43)$$

If the branch voltages and currents obey the hybrid relations†

$$\begin{bmatrix} \mathbf{J}_1 \\ \mathbf{V}_2 \end{bmatrix} = \mathbf{H}\begin{bmatrix} \mathbf{V}_1 \\ \mathbf{J}_2 \end{bmatrix} \qquad (11\text{-}44)$$

the right-hand side can be rewritten in the form

$$[\hat{\mathbf{V}}_1^T \quad \hat{\mathbf{J}}_2^T]\,\Delta\mathbf{H}\begin{bmatrix} \mathbf{V}_1 \\ \mathbf{J}_2 \end{bmatrix} \qquad (11\text{-}45)$$

as shown earlier [see Eq. (9-71)]. Hence, (11-42) can be rewritten

$$\Delta E_p = \text{Re}\left\{\sum_{i=1}^{m} [\hat{\mathbf{V}}_1^T(\omega_i)\,\hat{\mathbf{J}}_2^T(\omega_i)]\,\Delta\mathbf{H}(\omega_i)\begin{bmatrix} \mathbf{V}_1(\omega_i) \\ \mathbf{J}_2(\omega_i) \end{bmatrix}\right\} \qquad (11\text{-}46)$$

Here the parameter increments Δp_i are contained in $\Delta\mathbf{H}(\omega_i)$. The gradient component $\partial E_p/\partial p_k$ is then the coefficient of Δp_k on the right-hand side of (11-46).

---

† The hybrid branch matrix entering (11-44) is not to be confused with the Hessian matrix $\mathbf{H}^j$.

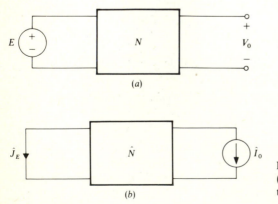

Figure 11-8 (a) Physical circuit N; (b) the adjoint network $\hat{N}$ excited by the function $\hat{I}_0$ of Eq. (11-41).

Example 11-7 We next return to the amplifier-equalizer example but assume that the output-voltage response $V_0(\omega)$ rather than the logarithmic gain $20 \log |V_0(\omega)/E|$ is specified. Since the branch relations are

$$
\begin{bmatrix} J_1 \\ J_2 \\ J_3 \\ J_4 \\ V_5 \end{bmatrix} = \begin{bmatrix} \dfrac{1}{R_0} & 0 & 0 & 0 & 0 \\ 0 & 0 & 0 & 0 & 0 \\ 0 & 0 & \dfrac{1}{r} & 0 & 0 \\ 0 & 0 & 0 & j\Omega c & 0 \\ 0 & -G & 0 & 0 & 0 \end{bmatrix} \begin{bmatrix} V_1 \\ V_2 \\ V_3 \\ V_4 \\ J_5 \end{bmatrix}
\tag{11-47}
$$

the square matrix on the right-hand side is **H**. If only r and c are designable, then

$$
\Delta \mathbf{H} = \begin{bmatrix} 0 & 0 & 0 & 0 & 0 \\ 0 & 0 & 0 & 0 & 0 \\ 0 & 0 & -\dfrac{\Delta r}{r^2} & 0 & 0 \\ 0 & 0 & 0 & j\Omega\,\Delta c & 0 \\ 0 & 0 & 0 & 0 & 0 \end{bmatrix}
\tag{11-48}
$$

and (11-46) gives

$$
\Delta E_p = \text{Re} \left[\sum_{i=1}^{m} \frac{-\hat{V}_3(\Omega_i)V_3(\Omega_i)}{r^2} \right] \Delta r + \text{Re} \left[\sum_{i=1}^{m} \hat{V}_4(\Omega_i)V_4(\Omega_i)j\Omega \right] \Delta c
\tag{11-49}
$$

Hence, the gradient vector is

$$
\mathbf{V} E_p = \begin{bmatrix} -\text{RE} \left[\sum_{i=1}^{m} \dfrac{\hat{V}_3(\Omega_i)V_3(\Omega_i)}{r^2} \right] \\ \text{Re} \left[\sum_{i=1}^{m} \hat{V}_4(\Omega_i)V_4(\Omega_i)j\Omega \right] \end{bmatrix}
\tag{11-50}
$$

Note that in evaluating $V_3(\Omega_i)$ and $V_4(\Omega_i)$ the network of Fig. 11-7a is to be analyzed; however, in calculating $\hat{V}_3(\Omega_i)$ and $\hat{V}_4(\Omega_i)$, the excitation of $\hat{N}$ is to be $\hat{I}_0(\Omega_i)$, as given in (11-41). Hence, the $\hat{V}_3$ and $\hat{V}_4$ obtained from Fig. 11-7b must be multiplied by the right-hand side of (11-41) before substituting into $\mathbf{V} E_p$.

We conclude that it is possible to find a relation, Eq. (11-23), giving E^{j+1} as a function of the circuit parameter increments Δp_i^j. The relation is only approximately true, for small parameter increments. It requires an evaluation of the gradient given by (11-21) and the Hessian matrix given by (11-22). The former can be calculated conveniently using the adjoint-network method; in fact, we have described two equivalent methods to carry out this calculation.

The Hessian matrix too can be calculated using the adjoint-network approach, as described in Sec. 9-4; however, this requires $n + 2$ circuit analyses, where n is the number of designable parameters.

In the next sections, some gradient methods of circuit optimization, i.e., methods based on (11-23), will be discussed. With only one exception (Newton's method, Sec. 11-5), these methods require only $\mathbf{V}E^j$ and not $\mathbf{H}^j$ for optimization.

11-4 THE STEEPEST-DESCENT METHOD

The magnitude† of the parameter-increment vector

$$\Delta\mathbf{p}^j \triangleq \begin{bmatrix} \Delta p_1^j \\ \Delta p_2^j \\ \vdots \\ \Delta p_n^j \end{bmatrix} \tag{11-51}$$

is defined by the relation

$$|\Delta\mathbf{p}^j| \triangleq \sqrt{\Delta\mathbf{p}^{j^T}\Delta\mathbf{p}^j} = \sqrt{(\Delta p_1^j)^2 + (\Delta p_2^j)^2 + \cdots + (\Delta p_n^j)^2} \tag{11-52}$$

For obvious reasons, $|\Delta\mathbf{p}^j|$ will be called the *step size*. We now answer the following question: for a given, very small step size $|\Delta\mathbf{p}^j| = \delta$, what is the optimum $\Delta\mathbf{p}^j$?

For $\delta \to 0$, (11-23) gives

$$E^{j+1}(\mathbf{p}^{j+1}) \approx E^j(\mathbf{p}^j) + \mathbf{V}E^j(\mathbf{p}^j)^T \Delta\mathbf{p}^j \tag{11-53}$$

Since we are trying to decrease E by the maximum amount, $\mathbf{V}E^{j^T}\Delta\mathbf{p}^j$ should be as large a negative number as possible. But, as well known from vector analysis, the scalar product $\mathbf{V}E^{j^T}\Delta\mathbf{p}^j$ can be written as $|\mathbf{V}E^j||\Delta\mathbf{p}^j|\cos\beta$, where β is the angle between the vectors $\mathbf{V}E^j$ and $\Delta\mathbf{p}^j$ (Fig. 11-9). The largest negative value, for given $|\mathbf{V}E^j|$ and $|\Delta\mathbf{p}^j| = \delta$, results for $\beta = \pi$. Then

$$E^{j+1}(\mathbf{p}^{j+1}) \approx E^j(\mathbf{p}^j) - |\mathbf{V}E^j(\mathbf{p}^j)|\delta \tag{11-54}$$

As Fig. 11-9 indicates, $\beta = \pi$ means that

$$\Delta\mathbf{p}^j = -k\,\mathbf{V}E^j(\mathbf{p}^j) \tag{11-55}$$

where

$$k = \frac{|\Delta\mathbf{p}^j|}{|\mathbf{V}E^j|} > 0 \tag{11-56}$$

† Sometimes called *euclidean norm*.

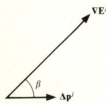

Figure 11-9 Illustration of the scalar product $\mathbf{V}E^{j^T}\Delta\mathbf{p}^j$.

Hence, for small step size $|\Delta p^j|$, *the step $\Delta \mathbf{p}^j$ should be taken in the direction of the negative gradient.* This is the direction in which E decreases most rapidly; it is called *the direction of the steepest descent.*

From (11-55), it follows that

$$\Delta p_i^j = -k \frac{\partial E^j}{\partial p_i^j} \qquad i = 1, 2, \ldots, n \tag{11-57}$$

Equation (11-57) can be used to find all Δp_i^j as soon as k has been chosen. A simple way to get a suitable value for k is by trial and error, as follows:

1. Calculate all $(\partial E^j / \partial p_i^j)/p_i^j$ for $i = 1, 2, \ldots, n$.
2. Assign the maximum relative change† in any component p_i^j in the first step; call this Δ_{max}.
3. The corresponding maximum value of k is then

$$k_{max} = \frac{\Delta_{max}}{\max_i \; |(\partial E^j / \partial p_i)/p_i|} \tag{11-58}$$

4. Evaluate

$$E(\mathbf{p}^j - k_{max} \, \nabla \mathbf{E}^j), \; E(\mathbf{p}^j - 2k_{max} \, \nabla \mathbf{E}^j), \; E(\mathbf{p}^j - 3k_{max} \, \nabla \mathbf{E}^j), \; \ldots \tag{11-59}$$

Stop incrementing $\mathbf{p}^j$ when E no longer decreases with each step. The optimum point (at which E was a minimum) is then designated as $\mathbf{p}^{j+1}$, the new gradient $\nabla \mathbf{E}^{j+1}$ is calculated there, and steps 1 to 4 are repeated to find $\mathbf{p}^{j+2}$.

The process given by steps 1 to 4, called a *linear search,* is illustrated in Fig. 11-10.

An alternative to the search is to use *polynomial interpolation* to find the minimum of E along the negative gradient. Again, k_{max} is found; but now instead of (11-59) the sequence

$$E(\mathbf{p}^j - k_{max} \, \nabla \mathbf{E}^j), \; E(\mathbf{p}^j - 2k_{max} \, \nabla \mathbf{E}^j), \; E(\mathbf{p}^j - 4k_{max} \, \nabla \mathbf{E}^j),$$

$$E(\mathbf{p}^j - 8k_{max} \, \nabla \mathbf{E}^j), \; \ldots, \; E(\mathbf{p}^j - 2^l k_{max} \, \nabla \mathbf{E}^j), \; \ldots \tag{11-60}$$

is calculated. Let the first increasing number E occur for

$$\Delta \mathbf{p}_0^j = -2^i \, k_{max} \nabla \mathbf{E}^j \tag{11-61}$$

Then the directional minimum, i.e., the minimum of $E(\mathbf{p})$ along the negative gradient line, is somewhere between $\mathbf{p}^j$ and $\mathbf{p}^j + \Delta \mathbf{p}_0^j$.‡ To find its approximate location, we assume that in this region the function $E(\mathbf{p}^j + \Delta \mathbf{p})$ can be approximated by a parabola of the form

$$E = E(\mathbf{p}^j) + a_1 \, |\Delta \mathbf{p}| + a_2 \, |\Delta \mathbf{p}|^2 \tag{11-62}$$

† Usually 1 to 50 percent.
‡ This process is called *bracketing the minimum.*

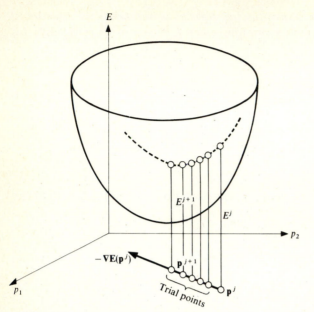

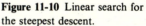

Figure 11-10 Linear search for the steepest descent.

As Fig. 11-10 illustrates, for a smooth error function near the minimum this assumption is reasonable. The coefficients a_1 and a_2 are found by matching the parabola to the two values $E(\mathbf{p}^j - 2^{i-1}k_{max}\nabla E^j) = E(\mathbf{p}^j + \Delta\mathbf{p}_0/2)$ and $E(\mathbf{p}^j - 2^i k_{max}\nabla E^j) = E(\mathbf{p}^j + \Delta\mathbf{p}_0)$, both already found during the bracketing operation. [They are the last two values in the sequence (11-60).]

The minimum of the parabola (11-62) occurs then for

$$\Delta\mathbf{p}^j = \frac{\Delta\mathbf{p}_0^j}{4}\frac{3E^j(\mathbf{p}^j) - 4E(\mathbf{p}^j + \Delta\mathbf{p}_0^j/2) + E(\mathbf{p}^j + \Delta\mathbf{p}_0^j)}{E^j(\mathbf{p}^j) - 2E(\mathbf{p}^j + \Delta\mathbf{p}_0^j/2) + E(\mathbf{p}^j + \Delta\mathbf{p}_0^j)} \tag{11-63}$$

The process is illustrated in Fig. 11-11.

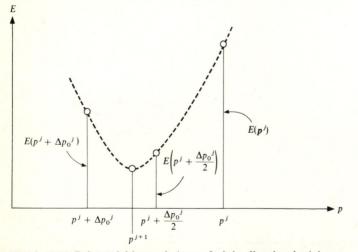

Figure 11-11 Polynomial interpolation to find the directional minimum.

Example 11-8 We illustrate the steepest-descent method first on a simple $E(\mathbf{p})$ function, which did not originate in circuit design but is useful for demonstrating the process.

Let

$$E(\mathbf{p}) = p_1^2 - 2p_1p_2 + 2p_2^2 \tag{11-64}$$

Since $E(\mathbf{p})$ can be rewritten in the form $(p_1 - p_2)^2 + p_2^2$, it is immediately obvious that the minimum $E = 0$ occurs for $p_1 = p_2 = 0$. Nevertheless, we can use this function to show how the various steps discussed above are carried out.

The gradient (which here can be found analytically) is

$$\nabla E(\mathbf{p}) = \begin{bmatrix} \dfrac{\partial E}{\partial p_1} \\ \dfrac{\partial E}{\partial p_2} \end{bmatrix} = \begin{bmatrix} 2(p_1 - p_2) \\ -2p_1 + 4p_2 \end{bmatrix} \tag{11-65}$$

If an initial approximation $\mathbf{p}^0 = (-4, 2)^T$ is assumed, the gradient is

$$\nabla E^0(\mathbf{p}^0) = \begin{bmatrix} -12 \\ 16 \end{bmatrix} \tag{11-66}$$

If we allow at most $\Delta_{max} = 50$ percent change in any parameter during one step, by (11-58),

$$k_{max} = \frac{\Delta_{max}}{\max_i \; |(\partial E/\partial p_i)/p_i|} = \frac{0.5}{8} = 0.0625 \tag{11-67}$$

Then, a linear search along the negative gradient direction with a step size

$$\Delta\mathbf{p} = -k_{max} \, \nabla E^0 = \begin{bmatrix} 0.75 \\ -1 \end{bmatrix} \tag{11-68}$$

gives the results shown in Table 11-5. Clearly, the correct choice for $\mathbf{p}^1$ is $(-1.75, -1)^T$. Then the new gradient is

$$\nabla E^1(\mathbf{p}^1) = \begin{bmatrix} -1.5 \\ -0.5 \end{bmatrix} \tag{11-69}$$

and the process can be repeated.

Table 11-5

i	$i\,\Delta\mathbf{p}$	$\mathbf{p}^0 + i\,\Delta\mathbf{p}$	$E(\mathbf{p}^0 + i\,\Delta\mathbf{p})$
0	$\begin{bmatrix} 0 \\ 0 \end{bmatrix}$	$\begin{bmatrix} -4 \\ 2 \end{bmatrix}$	40
1	$\begin{bmatrix} 0.75 \\ -1 \end{bmatrix}$	$\begin{bmatrix} -3.25 \\ 1 \end{bmatrix}$	19.0625
2	$\begin{bmatrix} 1.5 \\ -2 \end{bmatrix}$	$\begin{bmatrix} -2.5 \\ 0 \end{bmatrix}$	6.25
3	$\begin{bmatrix} 2.25 \\ -3 \end{bmatrix}$	$\begin{bmatrix} -1.75 \\ -1 \end{bmatrix}$	1.5625
4	$\begin{bmatrix} 3 \\ -4 \end{bmatrix}$	$\begin{bmatrix} -1 \\ -2 \end{bmatrix}$	5.00

Table 11-6

i	$2^i \, \Delta\mathbf{p}$	$\mathbf{p}^0 + 2^i \, \Delta\mathbf{p}$	$E(\mathbf{p}^0 + 2^i \, \Delta\mathbf{p})$
0	$\begin{bmatrix} 0.75 \\ -1 \end{bmatrix}$	$\begin{bmatrix} -3.25 \\ 1 \end{bmatrix}$	19.0625
1	$\begin{bmatrix} 1.5 \\ -2 \end{bmatrix}$	$\begin{bmatrix} -2.5 \\ 0 \end{bmatrix}$	6.25
2	$\begin{bmatrix} 3 \\ -4 \end{bmatrix}$	$\begin{bmatrix} -1 \\ -2 \end{bmatrix}$	5
3	$\begin{bmatrix} 6 \\ -8 \end{bmatrix}$	$\begin{bmatrix} 2 \\ -6 \end{bmatrix}$	100

Next, polynomial interpolation will be demonstrated. With $\Delta\mathbf{p}$ given in (11-68) as the initial step size, (11-60) gives the results shown in Table 11-6. Hence, we choose $\Delta\mathbf{p}_0 = (6, -8)^T$. Then (11-63) gives

$$\Delta\mathbf{p}^0 = \frac{1}{4} \frac{(3)(40) - (4)(5) + 100}{40 - (2)(5) + 100} \begin{bmatrix} 6 \\ -8 \end{bmatrix} \approx \begin{bmatrix} 2.3077 \\ -3.077 \end{bmatrix} \tag{11-70}$$

Hence,

$$\mathbf{p}^1 = \mathbf{p}^0 + \Delta\mathbf{p}^0 \approx \begin{bmatrix} -1.692 \\ -1.077 \end{bmatrix} \tag{11-71}$$

and $E^1(\mathbf{p}^1) \approx 1.538$.

Note that for this simple problem both the linear search and the polynomial interpolation were efficient. They decreased the error criterion from $E^0 = 40$ to $E^1 = 1.5$ to 1.6 in a single iteration. This is largely due to the inherently well-conditioned smooth error surface, which corresponds to the $E(\mathbf{p})$ given in (11-64). The general shape of the error surface is indicated in Fig. 11-12 and the corresponding contours in Fig. 11-13, which also shows the initial point $\mathbf{p}^0$, the initial

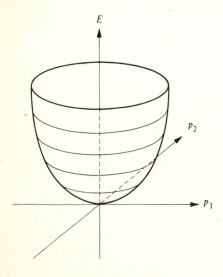

Figure 11-12 Error surface for $E = p_1^2 - 2p_1 p_2 + 2p_2^2$.

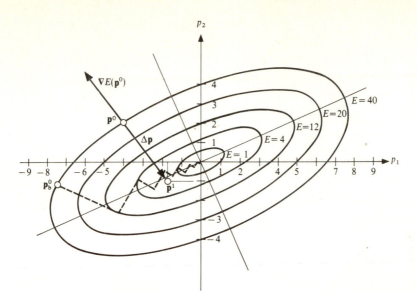

Figure 11-13 Error contours of $E = p_1^2 - 2p_1p_2 + 2p_2^2$.

gradient ∇E^0, and the result $\mathbf{p}^1$ of the first iteration using linear search (the result of the polynomial interpolation is very close to $\mathbf{p}^1$).

Example 11-9 For the amplifier-equalizer, using the initial values $r = 2.8$ and $c = 0.028$, as well as the error criterion of (11-25), we found the initial gradient ∇E^0 given in (11-34). Prescribing $\Delta_{max} = 2$ percent, (11-58) gives

$$k_{max} = \frac{\Delta_{max}}{\max_i |(\partial E/\partial p_i)/p_i|} \approx 2.36 \times 10^{-6} \tag{11-72}$$

Therefore, using the step size

$$\Delta \mathbf{p} = -k_{max} \nabla E^0 \approx \begin{bmatrix} 2.88 \times 10^{-6} \\ -5.6 \times 10^{-4} \end{bmatrix} \tag{11-73}$$

we obtain the results shown in Table 11-7, using linear search.†

Table 11-7

i	$i \Delta \mathbf{p}$	$\mathbf{p}_0 + i \Delta \mathbf{p}$	$E(\mathbf{p}^0 + i \Delta \mathbf{p})$
0	0.0	2.799999	0.139210
	0.0	0.028000	
1	0.000003	2.800002	0.047622
	−0.000560	0.027440	
2	0.000006	2.800005	0.023012
	−0.001120	0.026880	
3	0.000009	2.800008	0.059784
	−0.001680	0.026320	

† These calculations are best performed on a computer or programmable hand-held calculator, since the evaluation of $E(\mathbf{p}^0 + i \Delta \mathbf{p})$ is tedious.

Table 11-8

0	0.000003	2.800002	0.047622
	−0.000560	0.027440	
1	0.000006	2.800005	0.023012
	−0.001120	0.026880	
2	0.000012	2.800011	0.182302
	−0.002240	0.025760	

Clearly, the best result is obtained for $\Delta \mathbf{p}^0 \approx (6 \times 10^{-6}, -0.00112)^T$, giving $\mathbf{p}^1 \approx (2.8, 0.02688)^T$ and $E^1(\mathbf{p}^1) = 0.023012$.

The corresponding calculation using polynomial interpolation is illustrated in Table 11-8.

Formula (11-63) gives now for the increment vector

$$\Delta \mathbf{p}^0 \approx \frac{1}{4} \begin{bmatrix} 0.000012 \\ -0.00224 \end{bmatrix} \frac{(3)(0.13921) - (4)(0.023012) + 0.182302}{0.13921 - (2)(0.023012) + 0.182302}$$

$$\approx \begin{bmatrix} 0.00000553 \\ -0.0010324 \end{bmatrix} \tag{11-74}$$

so that
$$\mathbf{p}^1 \approx (2.8, 0.0269676)^T \tag{11-75}$$

and
$$E^1(\mathbf{p}^1) \approx 0.022972 \tag{11-76}$$

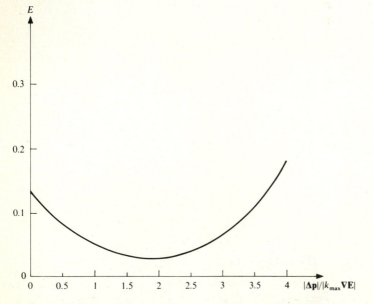

Figure 11-14 The variation of E in the steepest-descent direction.

We note that the computational effort for the polynomial interpolation is somewhat less and the result slightly better than for linear search. The computational saving is more pronounced for smaller step size; e.g., for $\Delta_{max} = 1$ percent, five steps are needed for the linear search but only three for polynomial interpolation. At the same time, the directional minimum value $E^1(\mathbf{p}^1)$ remains 0.023012 for the linear search and 0.022972 for the polynomial interpolation.

The success of the polynomial interpolation is explained by the shape of the E-vs.-$|\Delta p|$ curve, shown in Fig. 11-14. As the curve illustrates, the $E(|\Delta p|)$ function is indeed close to being a parabola of the form predicted by (11-62).

The main advantage of the steepest-descent method is its conceptual and computational simplicity; however, it is effective only if the error surface is smooth and the circuit parameters are scaled such that the error contours are close to being circular. To illustrate this last point, assume that for the $E(\mathbf{p})$ given in (11-64) the initial point is chosen to be $\mathbf{p}_b^0 = (-7.245, -1)^T$, on the same contour as the previous one (Fig. 11-13). Then the steepest-descent iteration approaches the minimum along a tortuous zigzag path,[†] illustrated in Fig. 11-13. Assume next that the scaling is such that the p_1 axis is expanded 1000 times; e.g., ohms rather than kilohms are used as the units of p_1. Then the elliptical contours become very elongated, and the zigzagging will be worsened for almost all initial points. By contrast, if the p_1 axis is compressed so that the contours become nearly circular, a single directional search will approach the exact optimum for any starting point.

The steepest descent obviously works well if the error surface is nearly planar, so that the gradient is approximately constant over a wide range of the parameters. This is likely to be the case far away from the optimum, i.e., in the early stages of the iteration. As the optimum is approached, zigzagging is likely to set in. Then other optimization methods become more efficient to use.

11-5 NEWTON'S METHOD

We recall that the steepest-descent method resulted from neglecting all but the constant and linear terms in (11-20) and (11-23). If we now include the quadratic terms as well, we obtain

$$E^{j+1}(\mathbf{p}^j + \Delta\mathbf{p}^j) \approx E(\mathbf{p}^j) + \sum_{i=1}^{n} \frac{\partial E^j}{\partial p_i^j} \Delta p_i^j + \frac{1}{2} \sum_{k=1}^{n} \sum_{l=1}^{n} \frac{\partial^2 E^j}{\partial p_k^j \partial p_l^j} \Delta p_k^j \Delta p_l^j \qquad (11\text{-}77)$$

The optimum Δp_i^j satisfy the extremum conditions

$$\frac{\partial E^{j+1}}{\partial \Delta p_i^j} = 0 \qquad i = 1, 2, \ldots, n \qquad (11\text{-}78)$$

† Note that the steepest-descent direction is orthogonal to the contour.

or, from (11-77),

$$\frac{\partial E^j}{\partial p_i^j} + \frac{1}{2} \sum_{l=1}^{n} \frac{\partial^2 E^j}{\partial p_i^j \, \partial p_l^j} \Delta p_l^j + \frac{1}{2} \sum_{k=1}^{n} \frac{\partial^2 E^j}{\partial p_k^j \, \partial p_i^j} \Delta p_k^j = 0 \qquad i = 1, 2, \ldots, n \qquad (11\text{-}79)$$

Since $\partial^2 E^j/(\partial p_i^j \, \partial p_l^j) = \partial^2 E^j/(\partial p_l^j \, \partial p_i^j)$, the two summations are equal. Hence, the conditions for a minimum E^{j+1} are

$$\frac{\partial E^j}{\partial p_i^j} + \sum_{k=1}^{n} \frac{\partial^2 E^j}{\partial p_k^j \, \partial p_i^j} \Delta p_k^j = 0 \qquad i = 1, 2, \ldots, n \qquad (11\text{-}80)$$

or, in vector form,

$$\nabla E^j(\mathbf{p}^j) + \mathbf{H}^j(\mathbf{p}^j) \, \Delta \mathbf{p}^j = \mathbf{0} \qquad (11\text{-}81)$$

Equation (11-80) represents a system of linear equations for the unknown Δp_i^j. They can be solved using gaussian elimination or, since $\mathbf{H}$ is a symmetric matrix, using Cholesky's method.[2] The latter is especially simple if $\mathbf{H}^j$ is a positive semidefinite matrix, i.e., if $\mathbf{x}^T \mathbf{H}^j \mathbf{x} \geq 0$ for any vector $\mathbf{x}$. Near the minimum M of the error surface (Fig. 11-15a) this is certainly true. For consider the behavior of $E(\mathbf{p}_M + \Delta \mathbf{p})$, where $\mathbf{p}_M$ is the parameter vector at the minimum and where $|\Delta \mathbf{p}|$ is very small. Using (11-23), we have, to a first-order approximation,

$$E(\mathbf{p}_M + \Delta \mathbf{p}) \approx E(\mathbf{p}_M) + \nabla E(\mathbf{p}_M)^T \, \Delta \mathbf{p} \qquad (11\text{-}82)$$

Since $\nabla E(\mathbf{p}_M)^T \, \Delta \mathbf{p} \leq 0$ if $\Delta \mathbf{p}$ points in the negative $\nabla E(\mathbf{p}_M)$ direction, we conclude that $E(\mathbf{p}_M + \Delta \mathbf{p}) < E(\mathbf{p}_M)$ unless $|\nabla E(\mathbf{p}_M)| = 0$. Since $E(\mathbf{p}_M)$ is the minimum value of $E(\mathbf{p})$, $E(\mathbf{p}_M + \Delta \mathbf{p})$ cannot be any smaller. We conclude that $\nabla E(\mathbf{p}_M) = \mathbf{0}$. Hence, again from (11-23),

$$E(\mathbf{p}_M + \Delta \mathbf{p}) \approx E(\mathbf{p}_M) + \tfrac{1}{2} \Delta \mathbf{p}^T \mathbf{H}(\mathbf{p}_M) \, \Delta \mathbf{p} \qquad (11\text{-}83)$$

Repeating our argument, since $E(\mathbf{p}_M + \Delta \mathbf{p}_M) \geq E(\mathbf{p}_M)$, the second term on the right-hand side must be nonnegative. We conclude hence that at the optimum point $\mathbf{p}_M$ the conditions

$$\nabla E(\mathbf{p}_M) = 0 \qquad \mathbf{x}^T \mathbf{H}(\mathbf{p}_M) \mathbf{x} \geq 0 \qquad \text{for any } \mathbf{x} \qquad (11\text{-}84)$$

must hold. These conditions are analogous to the relations $df(x)/dx = 0$ and $d^2 f(x)/dx^2 \geq 0$ which hold at the minimum of a scalar function $f(x)$. The analogy extends even further; it can be shown that the condition $\mathbf{x}^T \mathbf{H}(\mathbf{p}) \mathbf{x} > 0$ for all $\mathbf{x} \neq \mathbf{0}$ indicates that the error surface $E(\mathbf{p})$ is convex from the bottom, just as the condition $d^2 f(x)/dx^2 > 0$ indicates a convex curve for $f(x)$. These derivative conditions are illustrated in Fig. 11-15b.

Next, simple examples will illustrate the use of Eqs. (11-80) and (11-81) in optimization.

Example 11-10 For the function given in (11-64), if we start from $\mathbf{p}^0 = (-4, 2)^T$ and use the gradient $\nabla E^0(\mathbf{p}^0) = (-12, 16)^T$ found in (11-66) as well as the Hessian matrix

$$\mathbf{H}(\mathbf{p}) = \begin{bmatrix} 2 & -2 \\ -2 & 4 \end{bmatrix} \qquad (11\text{-}85)$$

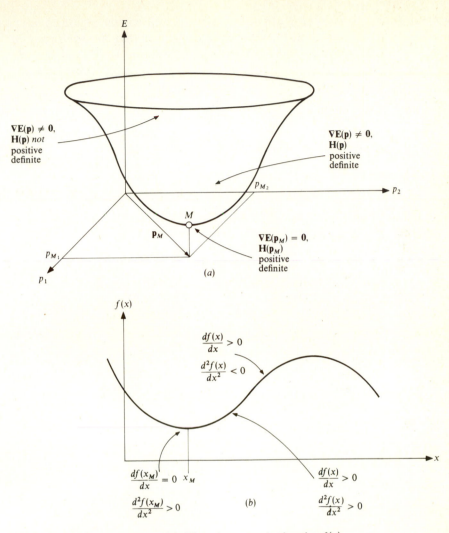

E

$\nabla E(\mathbf{p}) \neq 0,$
$\mathbf{H}(\mathbf{p})$ *not*
positive
definite

$\nabla E(\mathbf{p}) \neq 0,$
$\mathbf{H}(\mathbf{p})$
positive
definite

p_{M_2}

p_2

M

$\mathbf{p}_M$

$\nabla E(\mathbf{p}_M) = 0,$
$\mathbf{H}(\mathbf{p}_M)$
positive
definite

p_{M_1}

p_1

(a)

$f(x)$

$\dfrac{df(x)}{dx} > 0$

$\dfrac{d^2f(x)}{dx^2} < 0$

x

$\dfrac{df(x_M)}{dx} = 0$ x_M

$\dfrac{d^2f(x_M)}{dx^2} > 0$

(b)

$\dfrac{df(x)}{dx} > 0$

$\dfrac{d^2f(x)}{dx^2} > 0$

Figure 11-15 (a) Error surface $E(\mathbf{p})$; (b) analogous scalar function $f(x)$.

Equation (11-80) gives

$$2 \, \Delta p_1^0 - 2 \, \Delta p_2^0 = 12$$
$$-2 \, \Delta p_1^0 + 4 \, \Delta p_2^0 = -16$$

(11-86)

This can be easily solved to get $\Delta \mathbf{p}^0 = (4, \ -2)^T$. Hence, $\mathbf{p}^1 = \mathbf{p}^0 + \Delta \mathbf{p}^0 = (-4, 2)^T +$ $(4, \ -2)^T = (0, 0)^T$. The corresponding $E^1(\mathbf{p}^1)$ is zero. Hence, we obtained the true minimum in just one iteration. This is not too surprising, since our $E(\mathbf{p})$ is quadratic and hence all third- and higher-order derivatives of E with respect to the p_i are zero. Thus, (11-20), (11-23), and (11-80) all hold exactly, and no iterations are needed to make up for the omitted higher-order terms.

As long as $E(\mathbf{p})$ behaves nearly like a quadratic function in the region containing $\mathbf{p}$ and the minimum point $\mathbf{p}_M$, it is reasonable to expect that Newton's method will be effective. Its use will thus prevent the zigzagging which plagues the steepest-descent method near the optimum. Hence, it is possible to use a hybrid optimization process which applies steepest descent for the first few iterations and then switches to Newton's method once the minimum is approached (this is indicated by the failure of the steepest-descent method to decrease E rapidly).

Example 11-11 For the amplifier-equalizer, at the initial point $\mathbf{p}^0 = (2.8, 0.028)$ the gradient was found to be $\mathbf{V}E^0(\mathbf{p}^0) = (-1.22033, 236.924)^T$, as given in (11-34). Also, the Hessian matrix was found subsequently to be

$$\mathbf{H}^0(\mathbf{p}^0) = \begin{vmatrix} 8.42072 & -1637.87 \\ -1637.87 & 296{,}056 \end{vmatrix} \tag{11-87}$$

Hence, (11-81) now gives the linear equations

$$8.42072\, \Delta p_1^0 - 1637.87\, \Delta p_2^0 = 1.22033$$
$$-1637.87\, \Delta p_1^0 + 296{,}056\, \Delta p_2^0 = -236.924 \tag{11-88}$$

which can be solved to get

$$\Delta p_1^0 \approx 0.1411524 \qquad \Delta p_2^0 \approx -0.0000193706 \tag{11-89}$$

Hence
$$\mathbf{p}^1 = \mathbf{p}^0 + \Delta\mathbf{p}^0 \approx \begin{vmatrix} 2.8 \\ 0.028 \end{vmatrix} + \begin{vmatrix} 0.141152 \\ -1.937 \times 10^{-5} \end{vmatrix} \approx \begin{vmatrix} 2.941153 \\ 0.027981 \end{vmatrix} \tag{11-90}$$

and, from (11-25), $E(\mathbf{p}^1) \approx 0.03525$. Hence, the improvement is roughly comparable to that obtained using steepest descent. Note that neither linear search nor polynomial interpolation is needed to find the step size since (11-87) determines both the direction and the magnitude of $\Delta\mathbf{p}^0$.

From the limited experience gained from our two simple examples we can draw the following, generally valid, conclusions for the gradient methods described thus far:

1. Steepest descent combined with linear search is laborious, requires insight for choosing the step size of the search, and may zigzag if the initial point and/or the scaling is unlucky. Nonetheless, it guarantees by its very nature some improvement in each iteration.
2. Steepest descent with the step size obtained by polynomial interpolation is performed relatively easily. If the surface is well behaved, moreover, it will also be effective, although it is also subject to zigzagging. However, for badly conditioned surfaces it may give wildly wrong results, in which case linear search must be resorted to.
3. Newton's method is excellent for quadratic surfaces or for surfaces which are nearly quadratic in the sense that third and higher derivatives of E with respect to the p_i are small and $\mathbf{H}^j(\mathbf{p})$ is thus nearly constant. No zigzagging will be present in such a case, and the convergence will be quite fast. For badly condi-

tioned surfaces, however, Newton's method may also give a grossly erratic result and thus may lead to an increase, rather than decrease, of E. In addition, this method requires the calculation of the Hessian matrix $\mathbf{H}^j$ and the solution of a linear system of equations or the inversion of $\mathbf{H}^j$ in each iteration. This becomes quite a time-consuming task for larger n values (say, $n > 5$).

Performing linear search in the direction given by the Newton equation (11-80) (not discussed here) usually gives excellent results. The computational effort, however, now becomes quite substantial for each iteration.

In the vicinity of their minima, even ill-conditioned error surfaces tend to exhibit quadratic behavior. This observation can be used in developing hybrid techniques which employ steepest descent with linear search until the minimum is approached and then switch to Newton's method to take advantage of its faster convergence in the near-quadratic region.

A more sophisticated version of the hybrid approach is used in the popular *Fletcher-Powell process*. There the transition from the steepest-descent or linear search to Newton's method is gradual. In this technique, Eq. (11-81) is used, with an approximating matrix which changes from the unit matrix to the true inverse Hessian in a continuous manner.[4]

11-6 THE LEAST-pTH TAYLOR METHOD

As illustrated in Sec. 11-3, the calculation of the gradient $\mathbf{V}E^j(\mathbf{p}^j)$ can be accomplished accurately and relatively simply using the adjoint-network method. The calculation of the Hessian matrix $\mathbf{H}^j(\mathbf{p}^j)$, on the other hand, is a tedious task even when the adjoint-network concept is involved. Hence, it is useful to develop a process with most of the advantages of Newton's method without having to calculate the second derivatives of the response. For the important special case of the least-pth error criterion defined in Eq. (11-18), this can be achieved relatively simply.

When the simplified notation

$$e_i^j \triangleq e^j(\omega_i, \mathbf{p}^j) \triangleq w(\omega_i)[F(\omega_i) - F^j(\omega_i, \mathbf{p}^j)] \tag{11-91}$$

is introduced, (11-18) becomes simply

$$E_p^j = \sum_{i=1}^{m} |e_i^j|^p \tag{11-92}$$

If w, F, and F^j are all real and p is restricted to be an even integer, we can write

$$E_p^j = \sum_{i=1}^{m} (e_i^j)^p \tag{11-93}$$

and the kth component of $\mathbf{V}E^j$ is

$$\frac{\partial E_p^j}{\partial p_k} = p \sum_{i=1}^{m} (e_i^j)^{p-1} \frac{\partial e_i^j}{\partial p_k} \tag{11-94}$$

while the k, l component of the Hessian matrix becomes

$$\frac{\partial^2 E_p^j}{\partial p_k\,\partial p_l} = p\sum_{i=1}^{m}\left[(p-1)(e_i^j)^{p-2}\frac{\partial e_i^j}{\partial p_k}\frac{\partial e_i^j}{\partial p_l} + (e_i^j)^{p-1}\frac{\partial^2 e_i^j}{\partial p_k\,\partial p_l}\right] \quad (11\text{-}95)$$

The crucial step of the method discussed in this section is dropping the terms on the right-hand side of (11-95) which contain $\partial^2 e_i^j/(\partial p_k\,\partial p_l)$. The motivation is, of course, to avoid the difficult calculations needed to find the second derivatives. In addition, however, for circuits whose response does not vary too rapidly with changes in the element values, these terms are indeed much smaller than the corresponding terms containing first derivatives. Furthermore, the approximate Hessian matrix $\mathbf{H}^j$ obtained by dropping the second-derivative terms in (11-95) can be shown (Probs. 11-12 and 11-13) to be a symmetric positive definite matrix. This makes the numerical computations much more convenient.

Substituting (11-94) and the simplified form of (11-95) into Newton's relation (11-80) gives

$$\sum_{i=1}^{m}(e_i^j)^{p-1}\frac{\partial e_i^j}{\partial p_l} + (p-1)\sum_{k=1}^{n}\left[\sum_{i=1}^{m}(e_i^j)^{p-2}\frac{\partial e_i^j}{\partial p_k}\frac{\partial e_i^j}{\partial p_l}\right]\Delta p_k^j = 0$$
$$l = 1, 2\ldots, n \quad (11\text{-}96)$$

It is expedient at this stage to define the vector

$$\varepsilon^j \triangleq [(e_1^j)^{p-1} \quad (e_2^j)^{p-1} \quad \cdots \quad (e_m^j)^{p-1}]^T \quad (11\text{-}97)$$

as well as the matrices

$$\mathbf{A}^j \triangleq -\begin{bmatrix} \dfrac{\partial e_1^j}{\partial p_1^j} & \dfrac{\partial e_1^j}{\partial p_2^j} & \cdots & \dfrac{\partial e_1^j}{\partial p_n^j} \\[2mm] \dfrac{\partial e_2^j}{\partial p_1^j} & \dfrac{\partial e_2^j}{\partial p_2^j} & \cdots & \dfrac{\partial e_2^j}{\partial p_n^j} \\[2mm] \cdots\cdots\cdots\cdots\cdots\cdots \\[2mm] \dfrac{\partial e_m^j}{\partial p_1^j} & \dfrac{\partial e_m^j}{\partial p_2^j} & \cdots & \dfrac{\partial e_m^j}{\partial p_n^j} \end{bmatrix} \quad (11\text{-}98)$$

and

$$\mathbf{B}^j = \begin{bmatrix} (e_1^j)^{p-2} & 0 & 0 & \cdots & 0 \\ 0 & (e_2^j)^{p-2} & 0 & \cdots & 0 \\ \cdots\cdots\cdots\cdots\cdots\cdots \\ 0 & 0 & 0 & \cdots & (e_m^j)^{p-2} \end{bmatrix} \quad (11\text{-}99)$$

We note that ε^j and $\mathbf{B}^j$ contain simply the errors e_i^j at the m sample points, while $\mathbf{A}^j$ contains the first partial derivatives of the actual frequency response $F^j(\omega_i, \mathbf{p}^j)$; the k, l element of $\mathbf{A}^j$ is

$$-\frac{\partial e_k^j}{\partial p_l^j} = w(\omega_k)\frac{\partial F^j(\omega_k, \mathbf{p}^j)}{\partial p_l^j} \quad (11\text{-}100)$$

Hence, $\mathbf{A}^j$ can readily be obtained by analyzing the jth network N^j and its adjoint $\hat{N}^j$ at all m sample frequencies.

In terms ε^j, $\mathbf{A}^j$, and $\mathbf{B}^j$, from (11-94),

$$\mathbf{V}E^j(\mathbf{p}^j) = -p\mathbf{A}^{jT}\varepsilon^j \tag{11-101}$$

while from (11-95), neglecting the second-derivative term,

$$\mathbf{H}^j(\mathbf{p}^j) \approx p(p-1)\mathbf{A}^{jT}\mathbf{B}^j\mathbf{A}^j \tag{11-102}$$

Hence, the Newton relation (11-81) becomes

$$(p-1)\mathbf{A}^{jT}\mathbf{B}^j\mathbf{A}^j \, \Delta\mathbf{p}^j = \mathbf{A}^{jT}\varepsilon^j \tag{11-103}$$

Equation (11-103) is, of course, equivalent to Eq. (11-96), as can be verified by carrying out the matrix and vector multiplications indicated in (11-103).

In the jth iteration of the least-pth Taylor method, therefore, Eq. (11-103) or, equivalently, (11-96) is solved for the parameter increments Δp_k^j. Then $\mathbf{p}^j$ is incremented to obtain $\mathbf{p}^{j+1} = \mathbf{p}^j + \Delta\mathbf{p}^j$, the new $\mathbf{A}^{j+1}$, $\mathbf{B}^{j+1}$, and ε^{j+1} found, and the process repeated. The iteration proceeds until the errors e_i^j all become very small; then, by (11-97), $\varepsilon^j \approx \mathbf{0}$, and hence also the $\Delta\mathbf{p}^j$ found from (11-103) becomes nearly zero. This indicates convergence.

Example 11-12 For the amplifier-equalizer,

$$F^j(\omega_k, \mathbf{p}^j) = 10 \log [V_0(j\omega_k)V_0^*(j\omega_k)] \tag{11-104}$$

and hence the k, l component of $\mathbf{A}^j$ is

$$-\frac{\partial e_k^j}{\partial p_l^j} = \frac{\partial F^j(\omega_k, \mathbf{p}^j)}{\partial p_l^j} = \frac{10 \log e}{|V_0(j\omega_k)|^2} \frac{\partial [V_0(j\omega_k)V_0^*(j\omega_k)]}{\partial p_l^j}$$

$$= \frac{10 \log e}{|V_0(j\omega_k)|^2} \left[V_0(j\omega_k)\frac{V_0^*(j\omega_k)}{\partial p_l^j} + V_0^*(j\omega_k)\frac{V_0(j\omega_k)}{\partial p_l^j} \right]$$

$$= (20 \log e) \, \mathrm{Re} \left[\frac{\partial V_0(j\omega_k)/\partial p_l^j}{V_0(j\omega_k)} \right] \tag{11-105}$$

Here, $k = 1, 2, \ldots, 5$ and $l = 1, 2$. Substituting for V_0 and $\partial V_0/\partial p_l$ from Table 11-4, we obtain the $\mathbf{A}^0$ matrix

$$\mathbf{A}^0 = \begin{bmatrix} 0.605755 & -249.635 \\ 0.852818 & -224.929 \\ 1.095538 & -200.6565 \\ 1.322492 & -177.9611 \\ 1.527932 & -157.4172 \end{bmatrix}$$

ε^j and $\mathbf{B}^j$ can be obtained from (11-97) and (11-99). The numerical values of the e_k^0 have already been tabulated in Table 11-1; substitution gives

$$\varepsilon^0 = [4.9645 \times 10^{-5} \quad 0.090423 \quad 0.15447 \quad 0.04444 \quad 6.522 \times 10^{-6}]^T$$

$$\mathbf{B}^0 = \begin{bmatrix} 0.00135078 & 0 & 0 & 0 & 0 \\ 0 & 0.201458 & 0 & 0 & 0 \\ 0 & 0 & 0.28789 & 0 & 0 \\ 0 & 0 & 0 & 0.12546 & 0 \\ 0 & 0 & 0 & 0 & 0.00034909 \end{bmatrix}$$

Table 11-9

Iteration j	$p_1^j \equiv r$	$p_2^j \equiv c$	$E^j(\mathbf{p}^j)$
0	2.8	0.028	0.139208
1	2.82586	0.027368	$3.16543E-2$
2	2.89117	$2.73396E-2$	$2.02336E-2$
3	2.8841	$2.72736E-2$	$2.01255E-2$
4	2.88423	$2.72747E-2$	$2.01254E-2$
5	2.88423	$2.72747E-2$	$2.01254E-2$

Therefore, from (11-102),

$$\mathbf{H}^0(\mathbf{p}^0) \approx 12\mathbf{A}^{0T}\mathbf{B}^0\mathbf{A}^0 = \begin{vmatrix} 8.55342 & -1580.95 \\ -1580.95 & 310,198 \end{vmatrix}$$

This can be compared with the exact Hessian matrix

$$\mathbf{H}^0(\mathbf{p}^0) = \begin{vmatrix} 8.42072 & -1637.87 \\ -1637.87 & 296,056 \end{vmatrix}$$

obtained earlier; clearly, all matrix elements are accurate to within 5 percent, illustrating the validity of the approximation made in (11-102).

Next, setting up and solving the linear equations (11-103) for the parameter increment vector gives

$$\Delta\mathbf{p}^0 = \begin{vmatrix} 0.02586 \\ 0.000632 \end{vmatrix}$$

Hence, $\mathbf{p}^1 = [2.82586 \quad 0.027368]^T$. Evaluating E^1 from (11-25), we get $E^1 = 0.0316543$. Since $E^0 = 0.139208$, this represents a substantial improvement. Continuing the iteration, we obtain the results shown in Table 11-9.

Clearly, the process converges rapidly to the optimum. The final response is shown in Fig. 11-16. A comparison with the response of Fig. 11-4 (obtained by matching the desired response at 1 and 2 MHz) shows that there is significant improvement in the performance: the maximum error decreased from 0.54 dB to about 0.3 dB.

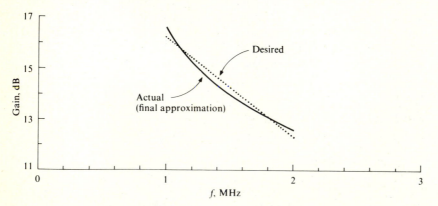

Figure 11-16 The frequency response of the optimized amplifier-equalizer.

At this stage, it appears that the least-pth Taylor method is a nearly ideal design tool; it requires only first derivatives $\partial F^j/\partial p_k$ which are readily obtained using the adjoint-network method; it is fast and yet has excellent convergence properties, as Example 11-12 illustrated. To demonstrate an important shortcoming of the method, we repeat the amplifier-equalizer optimization from a different and poorer initial approximation.

Example 11-13 Assume that instead of matching the desired response at 1 and 2 MHz, we obtain our initial values by choosing arbitrarily $r = 1$ and $c = 0.67$. A calculation similar to the preceding one gives then

$$
\mathbf{A}^0(\mathbf{p}^0) = \begin{bmatrix} 8.22195 & -0.692453 \\ 8.38314 & -0.451859 \\ 8.47339 & -0.317169 \\ 8.52874 & -0.234544 \\ 8.56506 & -0.180338 \end{bmatrix} \qquad \mathbf{\varepsilon}^0(\mathbf{p}^0) = \begin{bmatrix} 3915.63 \\ 3272.06 \\ 2681.23 \\ 2157.04 \\ 1701.85 \end{bmatrix}
$$

and

$$
\mathbf{B}^0(\mathbf{p}^0) = \begin{bmatrix} 248.428 & 0 & 0 & 0 & 0 \\ 0 & 220.401 & 0 & 0 & 0 \\ 0 & 0 & 192.999 & 0 & 0 \\ 0 & 0 & 0 & 166.945 & 0 \\ 0 & 0 & 0 & 0 & 142.544 \end{bmatrix}
$$

Setting up and solving Eq. (11-103) gives now

$$
\Delta\mathbf{p}^0 = \begin{bmatrix} 0.436575 \\ -2.53711 \end{bmatrix}
$$

Hence

$$
\mathbf{p}^1 = \mathbf{p}^0 + \Delta\mathbf{p}^0 = \begin{bmatrix} 1.14366 \\ -1.86711 \end{bmatrix}
$$

A danger signal here is the large value of Δp_2^0 and the resulting negative value of $p_2^1 = c$. Clearly, the process noted the very large value of p_2^0 (about 25 times the optimum value) and tried to decrease it drastically. In so doing, it overshot the goal and drove p_2^1 negative.

Proceeding with the calculation, we obtain the results shown in Table 11-10. Clearly, even though c varies in the right direction in each iteration, the values assumed by it oscillate with an increasing amplitude until overflow sets in. Thus, the process does not tolerate a poor initial approximation.

Table 11-10

Iteration j	$p_1^j \equiv r$	$p_2^j \equiv c$	$E^j(\mathbf{p}^j)$
0	1	0.67	195,731
1	1.4366	−1.867	76,224
2	1.898	95.12	29,583
3	2.337	-2.19×10^7	12,305
4		OVERFLOW	

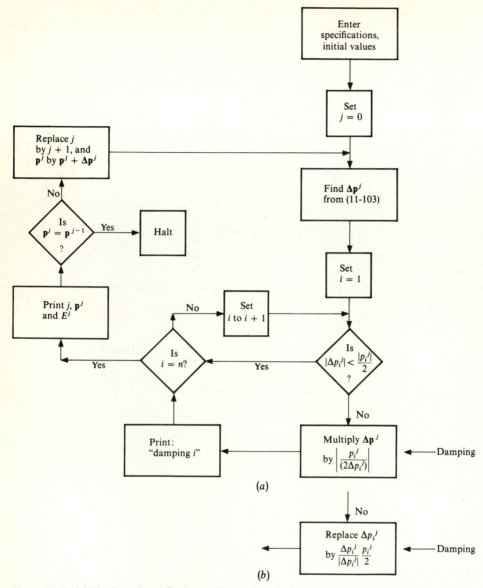

Figure 11-17 (a) The flow chart of a damped least-pth optimization process; (b) a simplified damping process.

To make this otherwise useful method more immune to divergence caused by bad initial values, a modification called *damping* can be implemented. In its simplest form, this can be achieved by reducing the size of $\Delta \mathbf{p}^j$, that is, by multiplying $\Delta \mathbf{p}^j$ by a scalar λ, where $0 < \lambda < 1$. If λ is sufficiently small, oscillation is prevented. Furthermore, it can be shown (see Prob. 11-14) that if $\lambda \to 0$, then $E(\mathbf{p}^j + \lambda \, \Delta \mathbf{p}^j) < E(\mathbf{p}^j)$ holds, provided $\Delta \mathbf{p}^j$ was obtained from (11-103). This is often expressed by saying that $\Delta \mathbf{p}^j$ points in the *downhill direction* on the error surface. Thus restricting the size but keeping the direction, i.e., the ratio of the components, of $\Delta \mathbf{p}^j$ results in a decreasing value of E^j, and also prevents large variations of the $\Delta \mathbf{p}^j$.

A simple damped optimization process therefore consists of the following steps:

1. $\Delta \mathbf{p}^j$ is found from Eq. (11-103).

Table 11-11

Iteration j	p_1^j	p_2^j	E^j
0	100	100	2.32547E + 6
DAMPING 1			
DAMPING 2			
1	50	50	820694.
DAMPING 1			
DAMPING 2			
2	25	25	201554
DAMPING 1			
DAMPING 2			
3	12.5	12.5	23669.4
DAMPING 2			
4	7.28228	6.25	1220.6
DAMPING 2			
5	5.42563	3.125	63.6175
DAMPING 2			
6	4.62996	1.5625	63.3766
DAMPING 2			
7	4.28685	0.78125	159.648
DAMPING 2			
8	4.24832	0.390625	170.914
DAMPING 2			
9	4.23035	0.195312	156.226
DAMPING 2			
10	4.20922	9.76562E − 2	94.2817
DAMPING 2			
11	4.14349	4.88281E − 2	9.05246
12	3.84559	3.46884E − 2	1.20578
13	3.50045	3.07937E − 2	0.369112
14	3.18007	0.028128	0.118496
15	3.03345	2.78271E − 2	3.37247E − 2
16	2.92275	2.73889E − 2	2.10625E − 2
17	2.8883	2.72897E − 2	2.01331E − 2
18	2.88432	2.72751E − 2	2.01254E − 2
19	2.88423	2.72747E − 2	2.01254E − 2
20	2.88423	2.72747E − 2	2.01254E − 2

2. All components Δp_i^j are tested against their specified limits. If none violates its limit, $\mathbf{p}^j$ is changed to $\mathbf{p}^j + \Delta\mathbf{p}^j$. Then, if the optimum has not been reached yet, the process returns to step 1.
3. If one (or more) of the components is larger than its limit, all components are reduced in size by the same factor until they are all within their limits.

Example 11-14 We shall illustrate the process by carrying out the damped least-fourth optimization of the amplifier-equalizer circuit from the initial values $p_1^0 = r^0 = 100$, $p_2^0 = c^0 = 100$. We know by our previous experience that these values represent a very poor initial approximation and would certainly lead to divergence if damping were not present.

The flow chart of the damping process is illustrated in Fig. 11-17a. As the figure illustrates, all components Δp_i^j are compared with $p_i^j/2$ in size; if any of them is larger, all components are multiplied by $|p_i^j/2 \Delta p_i^j|$ to decrease the size of $\Delta\mathbf{p}^j$, and the message "Damping i" is printed for the designer's information.

The program becomes somewhat simpler (and nearly as efficient) if the damping is performed by cutting the size of any Δp_i greater in modulus than $|p_i/2|$ back to $|p_i/2|$ (Fig. 11-17b). This damping method was programmed and used. The resulting output for our example is shown in Table 11-11.

As the table illustrates, both components needed damping for the first three iterations, while only $p_2^j \equiv c^j$ needed it for the next eight cycles. After $j = 10$ has been reached, both components were within a factor of 2 of their optimum values, and the process converged in 10 more iterations without any further damping.

It should be noted that the described damping method is not the only one available. A more sophisticated (and more useful) technique is due to Levenberg.[4,5]

Also, more involved schemes (such as the well-known *Fletcher-Powell* and *conjugate-gradient* methods) can be used to approximate the Hessian matrix using first-derivative data. Again, the reader is referred to the literature[4-6] for further information on these optimization methods.

11-7 THE LINEAR LEAST-SQUARES PROBLEM

An important special case of the least-pth approximation occurs when $p = 2$ and the actual response F^j is a linear function of the parameters p_i. Then

$$F^j(\omega, \mathbf{p}) = \sum_{i=1}^{n} p_i f_i(\omega) \tag{11-106}$$

and

$$E \triangleq \sum_{h=1}^{m} \left\{ w(\omega_h) \left[F(\omega_h) - \sum_{i=1}^{n} p_i f_i(\omega_h) \right] \right\}^2 \tag{11-107}$$

The minimum of E is obtained for

$$\frac{\partial E}{\partial p_k} = 0 \qquad k = 1, 2, \ldots, n \tag{11-108}$$

From (11-107) and (11-108), the linear system of equations

$$p_1 \sum_{h=1}^{m} [w(\omega_h)]^2 f_1(\omega_h) f_k(\omega_h) + p_2 \sum_{h=1}^{m} [w(\omega_h)]^2 f_2(\omega_h) f_k(\omega_h) + \cdots$$

$$+ p_n \sum_{h=1}^{m} [w(\omega_h)]^2 f_n(\omega_h) f_k(\omega_h) = \sum_{h=1}^{m} [w(\omega_h)]^2 F(\omega_h) f_k(\omega_h) \qquad k = 1, 2 \ldots, n$$

$$(11\text{-}109)$$

is obtained. This can be solved directly, *without* approximation or iteration, for the p_i giving the least-squares optimum.

In circuit realization, the response is never a linear function of the element values as (11-106) postulates. However, in solving the approximation problem, i.e., finding an appropriate network function, the form (11-106) often occurs.

Example 11-15 Find the impulse response of an *RC* circuit so as to approximate a specified time response

$$F(t) = 3 - \frac{t}{3}$$

in the range $0 \le t \le 9$. A least-squares approximation should be obtained. The natural modes of the circuit should be $s_1 = -\frac{1}{2}$ and $s_2 = -\frac{1}{5}$.

The actual response is in the form

$$F^j(t, p_1, p_2) = p_1 e^{-t/2} + p_2 e^{-t/5}$$

Choosing $m = 20$ sample points equidistantly in the $0 \le t \le 9$ range, we have

$$t_h = \frac{(h-1)9}{19} \qquad h = 1, 2, \ldots, 20$$

Hence, (11-109) gives now

$$p_1 \sum_{h=1}^{20} (e^{-t_h/2})^2 + p_2 \sum_{h=1}^{20} e^{-t_h/2} e^{-t_h/5} = \sum_{h=1}^{20} e^{-t_h/2} \left(3 - \frac{t_h}{3}\right)$$

$$p_1 \sum_{h=1}^{20} e^{-t_h/5} e^{-t_h/2} + p_2 \sum_{h=1}^{20} (e^{-t_h/5})^2 = \sum_{h=1}^{20} e^{-t_h/5} \left(3 - \frac{t_h}{3}\right)$$

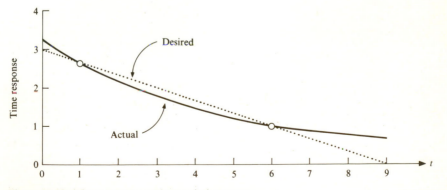

Figure 11-18 A least-squares optimum time response.

Carrying out the summations, we see that these equations become

$$2.65023p_1 + 3.53878p_2 = 11.4553 \qquad 3.53878p_1 + 5.66258p_2 = 18.5168$$

which can be solved to give

$$p_1 = -0.265905 \qquad p_2 = 3.4362$$

The optimized response and the desired one are compared in Fig. 11-18.

Example 11-16 Find the least-squares approximation of the desired frequency response $\exp\sqrt{\omega}$ with the actual response

$$F^j(\omega) = p_1 + p_2\omega^2 + p_3\omega^4$$

in the range $0 \le \omega \le 1$.

If we choose 10 sample points at

$$\omega_h = \frac{h-1}{9} \qquad h = 1, 2, \ldots, 10$$

Eq. (11-109) gives

$$p_1 \sum_{h=1}^{10} \omega_h^0 + p_2 \sum_{h=1}^{10} \omega_h^2 + p_3 \sum_{h=1}^{10} \omega_h^4 = \sum_{h=1}^{10} \exp\sqrt{\omega_h}$$

$$p_1 \sum_{h=1}^{10} \omega_h^2 + p_2 \sum_{h=1}^{10} \omega_h^4 + p_3 \sum_{h=1}^{10} \omega_h^6 = \sum_{h=1}^{10} \exp\left(\sqrt{\omega_h}\right)\omega_h^2$$

$$p_1 \sum_{h=1}^{10} \omega_h^4 + p_2 \sum_{h=1}^{10} \omega_h^6 + p_3 \sum_{h=1}^{10} \omega_h^8 = \sum_{h=1}^{10} \exp\left(\sqrt{\omega_h}\right)\omega_h^4$$

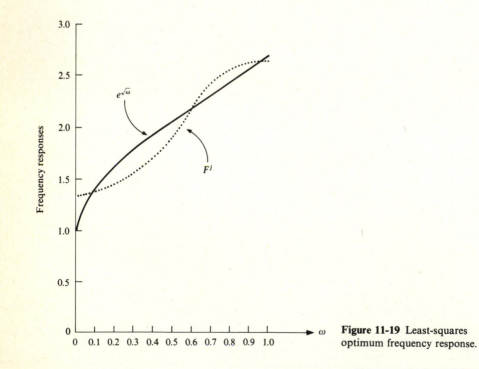

Figure 11-19 Least-squares optimum frequency response.

Table 11-12

ω	$\exp \sqrt{\omega}$	F^j	$\exp(\sqrt{\omega}) - F^j$
0.00	0.1000000E+01	0.1325458E+01	−0.3254576
0.05	0.1250579E+01	0.1332779E+01	−0.8220005E−01
0.10	0.1371943E+01	0.1354623E+01	0.1731968E−01
0.15	0.1472996E+01	0.1390626E+01	0.8236980E−01
0.20	0.1563948E+01	0.1440182E+01	0.1237659
0.25	0.1648721E+01	0.1502442E+01	0.1462784
0.30	0.1729309E+01	0.1576317E+01	0.1529922
0.35	0.1806890E+01	0.1660473E+01	0.1464176
0.40	0.1882226E+01	0.1753334E+01	0.1288919
0.45	0.1955841E+01	0.1853084E+01	0.1027575
0.50	0.2028114E+01	0.1957662E+01	0.7045269E−01
0.55	0.2099333E+01	0.2064766E+01	0.3456688E−01
0.60	0.2169716E+01	0.2171850E+01	−0.2134323E−02
0.65	0.2239439E+01	0.2276129E+01	−0.3668976E−01
0.70	0.2308642E+01	0.2374573E+01	−0.6593037E−01
0.75	0.2377442E+01	0.2463903E+01	−0.8646679E−01
0.80	0.2445934E+01	0.2540625E+01	−0.9469032E−01
0.85	0.2514199E+01	0.2600961E+01	−0.8676147E−01
0.90	0.2582307E+01	0.2640922E+01	−0.5861473E−01
0.95	0.2650317E+01	0.2656263E+01	−0.5946159E−02
1.00	0.2718282E+01	0.2642505E+01	0.7577705E−01

When the summations are evaluated the linear system

$$\begin{bmatrix} 0.1000000E+02 & 0.3518517E+01 & 0.2336990E+01 \\ 0.3518517E+01 & 0.2336990E+01 & 0.1841041E+01 \\ 0.2336990E+01 & 0.1841041E+01 & 0.1573437E+01 \end{bmatrix} \begin{bmatrix} p_1 \\ p_2 \\ p_3 \end{bmatrix} = \begin{bmatrix} 0.1979762E+02 \\ 0.8542876E+01 \\ 0.5954686E+01 \end{bmatrix}$$

is obtained. This can be solved for p_1, p_2, and p_3, giving

$$p_1 = 0.1325458E+01 \qquad p_2 = 0.2932744E+01 \qquad p_3 = -0.1615697E+01$$

The optimized response is compared with the desired values in Table 11-12 and in Fig. 11-19.

Evidently, a good approximation is obtained. Except near the frequency origin, the relative error is less than 10 percent over the whole range.

11-8 POINT MATCHING AND COEFFICIENT MATCHING

An obvious way of achieving a reasonably good approximation to a specified response is to force the actual circuit response to coincide with it at some selected points. This was the technique used in Sec. 11-1 to obtain the response shown in Fig. 11-4. To generalize the process, assume that we prescribe n points ω_1,

$\omega_2, \ldots, \omega_n$, where the desired response $F(\omega)$ and the actual response $F^j(\omega, \mathbf{p})$ must match. Then, the n equations

$$F^j(\omega_1, p_1, p_2, \ldots, p_n) = F(\omega_1)$$

$$F^j(\omega_2, p_1, p_2, \ldots, p_n) = F(\omega_2) \tag{11-110}$$

$$\cdots\cdots\cdots\cdots\cdots\cdots\cdots$$

$$F^j(\omega_n, p_1, p_2, \ldots, p_n) = F(\omega_n)$$

are to be solved for the parameters p_i. This is especially simple if $F^j(\omega, \mathbf{p})$ is a polynomial or a rational function in ω, with p_i being the coefficients. Then (11-110) results in a linear system of equations in the unknowns p_i. Note that the solution needs only to be found *once*: the method is not iterative.

Example 11-17 Approximate $F(\omega) = e^{\omega^2}$ with the rational function

$$F^j(\omega, \mathbf{p}) = \frac{1 + p_1\omega^2 + p_2\omega^4}{p_3 + p_4\omega^2}$$

in the range $0 \le \omega \le 1$.

Since $n = 4$ parameters p_i exist, $F(\omega)$ can be matched at four frequencies. Now (11-110) gives

$$\frac{1 + p_1\omega_i^2 + p_2\omega_i^4}{p_3 + p_4\omega_i^2} = e^{\omega_i^2} \qquad i = 1, 2, 3, 4$$

This can be rewritten in the form

$$\omega_i^2 p_1 + \omega_i^4 p_2 - e^{\omega_i^2}p_3 - e^{\omega_i^2}\omega_i^2 p_4 = -1 \qquad i = 1, 2, 3, 4$$

which is a linear set of equations in the unknowns p_1, p_2, p_3, and p_4.

When the matching frequencies are chosen at $\omega_1 = 0.2$, $\omega_2 = 0.4$, $\omega_3 = 0.6$, and $\omega_4 = 0.8$, the linear equations turn out to be

$$0.04p_1 + 0.0016p_2 - 1.04081p_3 - 0.041632p_4 = -1$$

$$0.16p_1 + 0.0256p_2 - 1.17351p_3 - 0.187762p_4 = -1$$

$$0.36p_1 + 0.1296p_2 - 1.43333p_3 - 0.515999p_4 = -1$$

$$0.64p_1 + 0.4096p_2 - 1.89648p_3 - 1.21375p_4 = -1$$

These can be solved to give

$$\mathbf{p} = \begin{bmatrix} 0.697131 \\ 0.204213 \\ 0.999978 \\ 0.302064 \end{bmatrix}$$

and the approximating function is therefore

$$F^j(\omega, \mathbf{p}) = \frac{1 + 0.697131\omega^2 + 0.204213\omega^4}{0.999978 - 0.302064\omega^2}$$

The quality of the approximation is illustrated by Table 11-13.

Table 11-13

ω	$F(\omega)$	$F^j(\omega, \mathbf{p})$	$F(\omega) - F^j(\omega, \mathbf{p})$
0	1	1.00002	$-2.22567E-05$
0.1	1.01005	1.01007	$-1.50818E-05$
0.2	1.04081	1.04081	$3.72529E-09$
0.3	1.09417	1.09417	$8.36141E-06$
0.4	1.17351	1.17351	$1.86265E-08$
0.5	1.28403	1.28404	$-1.45677E-05$
0.6	1.43333	1.43333	$8.56817E-08$
0.7	1.63232	1.63225	$6.25290E-05$
0.8	1.89648	1.89648	$3.87430E-07$
0.9	2.24791	2.24897	$-1.06078E-03$
1	2.71828	2.72432	$-6.04163E-03$

The match is very good over the whole $0 \le \omega \le 1$ range. This kind of optimization problem often arises in the design of loss equalizers, like the one discussed in Sec. 1-2, Eqs. (1-11) to (1-13).

In general, Eq. (11-110) represents a system of *nonlinear* equations in the unknowns p_i. Then an iterative process, also called *Newton's technique,*† can be used.

Assuming that a good initial approximation $p_1^0, p_2^0, \ldots, p_n^0$ is available, a truncated Taylor-series expansion of (11-110) around the initial values gives

$$F^0(\omega_k, \mathbf{p}^0) + \sum_{i=1}^{n} \frac{\partial F^0(\omega_k, \mathbf{p}^0)}{\partial p_i} \Delta p_i^0 = F(\omega_k) \qquad k = 1, 2, \ldots, n \quad (11\text{-}111)$$

which can be solved for $\Delta p_1^0, \Delta p_2^0, \ldots, \Delta p_n^0$. Then $\mathbf{p}^1 = \mathbf{p}^0 + \Delta \mathbf{p}^0$ is found and the process repeated. The general equation is

$$F^j(\omega_k, \mathbf{p}^j) + \sum_{i=1}^{n} \frac{\partial F^j(\omega_k, \mathbf{p}^j)}{\partial p_i^j} \Delta p_i^j = F(\omega_k)$$

$$k = 1, 2, \ldots, n \qquad j = 0, 1, 2 \ldots \qquad (11\text{-}112)$$

If the vectors

$$\mathbf{F}^j \triangleq [F^j(\omega_1, \mathbf{p}^j) \quad F^j(\omega_2, \mathbf{p}^j) \quad \cdots \quad F^j(\omega_n, \mathbf{p}^j)]^T$$
$$\mathbf{F} \triangleq [F(\omega_1) \quad F(\omega_2) \quad \cdots \quad F(\omega_n)]^T$$
$$(11\text{-}113)$$

are defined, along with the *Jacobian matrix* $\mathbf{J}^j$, an $n \times n$ matrix with its k, l element given by

$$J_{kl}^j \triangleq \frac{\partial F^j(\omega_k, \mathbf{p}^j)}{\partial p_l^j} \qquad (11\text{-}114)$$

† Not to be confused with the optimization method described in Sec. 11-5.

Table 11-14

ITERATION #	p_1	p_2	p_3
0	0.6667	3	0.333

TIME	DESIRED	ACTUAL	ERROR
1	2.66667	2.62739	3.92737E−02
5	1.33333	1.19644	0.13689
8	0.333333	0.579202	−0.245868

ITERATION #	p_1	p_2	p_3
1	3.2164	1.75292	0.673611

TIME	DESIRED	ACTUAL	ERROR
1	2.66667	2.53368	0.132985
5	1.33333	0.614531	0.718802
8	0.333333	0.125521	0.207812

ITERATION #	p_1	p_2	p_3
2	5.5019	−0.947949	0.537535

TIME	DESIRED	ACTUAL	ERROR
1	2.66667	2.66035	6.31325E−03
5	1.33333	1.80722	−0.473882
8	0.333333	0.584195	−0.250861

ITERATION #	p_1	p_2	p_3
3	5.50546	−0.692515	0.592036

TIME	DESIRED	ACTUAL	ERROR
1	2.66667	2.66252	4.14531E−03
5	1.33333	1.3903	−5.69644E−02
8	0.333333	0.380239	−4.69052E−02

ITERATION #	p_1	p_2	p_3
4	6.1848	−1.22207	0.6216

TIME	DESIRED	ACTUAL	ERROR
1	2.66667	2.66541	1.26158E−03
5	1.33333	1.32739	5.94427E−03
8	0.333333	0.334115	−7.81736E−04

ITERATION #	p_1	p_2	p_3
5	6.31249	−1.33427	0.624247

TIME	DESIRED	ACTUAL	ERROR
1	2.66667	2.66665	1.26585E−05
5	1.33333	1.33314	1.93499E−04
8	0.333333	0.333276	5.75250E−05

ITERATION #	p_1	p_2	p_3
6	6.31335	−1.33514	0.624241

TIME	DESIRED	ACTUAL	ERROR
1	2.66667	2.66667	3.72529E−09
5	1.33333	1.33333	−1.49012E−08
8	0.333333	0.33333	−3.25963E−09

then (11-112) can be rewritten in the simple form

$$\mathbf{J}^j \, \Delta \mathbf{p}^j = \mathbf{F} - \mathbf{F}^j \tag{11-115}$$

This is a system of linear equations which is to be solved for $\Delta \mathbf{p}^j$. Then, $\mathbf{p}^{j+1} = \mathbf{p}^j + \Delta \mathbf{p}^j$ is found, and the process is repeated until it converges, i.e., until $\Delta \mathbf{p}^j \approx 0$. Convergence may in some cases require damping; then the formula $\mathbf{p}^{j+1} = \mathbf{p}^j + \lambda \, \Delta \mathbf{p}^j,\, 0 < \lambda < 1$ can be used.

Example 11-18 Match the time response $F(t) = 3 - t/3$ at $t_1 = 1$, $t_2 = 5$, and $t_3 = 8$ with a realizable circuit impulse response in the form $F^j(t, \mathbf{p}^j) = (p_1^j t + p_2^j)e^{-p_3^j t}$.

In the absence of the realizability requirement, the problem has a trivial solution with $p_1^j = -\frac{1}{3}$, $p_2^j = 3$, and $p_3^j = 0$. This, however, corresponds to the s-domain function $(9s - 1)/3s^2$, which is not realizable either as a driving-point or a transfer function. Hence, in this case realizability constraints force us to use optimization techniques to handle an otherwise exactly solvable problem.

Realizability requires that all p_i^j be real and, in addition, that $p_3^j > 0$. To find a good initial approximation, we note that for $t = 0$, $F(0) = 3$ while $F^j(0, \mathbf{p}^j) = p_2^j$. Hence, $p_2^0 = 3$ can be chosen. Next, since $dF(t)/dt = -\frac{1}{3}$ while $dF^j(t, \mathbf{p}^j)/dt = (-p_1^j p_3^j + p_1^j - p_2^j p_3^j)e^{-p_3^j t}$, equating the two derivatives for $t = 0$ gives $p_1^0 - p_2^0 p_3^0 = -\frac{1}{3}$, so that $p_1^0 - 3p_3^0 = -\frac{1}{3}$. Choosing arbitrarily $p_3^0 = \frac{1}{3}$ then gives $p_1^0 = \frac{2}{3}$.

The Jacobian matrix is now

$$\mathbf{J}(\mathbf{p}^0) = \begin{bmatrix} t_1 e^{-p_3^0 t_1} & e^{-p_3^0 t_1} & -t_1(p_1^0 t_1 + p_2^0)e^{-p_3^0 t_1} \\ t_2 e^{-p_3^0 t_2} & e^{-p_3^0 t_2} & -t_2(p_1^0 t_2 + p_2^0)e^{-p_3^0 t_2} \\ t_3 e^{-p_3^0 t_3} & e^{-p_3^0 t_3} & -t_3(p_1^0 t_3 + p_2^0)e^{-p_3^0 t_3} \end{bmatrix}$$

For the given values of t_1, t_2, and t_3 as well as the chosen $\mathbf{p}^0$, substitution gives

$$\mathbf{J}(\mathbf{p}^0) = \begin{bmatrix} 0.716555 & 0.716555 & -2.62739 \\ 0.944535 & 0.188907 & -5.982222 \\ 0.556016 & 0.069502 & -4.63361 \end{bmatrix}$$

Setting up and solving the linear equations (11-115) then gives

$$\Delta \mathbf{p}^0 = \begin{bmatrix} 2.5497 \\ -1.24708 \\ 0.340411 \end{bmatrix}$$

Carrying out the iteration (without damping) gives the computer printout shown in Table 11-14. Evidently, the convergence is fast, and no damping is needed. The resulting circuit response is

$$F^j(t) = (6.31335t - 1.33514)e^{-0.624241t}$$

which is a realizable function. The desired response $F(t)$ is compared with $F^j(t)$ in Fig. 11-20. The largest error occurs at $t = 0$, and it is intuitively obvious that by shifting the matching points toward smaller values of t this maximum error can be decreased. Changing the matching points, therefore, to $t_1 = 0.2$, $t_2 = 3$, and $t_3 = 7$ and repeating the Newton process gives the realizable function

$$F^j(t) = (1.37002t + 2.91561)e^{-0.418807t}$$

This is compared with $F(t)$ in Fig. 11-21.

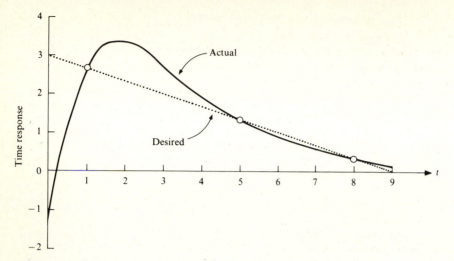

Figure 11-20 Time-response approximation by matching at $t = 1$, 5, and 8.

It is clear that by repeatedly adjusting the matching points and using Newton's process for obtaining the match we can decrease the maximum error. The approximation achieved when the maximum error is the smallest possible is characterized by

$$E_\infty^j \triangleq \max_{\omega_l \leq \omega \leq \omega_u} |w(\omega)[F(\omega) - F^j(\omega, \mathbf{p}^j)]| = \min \qquad (11\text{-}116)$$

This is the min-max error approximant mentioned in Sec. 11-2, Eqs. (11-14) and (11-15). In Sec. 11-9, an iterative method, which is more efficient than repeated point matching, will be described for obtaining the min-max approximation.

When the p_i^j are element values, the components of the Jacobian matrix, defined in (11-114), can be found efficiently using the adjoint-network method. The point-matching process is then performed in the following steps:

1. Suitable initial values p_i^0 are chosen for the element values.
2. The branch currents and/or voltages are found for the initial circuit and its adjoint network.

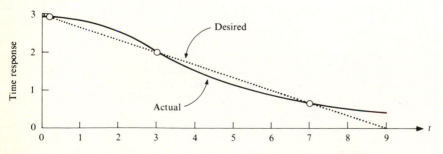

Figure 11-21 Time-response approximation by matching at $t = 0.2$, 3, and 7.

3. From the branch variables, the derivatives $\partial F^j(\omega_k, \mathbf{p}^j)/\partial p_i^j$ are found using the methods described in Chaps. 9 and 10.
4. Equations (11-115) are constructed and solved for $\Delta\mathbf{p}^j$.
5. $\mathbf{p}^j$ is replaced by $\mathbf{p}^{j+1} = \mathbf{p}^j + \Delta\mathbf{p}^j$, and (unless convergence is attained so that $\mathbf{p}^{j+1} \approx \mathbf{p}^j$) steps 1 to 4 are repeated.

Example 11-19 For the amplifier-equalizer, the desired response is

$$F(\Omega) = 20 - \frac{2}{\pi}\Omega \tag{11-117}$$

while the simplified expression for the actual response is

$$F^j(\Omega, r^j, c^j) = 10 \log \left(r^{j2} + \frac{1}{\Omega^2 c^{j2}}\right) \tag{11-118}$$

as given originally in Eqs. (11-5) and (11-7). A very good approximation can be obtained by matching $F(\Omega)$ to F^j at $\Omega_1 = 2.4\pi$ and $\Omega_2 = 3.6\pi$. In this simple example, the two nonlinear equations which result from (11-110) can actually be solved analytically, giving

$$c = \frac{1}{10}\sqrt{\frac{\Omega_1^{-2} - \Omega_2^{-2}}{10^{-\Omega_1/5\pi} - 10^{-\Omega_2/5\pi}}} \qquad r = 10\sqrt{\frac{\Omega_2^2 10^{-\Omega_2/5\pi} - \Omega_1^2 10^{-\Omega_1/5\pi}}{\Omega_2^2 - \Omega_1^2}} \tag{11-119}$$

which here result in

$$c = 0.026365321 \qquad r = 2.794244656$$

However, as an illustration of the general point-matching technique, the problem will be solved through repeated applications of (11-115), from the poor initial values $r^0 = c^0 = 1$. Straightforward application of (11-115) then leads to divergence and overflow in just four iterations. Hence, we must introduce damping, restricting the size of element increments by the condition $|\Delta p_i^j| \le |p_i^j/2|$, as in the flow chart of Fig. 11-17. This leads to the computer printout shown in Table 11-15. Evidently, damping is needed for the first five iterations; after that, the process converges smoothly.

Table 11-15

Iteration j	Δp_1^j	Δp_2^j	p_1^{j+1}	p_2^{j+1}
0	0.0209952	−0.50	1.02100	0.50
DAMPING 2				
1	0.0706629	−0.25	1.09166	0.25
DAMPING 2				
2	0.178960	−0.125	1.27062	0.125
DAMPING 2				
3	0.328148	−0.0625	1.59877	0.0625
DAMPING 2				
4	0.535757	−0.03125	2.13452	0.03125
DAMPING 2				
5	0.585459	−0.005337	2.71998	0.025913
6	0.0757897	0.0004439	2.79577	0.026357
7	−0.00150932	0.00000854	2.79426	0.026365
8	−0.00001464	0.00000006	2.79424	0.026365

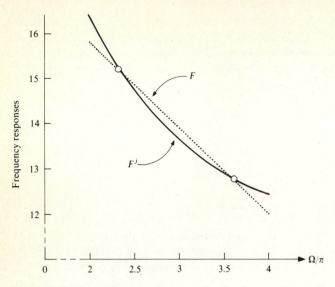

Figure 11-22 Frequency-response approximation by matching at $\Omega = 2.4\pi$ and 3.6π.

The resulting frequency response is shown in Fig. 11-22. The maximum error is obtained at $\Omega = 2\pi$, where $F - F^j \approx -0.459$ dB. The error at $\Omega = 3\pi$ is 0.197 dB, and at $\Omega = 4\pi$ it equals -0.283 dB.

For comparison, the iteration was also performed from the excellent starting point $r^0 = 2.8$ and $c^0 = 0.028$. It then converged without damping in just three iterations to the same result.

Closely related to the point-matching process is the realization technique known as *coefficient matching*.[7] Here, a transfer function $F(\omega)$ is prescribed as a rational function.

$$F(\omega) = \frac{a_0 + a_1\omega + a_2\omega^2 + \cdots + a_m\omega^m}{a_{m+1} + a_{m+2}\omega + a_{m+3}\omega^2 + \cdots + a_{m+k+1}\omega^k} \qquad (11\text{-}120)$$

(In practice, s rather than ω usually is the independent variable.) The actual transfer function after the jth iteration is in the form

$$F^j(\omega, p^j) = \frac{b_0^j + b_1^j\omega + b_2^j\omega^2 + \cdots + b_m^j\omega^m}{b_{m+1}^j + b_{m+2}^j\omega + b_{m+3}^j\omega^2 + \cdots + b_{m+k+1}^j\omega^k} \qquad (11\text{-}121)$$

where the coefficients b_i^j are known functions of the element values p_i^j. The two transfer functions will coincide if the equalities

$$b_0^j(p_1^j, p_2^j, \ldots, p_n^j) = Ka_0$$

$$b_1^j(p_1^j, p_2^j, \ldots, p_n^j) = Ka_1 \qquad (11\text{-}122)$$

$$\cdots\cdots\cdots\cdots\cdots\cdots\cdots\cdots$$

$$b_{m+k+1}(p_1^j, p_2^j, \ldots, p_n^j) = Ka_{m+k+1}$$

hold. Here, K is an arbitrary constant, which cancels in $F^j(\omega, \mathbf{p}^j)$.

Often, the number n of the circuit elements p_i^j equals $m + k + 1$. Then, the $m + k + 2$ equations (11-122) can be solved for the p_i^j and K. Since the relations between the b_i^j and the p_i^j are not linear, (11-122) represents a nonlinear system which can be solved using Newton's technique. For an initial approximation $\mathbf{p}^0$, Newton's technique applied to (11-122) gives

$$b_0^0(\mathbf{p}^0) + \sum_{i=1}^{n} \left(\frac{\partial b_0}{\partial p_i}\right)_{\mathbf{p}^0} \Delta p_i^0 = K^0 a_0$$

$$b_1^0(\mathbf{p}^0) + \sum_{i=1}^{n} \left(\frac{\partial b_1}{\partial p_i}\right)_{\mathbf{p}^0} \Delta p_i^0 = K^0 a_1 \qquad (11\text{-}123)$$

$$\cdots\cdots\cdots\cdots\cdots\cdots\cdots\cdots\cdots\cdots\cdots\cdots$$

$$b_{m+k+1}^0(\mathbf{p}^0) + \sum_{i=1}^{n} \left(\frac{\partial b_{m+k+1}}{\partial p_i}\right)_{\mathbf{p}^0} \Delta p_i^0 = K^0 a_{m+k+1}$$

Equation (11-123) is to be solved for the Δp_i^0 and K^0; then $\mathbf{p}^0$ is replaced by $\mathbf{p}^0 + \Delta\mathbf{p}^0$ and the process repeated until convergence is reached, i.e., until $\Delta\mathbf{p}^j \approx \mathbf{0}$ and (11-122) holds.

Sometimes $n \neq m + k + 1$, either because redundant elements are present $(n > m + k + 1)$ or some component values are prescribed $(n < m + k + 1)$. In the former case, we can postulate $n - (m + k + 1)$ relations between the element values; in the latter, we can define an error criterion of the form

$$E^j \triangleq \sum_{l=0}^{m+k+1} [b_l^j(\mathbf{p}^j) - K^j a_l]^2 \qquad (11\text{-}124)$$

and minimize E^j using either the least-pth Taylor process or some other gradient methods.

If the network functions $F(\omega)$ and $F^j(\omega)$ are polynomials rather than rational functions in ω (or s), K no longer cancels in $F^j(\omega, \mathbf{p}^j)$; hence, we must use $K = 1$.

Example 11-20 The circuit shown in Fig. 11-23 has the specified transfer function

$$\frac{V_1}{V_2} = 1.5s^3 + 3s^2 + 3s + 1.5 \qquad (11\text{-}125)$$

If $r = 0$, the circuit becomes a doubly terminated reactance two-port and the synthesis techniques described in Chap. 6 can be used to obtain the element values $R_1 = 1$, $C_1 = 1$, $C_2 = 0.5$, $L = 3$, and $R_2 = 2$. Assume now that the inductor is lossy. If its quality factor Q_L defined as $\omega L/r$ at $\omega = 1$ is 10, the losses can be represented by prescribing the relation

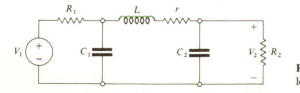

Figure 11-23 A low-pass filter with a lossy inductor.

Table 11-16

Iteration	L	C_1	C_2	R_2
0	3	1	0.5	2
1	3.51	0.9	0.465	2.54
2	3.63599	0.943663	0.435583	2.71598
3	3.64433	0.943069	0.436447	2.72881
4	3.64436	0.943081	0.436436	2.72887
5	3.64436	0.943081	0.436436	2.72887

$r = L/10$. Now our synthesis methods no longer apply, and optimization must be used to design the circuit. Since four coefficients must be matched, we can regard four element values, say C_1, C_2, L, and R_2, as variables. Then $R_1 = 1$ and $r = L/10$ determine the remaining element values.

Since the actual transfer function is now

$$\frac{V_1}{V_2} = R_1 LC_1 C_2 s^3 + \left[L\left(\frac{R_1}{R_2} C_1 + C_2\right) + rR_1 C_1 C_2 \right] s^2$$

$$+ \left[\frac{L}{R_2} + R_1(C_1 + C_2) + r\left(C_2 + \frac{R_1}{R_2} C_1\right) \right] s + 1 + \frac{R_1}{R_2} + \frac{r}{R_2} \qquad (11\text{-}126)$$

(11-123) gives, with $K^0 = 1$, the relations

$$R_1 LC_1 C_2 + R_1 C_1 C_2 \,\Delta L + R_1 LC_2 \,\Delta C_1 + R_1 LC_1 \,\Delta C_2 = 1.5$$

$$L\left(\frac{R_1}{R_2} C_1 + C_2 + 0.1R_1 C_1 C_2\right) + \left(\frac{R_1}{R_2} C_1 + C_2 + 0.1R_1 C_1 C_2\right) \Delta L$$

$$+ L\left(\frac{R_1}{R_2} + 0.1R_1 C_2\right) \Delta C_1 + L(1 + 0.1R_1 C_1) \,\Delta C_2 - L\frac{R_1 C_1}{R_2^2} \,\Delta R_2 = 3$$

$$\left[R_1(C_1 + C_2) + L\left(\frac{1}{R_2} + 0.1\frac{R_1}{R_2} C_1 + 0.1C_2\right) \right] + \left(\frac{1}{R_2} + 0.1\frac{R_1}{R_2} C_1 + 0.1C_2\right) \Delta L \qquad (11\text{-}127)$$

$$+ \left(R_1 + 0.1L\frac{R_1}{R_2}\right) \Delta C_1 + (R_1 + 0.1L) \,\Delta C_2 - L\frac{1 + 0.1R_1 C_1}{R_2^2} \,\Delta R_2 = 3$$

$$\left(1 + \frac{R_1}{R_2} + \frac{0.1\,L}{R_2}\right) + \frac{0.1}{R_2} \,\Delta L - \frac{R_1 + 0.1\,L}{R_2^2} \,\Delta R_2 = 1.5$$

where the relation $r = L/10$ was used to eliminate r.

When the lossless values $L = 3$, $C_1 = 1$, $C_2 = 0.5$, and $R_2 = 2$ are chosen as initial values, repeated solution of (11-127) leads to the iteration given in Table 11-16. In only five iterations, the desired solution is obtained. To demonstrate convergence from badly chosen initial values, next the initial values of C_1 and C_2 are interchanged. The results are then as shown in Table 11-17. The behavior of the element values is quite lively now during the iteration, but after a few iterations they settle down and in 12 iterations the correct values are obtained.

Table 11-17

Iteration	L	C_1	C_2	R_2
0	3	0.5	1	2
1	29.055	1.775	−10.235	6.47
2	22.3005	1.75103	−2.48848	6.4618
3	7.68056	1.67065	−1.70723	3.53534
4	4.92021	1.57697	−0.592401	2.98416
5	2.83807	1.58729	−5.34937E−02	2.5676
6	7.60952	−0.200509	0.362658	3.5219
7	6.26298	0.472734	0.298757	3.2526
8	3.63987	0.964179	0.321179	2.72797
9	3.6831	0.936182	0.432925	2.73662
10	3.64419	0.943157	0.436376	2.72884
11	3.64436	0.943081	0.436436	2.72887
12	3.64436	0.943081	0.436436	2.72887

11-9 THE REMEZ ALGORITHM

As mentioned in Sec. 11-8, a nearly equal-ripple error can be obtained by point matching at well-chosen values of ω. A highly effective iterative method for achieving exactly equal-ripple error is the *Remez algorithm*, consisting of the following operations:

1. An approximately equal-ripple error is attained by forcing the error $F(\omega) - F^j(\omega, \mathbf{p}^j)$ at $n + 1$ selected points $\omega_1^j, \omega_2^j, \ldots, \omega_{n+1}^j$ in the range $\omega_l \leq \omega \leq \omega_u$ of approximation† to take on the value E^j, with alternating signs. This is achieved by calculating the p_i^j as well as E^j from the equations

$$F^j(\omega_1^j, \mathbf{p}^j) - E^j = F(\omega_1^j)$$

$$F^j(\omega_2^j, \mathbf{p}^j) + E^j = F(\omega_2^j) \tag{11-128}$$

$$\ldots\ldots\ldots\ldots\ldots\ldots\ldots\ldots$$

$$F^j(\omega_{n+1}^j, \mathbf{p}^j) + (-1)^{n+1}E^j = F(\omega_{n+1}^j)$$

2. While the error is equal-ripple at the points ω_k^j, it may have larger local extrema elsewhere in the range (Fig. 11-24). Then, the abscissas of the largest $n + 1$ local error extrema must be found and designated as ω_k^{j+1}.
3. Next, step 1 is repeated at the new points ω_k^{j+1}, and (11-128) is solved for $\mathbf{p}^{j+1}$ and E^{j+1}, etc. The process is repeated until convergence is reached. This is indicated by the conditions

$$E^{j+1} = E^j \qquad \mathbf{p}^{j+1} = \mathbf{p}^j \qquad \omega_k^{j+1} = \omega_k^j \tag{11-129}$$

If $F^j(\omega, \mathbf{p}^j)$ is a linear function of the p_i^j, that is, if it is in the form given in (11-106), then (11-128) is a linear system of equations in all unknowns which can readily be solved.

† Usually $\omega_1^j = \omega_l$ and $\omega_{n+1}^j = \omega_u$, for all j.

$F(\omega) - F^j(\omega, \mathbf{p}^j)$

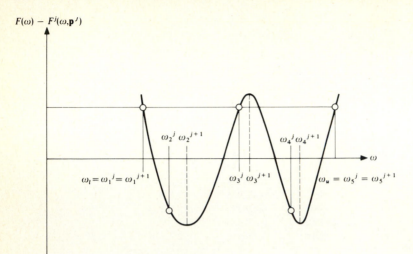

Figure 11-24 The Remez method, for $n = 4$.

Example 11-21 In Sec. 11-7, we carried out the least-squares approximation of the time response $F(t) = 3 - t/3$ with the actual circuit response $F^j(t) = p_1^j e^{-t/2} + p_2^j e^{-t/5}$. The resulting approximation was displayed in Fig. 11-18. The local error extrema are at $t_1^0 = 0$, $t_2^0 \approx 3$, and $t_3^0 = 9$. To obtain an equal-ripple error, we use (11-128) which now takes the form

$$p_1^j e^{-t_1^j/2} + p_2^j e^{-t_1^j/5} - E^j = 3 - \frac{t_1^j}{3}$$

$$p_1^j e^{-t_2^j/2} + p_2^j e^{-t_2^j/5} + E^j = 3 - \frac{t_2^j}{3} \qquad (11\text{-}130)$$

$$p_1^j e^{-t_3^j/2} + p_2^j e^{-t_3^j/5} - E^j = 3 - \frac{t_3^j}{3}$$

Substituting $j = 0$ and t_1^0, t_2^0, and t_3^0, as given above, leads to

$$1.00000p_1^0 + 1.00000p_2^0 - E^0 = 0$$

$$0.223130p_1^0 + 0.548812p_2^0 + E^0 = 2$$

$$0.011109p_1^0 + 0.165299p_2^0 - E^0 = 0$$

Then solving for p_1^0, p_2^0, and E^0 gives $p_1^0 \approx 0.926118$, $p_2^0 \approx 2.49691$, and $E^0 \approx 0.423024$. The extrema of the error $3 - t/3 - p_1^0 e^{-t/2} - p_2^0 e^{-t/5}$ are now at $t_1^1 = 0$, $t_2^1 \approx 3.44$, and $t_3^1 = 9$. Substituting these values into (11-130) gives

$$1.00000p_1^1 + 1.00000p_2^1 - E^1 = 3$$

$$0.179066p_1^1 + 0.502580p_2^1 + E^1 = 1.85333$$

$$0.011109p_1^1 + 0.165299p_2^1 - E^1 = 0$$

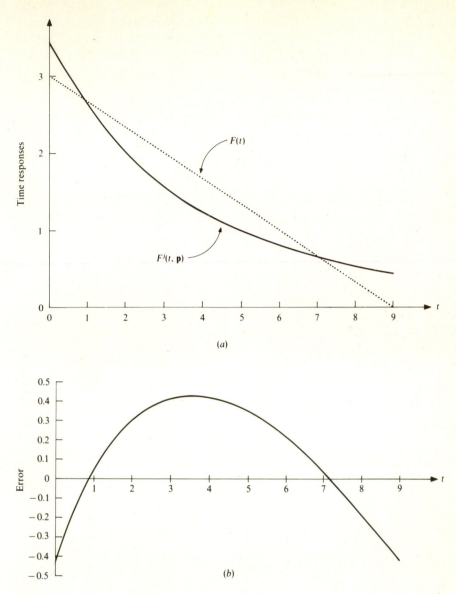

Figure 11-25 (a) An equal-ripple approximation of $F(t) = 3 - t/3$ using the approximant $F^j(t, p^j) = p_1^j e^{-t/2} + p_2^j e^{-t/5}$; (b) the error of the approximation.

which is again solved, resulting in $p_1^1 \approx 0.910177$, $p_2^1 \approx 2.51578$, and $E^1 \approx 0.425967$. The resulting approximation is illustrated in Fig. 11-25a, the error in Fig. 11-25b. The maximum errors are -0.42596 at $t_1^2 = 0$, $+0.42587$ at $t_2^2 \approx 3.5$, and -0.42597 at $t_3^2 = 9$. This indicates that very close approximation to the equal-ripple response was obtained in just two iterations.

If $F^j(\omega, \mathbf{p}^j)$ is not linear in the p_i^j, then (11-128) represents a nonlinear system of equations. This can be solved iteratively, using Newton's technique. Notice that since (11-128) must be solved in every iteration cycle of the Remez algorithm, we have here one iteration loop within another. To avoid an extravagant amount of computation, therefore, we usually perform only a few Newton iterations in each cycle.

The Newton equations which correspond to (11-128) can be obtained by Taylor-series expansion in (11-128)

$$\sum_{i=1}^{n} \frac{\partial F^j(\omega_1^j, \mathbf{p}^j)}{\partial p_i^j} \Delta p_i^j - E^j = F(\omega_1^j) - F^j(\omega_1^j, \mathbf{p}^j)$$

$$\sum_{i=1}^{n} \frac{\partial F^j(\omega_2^j, \mathbf{p}^j)}{\partial p_i^j} \Delta p_i^j + E^j = F(\omega_2^j) - F^j(\omega_2^j, \mathbf{p}^j) \qquad (11\text{-}131)$$

$$\cdots \cdots \cdots \cdots \cdots \cdots \cdots \cdots \cdots \cdots \cdots$$

$$\sum_{i=1}^{n} \frac{\partial F^j(\omega_{n+1}^j, \mathbf{p}^j)}{\partial p_i^j} \Delta p_i^j + (-1)^{n+1} E^j = F(\omega_{n+1}^j) - F^j(\omega_{n+1}^j, \mathbf{p}^j)$$

Note the close resemblance between (11-131) and the point-matching Newton equation (11-111).

Equation (11-131) must, in general, be solved repeatedly with updated derivatives to obtain a good solution of the nonlinear equations (11-128). Hence, this Newton iteration forms a loop within the loop of the Remez iteration.

Example 11-22 In Sec. 11-8, point matching at $t_1 = 0.2$, $t_2 = 3$, and $t_3 = 7$ was used to obtain an excellent realizable approximation in the form

$$F^j(t, \mathbf{p}^j) = (1.37002t + 2.91561)e^{-0.418807t}$$

to the unrealizable time response $F(t) = 3 - t/3$. The behavior of $F^j(t, \mathbf{p}^j)$ and $F(t)$ was illustrated in Fig. 11-21. Since the error, while small, is not equal-ripple, it is possible that its maximum value (which occurs at $t = 9$ and has the magnitude of 0.352) can be further decreased. Hence, an equal-ripple (min-max) approximation will next be obtained.

Since our approximant $F^j(t, \mathbf{p}^j) = (p_1^j t + p_2^j)e^{-p_3^j t}$ is nonlinear in the p_i^j, we shall have to use the Newton equation (11-131) in each iteration. To keep computational costs down, we shall permit only two Newton iterations in each cycle. In the first Remez cycle, we choose arbitrarily the abscissas $t_1^0 = 0$, $t_2^0 = 3$, $t_3^0 = 5$, and $t_4^0 = 9$ as the $n + 1$ points where equal-magnitude error will be forced.† Also, the previously obtained parameter vector $\mathbf{p}^0 = [1.37002 \quad 2.91561 \quad 0.418807]^T$ is used for initial values in the first Newton iteration. Equations (11-131) are now in the form

$$t_k^j \exp\left(-p_3^j t_k^j\right) \Delta p_1^j + \exp\left(-p_3^j t_k^j\right) \Delta p_2^j - t_k^j(p_1^j t_k^j + p_2^j) \exp\left(-p_3^j t_k^j\right) \Delta p_3^j$$

$$+ (-1)^k E^j = 3 - \frac{t_k^j}{3} - (p_1^j t_k^j + p_2^j) \exp\left(-p_3^j t_k^j\right)$$

$$k = 1, 2, 3, 4 \qquad j = 0, 1, 2, \ldots \qquad (11\text{-}132)$$

---

† If the more suitable points $t = 0, 1, 5$, and 9 (which can be immediately guessed from Fig. 11-21) are used, the convergence is much faster—too fast for purposes of demonstration.

Table 11-18 Remez cycle 1

ITERATION #	p_1	p_2	p_3
0	1.37002	2.91561	0.418807

$E = 0$

TIME	DESIRED	ACTUAL	ERROR
0	3	2.91561	0.08439
3	2	2	$-4.80190E-06$
5	1.33333	1.20303	0.130305
9	0	0.351706	-0.351706

ITERATION #	p_1	p_2	p_3
1	2.50074	2.808	0.542644

$E = -0.191997$

TIME	DESIRED	ACTUAL	ERROR
0	3	2.808	0.191997
3	2	2.02426	$-2.42586E-02$
5	1.33333	1.01552	0.317817
9	0	0.191587	-0.191587

ITERATION #	p_1	p_2	p_3
2	2.83821	2.78739	0.543694

$E = -0.212607$

TIME	DESIRED	ACTUAL	ERROR
0	3	2.78739	0.212607
3	2	2.212	-0.212004
5	1.33333	1.12016	0.213172
9	0	0.2124	-0.2124

The computer printout of the first Remez cycle is shown in Table 11-18. It will be noted that at the end of the second Newton iteration, the absolute errors at the t_k are nearly equal to each other and to E. It is not worthwhile improving the match any further (although about two more iterations would make it perfect), since the matching itself is only an intermediate step in the overall process. Hence, we proceed to step 2. The error between $F(t)$ and the approximant

$$F^1(t) = (2.83821t + 2.78739)e^{-0.543694t}$$

is calculated (in $\Delta t = 0.25$ step) over the range $0 \le t \le 9$, and error extrema are located. They are found to be at $t_1^1 = 0$, $t_2^1 = 1.25$, $t_3^1 = 5.75$, and $t_4^1 = 9$. Hence, the error will next be forced to take on values of (nearly) equal amplitudes but alternating signs at these points. The results are shown in Table 11-19. Next, the error extrema for $F^i(t) = (1.98248t + 2.75374)e^{-0.419622t}$ are found, at $t_1^2 = 0$, $t_2^2 = 1.25$, $t_3^2 = 5.24$, and $t_4^2 = 9$. Matching $F(t) \pm E^{j+1}$ at these points gives the results in Table 11-20. Evidently, an excellent match was obtained, and the maximum residual error $E \approx 0.249$ is quite an improvement over the already good initial value 0.352. The final response and its error function are illustrated in Fig. 11-26. The error was decreased at the upper end of the range (where the error was worst before) and increased elsewhere, as compared with the approximation of Fig. 11-21. It depends on the actual application whether this change can be regarded as an improvement.

Table 11-19 Remez cycle 2

ITERATION #	p_1	p_2	p_3
0	2.83821	2.78739	0.543694
$E = 0$			
TIME	DESIRED	ACTUAL	ERROR
0	3	2.78739	0.21261
1.25	2.58333	3.21072	−0.62739
5.75	1.08333	0.838467	0.244867
9	0	0.21401	−0.212401
ITERATION #	p_1	p_2	p_3
1	1.95719	2.75762	0.496798
$E = -0.24238$			
TIME	DESIRED	ACTUAL	ERROR
0	3	2.75762	0.24238
1.25	2.58333	2.79673	−0.213399
5.75	1.08333	0.805163	0.27817
9	0	0.232934	−0.232934
ITERATION #	p_1	p_2	p_3
2	1.98248	2.75374	0.491662
$E = -0.246259$			
TIME	DESIRED	ACTUAL	ERROR
0	3	2.75374	0.246259
1.25	2.58333	2.82975	−0.246413
5.75	1.08333	0.837673	0.24566
9	0	0.246633	−0.246633

Table 11-20 Remez cycle 3

ITERATION #	p_1	p_2	p_3
0	1.98248	2.75374	0.491662
$E = 0$			
TIME	DESIRED	ACTUAL	ERROR
0	3	2.75374	0.24626
1.25	2.58333	2.82974	−0.246409
5.25	1.25	0.996098	0.253902
9	0	0.246632	−0.246632
ITERATION #	p_1	p_2	p_3
1	1.983	2.75119	0.490691
$E = -0.248812$			
TIME	DESIRED	ACTUAL	ERROR
0	3	2.75119	0.248812
1.25	2.58333	2.83215	−0.248813
5.25	1.25	1.0012	0.248799
9	0	0.248822	−0.248822
ITERATION #	p_1	p_2	p_3
2	1.98303	2.75119	0.490696
$E = -0.248814$			
TIME	DESIRED	ACTUAL	ERROR
0	3	2.75119	0.248814
1.25	2.58333	2.83215	−0.248814
5.25	1.25	1.00119	0.248814
9	0	0.248814	−0.248814

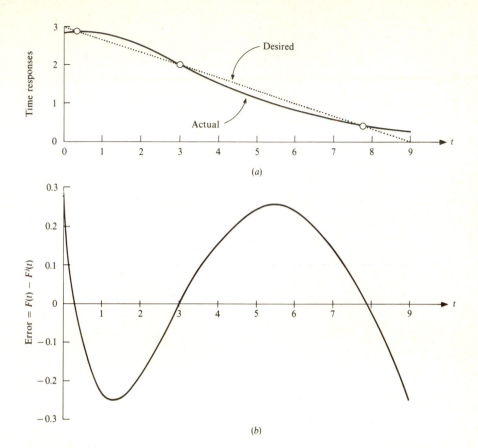

Figure 11-26 (a) An equal-ripple approximation of $F(t) = 3 - t/3$ using the approximant $F^j(t, p^j) = (p_1^j t + p_2^j)\exp(-p_3^j t)$; (b) the error of the approximation.

In Examples 11-21 and 11-22 the Remez method was used to solve the *approximation* problem, i.e., to find the constants in the network function. As Example 11-23 will illustrate, it can also be used to find the element values of the circuit, i.e., for *realization*. If the p_i^j are element values, the derivatives $\partial F^j/\partial p_i^j$ can be found efficiently using the adjoint-network approach.

Example 11-23 In Sec. 11-8, the amplifier-equalizer was designed to match a desired response $F(\Omega) = 20 - (2/\pi)\Omega$ at $\Omega_1 = 2.4\pi$ and $\Omega_2 = 3.6\pi$. The resulting frequency characteristics, shown in Fig. 11-22, were not equal-ripple: the local extrema of the error were

$$F - F^j = \begin{cases} -0.459 \text{ dB} & \text{at } \Omega_1 = 2\pi \\ +0.197 \text{ dB} & \text{at } \Omega_2 = 3\pi \\ -0.283 \text{ dB} & \text{at } \Omega_3 = 4\pi \end{cases}$$

Next, the parameter values $r = 2.79424$ and $c = 0.026365$ were used as initial values for a Remez iteration aimed at making the response exactly equal-ripple. The computer print-out for the Remez iterations is shown in Table 11-21; the resulting frequency responses are

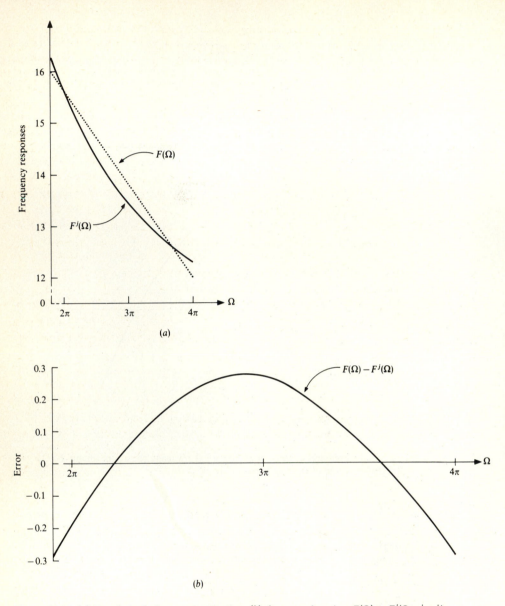

Figure 11-27 (a) Equal-ripple frequency response; (b) the error function $F(\Omega) - F^j(\Omega, p_1^j, p_2^j)$.

shown in Fig. 11-27a and b. The maximum error decreased from 0.459 to 0.286 dB; the final element values are $r = 2.89778$ and $c = 0.0272451$.

Note that in these conditions we permitted only *one* Newton iteration in each Remez cycle; the steps performed were therefore the following:

1. We set $j = 0$, $p_1^j \equiv r^j = 2.79424$, $p_2^j \equiv c^j = 0.026365$, $\Omega_1^j = 2\pi$, $\Omega_2^j = 3\pi$, and $\Omega_3^j = 4\pi$.

Table 11-21

Iteration j	p_1^j	p_2^j	Ω_1^j
0	2.794240D 00	2.636500D−02	6.283185D 00
1	2.900515D 00	2.723103D−02	6.283185D 00
2	2.897130D 00	2.725115D−02	6.283185D 00
3	2.987780D 00	2.724506D−02	6.283185D 00

Iteration j	Ω_2^j	Ω_3^j	E^j
0	9.424778D 00	1.256637D−01	2.822918D−01
1	9.299069D 00	1.256628D+01	2.841431D−01
2	9.047746D 00	1.256628D+01	2.860868D−01
3	9.047746D 00	1.256628D+01	2.860868D−01

2. The adjoint-network method is used† to find the Jacobian matrix elements $\partial F^j(\Omega, p_1, p_2)/\partial p_l$ for $\Omega = \Omega_k^j$, $p_1 = p_1^j$, and $p_2 = p_2^j$ and $k = 1, 2, 3, l = 1, 2$.
3. Equation (11-131) is set up and solved for Δp_1^j, Δp_2^j, and E^j but only once for each j.
4. The parameter values are updated, so that $p_l^{j+1} = p_l^j + \Delta p_l^j$. If $|\Delta p_l^j| > 10^{-5} |p_l^j|$ for $l = 1$ or 2, we set j to $j + 1$, find the local extrema of the new error function $F(\Omega) - F^j(\Omega, p_1^j, p_2^j)$, and designate them as Ω_k^j, $k = 1, 2, 3$. Then the process returns to step 2.
5. If the convergence criteria $|\Delta p_l^j| \leq 10^{-5} |p_l^j|$ are satisfied for both $l = 1$ and 2, the final responses $F^j(\Omega, p_1^j, p_2^j)$, $F(\Omega)$, and $F(\Omega) - F^j(\Omega, p_1^j, p_2^j)$ are tabulated and plotted and the program halts.

As Table 11-21 illustrates, the above process converged rapidly (in just three iterations) without any need for damping to the optimum solution.

11-10 SUMMARY

Chapter 11 describes a variety of techniques for the iterative design of circuits which, for one reason or another, could not be synthesized using the step-by-step methods of Chaps. 3 to 8. The main subjects discussed were as follows:

1. The conditions under which circuit synthesis is possible and the description of a counterexample, the design of an amplifier-equalizer (Sec. 11-1).
2. The steps involved in iterative optimization: establishing the specifications, deriving an initial approximation, error evaluation, and the readjustment of circuit parameters (Sec. 11-2).
3. Gradient methods of optimization, the error gradient, the Hessian matrix, the use of the adjoint-network method for calculating the derivatives of the performance index (Sec. 11-3).
4. The steepest-descent optimization method using linear search or polynomial interpolation; the limitations of steepest descent (Sec. 11-4).

† In this simple problem, analytical differentiation can also be used.

5. Optimization using Newton's method; a comparison of steepest descent and Newton's method; hybrid methods (Sec. 11-5).
6. The least-pth Taylor method: approximation to the Hessian matrix, divergence problems, damping (Sec. 11-6).
7. The solution of linear least-squares problems (Sec. 11-7).
8. Optimization using point matching combined with Newton's technique (Sec. 11-8).
9. Circuit optimization using coefficient matching (Sec. 11-8).
10. Equal-ripple approximation using the Remez algorithm, for linear and nonlinear design and for approximation and realization problems (Sec. 11-9).

We still owe the reader a discussion of how the network functions, assumed to be heaven-sent in Chaps. 3 to 8 and sometimes also in Chap. 11, can be found in practical situations. Finding these functions is the subject of approximation theory, discussed in Chap. 12.

PROBLEMS

11-1 Repeat the matching of $F(\Omega)$ and $F_1(\Omega, r, c)$ [compare Eqs. (11-5) to (11-8)] at Ω_1 and Ω_2, using the exact expression (11-6) for $F_1(\Omega, r, c)$. How much do the values of r and c change? How much more complicated are the calculations needed to obtain them?

11-2 Repeat the matching process, using the simplified formula (11-7) for $F_1(\Omega, r, c)$ but equating the two curves at $\Omega_1 = 2.4\pi$ and 2.8π. Find r and c and plot $F_1(\Omega, r, c)$ as well as $F(\Omega)$. How does the maximum error compare with that obtained earlier (Fig. 11-4)? Why?

11-3 (a) Show that the Hessian matrix H^j is symmetric.
(b) How many different elements can H^j have?
(c) How many evaluations of E^j are required to calculate all elements of H^j using (1) forward-difference quotients; (2) central-difference quotients?

11-4 Calculate the $(1, 1)$ component of H^j from the E_4^j given in (11-25) by analytic differentiation.

11-5 Find a_1 and a_2 of Eq. (11-62) in terms of $E^j(\mathbf{p}^j)$, $E(\mathbf{p}^j + \Delta \mathbf{p}_0^j/2)$, and $E(\mathbf{p}^j + \Delta \mathbf{p}_0^j)$. Show that Eq. (11-63) holds.

11-6 In Table 11-5 linear search was used in the steepest-descent direction to find an improved parameter vector $\mathbf{p}^1 = [-1.75 \quad -1]^T$ from the initial vector $\mathbf{p}^0 = [-4 \quad 2]^T$. Perform two more searches to find $\mathbf{p}^2$ and $\mathbf{p}^3$.

11-7 Repeat the calculations of Table 11-5 to find $\mathbf{p}^1$ from the initial point $\mathbf{p}^0 = [-7.245 \quad -1]^T$.

11-8 In Table 11-6 and Eq. (11-70), polynomial interpolation was used to find the improved parameter vector $\mathbf{p}^1 \approx [-1.692 \quad -1.077]^T$ from the initial vector $\mathbf{p}^0 = [-4 \quad 2]^T$. Perform two more searches to find $\mathbf{p}^2$ and $\mathbf{p}^3$.

11-9 Repeat the calculations of Table 11-6 and Eqs. (11-70) and (11-71) from the new initial vector $\mathbf{p}^0 = [-7.245 \quad -1]^T$.

11-10 From which initial points on the $E = 40$ contour in Fig. 11-13 will the global optimum be obtained in a single steepest-descent iteration?

11-11 Replot the contours of Fig. 11-13 when the new parameter $p_1' = p_1/2$ replaces p_1. Is this a better or worse choice of variables for steepest-descent optimization?

11-12 Show that the approximate Hessian matrix given by (11-102) is a positive definite matrix, i.e., that $\mathbf{x}^T H^j \mathbf{x} > 0$ for any vector $\mathbf{x} \neq \mathbf{0}$. *Hint:* Substitute H^j from (11-102) and carry out the vector and matrix multiplications indicated.

11-13 Show that the H^j given by (11-102) is a symmetric matrix.

11-14 Prove that for sufficiently small λ, $E(\mathbf{p}^j + \lambda \Delta \mathbf{p}^j) < E(\mathbf{p}^j)$ if $\Delta \mathbf{p}^j$ is given by (11-103). *Hint:* First show, as in (11-53), that $E(\mathbf{p}^j + \lambda \Delta \mathbf{p}^j) - E(\mathbf{p}^j) \approx \lambda \, \nabla E^j(\mathbf{p}^j)^T \Delta \mathbf{p}^j$. From (11-101) to (11-103), if we define $\mathbf{x} \neq \mathbf{0}$ by the relation $\nabla E^j = H^j \mathbf{x}$, $\nabla E^{j^T} \Delta \mathbf{p}^j = -\nabla E^{j^T}(H^j)^{-1} \nabla E^j = -\mathbf{x}^T H^j(H^j)^{-1} H^j \mathbf{x} = -\mathbf{x}^T H^j \mathbf{x} < 0$ since H^j is positive definite as shown in Prob. 11-12.

11-15 For the error criterion

$$E = \sin^2 p_1 + e^{p_2} \cosh p_1$$

calculate the gradient vector and the Hessian matrix. Perform a steepest-descent step, using linear search and permitting a maximum parameter change of 5 percent, from $p_1 = p_2 = 1$.

11-16 Repeat the steepest-descent step of Prob. 11-15 using polynomial interpolation. Compare the computational effort needed and the results achieved by the two processes.

11-17 Using the Hessian matrix and gradient vector found in Prob. 11-15 perform a Newton iteration (using the equations of Sec. 11-5) on the error criterion $E = \sin^2 p_1 + e^{p_2} \cosh p_1$. Is Newton's method more or less effective than steepest descent for this problem?

11-18 The circuit shown in Fig. 11-28 has the specified response $F(\omega) \triangleq |V_0(j\omega)/E| = 1 - \omega/40\pi$ for $0 \le \omega \le 40\pi$.

(a) By matching $F(\omega)$ at $\omega = 20\pi$ and 40π, find suitable initial values for L and C.

(b) Using a least-squares error criterion and 10 sample points, perform a steepest-descent step with linear search.

(c) Repeat the process using polynomial interpolation.

11-19 Provide a complete solution to Prob. 11-18 using the least-squares Taylor method described in Sec. 11-6.

11-20 Prove that the direction of steepest descent is in the negative gradient direction, using conventional argument rather than the geometric interpretation of Fig. 11-9. *Hint:* For a given step size δ,

$$\sum_{i=1}^{n} (\Delta p_i)^2 = \delta^2$$

Also, from (11-53),

$$E^{j+1} - E^j = \mathbf{V}\mathbf{E}^{j^T} \Delta \mathbf{p}^j = \sum_{i=1}^{n} \frac{\partial E^j}{\partial p_i} \Delta p_i$$

For a minimum

$$\frac{\partial(E^{j+1} - E^j)}{\partial p_k} = 0 \qquad k = 0, 1, 2, \ldots, n-1$$

where Δp_n can be expressed in terms of $\delta, \Delta p_0, \ldots, \Delta p_{n-1}$.

11-21 Prove that the direction of the steepest descent, given by (11-55) is downhill, i.e., that

$$E(\mathbf{p}^j + \lambda \Delta \mathbf{p}^j) < E(\mathbf{p}^j)$$

if $|\Delta \mathbf{p}^j| > 0$ and $\lambda \to 0$.

11-22 (a) Prove that if the Hessian matrix $\mathbf{H}^j$ is positive definite, then so is $(\mathbf{H}^j)^{-1}$, and vice versa.

(b) Prove that the direction of the increment vector given by (11-81) is downhill if $|\Delta \mathbf{p}^j| > 0$ and if $\mathbf{H}^j(\mathbf{p}^j)$ is positive definite.

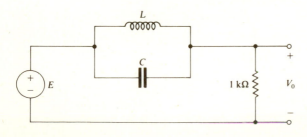

Figure 11-28 Circuit to be optimized for the frequency response $|V_0/E| = 1 - \omega/(40\pi)$.

11-23 Carry out the steepest descent and Newton optimization of

$$E(\mathbf{p}) = (p_1 + 2p_2 - 7)^2 + (2p_1 + p_2 - 5)^2$$

from $p_1^0 = p_2^0 = 0$. Draw the error contours and indicate the minimum.

11-24 Repeat the calculations of Prob. 11-23 for the error criterion

$$E(\mathbf{p}) = p_1 + \frac{p_2}{4} + \frac{4}{p_1} + \frac{4}{p_2}$$

using $p_1^0 = 4.5$ and $p_2^0 = 14$. Why is the convergence much worse for this example?

11-25 (a) Use the least-fourth Taylor method to optimize the circuit of Fig. 11-29a for the desired response (Fig. 11-29b)

$$F(\omega) \triangleq 20 \log |Y(j\omega)| = \frac{2.5}{\pi}\omega + 5$$

where ω is in kiloradians per second.

(b) Repeat your calculations for the simplified circuit shown in Fig. 11-30. Are the results significantly worse?

11-26 (a) Match the function $(p_1 + p_2\omega^2)/(1 + 0.4\omega^2)$ in the least-squares sense to the desired response $\exp\sqrt{\omega}$ in the $0.1 \le \omega \le 1$ range.

(b) Regard the p_1, p_2 values obtained in part (a) as fixed numbers, denoted by $p_1 = a$, $p_2 = b$.

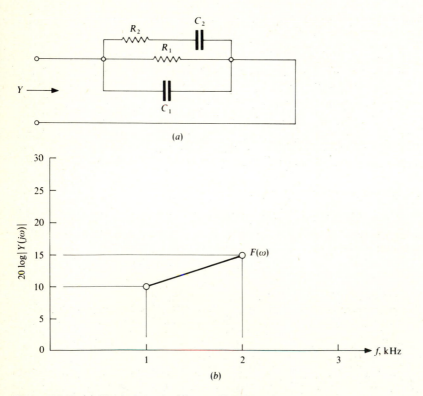

(a)

(b)

Figure 11-29 (a) Physical circuit; (b) prescribed frequency response.

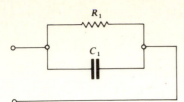

Figure 11-30 Simplified physical circuit.

Next, match the reciprocal function $(1 + p_1 \omega^2)/(a + b\omega^2)$ (in which the constant 0.4 was replaced by the variable p_1) to $\exp(-\sqrt{\omega})$ in the least-squares sense for $0.1 \leq \omega \leq 1$.

(c) On the basis of parts (a) and (b) attempt to formulate an iterative scheme for the least-squares optimization of rational functions

$$\frac{p_1 + p_2 f_2(\omega) + \cdots + p_k f_k(\omega)}{1 + p_{k+1} f_{k+1}(\omega) + \cdots + p_n f_n(\omega)}$$

11-27 Let the desired curve be given by $F(\omega) = 10^{-C - D\sqrt{\omega}}$. $F(\omega)$ goes through the points $F(0.075) = 10^4$ and $F(1) = 2.5$.

(a) Find C and D.

(b) Find the coefficients p_1 and p_2 in the least-squares approximant

$$F^j(\omega) = \frac{1 + p_1 \omega^2 + p_2 \omega^4}{3 \times 10^{-5} + 0.01\omega^2 + 0.125\omega^4}$$

The approximation should be valid for $0.075 \leq \omega \leq 1$.

(c) Plot the desired and the actual curves.

(d) What conclusions can be drawn from these results?

11-28 Approximate the desired response

$$F(\omega) = 10^{0.15\omega^2 - 0.6\sqrt{\omega}}$$

in the range $0.003 \leq \omega \leq 1$ by the function

$$\frac{\omega^6 + p_3\omega^4 + p_2\omega^2 + p_1}{\omega^6 + 0.088\omega^4 + 2 \times 10^{-4}\omega^2 + 5.6 \times 10^{-3}}$$

in the least-squares sense. Plot the desired and actual responses and evaluate the quality of the approximation.

11-29 A coaxial cable has an attenuation response (due to skin effect) of the form

$$\alpha(f) = C + D\sqrt{f}$$

where α is in decibels and f is the frequency. The frequency range of interest is $f_l \leq f \leq f_u$, with $f_l = 60$ kHz, $f_u = 800$ kHz. From measurements,

$$\alpha(f_l) = 4 \text{ dB} \qquad \alpha(f_u) = 40 \text{ dB}$$

(a) Find C and D.

(b) A loss equalizer must be designed for the system. The attenuation of the equalizer is given by the equation

$$\alpha_e(f) = \frac{1 + p_1 x + p_2 x^2}{p_3 + p_4 x + p_5 x^2}$$

where $x = (f/f_u)^2$. It must satisfy

$$\alpha(f) + \alpha_e(f) \approx \text{const} = 44 \text{ dB}$$

for $f_l \leq f \leq f_u$. Find $p_1^0, p_2^0, \ldots, p_5^0$ by matching $\alpha_e(f)$ to $44 - \alpha(f)$ at $x_1 = 0.01$, $x_2 = 0.03$, $x_3 = 0.1$, $x_4 = 0.3$, and $x_5 = 1$.

(c) Plot the error function on a logarithmic x scale for the p^0 obtained in part (b).

11-30 Redesign the equalizer obtained in Prob. 11-29 for a Chebyshev (equal-ripple) error function. Use the p^0 obtained in Prob. 11-29 as initial approximation for the Remez-Newton procedure.

11-31 Repeat the coefficient-matching procedure carried out in Sec. 11-8 for the circuit of Fig. 11-23, with the specified function $V_1/V_2 = \frac{4}{3}(s^3 + 2s^2 + 2s + 1)$, $r = 0.1L$, and initial values $r = 0$ as well as:

(a) $R_1 = C_1 = 1$, $L = 4$, $C_2 = \frac{1}{3}$, $R_2 = 3$

(b) $R_1 = C_1 = L = C_2 = R_2 = 1$

Interpret your results.

11-32 Repeat the solution of Prob. 11-31 for the function

$$\frac{V_1}{V_2} = 2(s^3 + 2s^2 + 2s + 1) \qquad r = 0.1L$$

with initial values

(a) $R_1 = C_1 = C_2 = R_2 = 1$, $L = 2$

(b) $R_1 = C_1 = 1$, $L = 3$, $C_2 = 0.5$, $R_2 = 2$

Explain the difficulties you encounter. *Hint:* Examine the coefficient matrix of the Newton equations for the initial values given in part (a). These initial values satisfy the given V_1/V_2 for $r = 0$; hence the solution must lie close to them.

REFERENCES

1. Director, S. W., and R. A. Rohrer: Automated Network Design: The Frequency Domain Case, *IEEE Trans. Circuit Theory*, August 1969, pp. 330–337.
2. Westlake, Joan: "A Handbook of Numerical Matrix Inversion and Solution of Linear Equations," Wiley, New York, 1968.
3. Aaron, M. R.: The Use of Least Squares in System Design, *IRE Trans. Circuit Theory*, December 1956, pp. 224–231.
4. Temes, G. C.: Optimization Methods in Circuit Design, in F. F. Kuo and W. G. Magnuson, Jr. (eds.): "Computer Oriented Circuit Design," Prentice-Hall, Englewood Cliffs, N.J., 1969.
5. Temes, G. C., and D. A. Calahan: Computer-aided Network Optimization: The State of the Art, *Proc. IEEE*, November 1967, pp. 1832–1363.
6. Aoki, M.: "Introduction to Optimization Techniques," Macmillan, New York, 1971.
7. Calahan, D. A.: Computer Design of Linear Frequency-selective Networks, *Proc. IEEE*, November 1965, pp. 1701–1706.

CHAPTER TWELVE

APPROXIMATION THEORY

12-1 THE APPROXIMATION PROBLEM

In all previous chapters, it was tacitly assumed that a realizable network function (driving-point or transfer) was given at the outset of the design procedure, presumably as part of the specifications. In practice, of course, this is almost never the case; normally, the desired circuit performance is prescribed either in the form of a chart or a graph or, more often, as a set of inequalities. A typical set of specifications may be in the form given in the following example.

Example 12-1 Design a doubly terminated reactance two-port with both terminations equal to 50 Ω and with the following restrictions on the transducer loss α:

$$\alpha \leq 1 \text{ dB} \quad \text{for} \quad 0 \leq f \leq 1.8 \text{ MHz}$$
$$\alpha \geq 50 \text{ dB} \quad \text{for} \quad 7 \text{ MHz} \leq f \leq \infty$$

The solution of this problem will be discussed later in the chapter. Here, we merely note that the specified circuit is evidently intended to pass low frequencies (below 1.8 MHz) essentially unattenuated and suppress high frequencies (above 7 MHz) strongly.

Such frequency-selective two-ports are called *filters*. Depending on the frequency bands passed (or suppressed), they can be classified as low-pass, or high-pass, bandpass, or bandstop, etc., filters. The circuit specified above is clearly a low-pass filter. The frequency region in which the loss must be low ($0 \leq f \leq$ 1.8 MHz in the above example) is called the *passband* of the filter; the high-loss region (7 MHz $\leq f \leq \infty$ in the example) is called the *stopband*. The frequency limits of these bands are called *passband limit* and *stopband limit*, respectively. In the above example, the passband limit is $f_p = 1.8$ MHz, and the stopband limit is $f_s = 7$ MHz.

The branch of circuit theory which deals with finding a realizable network function meeting such practical specifications is called *approximation theory*. Its fundamentals will be discussed in this chapter.

It will be demonstrated next that the approximation problem is especially

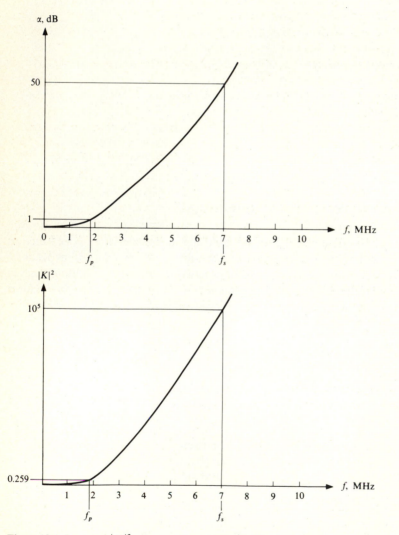

Figure 12-1 Loss and $|K|^2$ responses for a low-pass filter.

conveniently solved in terms of the transducer function $H(s)$ and the characteristic function $K(s)$ [introduced in Chap. 6 via Eqs. (6-5) and (6-21), respectively]. Specifically, if only the behavior of the transducer loss $\alpha(\omega)$ is specified, it is advantageous to find $|K(j\omega)|^2$; if the phase or time response is also prescribed, $H(s)$ should be calculated. The reason for these preferences is that, by Eqs. (6-8) and (6-25), the transducer loss for $s = j\omega$ is

$$\alpha = 10 \log |H|^2 = 10 \log (|K|^2 + 1) \qquad (12\text{-}1)$$

Consequently, whenever $\alpha \approx 0$, $|K|^2 \approx 0$; and whenever $\alpha \to \infty$, $|K|^2 \to \infty$. Similarly, if $\alpha(\omega)$ is increasing (decreasing), $|K(j\omega)|^2$ must also be increasing (decreasing). In conclusion, the loss $\alpha(\omega)$ and the squared modulus of $K(j\omega)$ vary exactly the same way with frequency, except that there is of course a difference between their actual numerical values. This point is illustrated in Fig. 12-1, which compares the two responses for the low-pass filter specified in the above example. Obviously, neither curve is drawn to a true scale.

Another advantage of using $|K|^2$ for approximating a prescribed filter loss response is based on the fact (to be demonstrated) that it is efficient to place all zeros of the $\alpha(\omega)$ function in those frequency bands where α must be small, i.e., in the *passbands*, and, similarly, that all poles of $\alpha(\omega)$ should be in the frequency bands of specified high attenuation, i.e., in the *stopbands*. By Eq. (12-1), the zeros and poles of $|K|^2$ are located at the same frequencies as those of α. Hence, the approximate locations of the zeros and poles of $|K|^2$ are usually known in advance, and these poles and zeros are normally all on the $j\omega$ axis of the s plane. By contrast, the zeros of $H(s)$ are in the inside of the left half s-plane. Finding the zeros of $H(s)$ is therefore a two-dimensional problem, while the search for those of $K(s)$ is usually restricted to the $j\omega$ axis, i.e., is in one dimension only.

Example 12-2 To illustrate how easily a filter-type response can be obtained by properly choosing the location of the zeros and poles of $|K|^2$, assume that we wish α to be small for $0 \le \omega \le 1$ and as large as possible for $2 \le \omega \le 4$. Let $K(s)$ be of degree 4; then in general

$$K(s) = C \, \frac{(s - s_1)(s - s_2)(s - s_3)(s - s_4)}{(s - p_1)(s - p_2)(s - p_3)(s - p_4)}$$

Since $K(s)$ is a *real* rational function, the s_i and p_i must occur in conjugate pairs. Furthermore, we anticipate that distributing the *zeros* s_i along the $j\omega$ axis in the $0 \le \omega \le 1$ range, the values of $|K|$ and α will be small in that range. Finally, placing the *poles* p_i on the $j\omega$ axis in the $2 \le \omega \le 4$ range should make $|K|$ and hence also α large in that frequency range. Hence, choosing the zeros and poles arbitrarily at the plausible locations given by

$$s_1 = s_2^* = j0.3 \qquad s_3 = s_4^* = j0.7 \qquad p_1 = p_2^* = j2.5 \qquad p_3 = p_4^* = j3.5$$

and selecting $C = 100$ gives the resulting $K(j\omega)$ in the form

$$K(j\omega) = 100 \, \frac{(\omega^2 - 0.3^2)(\omega^2 - 0.7^2)}{(\omega^2 - 2.5^2)(\omega^2 - 3.5^2)}$$

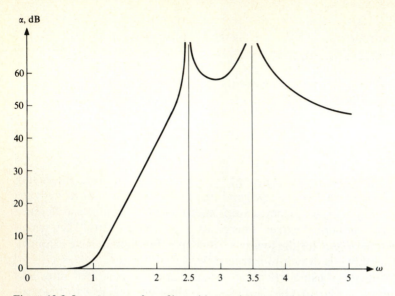

Figure 12-2 Loss response for a filter with $j\omega$-axis zeros and poles.

The loss response $\alpha(\omega)$ thus obtained is shown in Fig. 12-2. It confirms the effectiveness of distributing the zeros and poles in the passband and stopband, respectively: the passband loss is small (less than 0.1 dB over 80 percent of the passband) and the stopband loss is large.

The situation changes, however, if we are interested in the phase or delay response of our circuit. By Eq. (6-7), the phase lag β between the output and input voltages is simply the phase of the complex function $H(j\omega)$:

$$\beta(\omega) = \tan^{-1} \frac{\text{Im } H(j\omega)}{\text{Re } H(j\omega)} \tag{12-2}$$

Hence, the phase as well as the *phase delay*

$$T_{\text{ph}}(\omega) = \frac{\beta(\omega)}{\omega} \tag{12-3}$$

and the *group delay*

$$T_g(\omega) = \frac{d\beta(\omega)}{d\omega} \tag{12-4}$$

are closely related to the transducer function $H(s)$ and are easily expressed from it. There is no direct connection between these quantities and $K(s)$. Hence, if $\beta(\omega)$, $T_{\text{ph}}(\omega)$, or $T_g(\omega)$ is specified, it is more expedient to calculate in terms of $H(s)$.

Similarly, the time response of a two-port is given by

$$v_2(t) = \mathscr{L}^{-1}[V_2(s)] = \frac{1}{2}\sqrt{\frac{R_L}{R_G}} \mathscr{L}^{-1}\left[\frac{E_G(s)}{H(s)}\right] \tag{12-5}$$

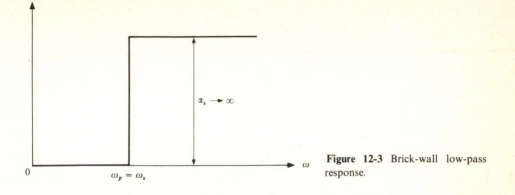

Figure 12-3 Brick-wall low-pass response.

where $\mathscr{L}^{-1}$ denotes the inverse Laplace transform and Eq. (6-5) was utilized. There is no comparable relation in terms of $K(s)$. Therefore, only $H(s)$ is useful for satisfying time-response specifications.

In most design problems, the desired ideal characteristics cannot be exactly achieved. For example, an ideal low-pass filter would have to exhibit the infinitely sharp "brick-wall" loss response illustrated in Fig. 12-3. Such a characteristic is not realizable with any circuit built from a finite number of elements (refer to Probs. 12-1 and 12-2). Hence, the desired loss function must be *approximated* by functions which are realizable. Obviously, we want to achieve the best possible agreement between the desired and the actual performance. Two widely used procedures for measuring this agreement are:

(a) To compare the two functions and their first $n - 1$ derivatives at one single value of the independent variable.†
(b) Or to evaluate the maximum deviation between the two functions in a range of the independent variable.

Criterion *a* requires therefore that

$$F_{\text{spec}}(\omega) = F_{\text{act}}(\omega)$$

$$\frac{dF_{\text{spec}}}{d\omega} = \frac{dF_{\text{act}}}{d\omega}$$

$$\frac{d^2 F_{\text{spec}}}{d\omega^2} = \frac{d^2 F_{\text{act}}}{d\omega^2} \tag{12-6}$$

$$\cdot \cdot \cdot \cdot \cdot \cdot \cdot \cdot \cdot \cdot \cdot \cdot \cdot \cdot \cdot$$

$$\frac{d^{n-1} F_{\text{spec}}}{d\omega^{n-1}} = \frac{d^{n-1} F_{\text{act}}}{d\omega^{n-1}}$$

for some $\omega = \omega_0$. Here, $F_{\text{spec}}(\omega)$ is the specified response while $F_{\text{act}}(\omega)$ is the actual one. Obviously, $F_{\text{act}}(\omega)$ must have n free parameters, e.g., coefficients, or

† Usually the frequency ω is the independent variable.

zeros and poles if F_{act} is a rational function of ω, in order to satisfy the n equations in (12-6). If F_{act} does satisfy Eq. (12-6), the error between F_{spec} and F_{act} is called *maximally flat* and F_{act} is a *maximally flat approximation* of F_{spec}.

Criterion b suggests the maximum absolute error in the $\omega_l \le \omega \le \omega_u$ range:

$$E = \max_{\omega_l \le \omega \le \omega_u} \left| F_{spec}(\omega) - F_{act}(\omega) \right| \tag{12-7}$$

be minimized. Assuming again that F_{act} has n adjustable parameters, this will

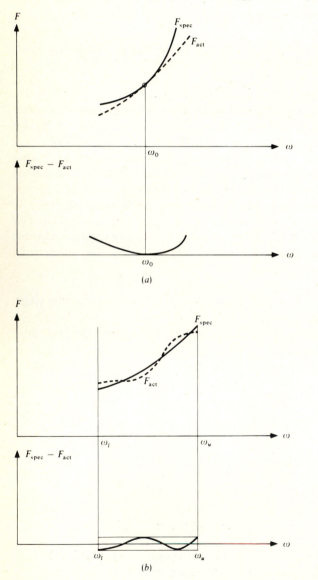

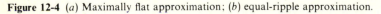

Figure 12-4 (*a*) Maximally flat approximation; (*b*) equal-ripple approximation.

usually happen if the error $F_{spec}(\omega) - F_{act}(\omega)$ has $n + 1$ equal alternating extrema in the frequency range between ω_l and ω_u. Since such error is called equal-ripple, the approximation itself will be called *equal-ripple* or *min-max approximation*. Responses representing maximally flat and equal-ripple approximations are illustrated in Fig. 12-4a and b, respectively.

In the following, we shall obtain maximally flat or equal-ripple approximations to some idealized responses, such as the brick-wall response illustrated in Fig. 12-3. From the discussions of Chap. 6 we know that the following information is required for the circuit realization process:

1. The degree n of the circuit, i.e., the number of its natural frequencies
2. The characteristic function $K(s)$ and/or the transducer factor $H(s)$

$K(s)$ and $H(s)$ can be described by the polynomials $E(s)$, $F(s)$, and $P(s)$ or by the critical frequencies, i.e., natural modes, loss poles (transmission zeros), and zero-loss points (reflection zeros) of the circuit.

The next sections will describe the calculation of these functions and parameters for some simple but important responses. Initially, some approximations to the ideal low-pass response of Fig. 12-3 will be discussed. Then it will be shown how transformation of the frequency variable can be used to extend these approximations to the design of high-pass, bandpass, and bandstop filters. After that, the maximally flat approximation of a linear phase, i.e., constant group delay, will be described for low-pass filters. Finally, some simple solutions of the approximation problem for bandpass filters will be discussed.

12-2 BUTTERWORTH APPROXIMATION

One of the simplest techniques for finding a useful realizable function which simulates the brick-wall loss function of Fig. 12-3 is to carry out a maximally flat approximation in the vicinity of $\omega_0 = 0$, where α and all its derivatives vanish. Choosing a *polynomial* rather than rational $K(s)$, the function

$$|K(j\omega)|^2 = K(j\omega)K(-j\omega) \tag{12-8}$$

will also be a polynomial in ω^2:

$$|K(j\omega)|^2 = C_n\omega^{2n} + C_{n-1}\omega^{2(n-1)} + \cdots + C_0 \tag{12-9}$$

Next, observe that α and $|K|^2$ satisfy

$$\alpha = 10 \log (1 + |K|^2)$$

$$\frac{d\alpha}{d(\omega^2)} = \frac{10 \log e}{1 + |K|^2} \frac{d|K|^2}{d(\omega^2)} \tag{12-10}$$

$$\frac{d^2\alpha}{d(\omega^2)^2} = \frac{10 \log e}{1 + |K|^2} \left| - \frac{1}{1 + |K|^2} \left[\frac{d|K|^2}{d(\omega^2)} \right]^2 + \frac{d^2|K|^2}{d(\omega^2)^2} \right|$$

From Eq. (12-10), it follows that whenever the conditions

$$|K|^2 = 0$$

$$\frac{d|K|^2}{d(\omega^2)} = 0 \tag{12-11}$$

$$\frac{d^2|K|^2}{d(\omega^2)^2} = 0$$

$$\cdots\cdots\cdots$$

hold for some $\omega = \omega_0$, the same conditions will be valid for α, $d\alpha/d(\omega^2)$, $d^2\alpha/d(\omega^2)^2$, etc., and vice versa. Hence, the maximally flat approximation of the brick-wall response of Fig. 12-3 requires that $|K|^2$ satisfy relations (12-11).

Applying, accordingly, the equations of (12-11) to the function of Eq. (12-9) one by one, we obtain the conditions

$$C_0 = C_1 = C_2 = \cdots C_{n-1} = 0 \tag{12-12}$$

The nth derivative of $|K|^2$ can only be matched to zero by setting C_n equal to zero. This, however, would result in $|K|^2 \equiv 0$ and hence $\alpha \equiv 0$ for all ω. Such an allpass network cannot be regarded as a useful approximation of the characteristic of Fig. 12-3. Hence, we keep $C_n > 0$† and obtain, for $s = j\omega$,

$$|K|^2 = C_n\omega^{2n} \qquad K(s) = \pm\sqrt{C_n}\, s^n \tag{12-13}$$

This means that by the Feldtkeller equation (6-29)

$$H(s)H(-s) = K(s)K(-s) + 1 = C_n(-1)^n s^{2n} + 1 \tag{12-14}$$

Therefore the zeros of $H(s)H(-s)$, that is, the natural modes and their mirror images with respect to the origin, satisfy

$$s_k^{2n} = (-1)^{n-1}C_n^{-1} = \frac{e^{j\pi(n-1+2k)}}{C_n} \qquad k = 1, 2, \ldots, 2n \tag{12-15}$$

or

$$s_k = C_n^{-1/2n}e^{j\pi(n-1+2k)/2n} \qquad k = 1, 2, \ldots, 2n \tag{12-16}$$

Thus, the s_k lie on a circle in the s plane, with a radius of $C_n^{-1/2n}$, at angles $\pi(n-1+2k)/2n$. The natural modes are, of course, those s_k which lie in the LHP, that is, $s_1, s_2, \ldots, s_n$. Figure 12-5 illustrates the $n = 4$ case.

The approximation represented by Eqs. (12-13) to (12-16) is called (after its first proponent) the *Butterworth approximation*. Filters realized using this process are called *Butterworth filters*.

At this stage, we see from Eq. (12-13) that all reflection zeros are at $s = 0$ and that all loss poles are at $s \to \infty$. We also see from Eq. (12-16) that the natural modes are at

$$s_k = C_n^{-1/2n}e^{j\pi(n-1+2k)/2n} \qquad k = 1, 2, \ldots, n \tag{12-17}$$

---

† $C_n < 0$ would make $|K|^2 < 0$, which is mathematically meaningless.

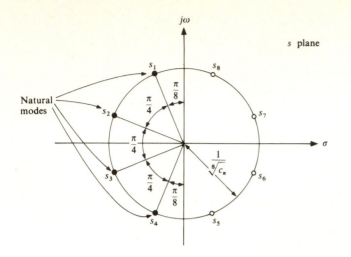

Figure 12-5 Natural modes and their mirror images for a Butterworth filter of degree $n = 4$.

We have yet to find C_n and the necessary degree n of the network. These parameters will depend on the actual specifications for the network. As mentioned earlier, they are usually of the form

$$
\begin{array}{ll}
\alpha \le \alpha_p & \text{for } 0 \le f \le f_p \\
\alpha \ge \alpha_s & \text{for } f_s \le f \le \infty
\end{array}
\tag{12-18}
$$

where α_p, α_s, f_p, and f_s are prescribed. These conditions can be illustrated graphically, as shown in Fig. 12-6. In order to meet the specifications, the $\alpha(\omega)$ curve must stay *below* the shaded barrier for $0 \le f \le f_p$ and *above* the barrier for $f_s \le f \le \infty$.

Consider now the Butterworth function given in (12-13). By Eq. (12-1),

$$
\alpha(\omega) = 10 \log (1 + |K|^2) = 10 \log (1 + C_n \omega^{2n})
\tag{12-19}
$$

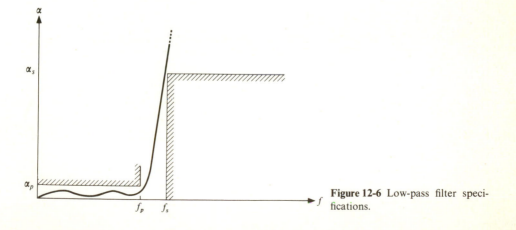

Figure 12-6 Low-pass filter specifications.

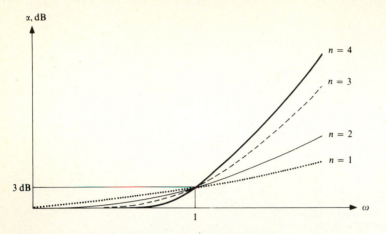

Figure 12-7 Loss responses of Butterworth filters for $n = 1, 2, 3,$ and 4 and $C_n = 1$.

The loss response given by Eq. (12-19) is illustrated in Fig. 12-7 for $C_n = 1$ and for various low values of n. Clearly, the greater n is, the lower the loss in the passband (if we assume $f_p \leq 1$) and the higher in the stopband (for $f_s \geq 1$). Because of the monotonic shape of the $\alpha(\omega)$ curve, if the inequalities (12-18) are satisfied at f_p and at f_s, they must be satisfied for all f. Hence, assuring

$$\alpha(f_p) \leq \alpha_p \qquad \alpha(f_s) \geq \alpha_s \tag{12-20}$$

is sufficient to guarantee that the conditions in Eq. (12-18) are met. Using Eqs. (12-19) and (12-20) gives

$$(2\pi f_p)^{2n} \leq \frac{10^{\alpha_p/10} - 1}{C_n} \qquad (2\pi f_s)^{-2n} \leq \frac{C_n}{10^{\alpha_s/10} - 1} \tag{12-21}$$

Since both sides in both inequalities are positive, we can multiply them together to obtain the new inequality

$$\left(\frac{f_p}{f_s}\right)^{2n} \leq \frac{10^{\alpha_p/10} - 1}{10^{\alpha_s/10} - 1} \tag{12-22}$$

At this stage, it is expedient to introduce the *selectivity parameter*

$$k \triangleq \frac{f_p}{f_s} < 1 \tag{12-23}$$

and the *discrimination parameter*

$$k_1 \triangleq \sqrt{\frac{10^{\alpha_p/10} - 1}{10^{\alpha_s/10} - 1}} \approx \frac{\sqrt{0.23\alpha_p}}{10^{\alpha_s/20}} \ll 1 \tag{12-24}$$

As Fig. 12-6 illustrates, the *larger* k is, the more selective, i.e., steeper, the response. Also, the *smaller* k_1 is, the greater the difference between passband and stopband loss.

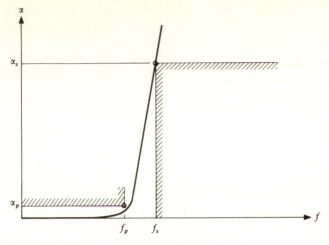

Figure 12-8 Filter response with design margin at f_p.

From Eqs. (12-22) to (12-24), the degree n satisfies

$$n \geq \frac{|\log k_1|}{|\log k|} = \frac{\log (1/k_1)}{\log (1/k)} \tag{12-25}$$

Obviously, in almost all cases the right-hand side of Eq. (12-25) gives a fractional (noninteger) number. Then n should be chosen as the next *higher* integer. Choosing n higher than (12-25) means that the requirements (12-20) will be surpassed. We can, for example, satisfy the specifications at f_s exactly and obtain a safety margin at f_p (Fig. 12-8).† Thus,

$$\alpha(f_s) = \alpha_s \qquad \alpha(f_p) < \alpha_p \tag{12-26}$$

With this choice, from Eq. (12-19),

$$C_n = \frac{10^{\alpha_s/10} - 1}{(2\pi f_s)^{2n}} \tag{12-27}$$

and hence, from Eq. (12-13)

$$|K|^2 = (10^{\alpha_s/10} - 1)\left(\frac{f}{f_s}\right)^{2n} \tag{12-28}$$

Therefore, by (12-1), the loss response is

$$\alpha = 10 \log \left[1 + (10^{\alpha_s/10} - 1)\left(\frac{f}{f_s}\right)^{2n}\right] \tag{12-29}$$

† The response is usually most sensitive to element dissipation, tolerances, etc., around f_p. Hence, we normally try to obtain some design margin there.

Example 12-3 Obtain the necessary degree n, the constant C_n, and the safety margin at f_p for the low-pass filter specified in Sec. 12-1. Find the characteristic function $K(s)$ and the transducer function $H(s)$. Design the circuit.

From Eqs. (12-23) to (12-25),

$$k = \frac{f_p}{f_s} = \frac{1.8}{7} \approx 0.257143$$

$$k_1 = \sqrt{\frac{10^{\alpha_p/10} - 1}{10^{\alpha_s/10} - 1}} = \sqrt{\frac{10^{0.1} - 1}{10^5 - 1}} \approx 1.60912 \times 10^{-3}$$

$$n \geq \frac{-\log k_1}{-\log k} \approx 4.736$$

Hence we choose $n = 5$. If we require (12-26) to hold, then, by (12-27),

$$C_n = \frac{10^{\alpha_s/10} - 1}{(2\pi f_s)^{2n}} = \frac{10^5 - 1}{(2\pi 7 \times 10^6)^{10}} \approx 3.7 \times 10^{-72}$$

This extremely small number and other very small or large numbers which would occur in the remainder of the calculation result from carrying out the calculations without using normalization. To improve the situation, we choose as our frequency unit the half-power (3-dB) radian frequency ω_0 of the response. This frequency is given by

$$|H(j\omega_0)|^2 = \frac{P_{max}}{P_2} = 2$$

$$|K(j\omega_0)|^2 = C_n \omega_0^{2n} = |H(j\omega_0)|^2 - 1 = 1 \qquad (12\text{-}30)$$

$$\omega_0 = C_n^{-1/(2n)}$$

Therefore, in terms of $\Omega = \omega/\omega_0$,

$$|K(j\Omega)|^2 = C_n(\omega_0\Omega)^{2n} = \Omega^{2n} \qquad (12\text{-}31)$$

A comparison of Eqs. (12-31) and (12-13) shows that the normalization results in replacing C_n by 1. Therefore, if we use Ω and $S = j\Omega$, for this specific example

$$|K|^2 = \Omega^{10} \qquad \text{and} \qquad K(S) = \pm S^5$$

The natural modes are then, by Eq. (12-16), at

$$S_k = e^{j\pi(4 + 2k)/10} = \cos\left[\pi(0.4 + 0.2k)\right] + j \sin\left[\pi(0.4 + 0.2k)\right] \qquad k = 1, 2, \ldots, 5$$

Hence

$$H(S) = \pm \prod_{k=1}^{5} (S - S_k) = \pm(S + 1)(S^2 + 0.618034S + 1)(S^2 + 1.618034S + 1)$$

$$= \pm(S^5 + a_4 S^4 + a_3 S^3 + a_2 S^2 + a_1 S + 1)$$

with $a_1 = a_4 = 3.23607$ and $a_2 = a_3 = 5.23607$. If the positive sign in both $K(S)$ and $H(S)$ is chosen and impedance normalization is used so that $R_G = R_L = 1$, Eq. (6-62) gives

$$z_{11} = \frac{H_e - K_e}{H_o + K_o} = \frac{a_4 S^4 + a_2 S^2 + 1}{2S^5 + a_3 S^3 + a_1 S}$$

with the a_k given above. Developing z_{11} into a ladder circuit using the techniques learned in

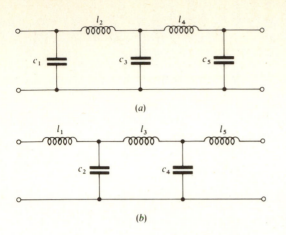

(a)

(b)   **Figure 12-9** Butterworth filters for $n = 5$.

Chaps. 5 and 6, the network of Fig. 12-9a results, with the normalized element values $c_1 = c_5 \approx 0.618034$, $c_3 = 2$ and $l_2 = l_4 \approx 1.618034$. The reader should carry out the details as an exercise.

Next, assume that the negative sign is chosen in $K(S)$ so that $K(S) = -S^5$. Then, from Eq. (6-65),

$$y_{11} = \frac{H_e + K_e}{H_o - K_o} = \frac{a_4 S^4 + a_2 S^2 + 1}{2S^5 + a_3 S^3 + a_1 S}$$

Ladder expansion gives now the circuit of Fig. 12-9b with $l_1 = l_5 \approx 0.618034$, $l_3 = 2$, and $c_2 = c_4 \approx 1.618034$. Clearly, the two circuits of Fig. 12-9 are duals of each other. Both are normalized in impedance (scaled down by 50) and frequency normalized to the still undetermined 3-dB radian frequency ω_0. To obtain the physical element values, ω_0 must be found. A simple way is provided by Eqs. (12-19) and (12-26), which give

$$\alpha_s = 10 \log \left[1 + \left(\frac{\omega_s}{\omega_0} \right)^{2n} \right]$$

$$\omega_0 = \omega_s (10^{\alpha_s/10} - 1)^{-1/2n} \tag{12-32}$$

$$= (2\pi 7 \times 10^6)(10^5 - 1)^{-0.1} \approx 1.39084 \times 10^7 \text{ rad/s}$$

Hence, the physical-element values can be obtained by multiplying the terminating resistors by $R_0 = 50$ Ω, all inductors by $L_0 = R_0/\omega_0 \approx 3.59494$ μH, and all capacitors by $C_0 = 1/R_0 \omega_0 \approx 1.43798$ nF. This routine task is left to the reader as an exercise.

We are also often interested in the behavior of the loss response for very small and very large values of ω. For $\Omega \ll 1$, by series expansion,

$$\alpha = 10 \log (1 + \Omega^{2n}) \approx (10 \log e) \Omega^{2n} \tag{12-33}$$

while for $\Omega \gg 1$

$$\alpha \approx 20n \log \Omega \tag{12-34}$$

Table 12-1 Zeros of normalized Butterworth polynomials†

$n = 1$	$n = 2$	$n = 3$	$n = 4$	$n = 5$	$n = 6$	$n = 7$	$n = 8$	$n = 9$	$n = 10$
-1.0000000	-0.7071068 $\pm j0.7071068$	-1.0000000	-0.3826834 $\pm j0.9238795$	-1.0000000	-0.2588190 $\pm j0.9659258$	-1.0000000	-0.1950903 $\pm j0.9807853$	-1.0000000	-0.1564345 $\pm j0.9876883$
		-0.5000000 $\pm j0.8660254$	-0.9238795 $\pm j0.3826834$	-0.3090170 $\pm j0.9510565$	-0.7071068 $\pm j0.7071068$	-0.2225209 $\pm j0.9749279$	-0.5555702 $\pm j0.8314696$	-0.1736482 $\pm j0.9848078$	-0.4539905 $\pm j0.8910065$
				-0.8090170 $\pm j0.5877852$	-0.9659258 $\pm j0.2588190$	-0.6234898 $\pm j0.7818315$	-0.8314696 $\pm j0.5555702$	-0.5000000 $\pm j0.8660254$	-0.7071068 $\pm j0.7071068$
						-0.9009689 $\pm j0.4338837$	-0.9807853 $\pm j0.1950903$	-0.7660444 $\pm j0.6427876$	-0.8910065 $\pm j0.4539905$
								-0.9396926 $\pm j0.3420201$	-0.9876883 $\pm j0.1564345$

† Reproduced by permission from L. Weinberg, "Network Analysis and Synthesis." p. 495, McGraw-Hill, New York, 1962; reprinted by Robert E. Krieger Publishing Co., Inc., Huntington, N.Y., 1975.

Table 12-2 Coefficients of normalized Butterworth polynomials ($a_0 = a_n = 1$ for all n)†

n	a_1	a_2	a_3	a_4	a_5	a_6	a_7	a_8	a_9
1									
2	1.4142136								
3	2.0000000	2.0000000							
4	2.6131259	3.4142136	2.6131259						
5	3.2360680	5.2360680	5.2360680	3.2360680					
6	3.8637033	7.4641016	9.1416202	7.4641016	3.8637033				
7	4.4939592	10.0978347	14.5917939	14.5917939	10.0978347	4.4939592			
8	5.1258309	13.1370712	21.8461510	25.6883559	21.8461510	13.1370712	5.1258309		
9	5.7587705	16.5817187	31.1634375	41.9863857	41.9863857	31.1634375	16.5817187	5.7587705	
10	6.3924532	20.4317291	42.8020611	64.8823963	74.2334292	64.8823963	42.8020611	20.4317291	6.3924532

† Reproduced by permission from L. Weinberg, "Network Analysis and Synthesis," p. 495, McGraw-Hill, New York, 1962; reprinted by Robert E. Krieger Publishing Co., Inc., Huntington, N.Y., 1975.

Table 12-3 Element values of Butterworth filters for $R_G = R_L = 1\,\Omega$ and $\omega_0 = 1$ rad/s†

Value of n	c_1 or l'_1	l_2 or c'_2	c_3 or l'_3	l_4 or c'_4	c_5 or l'_5	l_6 or c'_6	c_7 or l'_7	l_8 or c'_8	c_9 or l'_9	l_{10} or c'_{10}
1	2.0000									
2	1.4142	1.4142								
3	1.0000	2.0000	1.0000							
4	0.7654	1.8478	1.8478	0.7654						
5	0.6180	1.6180	2.0000	1.6180	0.6180					
6	0.5176	1.4142	1.9319	1.9319	1.4142	0.5176				
7	0.4450	1.2470	1.8019	2.0000	1.8019	1.2470	0.4450			
8	0.3902	1.1111	1.6629	1.9616	1.9616	1.6629	1.1111	0.3902		
9	0.3473	1.0000	1.5321	1.8794	2.0000	1.8794	1.5321	1.0000	0.3473	
10	0.3129	0.9080	1.4142	1.7820	1.9754	1.9754	1.7820	1.4142	0.9080	0.3129

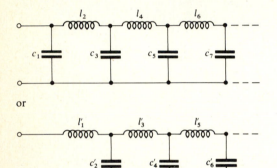

or

† Reproduced by permission from L. Weinberg, "Network Analysis and Synthesis," p. 605, McGraw-Hill, New York, 1962; reprinted by Robert E. Krieger Publishing Co., Inc., Huntington, N.Y., 1975.

Hence, doubling Ω results in an increase of

$$\Delta\alpha \approx 6.02n \text{ dB} \tag{12-35}$$

in the value of α. Therefore, on a logarithmic frequency scale, the α–vs.–log Ω curve tends asymptotically to a straight line, with a slope of 6.02n dB/octave.†

In the available literature, the normalized roots and coefficients of $H(S)$ are tabulated, as are the element values, for Butterworth filters. Some of these tables are reproduced in Tables 12-1 to 12-3. Our painfully computed circuits (Fig. 12-9) could readily have been found from Table 12-3. The calculation of n and ω_0, however, is necessary even if Table 12-3 is used. As a matter of interest, it should be noted that explicit formulas are also available[1,9] for the element values of Butterworth filters.

† An octave is the interval between two frequencies which have a ratio 1 : 2.

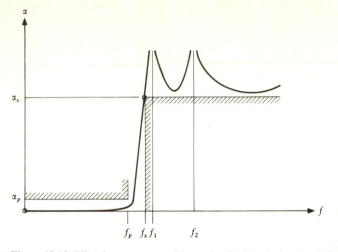

Figure 12-10 Filter loss response with maximally flat passband and finite loss poles.

A minor, but useful, modification of the Butterworth response is obtained if some (or all) of the loss poles are shifted from $\omega \to \infty$ to some finite frequencies ω_i. This results in a rational characteristic function $K(s)$ of the form

$$K(s) = \pm \sqrt{C_n} \frac{s^n}{\displaystyle\prod_{i=1}^{k} (s^2 + \omega_i^2)} \qquad k \le \frac{n}{2} \qquad (12\text{-}36)$$

The resulting loss response is still "maximally flat" at the frequency origin; this can easily be verified by repeatedly differentiating the loss function

$$\alpha = 10 \log \left[1 + C_n \frac{\omega^{2n}}{\displaystyle\prod_{i=1}^{k} (\omega_i^2 - \omega^2)^2} \right] \qquad (12\text{-}37)$$

The resulting new $\alpha(\omega)$ function is illustrated in Fig. 12-10. It is evident that the selectivity has improved, compared with the response of Fig. 12-8. The price paid is an increase of the number of elements, since the k finite-loss poles require tuned resonant circuits in k of the branches. Also, the previously established design formulas and tables are not applicable to these circuits. Their design can be performed using computer-aided techniques[2] to calculate the optimum values of the ω_i. These are beyond the scope of the present discussion. Only one class of this filter type, the *inverse Chebyshev filter*, to be discussed later, can be designed using straightforward analytical methods.

12-3 CHEBYSHEV FILTERS

An alternative way of approximating the ideal characteristics of Fig. 12-3 is to obtain an *equal-ripple* error in the range $0 \leq \omega \leq \omega_p$. It is convenient to introduce immediately frequency normalization, with

$$\omega_0 = 2\pi f_p \tag{12-38}$$

as the frequency unit; i.e., we are normalizing to the passband limit. The desired response is then that shown in Fig. 12-11. The oscillatory response in the pass-band immediately suggests a squared and horizontally compressed trigonometric function. Hence, we attempt to find the solution for $|K|^2$ in the form[†]

$$|K|^2 = k_p^2 \cos^2 nu(\Omega) \tag{12-39}$$

where $u(\Omega)$ is some function of Ω. If we choose

$$u(\Omega) = \cos^{-1} \Omega \tag{12-40}$$

the following conclusions can be drawn:

1. $|K|^2$ is a polynomial in Ω^2, since

$$\cos nu = \mathrm{Re}\,(e^{ju})^n = \mathrm{Re}\,(\cos u + j \sin u)^n$$

$$= \cos^n u - \binom{n}{2} \cos^{n-2} u \,(1 - \cos^2 u) + \binom{n}{4} \cos^{n-4} u \,(1 - \cos^2 u)^2 - + \cdots \tag{12-41}$$

[†] Of course, this solution can also be found in a less heuristic way. See, for example, N. Balabanian, "Network Synthesis," Prentice-Hall, Englewood Cliffs, N.J., 1958.

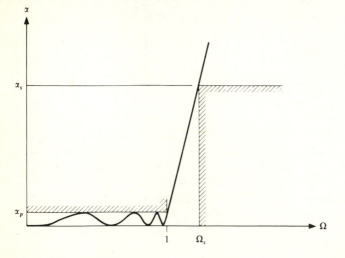

Figure 12-11 Chebyshev passband loss response.

where $\binom{n}{k} = n!/[k!(n-k)!]$. From Eq. (12-41) it is obvious that for n even (odd), $\cos nu$ is a pure even (odd) polynomial in $\cos u$. From this fact and Eqs. (12-39) and (12-40) our statement follows.

2. $|K|^2$ oscillates between 0 and k_p^2 for $-1 \le \Omega \le +1$. This is true, since by Eq. (12-40) in this range u is real and as Ω grows from -1 to $+1$, u can be considered to grow from $-\pi$ to 0. (In fact, $\cos^{-1} \Omega$ is multivalued, and hence this choice is arbitrary.) Hence, $|K|^2$ oscillates between zero and k_p^2, taking on the value 0 a total of n times and the value k_p^2 a total of $n+1$ times between $-\Omega_p = -1$ and $\Omega_p = +1$.

3. For values of Ω greater than 1, $|K|^2$ tends monotonically to infinity. This can be seen from the relation

$$\cos nu = \tfrac{1}{2}(e^{jnu} + e^{-jnu})$$
$$= \tfrac{1}{2}[(\cos u + \sqrt{\cos^2 u - 1})^n + (\cos u + \sqrt{\cos^2 u - 1})^{-n}]$$
$$= \tfrac{1}{2}[(\Omega + \sqrt{\Omega^2 - 1})^n + (\Omega + \sqrt{\Omega^2 - 1})^{-n}] \tag{12-42}$$

For $\Omega \to \infty$, $\cos nu \to 2^{n-1}\Omega^n$, and hence $|K|^2 \to k_p^2 2^{2n-2}\Omega^{2n}$.

Due to the described properties, the filter defined by Eqs. (12-39) and (12-40) does indeed have the behavior shown in Fig. 12-11 (which illustrates the $n=7$ case). The maximum passband loss α_p and k_p are related, from Eq. (12-1), by

$$\alpha_p = 10 \log (1 + k_p^2) \tag{12-43}$$

Filters which have $|K|^2$ given by (12-39) and (12-40) are called *Chebyshev filters* after the mathematician first analyzing the properties of the polynomials $\cos (n \cos^{-1} x)$, called *Chebyshev polynomials*.

Let us now compare the stopband responses of a Butterworth and a Chebyshev filter with the same passband limit, say $\Omega_p = 1$, and the same maximum passband loss α_p. In the stopband, typically $\alpha > 30$ dB, and hence by (12-1) $|K|^2 \gg 1$; therefore

$$\alpha \approx 20 \log |K| \tag{12-44}$$

Then, for the loss α_B of the Butterworth filter, using Eq. (12-34), we get

$$\alpha_B \approx 10 \log k_p^2 \Omega^{2n} = 20 \log k_p + 20n \log \Omega \tag{12-45}$$

while for the loss α_{Ch} of the Chebyshev filter of the same degree n,

$$\alpha_{Ch} \approx 20 \log k_p + (10)(2n - 2) \log 2 + 20n \log \Omega \tag{12-46}$$

Hence, at the same stopband frequency Ω,

$$\alpha_{Ch} \approx \alpha_B + 6.02(n - 1) \tag{12-47}$$

For even moderate degrees, the additional stopband loss $6.02(n-1)$ dB of the Chebyshev filter is significant. For example, for $n=5$, the added loss is over 24 dB. This illustrates that the Chebyshev filter is significantly more efficient in its loss characteristics than the Butterworth filter.

Since $|K|^2$ is a polynomial function of Ω^2, as we have already seen from

Eqs. (12-39) to (12-41), all poles of $|K|^2$, that is, all loss poles, lie at $\Omega \to \infty$. The reflection zeros, by Eq. (12-39), are located at values of u satisfying

$$nu_k^{(r)} = \frac{2k-1}{2}\pi \qquad k = 1, 2, \ldots, n \tag{12-48}$$

or, using Eq. (12-40),

$$\Omega_k^{(r)} = \cos u_k^{(r)} = \cos\frac{2k-1}{2n}\pi \qquad k = 1, 2, \ldots, n \tag{12-49}$$

The natural modes S_k and their mirror images can be found by extending Eqs. (12-39) and (12-40), which are valid on the $j\Omega$ axis only. Hence, replacing Ω by S/j, we get

$$|K|^2 + 1 = k_p^2 \cos^2 nu_k + 1 = 0 \qquad S_k = j \cos u_k \qquad k = 1, 2, \ldots, 2n \tag{12-50}$$

We have to select the S_k in the LHP as the natural modes.

Anticipating complex solutions for both u_k and S_k, we have by analytic continuation

$$u_k = v_k + jw_k \qquad S_k = \Sigma_k + j\Omega_k \tag{12-51}$$

where v_k, w_k, Σ_k, and Ω_k are all real. Then, using the familiar identity

$$\cos(x + jy) = \cos x \cos jy - \sin x \sin jy = \cos x \cosh y - j \sin x \sinh y \tag{12-52}$$

by the first relation in Eq. (12-50) we have

$$\cos nu_k = \cos nv_k \cosh nw_k - j \sin nv_k \sinh nw_k = \pm\frac{j}{k_p} \tag{12-53}$$

Equating real and imaginary parts on both sides, we get

$$\cos nv_k \cosh nw_k = 0 \qquad \sin nv_k \sinh nw_k = \pm\frac{1}{k_p} \tag{12-54}$$

Since $\cosh nw_k > 0$, we can write

$$nv_k = \pm\frac{2k-1}{2}\pi \qquad k = 1, 2, \ldots \tag{12-55}$$

and hence $\sin nv_k = \pm 1$ and

$$\sinh nw_k = \pm\frac{1}{k_p} \tag{12-56}$$

$$w_k = \pm\frac{1}{n}\sinh^{-1}\frac{1}{k_p} = \pm\frac{1}{n}\ln\left(\frac{1}{k_p} + \sqrt{\frac{1}{k_p^2} + 1}\right)$$

At this stage, v_k and w_k are known, from Eqs. (12-55) and (12-56). Substituting into Eq. (12-50) gives

$$S_k = \Sigma_k + j\Omega_k = j \cos u_k = j \cos (v_k + jw_k)$$

$$\Sigma_k + j\Omega_k = j \cos v_k \cosh w_k + \sin v_k \sinh w_k$$

(12-57)

Hence, $\qquad \Sigma_k = \sin v_k \sinh w_k \qquad \Omega_k = \cos v_k \cosh w_k$ (12-58)

or, from Eqs. (12-55) and (12-56),

$$\Sigma_k = -\sin \left(\frac{2k-1}{n}\frac{\pi}{2}\right)\frac{1}{2}(a^{1/n} - a^{-1/n})$$

$$\Omega_k = \cos \left(\frac{2k-1}{n}\frac{\pi}{2}\right)\frac{1}{2}(a^{1/n} + a^{-1/n})$$

(12-59)

In Eq. (12-59), $k = 1, 2, \ldots, n$, and

$$a \triangleq \frac{1}{k_p} + \sqrt{\frac{1}{k_p^2} + 1}$$

(12-60)

The locus of the S_k in the S plane can be found by solving the two equations in (12-59) for $\sin [(2k-1)\pi/2n]$ and $\cos [(2k-1)\pi/2n]$, respectively, and then squaring and adding the two relations. This gives

$$\frac{\Sigma_k^2}{[(a^{1/n} - a^{-1/n})/2]^2} + \frac{\Omega_k^2}{[(a^{1/n} + a^{-1/n})/2]^2} = \sin^2 \left(\frac{2k-1}{n}\frac{\pi}{2}\right) + \cos^2 \left(\frac{2k-1}{n}\frac{\pi}{2}\right)$$

$$= 1$$

(12-61)

Hence, the locus is an *ellipse*, with half axes $(a^{1/n} + a^{-1/n})/2$ and $(a^{1/n} - a^{-1/n})/2$. The natural modes for the $n = 4$ case are illustrated in Fig. 12-12.

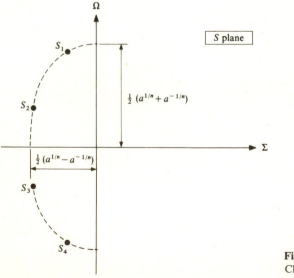

Figure 12-12 Natural modes of a Chebyshev filter for $n = 4$.

Next, the degree n necessary to satisfy a given set of specifications will be derived. For requirements of the form shown in Eq. (12-20), from Eqs. (12-39) and (12-40), we find that the passband requirement is always satisfied if k_p is obtained from (12-43) and if the normalization

$$\Omega = \frac{\omega}{\omega_0} = \frac{2\pi f}{2\pi f_p} = \frac{f}{f_p} \tag{12-62}$$

is used. To satisfy the stopband specification (Fig. 12-11), we must have

$$\alpha(\Omega_s) = 10 \log \left[|K(j\Omega_s)|^2 + 1 \right] \geq \alpha_s \tag{12-63}$$

By Eqs. (12-39) and (12-40), therefore,

$$k_p^2 \cos^2 nu(\Omega_s) \geq 10^{\alpha_s/10} - 1 \qquad u(\Omega_s) = \cos^{-1} \Omega_s \tag{12-64}$$

Now since $\Omega_s > 1$, $u(\Omega_s)$ is imaginary, i.e.,

$$u(\Omega_s) = jw_s \tag{12-65}$$

but
$$\qquad w_s = -ju(\Omega_s) = -j \cos^{-1} \Omega_s = \cosh^{-1} \Omega_s \tag{12-66}$$

is real. By Eq. (12-64),

$$k_p \cos njw_s = k_p \cosh nw_s \geq \sqrt{10^{\alpha_s/10} - 1} \tag{12-67}$$

Hence, from Eqs. (12-66) and (12-67),

$$nw_s = n \cosh^{-1} \Omega_s \geq \cosh^{-1} \frac{\sqrt{10^{\alpha_s/10} - 1}}{k_p} \tag{12-68}$$

Next, we can express k_p from (12-43)

$$k_p = \sqrt{10^{\alpha_p/10} - 1} \tag{12-69}$$

and Ω_s from (12-23)

$$\Omega_s = \frac{\Omega_p}{k} = \frac{1}{k} \tag{12-70}$$

Substituting into (12-68), and using (12-24), we obtain

$$n \geq \frac{\cosh^{-1} \sqrt{(10^{\alpha_s/10} - 1)/(10^{\alpha_p/10} - 1)}}{\cosh^{-1} \Omega_s} = \frac{\cosh^{-1} (1/k_1)}{\cosh^{-1} (1/k)} \tag{12-71}$$

which is the desired result. It is of some interest to note the similarity of Eqs. (12-25) and (12-71). In both, n can be expressed exclusively in terms of k and k_1.

Example 12-4 Design a Chebyshev low-pass filter for the specifications

$$\alpha_p = 1 \text{ dB} \qquad f_p = 1.8 \text{ MHz} \qquad \alpha_s = 50 \text{ dB} \qquad f_s = 7 \text{ MHz} \qquad R_G = R_L = 50 \ \Omega$$

satisfied earlier by a Butterworth filter.

As found before, $k = 0.257143$ and $k_1 = 1.60912 \times 10^{-3}$. Hence, from (12-71),

$$n \geq \frac{\cosh^{-1}(1/k_1)}{\cosh^{-1}(1/k)} \approx 3.5025$$

Hence, $n = 4$ can be chosen, one lower than for the Butterworth filter. From (12-41) and (12-40) then

$$\cos 4u = \cos^4 u - \binom{4}{2} \cos^2 u(1 - \cos^2 u) + \binom{4}{4}(1 - \cos^2 u)^2 = 8\Omega^4 - 8\Omega^2 + 1$$

Hence, by (12-39),

$$|K|^2 = k_p^2(8\Omega^4 - 8\Omega^2 + 1)^2$$

where now from (12-69)

$$k_p = \sqrt{10^{\alpha_p/10} - 1} = \sqrt{10^{0.1} - 1} \approx 0.508847$$

Hence, we can choose

$$K(S) = \pm k_p(8S^4 + 8S^2 + 1) \approx \pm(4.070776S^4 + 4.070776S^2 + 0.508847)$$

The natural modes can be obtained, say from (12-58), where now, by (12-55),

$$v_k = \frac{\pm(k - \frac{1}{2})\pi}{4} \qquad k = 1, 2, 3, 4$$

and, by (12-56),

$$w_k = \pm\frac{1}{4}\sinh^{-1}\frac{1}{k_p} = \pm0.356994$$

Hence, for the first natural mode

$$\Sigma_1 = -\sin\frac{\pi}{8}\sinh 0.356994 \approx -0.139536 \qquad \Omega_1 = \cos\frac{\pi}{8}\cosh 0.356994 \approx 0.983379$$

Proceeding this way, we can find all $S_k = \Sigma_k + j\Omega_k$. We obtain $S_4 = S_1^*$ and $S_2 = S_3^* = -0.33687 + j0.407329$. Then

$$H(S) = C\prod_{i=1}^{4}(S - S_i)$$

Here the constant factor C is the coefficient of S^4 in $H(S)$. By the Feldtkeller equation, the coefficients of S^4 in $K(S)$ and $H(S)$ must have the same absolute values. Hence $C = \pm4.070776$. The overall function is thus found to be

$$H(S) = \pm4.070776(S^2 - 2\Sigma_1 S + \Sigma_1^2 + \Omega_1^2)(S^2 - 2\Sigma_2 S + \Sigma_2^2 + \Omega_2^2)$$

$$= \pm(a_4 S^4 + a_3 S^3 + a_2 S^2 + a_1 S + a_0)$$

where $\qquad a_4 = 4.070776 \qquad a_3 = 3.878684 \qquad a_2 = 5.9186$

$$a_1 = 3.02304 \qquad a_0 = 1.12202$$

Hence, when we use impedance normalization and choose the positive sign in both $K(S)$ and $H(S)$, Eq. (6-65) gives

$$y_{11} = \frac{H_e + K_e}{H_o - K_o} = \frac{8.141552S^4 + 9.98938S^2 + 1.630867}{3.87868S^3 + 3.02304S}$$

Table 12-4 Natural modes for Chebyshev filters with $\alpha_p = 0.5$ dB and $\alpha_p = 1$ dB†

$\tfrac{1}{2}$-dB ripple

$n=1$	$n=2$	$n=3$	$n=4$	$n=5$	$n=6$	$n=7$	$n=8$	$n=9$	$n=10$
−2.8627752	−0.7128122 ±j1.0040425	−0.6264565	−0.1753531 ±j1.0162529	−0.3623196	−0.0776501 ±j1.0084608	−0.2561700	−0.0436201 ±j1.0050021	−0.1984053	−0.0278994 ±j1.0032732
		−0.3132282 ±j1.0219275	−0.4233398 ±j0.4209457	−0.1119629 ±j1.0115574	−0.2121440 ±j0.7382446	−0.0570032 ±j1.006405	−0.1242195 ±j0.8519996	−0.0344527 ±j1.0040040	−0.0809672 ±j0.9050658
				−0.2931227 ±j0.6251768	−0.2897940 ±j0.2702162	−0.1597194 ±j0.8070770	−0.1859076 ±j0.5692879	−0.0992026 ±j0.8829063	−0.1261094 ±j0.7182643
						−0.2308012 ±j0.4478939	−0.2192929 ±j0.1999073	−0.1519873 ±j0.6553170	−0.1589072 ±j0.4611541
								−0.1864400 ±j0.3486869	−0.1761499 ±j0.1589029

1-dB ripple

$n=1$	$n=2$	$n=3$	$n=4$	$n=5$	$n=6$	$n=7$	$n=8$	$n=9$	$n=10$
−1.9652267	−0.5488672 ±j0.8951286	−0.4941706	−0.1395360 ±j0.9833792	−0.2894933	−0.0621810 ±j0.9934115	−0.2054141	−0.0350082 ±j0.9964513	−0.1593305	−0.0224144 ±j0.9977755
		−0.2470853 ±j0.9659987	−0.3368697 ±j0.4073290	−0.0894584 ±j0.9901071	−0.1698817 ±j0.7272275	−0.0457089 ±j0.9952839	−0.0996950 ±j0.8447506	−0.0276674 ±j0.9972297	−0.1013166 ±j0.7143284
				−0.2342050 ±j0.6119198	−0.2320627 ±j0.2661837	−0.1280736 ±j0.7981557	−0.1492041 ±j0.5644443	−0.0796652 ±j0.8769490	−0.0650493 ±j0.9001063
						−0.1850717 ±j0.4429430	−0.1759983 ±j0.1982065	−0.1220542 ±j0.6508954	−0.1276664 ±j0.4586271
								−0.1497217 ±j0.3463342	−0.1415193 ±j0.1580321

† Reproduced by permission from L. Weinberg, "Network Synthesis and Analysis," p. 514, McGraw-Hill, New York, 1962; reprinted by Robert E. Krieger Publishing Co., Inc., Huntington, N.Y., 1975.

Table 12-5 Coefficients of the natural-mode polynomial for Chebyshev filters with $\alpha_p = 0.5$ dB and $\alpha_p = 1$ dB†

All coefficients are divided by $2^{n-1}k_p$. Hence $H(S) = 2^{n-1}k_p\left(S^n + \sum_{k=0}^{n-1} a_k S^k\right)$, where the a_k are the coefficients in the table.

n	a_0	a_1	a_2	a_3	a_4	a_5	a_6	a_7	a_8	a_9
					½-dB ripple					
1	2.8627752									
2	1.5162026	1.4256245								
3	0.7156938	1.5348954	1.2529130							
4	0.3790506	1.0254553	1.7168662	1.1973856						
5	0.1789234	0.7525181	1.3095747	1.9373675	1.1724909					
6	0.0947626	0.4323669	1.1718613	1.5897635	2.1718446	1.1591761				
7	0.0447309	0.2820722	0.7556511	1.6479029	1.8694079	2.4126510	1.1512176			
8	0.0236907	0.1525444	0.5735604	1.1485894	2.1840154	2.1492173	2.6567498	1.1460801		
9	0.0111827	0.0941198	0.3408193	0.9836199	1.6113880	2.814990	2.4293297	2.9027337	1.1425705	
10	0.0059227	0.0492855	0.2372688	0.6269689	1.5274307	2.1442372	3.4409268	2.7097415	3.1498757	1.1400664
					1-dB ripple					
1	1.9652267									
2	1.1025103	1.0977343								
3	0.4913067	1.2384092	0.9883412							
4	0.2756276	0.7426194	1.4539248	0.9528114						
5	0.1228267	0.5805342	0.9743961	1.6888160	0.9368201					
6	0.0689069	0.3070808	0.9393461	1.2021409	1.9308256	0.9282510				
7	0.0307066	0.2136712	0.5486192	1.3575440	1.4287930	2.1760778	0.9231228			
8	0.0172267	0.1073447	0.4478257	0.8468243	1.8369024	1.6551557	2.4230264	0.9198113		
9	0.0076767	0.0706048	0.2441864	0.7863109	1.2016071	2.3781188	1.8814798	2.6709468	0.9175476	
10	0.0043067	0.0344971	0.1824512	0.4553892	1.2444914	1.6129856	2.9815094	2.1078524	2.9194657	0.9159320

† Reproduced by permission from L. Weinberg, "Network Synthesis and Analysis," p. 516, McGraw-Hill, New York, 1962; reprinted by Robert E. Krieger Publishing Co., Inc., Huntington, N.Y., 1975.

Table 12-6 Element values for Chebyshev filters with $\alpha_p = 0.5$ dB†

The transformer ratio t is prescribed for each circuit

Value of n	c_1 or l'_1	l_2 or c'_2	c_3 or l'_3	l_4 or c'_4	c_5 or l'_5	l_6 or c'_6	c_7 or l'_7	l_8 or c'_8	c_9 or l'_9	l_{10} or c'_{10}
					$t^2 = 3$					
1	1.3972									
2	2.8282	0.3109								
3	4.3200	0.4405	2.9371							
4	3.6172	0.6399	4.1985	0.3620						
5	4.7896	0.5293	5.8898	0.4809	3.1130					
6	3.7922	0.6851	4.8770	0.6852	4.3536	0.3722				
7	4.9305	0.5495	6.2770	0.5603	6.0535	0.4901	3.1632			
8	3.8560	0.6990	5.0230	0.7235	4.9937	0.6953	4.4022	0.3759		
9	4.9901	0.5572	6.3947	0.5770	6.4061	0.5671	6.1064	0.4936	3.1841	
10	3.8860	0.7051	5.0780	0.7348	5.1229	0.7314	5.0307	0.6993	4.4237	0.3776

$t^2 = 2$

n	1	2	3	4	5	6	7	8	9	10
1	1.0479									
2	1.5132	0.6538								
3	2.9431	0.6503	2.1903							
4	1.8158	1.1328	2.4881	0.7732						
5	3.2228	0.7645	4.1228	0.7116	2.3197					
6	1.8786	1.1884	2.7589	1.2403	2.5976	0.7976				
7	3.3055	0.7899	4.3575	0.8132	4.2419	0.7252	2.3566			
8	1.9012	1.2053	2.8152	1.2864	2.8479	1.2628	2.6310	0.8063		
9	3.3403	0.7995	4.4283	0.8341	4.4546	0.8235	4.2795	0.7304	2.3719	
10	1.9117	1.2127	2.8366	1.2999	2.8964	1.3054	2.8744	1.2714	2.6456	0.8104

$t^2 = 1$

n	1	2	3	4	5	6	7	8	9
1	0.6986								
2	1.5963	1.0967							
3	1.7058	1.2296	1.5963						
4									
5	1.7058	1.2296	2.5408	1.2296	1.7058				
6									
7	1.7373	1.2582	2.6383	1.3443	2.6383	1.2582	1.7373		
8									
9	1.7504	1.2690	2.6678	1.3673	2.7239	1.3673	2.6678	1.2690	1.7504
10									

† Reproduced by permission from L. Weinberg, "Network Synthesis and Analysis," p. 611, McGraw-Hill, New York, 1962; reprinted by Robert E. Krieger Publishing Co., Inc., Huntington, N.Y., 1975.

Table 12-7 Element values for Chebyshev filters with $\alpha_p = 1$ dB†

The transformer ratio t is prescribed for each circuit

Value of n	c_1 or l'_1	l_2 or c'_2	c_3 or l'_3	l_4 or c'_4	c_5 or l'_5	l_6 or c'_6	c_7 or l'_7	l_8 or c'_8	c_9 or l'_9	l_{10} or c'_{10}
					$t^2 = 3$					
1	2.0354									
2	2.5721	0.4702								
3	4.9893	0.4286	3.8075							
4	3.0355	0.7929	3.7589	0.5347						
5	5.3830	0.4915	6.6673	0.4622	3.9944					
6	3.1307	0.8287	4.1451	0.8467	3.8812	0.5475				
7	5.4978	0.5050	6.9839	0.5177	6.8280	0.4696	4.0473			
8	3.1647	0.8395	4.2237	0.8764	4.2404	0.8580	3.9186	0.5520		
9	5.5459	0.5101	7.0783	0.5288	7.1141	0.5232	6.8785	0.4724	4.0693	
10	3.1806	0.8442	4.2532	0.8851	4.3088	0.8857	4.2691	0.8623	3.9349	0.5541

$$t^2 = 2$$

1	1.5265							
2	3.4774	0.6153						
3	3.7211	0.6949	2.8540					
4	3.7916	0.7118	4.7448	0.6650				
5	3.8210	0.7182	4.9425	0.7348	2.9936			
6			5.0013	0.7485	4.8636	0.6757		
7					5.0412	0.7429	3.0331	
8							4.9004	0.6797
9									3.0495
10								

$$t^2 = 1$$

1	1.0177							
2	2.0236	0.9941						
3	2.1349	1.0911	2.0236					
4	2.1666	1.1115	3.0009	1.0911				
5	2.1797	1.1192	3.0936	1.1735	2.1349			
6			3.1214	1.1897	3.0936	1.1115		
7					3.1746	1.1897	2.1666	
8							3.1214	1.1192
9									2.1797
10								

† Reproduced by permission from L. Weinberg, "Network Synthesis and Analysis," McGraw-Hill, New York, 1962; reprinted by Robert E. Krieger Publishing Co., Inc., Huntington, N.Y., 1975.

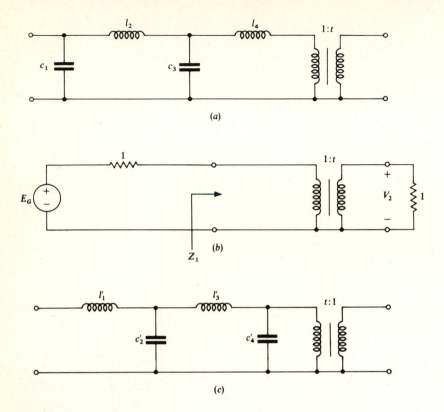

Figure 12-13 (a) Chebyshev filter; (b) equivalent circuit for $\omega = 0$; (c) dual filter.

Developing y_{11} into a ladder gives the normalized circuit of Fig. 12-13a. The element values can be found the usual way to be $c_1 = 2.09905$, $l_2 = 1.06444$, $c_3 = 2.831$, and $l_4 = 0.7892$. The transformer ratio can be found, for example, by developing y_{22}; it is much simpler to note, however, that at zero frequency, by Fig. 12-13b, the input impedance is $Z_1(0) = t^{-2}$. Hence, by (6-10), at zero frequency

$$\rho_1(0) = \frac{R_G - Z_1(0)}{R_G + Z_1(0)} = \frac{1 - t^{-2}}{1 + t^{-2}} = \frac{t^2 - 1}{t^2 + 1}$$

Also, by (6-21),

$$\rho_1(0) = \frac{K(0)}{H(0)} = \frac{0.508847}{1.12202} \approx 0.453510$$

Equating the two expressions and solving for t gives $t = \pm 1.630864$ (the negative sign means an inverting transformer).

Using the negative sign in $K(S)$ merely replaces the circuit by its dual. This is shown in Fig. 12-13c. The elements are $l'_1 = c_1$, $c'_2 = l_2$, $l'_3 = c_2$, and $c'_4 = l_4$.

As Eq. (12-50) shows, the natural modes S_k of the Chebyshev filter depend on k_p and hence on α_p. Therefore, any tabulation of the S_k, and hence of $H(S)$ and the

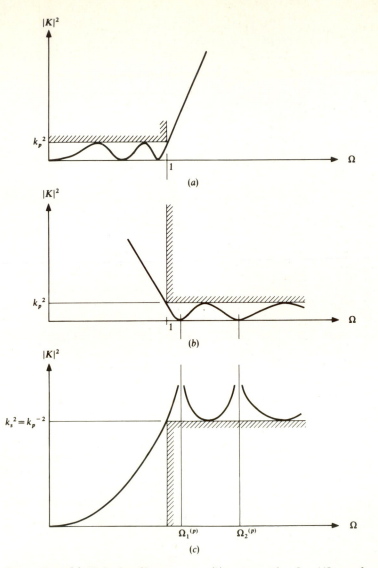

Figure 12-14 (a) Chebyshev filter response; (b) response after $\Omega \to 1/\Omega$ transformation; (c) response after $|K|^2 \to 1/|K|^2$ transformation.

element values, must have α_p as a parameter, and a different table will apply for each α_p. As an illustration, Tables 12-4 to 12-7 give the natural modes, the coefficients of $H(S)$, and the element values for Chebyshev filters with $\alpha_p = 0.5$ dB and $\alpha_p = 1$ dB. The circuit configurations are the same as in Fig. 12-13a and c, for unprimed and primed element values, respectively.

A filter response closely related to the Chebyshev response just discussed is the *inverse Chebyshev* (or *Chebyshev stopband*) characteristics. This can be obtained from the Chebyshev function in the following steps (Fig. 12-14):

1. Replace Ω by Ω^{-1} in Eq. (12-39). This will turn the $|K|^2$-vs.-Ω curve around, with the $\Omega = 1$ point as the pivot (Fig. 12-14b).
2. Replace $|K|^2$ by $|K|^{-2}$; this will result in the equal-ripple stopband characteristics shown in Fig. 12-14c.

Accordingly, the squared modulus of the characteristic function for the inverse Chebyshev function will be given by the expressions

$$|K|^2 = \frac{k_s^2}{\cos^2 nu(\Omega)} \tag{12-72}$$

and

$$u(\Omega) = \cos^{-1} \frac{1}{\Omega} \tag{12-73}$$

where

$$k_s^2 = 10^{\alpha_s/10} - 1 \triangleq \frac{1}{k_p^2} \tag{12-74}$$

From the steps leading to Eqs. (12-72) to (12-74), it is evident that the unit frequency is now the *stopband* limit frequency (Fig. 12-14). It can also be shown that Eq. (12-71), giving n in terms of k and k_1, remains valid for inverse Chebyshev filters. The proof is left as an exercise (Prob. 12-13). By the construction of $|K|^2$ it also follows that the loss poles are located at the reciprocal frequencies of the Chebyshev-filter reflection zeros. Hence, by (12-49),

$$\Omega_k^{(p)} = \frac{1}{\cos[(2k-1)/2\pi n]} \qquad k = 1, 2, \ldots, n \tag{12-75}$$

(Fig. 12-14c). The calculation of the natural modes is left to the reader (Prob. 12-15).

12-4 EQUAL-RIPPLE PASSBAND AND GENERAL STOPBAND FILTERS

A generalization of Chebyshev filters, important in practical problems, is provided by the class of filters with equal-ripple passbands and prescribed finite loss poles. A typical response is shown in Fig. 12-15a. These circuits are also often called (somewhat imprecisely) *general-parameter filters*.

The calculation of such a response with prescribed values of k_p (or, equivalently, α_p) and of the loss poles $\Omega_1, \Omega_2, \ldots$ requires some manipulations. Consider the transformation

$$Z \triangleq \sqrt{1 + S^{-2}} \qquad \text{Re } Z \geq 0 \tag{12-76}$$

Since in the passband $S = j\Omega$, $|\Omega| \leq 1$, the values of S in the passband transform into imaginary values jY of $Z = X + jY$. In the stopband $S = j\Omega$, $|\Omega| > 1$,

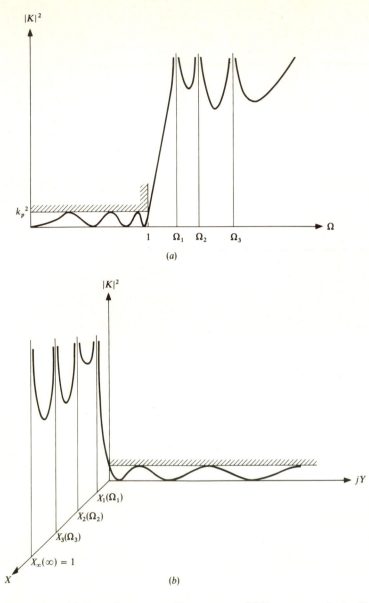

Figure 12-15 (a) General-parameter filter response; (b) filter response in the Z domain.

and hence Z is now real; that is, $Z = X$, where X, by (12-76), is nonnegative. The transformation is illustrated in Fig. 12-15b by showing the characteristic of Fig. 12-15a in the Z plane. A loss pole at $\Omega_i > 1$ is transformed to the real value

$$X_i = +\sqrt{1 - \Omega_i^{-2}} \qquad (12\text{-}77)$$

and hence clearly $0 < X_i < 1$. Any loss pole at $\Omega \to \infty$ is transformed to $X_\infty = 1$.

Next we define the polynomial

$$P(Z) \triangleq \prod_{i=1}^{n} (Z + X_i) \tag{12-78}$$

All desired loss poles X_i (including the X_∞) must be included in $P(Z)$; since poles at $+j\Omega_i$ and $-j\Omega_i$ map to the same X_i, such finite poles contribute a squared factor $(Z + X_i)^2$.

Let us now analyze the properties of the function

$$|K|^2 = k_p^2 \frac{[P_e(Z)]^2}{P(Z)P(-Z)} = k_p^2 \frac{[P_e(Z)]^2}{[P_e(Z)]^2 - [P_o(Z)]^2} = \frac{k_p^2}{1 - [P_o(Z)/P_e(Z)]^2} \tag{12-79}$$

Here, as before, P_e denotes the even part and P_o the odd part of $P(Z)$. Clearly, $|K|^2$ is an even rational function in Z, that is, a rational function of Z^2. By (12-76) therefore it is a rational function of S^2. Since $P(-Z) = \prod_{i=1}^{n} (-Z + X_i)$ is contained in the denominator, $|K|^2$ possesses the desired poles X_i. Finally, in terms of the transformed variable Z, $P(Z)$ is by (12-78) a strictly Hurwitz polynomial since all $X_i > 0$. Hence, $P_o(Z)/P_e(Z)$ is a reactance function in Z. Therefore, for $Z = jY$, that is, for values of Z in the transformed passband, the function $P_o(jY)/P_e(jY)$ is pure imaginary and acts as a reactance (Fig. 12-16a). Hence, $-(P_o/P_e)^2$ is a nonnegative function which oscillates between 0 and ∞ as Y increases from 0 to ∞. Now, as the last expression in (12-79) shows, $|K|^2 \to 0$ when $-(P_o/P_e)^2 \to \infty$; $|K|^2 = k_p^2$ when $-(P_o/P_e)^2 = 0$; and $0 < |K|^2 < k_p^2$ when $0 < -(P_o/P_e)^2 < \infty$. Thus, $|K|^2$ oscillates between 0 and k_p^2 as Y increases from 0 to ∞.

The above results show that $|K|^2$ as given in (12-79) does indeed have the desired response illustrated in Figs. 12-14 and 12-15 and is also a realizable function.†

It is possible to bring (12-79) to a form which indicates clearly that it is an extension of the Chebyshev filter function defined in Eqs. (12-39) and (12-40). It is easy to see that

$$|K|^2 = \frac{k_p^2}{4} \frac{[P(Z) + P(-Z)]^2}{P(Z)P(-Z)} = \frac{k_p^2}{2} \left[1 + \frac{1}{2} \frac{P(Z)}{P(-Z)} + \frac{1}{2} \frac{P(-Z)}{P(Z)} \right] \tag{12-80}$$

Let us now define

$$u_i \triangleq \tan^{-1} \frac{Z}{jX_i} \tag{12-81}$$

---

† See also Prob. 12-17, which proves that $|K|^2 > 0$ in the stopband as well as in the passband.

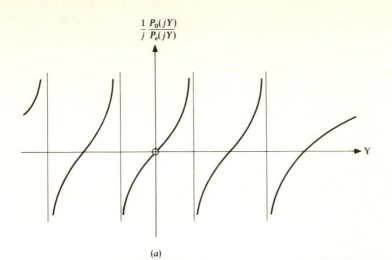

(a)

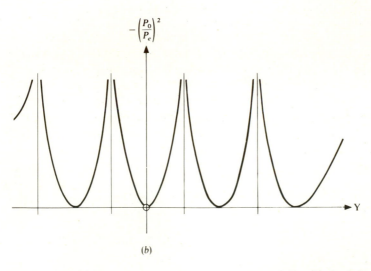

(b)

Figure 12-16 (a) The response of the quasi reactance P_o/P_e; (b) the behavior of $-(P_o/P_e)^2$ for $Z = jY$.

Then
$$\frac{Z + X_i}{-Z + X_i} = \frac{1 + j \tan u_i}{1 - j \tan u_i} = \frac{\cos u_i + j \sin u_i}{\cos u_i - j \sin u_i} = e^{j2u_i} \qquad (12\text{-}82)$$

Hence, by (12-78),

$$\frac{P(Z)}{P(-Z)} = \prod_{i=1}^{n} \frac{Z + X_i}{-Z + X_i} = \exp\left(j2 \sum_{i=1}^{n} u_i\right) \qquad (12\text{-}83)$$

and (12-80) gives

$$|K|^2 = \frac{k_p^2}{2}\left[1 + \frac{\exp\left(j2\sum_{i=1}^{n} u_i\right) + \exp\left(-j2\sum_{i=1}^{n} u_i\right)}{2}\right]$$

$$= \frac{k_p^2}{2}\left[1 + \cos\left(2\sum_{i=1}^{n} u_i\right)\right] = k_p^2 \cos^2\left(\sum_{i=1}^{n} u_i\right) \tag{12-84}$$

Equation (12-84), together with the defining equation† of the u_i

$$u_i \triangleq \tan^{-1}\frac{Z}{jX_i} = \cos^{-1}\left\{\left[1 - \left(\frac{Z}{X_i}\right)^2\right]^{-1/2}\right\} = \cos^{-1}\left(\Omega\sqrt{\frac{\Omega_i^2 - 1}{\Omega_i^2 - \Omega^2}}\right) \tag{12-85}$$

is compared in (12-86) with the basic equations (12-39) and (12-40) of Chebyshev filters:
Chebyshev filters:

$$|K|^2 = k_p^2 \cos^2 nu(\Omega) \qquad \text{where } u(\Omega) \triangleq \cos^{-1}\Omega \tag{12-86a}$$

General-parameter filters:

$$|K|^2 = k_p^2 \cos^2\left[\sum_{i=1}^{n} u_i(\Omega)\right] \qquad \text{where } u_i(\Omega) \triangleq \cos^{-1}\left(\Omega\sqrt{\frac{\Omega_i^2 - 1}{\Omega_i^2 - \Omega^2}}\right) \tag{12-86b}$$

It is clear that if all $\Omega_i \to \infty$, then $u_i(\Omega) \to \cos^{-1}\Omega$ for all i and $\sum_{i=1}^{n} u_i(\Omega) \to nu(\Omega)$. Hence, the Chebyshev filter is a special case of the general-parameter one.

Equation (12-86) can be also regarded as the defining equation of a general-parameter filter, and all properties of $|K|^2$ can be derived from it (see, for example, Probs. 12-19 to 12-21).

The design procedure to be used when the parameters α_p, Ω_1, Ω_2, ..., Ω_n are prescribed can be readily found from Eqs. (12-76) to (12-79). The steps are the following:

1. Calculate from (12-69)

$$k_p^2 = 10^{\alpha_p/10} - 1$$

and from (12-77)

$$X_i = +\sqrt{1 - \Omega_i^{-2}} \qquad i = 1, 2, \ldots, n$$

† See Prob. 12-18 for the detailed derivation of Eq. (12-85).

Note that if there are n_∞ loss poles specified at $\Omega \to \infty$, n_∞ of the X_i will equal 1. Note also that a finite pole pair at $\pm j\Omega_i$ will result in two equal X_i values.

2. Calculate the coefficients of the polynomial

$$P(Z) = \prod_{i=1}^{n} (Z + X_i) = (Z + 1)^{n_\infty} \prod_{i=1}^{(n-n_\infty)/2} (Z + X_i)^2 \qquad (12\text{-}87)$$

Here, on the right-hand side, the loss poles at infinity have been separated and collected in the first factor. Each of the remaining factors corresponds to a $\pm j\Omega_i$ pole pair; hence each X_i enters only one (squared) factor.

3. From (12-79), calculate the coefficients of the rational function

$$|K|^2 = k_p^2 \frac{\left| \mathrm{Ev} \left[(Z + 1)^{n_\infty} \prod_{i=1}^{(n-n_\infty)/2} (Z + X_i)^2 \right] \right|^2}{(1 - Z^2)^{n_\infty} \prod_{i=1}^{(n-n_\infty)/2} (X_i^2 - Z^2)^2} \qquad (12\text{-}88)$$

where Ev stands for "even part of." Only even powers of Z enter $|K|^2$.

4. Using (12-76), replace Z^2 by $1 + S^{-2}$. The result is a rational function of S^2.
5. Calculate $K(S)$ and $H(S)$ in the usual manner and complete the synthesis.

Example 12-5 Find $K(S)$ and $H(S)$ and design the filter from the following specifications:

1. $\alpha \le 0.28$ dB for $|f| \le 10$ kHz.
2. Desired loss poles: $f_1 = 26$ kHz and one pole at infinity.
3. Both terminations are to be 100 Ω.

Using $\omega_0 = 2\pi f_p = 2\pi 10^4$ rad/s as the unit radian frequency, we get $\Omega_1 = 2.6$, $\Omega_2 = -2.6$, and $\Omega_3 \to \infty$. Hence, by (12-77), $X_1 = X_2 = \sqrt{1 - \Omega_1^{-2}} \approx 0.923077$. Also, $X_3 = 1$. From (12-69),

$$k_p^2 = 10^{\alpha_p/10} - 1 = 10^{0.028} - 1 \approx 0.0665961$$

Next, from (12-87),

$$P(Z) = (Z + 1)(Z + X_1)^2 = Z^3 + a_2 Z^2 + a_1 Z + a_0$$

where $a_2 = 2.846154$, $a_1 = 2.698225$, and $a_0 = 0.852071$. Substituting into (12-88), we get

$$|K|^2 = \frac{k_p^2 (a_2 Z^2 + a_0)^2}{(1 - Z^2)(X_1^2 - Z^2)^2}$$

Next, replacing Z^2 by $1 + 1/S^2$, we obtain

$$|K|^2 = \frac{-(b_3 S^3 + b_1 S)^2}{(S^2 + \Omega_1^2)^2}$$

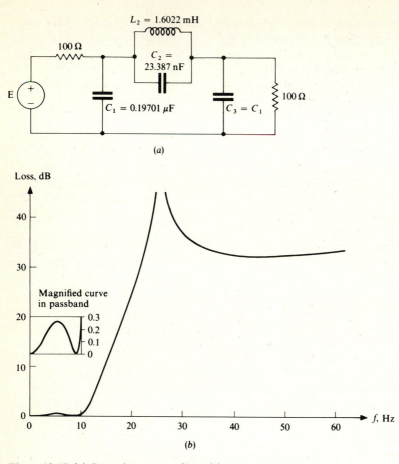

Figure 12-17 (*a*) General-parameter filter; (*b*) its loss response.

where $b_3 = 6.451555$ and $b_1 = 4.965117$. Since

$$|K|^2 = K(S)K(-S) = \frac{(b_3 S^3 + b_1 S)(-b_3 S^3 - b_1 S)}{(S^2 + \Omega_1^2)(S^2 + \Omega_1^2)}$$

clearly we can choose

$$K(S) = \pm \frac{b_3 S^3 + b_1 S}{S^2 + \Omega_1^2}$$

By the Feldtkeller equation (6-42), $E(S)E(-S) = -(b_3 S^3 + b_1 S)^2 + (S^2 + \Omega_1^2)^2$, and the Hurwitz factor is

$$E(S) = 6.451555S^3 + 9.423913S^2 + 11.77046S + 6.76$$

Hence, from Table 6-1, choosing the plus sign for $K(S)$, we obtain

$$z_{11} = z_{22} = R_1 \frac{E_e}{E_o + F_o} = \frac{942.3913S^2 + 676}{12.90311S^3 + 16.735637S}$$

Synthesizing the two-port using ladder development gives the circuit of Fig. 12-17a. Its loss response is shown in Fig. 12-17b.

In practical filter design, it is often the minimum permissible value of the loss in the stopband which is specified, rather than the location of the loss poles. Then an iterative design procedure, based on Eq. (12-80), can be used. This procedure is beyond the scope of our book; the reader should consult Refs. 2 and 3.

Another practical aspect concerns the numerical accuracy of filter design calculations. It can be shown[2,3] that the use of the transformed variable Z defined in (12-76) is very advantageous in that it preserves the accuracy of the calculations even for filters of very high order. This is only true, however, if *all* calculations from the construction of $|K|^2$ on all the way to the calculation of the element values are done in terms of Z rather than S. Again, the reader should consult Refs. 2 and 3 for the details of this process.

A special case of the class of general-parameter filters is obtained when the loss poles are located in such a manner that the stopband as well as the passband is equal-ripple (Fig. 12-18). Since these filters can be treated purely analytically and their $K(S)$ constructed in terms of elliptic functions, they are called *elliptic filters*. A detailed analysis[5] is quite lengthy and is therefore not included here. An important property of elliptic filters is that for given α_p, α_s, f_p, and f_s they require the lowest possible degree of all lumped linear filters. They are thus very economical. Since their design is complicated, industrial filter designers often rely on the many excellent reference works[1,6–8] listing the zeros, poles, and element values of elliptic filters. Since the actual values depend on α_p, α_s, k, and n, these tables tend to be voluminous. Table 12-8 contains a tabulation of the natural modes, loss poles, and element values of elliptic filters with $n = 3$, $\Omega_p = 1$, $R_G = R_L = 1\ \Omega$, and a maximum passband reflection factor of 20 percent. The first column contains the values of $\theta \triangleq \sin^{-1} k$; the second Ω_s; the third $A_{\min} \equiv \alpha_s$; the next four the zeros and poles, and the last three the element values.

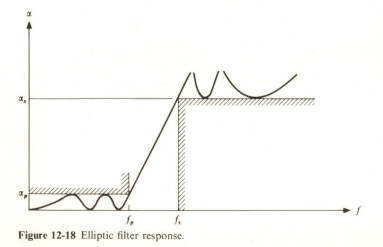

Figure 12-18 Elliptic filter response.

Table 12-8 Poles, zeros, and element values for third-order elliptic filters with 20 percent passband reflection coefficient, $\Omega_p = 1$, $n = 3$, and equal terminations of 1 Ω†

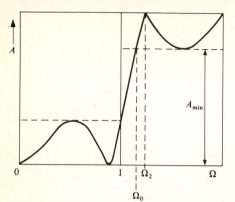

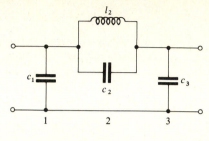

θ	Ω_s	$A_{min,}$ dB	σ_0	σ_1	Ω_1	Ω_2	$c_1 = c_3$	c_2	l_2
1	57.2987	115.77	0.84082	0.42027	1.13143	66.1616	1.1893	0.0002	1.1540
2	28.6537	97.70	0.84114	0.42001	1.13150	33.0839	1.1889	0.0008	1.1533
3	19.1073	87.13	0.84169	0.41958	1.13160	22.0595	1.1881	0.0018	1.1522
4	14.3356	79.63	0.84246	0.41899	1.13175	16.5483	1.1870	0.0032	1.1507
5	11.4737	73.81	0.84345	0.41822	1.13194	13.2424	1.1856	0.0050	1.1488
6	9.5668	69.05	0.84466	0.41728	1.13218	11.0392	1.1839	0.0072	1.1464
7	8.2055	65.03	0.84609	0.41617	1.13245	9.4661	1.1819	0.0098	1.1436
8	7.1853	61.54	0.84776	0.41489	1.13276	8.2868	1.1796	0.0128	1.1404
9	6.3925	58.46	0.84965	0.41344	1.13311	7.3700	1.1770	0.0162	1.1367
10	5.7588	55.70	0.85177	0.41182	1.13350	6.6370	1.1740	0.0200	1.1326
11	5.2408	53.20	0.85413	0.41003	1.13392	6.0377	1.1708	0.0243	1.1281
12	4.8097	50.92	0.85673	0.40807	1.12438	5.5386	1.1672	0.0290	1.1231
13	4.4454	48.82	0.85957	0.40593	1.13466	5.1166	1.1634	0.0342	1.1177
14	4.1336	46.87	0.86266	0.40363	1.13538	4.7552	1.1592	0.0398	1.1119
15	3.8637	45.05	0.86600	0.40115	1.13592	4.4423	1.1547	0.0458	1.1057
16	3.6280	43.35	0.86959	0.39851	1.13649	4.1088	1.1500	0.0524	1.0990
17	3.4203	41.75	0.87345	0.39569	1.13709	3.9277	1.1449	0.0594	1.0919
18	3.2361	40.23	0.87759	0.39270	1.13770	3.7137	1.1595	0.0669	1.0844
19	3.0716	38.80	0.88199	0.38954	1.13833	3.5224	1.1338	0.0749	1.0764
20	2.9238	37.44	0.88668	0.38621	1.13897	3.3505	1.1278	0.0834	1.0681
21	2.7904	36.14	0.89167	0.38272	1.13963	3.1951	1.1215	0.0925	1.0593
22	2.6695	34.90	0.89695	0.37905	1.14029	3.0541	1.1149	0.1021	1.0500
23	2.5593	33.71	0.90254	0.37521	1.14096	2.9256	1.1080	0.1123	1.0404
24	2.4586	32.57	0.90845	0.37120	1.14162	2.8079	1.1008	0.1231	1.0303
25	2.3662	31.47	0.91469	0.36702	1.14228	2.6999	1.0933	0.1345	1.0199
26	2.2812	30.41	0.92127	0.36268	1.14294	2.6003	1.0855	0.1466	1.0090
27	2.2027	29.39	0.92820	0.35817	1.14358	2.5083	1.0773	0.1593	0.9976
28	2.1301	28.41	0.93550	0.35349	1.14420	2.4231	1.0682	0.1728	0.9850
29	2.0627	27.45	0.94318	0.34864	1.14480	2.3438	1.0602	0.1869	0.9733
30	2.0000	26.53	0.95125	0.34364	1.14538	2.2701	1.0512	0.2019	0.9612
θ	Ω_s	$A_{min,}$ dB	σ_0	σ_1	Ω_1	Ω_2	$l'_1 = l'_3$	l'_2	c'_2

θ	Ω_s	$A_{min,}$ dB	σ_0	σ_1	Ω_1	Ω_2	$c_1 = c_3$	c_2	l_2
31	1.9416	25.63	0.95973	0.33847	1.14592	2.2012	1.0420	0.2176	0.9483
32	1.8871	24.76	0.96863	0.33313	1.14643	2.1368	1.0324	0.2343	0.9349
33	1.8361	23.92	0.97799	0.32764	1.14689	2.0765	1.0225	0.2518	0.9212
34	1.7883	23.09	0.98780	0.32199	1.14730	2.0199	1.0123	0.2702	0.9070
35	1.7434	22.29	0.99810	0.31619	1.14766	1.9666	1.0019	0.2897	0.8925
36	1.7013	21.51	1.00890	0.31023	1.14796	1.9165	0.9912	0.3103	0.8776
37	1.6616	20.74	1.02024	0.30412	1.14819	1.8602	0.9802	0.3320	0.8623
38	1.6243	20.00	1.03213	0.29786	1.14835	1.8245	0.9689	0.3549	0.8466
39	1.5890	19.27	1.04460	0.29147	1.14842	1.7823	0.9573	0.3791	0.8305
40	1.5557	18.56	1.05768	0.28493	1.14841	1.7423	0.9455	0.4047	0.8141
41	1.5243	17.86	1.07140	0.27825	1.14830	1.7044	0.9334	0.4318	0.7973
42	1.4945	17.18	1.08579	0.27145	1.14810	1.6684	0.9210	0.4605	0.7801
43	1.4663	16.52	1.10089	0.26452	1.14778	1.6343	0.9084	0.4909	0.7627
44	1.4396	15.86	1.11673	0.25747	1.14735	1.6018	0.8055	0.5232	0.7448
45	1.4142	15.22	1.13336	0.25031	1.14679	1.5710	0.8823	0.5576	0.7267
46	1.3902	14.60	1.15082	0.24304	1.14611	1.5415	0.8689	0.5942	0.7082
47	1.3673	13.98	1.16915	0.23567	1.14528	1.5135	0.8553	0.6331	0.6895
48	1.3456	13.38	1.18840	0.22821	1.14432	1.4868	0.8415	0.6747	0.6705
49	1.3250	12.79	1.20862	0.22067	1.14320	1.4613	0.8274	0.7192	0.6511
50	1.3054	12.22	1.22988	0.21306	1.14192	1.4369	0.8131	0.7668	0.6316
51	1.2868	11.65	1.25221	0.20539	1.14048	1.4137	0.7986	0.8179	0.6118
52	1.2690	11.10	1.27570	0.19766	1.13887	1.3914	0.7839	0.8728	0.5918
53	1.2521	10.56	1.30040	0.18990	1.13709	1.3702	0.7600	0.9319	0.5716
54	1.2361	10.03	1.32639	0.18211	1.13512	1.3498	0.7539	0.9958	0.5512
55	1.2208	9.51	1.35374	0.17431	1.13297	1.3303	0.7387	1.0648	0.5306
56	1.2062	9.01	1.38253	0.16652	1.13064	1.3117	0.7233	1.1397	0.5100
57	1.1924	8.51	1.41284	0.15873	1.12811	1.2938	0.7078	1.2210	0.4892
58	1.1792	8.03	1.44478	0.15098	1.12540	1.2767	0.6921	1.3097	0.4684
59	1.1666	7.57	1.47842	0.14328	1.12249	1.2603	0.6764	1.4065	0.4476
60	1.1547	7.11	1.51387	0.13565	1.11939	1.2446	0.6606	1.5127	0.4268
θ	Ω_s	$A_{min,}$ dB	σ_0	σ_1	Ω_1	Ω_2	$l_1 = l_3$	l_2'	c_2'

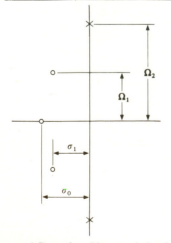

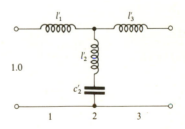

† Reproduced by permission from A. J. Zverev, "Handbook of Filter Synthesis," p. 177, Wiley, New York, 1967.

12-5 REACTANCE TRANSFORMATIONS AND MOEBIUS MAPPINGS

All filter types described thus far represent approximations to the ideal low-pass brick-wall response shown in Fig. 12-3. With a simple manipulation, however, all previous results can be applied to a variety of other problems. Assume, for example, that in the filter of Fig. 12-17a the inductor L_2 is replaced by a capacitor C'_2 and, vice versa, all capacitors C_1, C_2, C_3 are replaced by inductors L'_1, L'_2, L'_3, respectively (Fig. 12-19a). Assume furthermore that the new element values satisfy the relations

$$C'_2 = \frac{1}{AL_2} \qquad L'_i = \frac{1}{AC_i} \qquad i = 1, 2, 3 \tag{12-89}$$

What will the response of this new filter be?

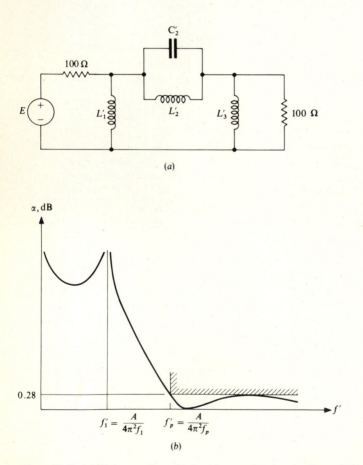

(a)

(b)

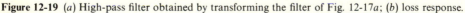

Figure 12-19 (a) High-pass filter obtained by transforming the filter of Fig. 12-17a; (b) loss response.

In the calculation of the original filter response, ω enters only through the reactive immittances $j\omega L_2$, $j\omega C_1$, $j\omega C_2$, and $j\omega C_3$. In the new filter, these immittances are replaced, according to Fig. 12-19a and Eq. (12-89), as follows:

$$j\omega L_i \rightarrow \frac{1}{j\omega' C_i'} = \frac{-jAL_i}{\omega'} \qquad i = 2$$

$$j\omega C_i \rightarrow \frac{1}{j\omega' L_i'} = \frac{-jAC_i}{\omega'} \qquad i = 1, 2, 3 \tag{12-90}$$

where ω' is the radian frequency variable of the new filter. Hence, in effect, the variable ω has been replaced by the new variable ω' through the relation

$$\omega = -\frac{A}{\omega'} \tag{12-91}$$

For example, $A = 2 \times 10^8$. Then the loss value 0.28 dB, which the original filter had at its passband limit $f_p = 10$ kHz (Fig. 12-17), will be obtained for the new filter at

$$\omega' = -\frac{A}{\omega_p} = -\frac{2 \times 10^8}{2\pi 10^4} = -\frac{10^4}{\pi} \text{ rad/s} \tag{12-92}$$

or

$$f' = -\frac{10^4}{2\pi^2} \approx -0.5066 \text{ kHz} \tag{12-93}$$

Since the loss response is an even function of f', the loss will be the same at $f_p' \triangleq +0.5066$ kHz. Equation (12-91) also shows that loss values in the original passband $|f| \leq f_p$ will appear in the range $|f'| \geq f_p'$ for the new filter (Fig. 12-19b), and vice versa. Thus, the filter obtained through the transformation described by Eqs. (12-90) and (12-91) from a low-pass filter is a *high-pass filter*. The frequency values related by the transformation (12-91) can be easily visualized if we plot the ω-vs.-ω' curve (Fig. 12-20a). Figure 12-20b illustrates schematically† the change in the loss response due to the transformation.

Assume now that a high-pass filter with specified values of ω_p', ω_s', α_p, and α_s must be designed. The techniques of Secs. 12-2 to 12-4 enable us to design a *low-pass* filter from which the final high-pass circuit is obtainable using (12-90) and (12-91). Hence, we need merely to find the parameters ω_p, ω_s of the low-pass filter and the transformation constant A. The selectivity parameter k of the low-pass filter satisfies, by (12-91),

$$k \triangleq \frac{\omega_p}{\omega_s} = \frac{-A/\omega_p'}{-A/\omega_s'} = \frac{\omega_s'}{\omega_p'} \tag{12-94}$$

and is thus known. When ω_p is chosen arbitrarily (say at $\omega_p = 1$), the transformation constant A is given, from (12-92), by

$$A = \omega_p \omega_p' \tag{12-95}$$

† For a Chebyshev filter.

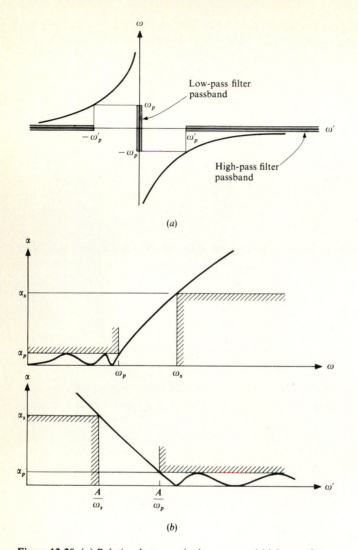

Figure 12-20 (*a*) Relation between the low-pass and high-pass frequency variables; (*b*) the resulting transformation of the loss response.

We conclude that the high-pass filter design using a low-pass prototype filter can be carried out in the following steps:

1. From the specified high-pass filter parameters α_p, α_s, f'_p, f'_s, the selectivity $k = f'_s/f'_p < 1$ of the low-pass prototype is found, and ω_p is chosen.
2. From ω_p, $\omega_s = \omega_p/k$, α_p, and α_s, the low-pass filter is designed.†

† Note that α_p and α_s remain the same for the low-pass and high-pass filters since our transformation affects only the ω axis.

3. From the elements of the low-pass filter, those of the desired high-pass filter can be obtained using the relations

$$C_i' = \frac{1}{AL_i} = \frac{1}{\omega_p \omega_p' L_i} \qquad i = 1, 2, \ldots$$

$$L_i' = \frac{1}{AC_i} = \frac{1}{\omega_p \omega_p' C_i} \qquad i = 1, 2, \ldots$$

(12-96)

Example 12-6 Design a high-pass filter satisfying the following specifications:

$$\alpha \le 0.1 \text{ dB} \qquad \text{for } f \ge 15 \text{ kHz}$$
$$\alpha \ge 40 \text{ dB} \qquad \text{for } f \le 2.5 \text{ kHz}$$

Terminating resistors: 600 Ω

Following the design steps outlined above, we find for the low-pass prototype filter the selectivity parameter

$$k = \frac{\omega_p}{\omega_s} = \frac{f_s'}{f_p'} = \frac{2.5 \times 10^3}{15 \times 10^3} = \frac{1}{6}$$

and, from (12-24), the discrimination parameter

$$k_1 = \sqrt{\frac{10^{\alpha_p/10} - 1}{10^{\alpha_s/10} - 1}} \approx 1.52628 \times 10^{-3}$$

Hence, choosing a Chebyshev filter, by (12-71) the degree must satisfy

$$n \ge \frac{\cosh^{-1}(1/k_1)}{\cosh^{-1}(1/k)} \approx 2.896$$

Using $n = 3$ and postulating (for a change)

$$\alpha(f_p) = \alpha_p \qquad \alpha(f_s) > \alpha_s$$

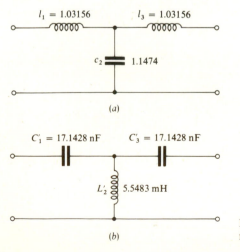

$l_1 = 1.03156$ $l_3 = 1.03156$

C_2 ▮ 1.1474

(a)

$C_1' = 17.1428$ nF $C_3' = 17.1428$ nF

L_2' 5.5483 mH

(b)

Figure 12-21 (a) Normalized low-pass prototype filter; (b) final high-pass filter circuit.

we get an increased stopband loss of 42.218 dB. Proceeding with the calculation of $K(S)$, $H(S)$, and the element values, as discussed in Sec. 12-3, we obtain the circuit shown in Fig. 12-21a. This circuit is both frequency- and impedance-normalized: it is designed with $\omega_p = 1$ and $R_G = R_L = 1$.

Next, the circuit is transformed, element by element, into an impedance-normalized high-pass filter. Using (12-95), we obtain

$$A = \omega_p \omega_p' = (1)(2\pi)(15 \times 10^3) \approx 9.424778 \times 10^4$$

and hence, by (12-96),

$$c_1' = c_3' = \frac{1}{Al_1} \approx 1.02857 \times 10^{-5} \qquad \text{and} \qquad l_2' = \frac{1}{Ac_2} = 0.924728 \times 10^{-5}$$

Finally, impedance denormalization is accomplished by multiplying l_2' by 600 and dividing c_1' as well as c_3' by 600. This gives the final circuit shown in Fig. 12-21b.

It is clear that the transformation $\omega = -A/\omega'$ given in (12-91) performs two functions: (1) it replaces a low-pass frequency response by a high-pass one; (2) it replaces the element immittances in the low-pass prototype by realizable immittances in the high-pass circuit, as Eqs. (12-89) and (12-90) illustrate. Next, consider the transformation

$$\omega = -A\frac{\omega_1^2 - \omega'^2}{\omega'} \tag{12-97}$$

This replaces an inductance in the prototype network by an impedance according to the relation

$$j\omega L_i = -jA\frac{\omega_1^2 L_i}{\omega'} + jAL_i\omega' = \frac{1}{j\omega'C_i'} + jL_i'\omega' \tag{12-98}$$

where

$$C_i' \triangleq \frac{1}{A\omega_1^2 L_i} \qquad L_i' \triangleq AL_i \tag{12-99}$$

Clearly, L_i becomes a series resonant circuit (Fig. 12-22a) in the transformed network. The resonant frequency is

$$\omega_r = \frac{1}{\sqrt{L_i'C_i'}} = \omega_1 \tag{12-100}$$

Similarly, a capacitor is replaced as suggested by

$$j\omega C_i = -jA\frac{\omega_1^2 C_i}{\omega'} + jAC_i\omega' = \frac{1}{j\omega'L_i'} + j\omega'C_i' \tag{12-101}$$

The capacitor C_i is thus transformed into a parallel resonant circuit (Fig. 12-22b). The element values are, from (12-101),

$$L_i' \triangleq \frac{1}{A\omega_1^2 C_i} \qquad C_i' \triangleq AC_i \tag{12-102}$$

The resonant frequency is again given by (12-100).

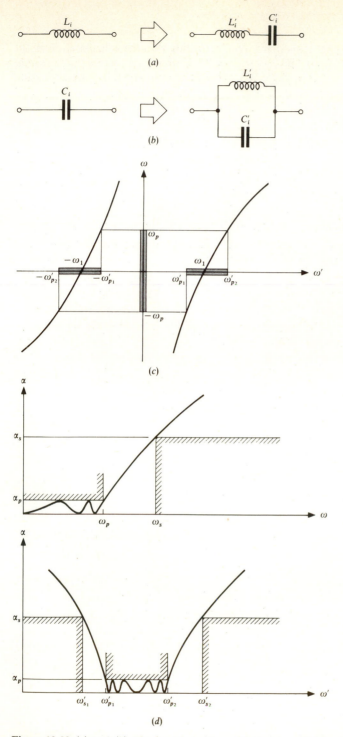

Figure 12-22 (a) and (b). The transformation of low-pass prototype elements into bandpass filter impedances; (c) the corresponding transformation of the frequency variable; (d) the effect of the transformation on the loss response.

To find the effect of the transformation (12-97) on the frequency response, the ω-vs.-ω' curve can be plotted. The result (Fig. 12-22c) demonstrates that now the frequency range $-\omega_p \leq \omega \leq \omega_p$ is transformed into the ranges $-\omega'_{p2} \leq \omega' \leq -\omega'_{p1}$ and $\omega'_{p1} \leq \omega' \leq \omega'_{p2}$. Hence, if the prototype circuit was a low-pass filter, the transformed circuit will be a *bandpass* one. Figure 12-22d illustrates the bandpass-filter response obtained. The passband limits ω'_{p1} and ω'_{p2} can be obtained from (12-97). As Fig. 12-22c shows, ω_p transforms into $-\omega'_{p1}$ and ω'_{p2}; hence, the latter frequencies are the solutions of the equation

$$\omega_p = -A\frac{\omega_1^2 - \omega'^2}{\omega'} \qquad \omega'^2 - \frac{\omega_p}{A}\omega' - \omega_1^2 = 0 \qquad (12\text{-}103)$$

Hence, ω'_{p1} and ω'_{p2} satisfy the relations

$$(\omega' + \omega'_{p1})(\omega' - \omega'_{p2}) = \omega'^2 - \frac{\omega_p}{A}\omega' - \omega_1^2$$

$$\omega'_{p2} - \omega'_{p1} = \frac{\omega_p}{A} \qquad \omega'_{p1}\omega'_{p2} = \omega_1^2 \qquad (12\text{-}104)$$

In the same way, the stopband limit ω_s of the low-pass filter transforms into the high-pass filter stopband limit frequencies $-\omega'_{s1}$ and ω'_{s2}. Performing an analysis exactly analogous to that giving (12-104) gives the results

$$\omega'_{s2} - \omega'_{s1} = \frac{\omega_s}{A} \qquad \omega'_{s1}\omega'_{s2} = \omega_1^2 \qquad (12\text{-}105)$$

As Eqs. (12-104) and (12-105) show, if the low-pass filter has a loss α at frequency ω, the same loss will be obtained for the bandpass filter at the positive frequencies ω'_a and ω'_b; these frequencies will have a geometric symmetry around ω_1 so that $\omega'_a\omega'_b = \omega_1^2$. Hence, the loss response of the bandpass filter will have a geometric symmetry around ω_1.

At this stage, we can piece together the design procedure for a bandpass filter using a low-pass prototype. The design steps are the following:

1. We check the given bandpass-filter parameters ω'_{p1}, ω'_{p2}, ω'_{s1}, and ω'_{s2} to see whether the geometric symmetry condition

$$\omega'_{p1}\omega'_{p2} = \omega'_{s1}\omega'_{s2} \qquad (12\text{-}106)$$

[which follows from Eqs. (12-104) and (12-105)] is met. If not, one of the parameters can be readjusted to restore symmetry *and* introduce some safety margin. If, for example,

$$\omega'_{p1}\omega'_{p2} > \omega'_{s1}\omega'_{s2} \qquad (12\text{-}107)$$

then ω'_{p1} can be lowered to $\omega'_{s1}\omega'_{s2}/\omega'_{p2}$.

2. The selectivity of the low-pass filter prototype can be found from (12-104) and (12-105):

$$k \triangleq \frac{\omega_p}{\omega_s} = \frac{A(\omega'_{p2} - \omega'_{p1})}{A(\omega'_{s2} - \omega'_{s1})} = \frac{\omega'_{p2} - \omega'_{p1}}{\omega'_{s2} - \omega'_{s1}} \qquad (12\text{-}108)$$

From k, α_p, and α_s, the low-pass prototype filter can be designed. Since in the design process the degree n is rounded *up*, k will actually be higher than the value given by (12-108).

3. To obtain the element values of the bandpass filter from those of the prototype, the transformation constants A and ω_1^2 are needed. They in turn can be obtained from the limit frequencies of the filters. From (12-108) and (12-106)

$$\omega'_{p2} - \omega'_{p1} = k(\omega'_{s2} - \omega'_{s1}) \qquad \omega'_{p2}\omega'_{p1} = \omega'_{s2}\omega'_{s1} \qquad (12\text{-}109)$$

where k is the actual (increased) selectivity of the low-pass filter. Keeping, say, ω'_{s1} and ω'_{s2} at their specified values, we obtain a second-degree equation

$$(\omega'_{p2})^2 - k(\omega'_{s2} - \omega'_{s1})\omega'_{p2} - \omega'_{s1}\omega'_{s2} = 0 \qquad (12\text{-}110)$$

for ω'_{p2}. Hence†

$$\omega'_{p2} = \frac{k}{2}(\omega'_{s2} - \omega'_{s1}) + \sqrt{\frac{k^2}{4}(\omega'_{s2} - \omega'_{s1})^2 + \omega'_{s1}\omega'_{s2}} \qquad (12\text{-}111)$$

while

$$\omega'_{p1} = \frac{\omega'_{s2}\omega'_{s1}}{\omega'_{p2}} = \omega'_{p2} - k(\omega'_{s2} - \omega'_{s1}) \qquad (12\text{-}112)$$

Next, from (12-104) and (12-105),

$$A = \frac{\omega_p}{\omega'_{p2} - \omega'_{p1}} = \frac{\omega_s}{\omega'_{s2} - \omega'_{s1}} \qquad \omega_1^2 = \omega'_{s1}\omega'_{s2} = \omega'_{p1}\omega'_{p2} \qquad (12\text{-}113)$$

Finally, the bandpass-filter-element values can be found from Fig. 12-22a and b, as well as Eqs. (12-99) and (12-102).

Example 12-7 Design a bandpass filter satisfying the following specifications:

$$\text{For } 4.82 \text{ MHz} \leq f \leq 5.18 \text{ MHz} \qquad \alpha \leq 0.2 \text{ dB}$$

$$\text{for } f \leq 4.34 \text{ MHz and } f \geq 5.66 \text{ MHz} \qquad \alpha \geq 36 \text{ dB}$$

Both terminations must be 150 Ω.

Since (calculating in megahertz)

$$f'_{p1}f'_{p2} = 24.9676 > f'_{s1}f'_{s2} = 24.5644$$

we have to readjust f'_{p1} to

$$\frac{f'_{s1}f'_{s2}}{f'_{p2}} \approx 4.74216 \text{ MHz}$$

By (12-108), the minimum value of the low-pass filter selectivity is

$$k = \frac{f'_{p2} - f'_{p1}}{f'_{s2} - f'_{s1}} \approx \frac{0.43784}{1.32} \approx 0.3317$$

† The solution containing the negative sign before the square root yields $-\omega'_{p1}$.

When an elliptic-filter prototype is chosen from Table 12-8, the filter with $\theta = 21°$ may be selected. This filter has a maximum passband reflection factor $\rho_{max} = 20$ percent, corresponding to

$$\alpha_p = -10 \log \frac{P_2}{P_{max}} = -10 \log \frac{P_{max} - P_r}{P_{max}}$$

$$\alpha_p = -10 \log (1 - \rho_{max}^2) \approx 0.17729 \text{ dB}$$

where Eqs. (6-16) to (6-20) have been utilized. Since for this filter $\alpha_p < 0.2$ dB, $\alpha_s = 36.14$ dB > 36 dB, and $k = 1/\Omega_s \approx 0.35837 > 0.3317$, it meets all requirements. The circuit diagram is shown in Fig. 12-23a. The element values are obtained from Table 12-8 as $C_1 = 1.1215$, $C_2 = 0.0925$, $L_2 = 1.0593$, with a normalization such that $\omega_p = 1$.

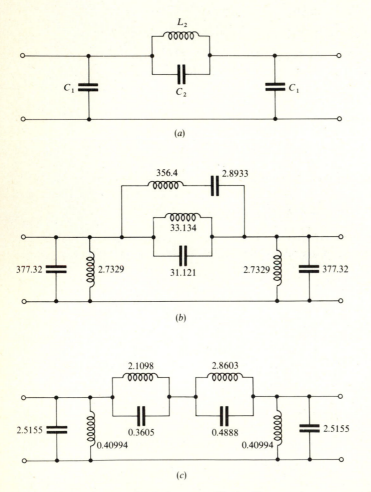

(a)

(b)

(c)

Figure 12-23 (a) Normalized low-pass elliptic filter, used as a prototype; (b) transformed-impedance–normalized-bandpass filter (element values in nanofarads and nanohenrys); (c) final bandpass filter (element values in nanofarads and microhenrys).

Keeping f'_{s_1} and f'_{s_2} unchanged and working in terms of f rather than ω, we see that Eq. (12-111) gives the readjusted passband limit

$$f'_{p_2} = \frac{k}{2}(f'_{s_2} - f'_{s_1}) + \sqrt{\frac{k^2}{4}(f'_{s_2} - f'_{s_1})^2 + f'_{s_1}f'_{s_2}} \approx 5.198365 \text{ MHz}$$

and from (12-112)

$$f'_{p_1} = \frac{f'_{s_1}f'_{s_2}}{f'_{p_2}} \approx 4.7254 \text{ MHz}$$

We note that the original specifications are met with some safety margin.

The transformation constants can be found from (12-113):

$$A = \frac{\omega_s}{\omega'_{s_2} - \omega'_{s_1}} = \frac{\Omega_s}{2\pi(f'_{s_2} - f'_{s_1})} \approx \frac{2.7904}{(2\pi)(1.32 \times 10^6)}$$

$$A \approx 0.336444 \times 10^{-6} \qquad \omega_1^2 = (2\pi)^2 f'_{s_1} f'_{s_2} \approx 969.76364 \times 10^{12}$$

Hence, by Fig. 12-22b and Eq. (12-102), C_1 becomes the parallel combination of a capacitance

$$C'_1 = AC_1 \approx 0.37732 \ \mu\text{F}$$

and an inductance

$$L'_1 = \frac{1}{C'_1 \omega_1^2} \approx 2.7329 \text{ nH}$$

Similarly, L_2 is replaced by a series resonant circuit. The element values are, by (12-99),

$$L'_2 = AL_2 \approx 356.4 \text{ nH} \qquad C'_2 = \frac{1}{L'_2 \omega_1^2} \approx 2.8933 \text{ nF}$$

Finally, C_2 is replaced by a parallel tuned circuit with element values $AC_2 \approx 31.121$ nF and $1/\omega_1^2 AC_2 \approx 33.134$ nH (Fig. 12-23b).

The circuit of Fig. 12-23b is still impedance-normalized, since the prototype filter has 1-Ω terminations. Furthermore, the element values (although positive and thus theor-

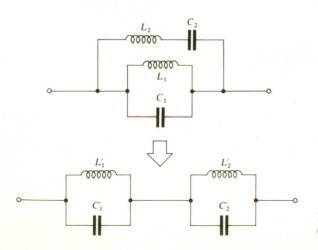

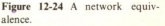

Figure 12-24 A network equivalence.

etically realizable) are too widely spread for easy practical construction; the ratios L_2'/L_1' and C_1'/C_2' are over 130.

It is known from experience that this phenomenon is often encountered for *narrow-band* bandpass filters, i.e., for filters where $\omega_{p_1}' - \omega_{p_2}' \ll \sqrt{\omega_{p_1}' \omega_{p_2}'}$, as is the case here. To remedy the situation, the circuit equivalence shown in Fig. 12-24 may be used. The circuits shown are equivalent if the following relations hold:

$$L_1' = L_1 \frac{1-y}{2} \qquad C_1' = C_2 x \frac{1-z}{2}$$

$$L_2' = L_1 \frac{1+y}{2} \qquad C_2' = C_2 x \frac{1+z}{2} \tag{12-114}$$

where

$$x \triangleq \left(1 + \frac{C_1}{C_2} + \frac{L_2}{L_1}\right)^2 - 4\frac{C_1 L_2}{C_2 L_1}$$

$$y \triangleq \sqrt{1 - \frac{4 L_2}{x L_1}} \qquad z \triangleq \sqrt{1 - \frac{4 C_1}{x C_2}} \tag{12-115}$$

For the circuit of Fig. 12-23b, $C_1/C_2 = L_2/L_1$ and hence (12-115) gives

$$x = 44.0291 \qquad y = z = 0.15102$$

Hence, using (12-114) and denormalizing the impedance level, i.e., multiplying all inductances and dividing all capacitances by $R_0 = 150\ \Omega$, gives the element values indicated in Fig. 12-23c. The spread of element values is now less than 7.

Next, consider the frequency transformation

$$\omega = A \frac{\omega'}{\omega_1^2 - \omega'^2} \tag{12-116}$$

Proceeding as we did before with (12-97), we can readily derive the corresponding element transformations (Fig. 12-25a and b) and the ω-vs.-ω' curve (Fig. 12-25c). The latter makes it obvious that Eq. (12-116) transforms a low-pass filter into a bandstop one. Figure 12-25d illustrates (for a Chebyshev filter) the resulting mapping of the loss response.

Since the analysis of this transformation is a close parallel of that of the lowpass-to-bandpass transformation, the detailed calculations are left to the reader as an exercise (see Probs. 12-31 to 12-34).

A review of Eqs. (12-91), (12-97), and (12-116) reveals that each of these relations replaces ω by a *reactance function* $f(\omega')$ of ω'. This makes it possible to replace the immittances ωL_i and ωC_i of the lowpass prototype filter, one by one, by realizable reactances to obtain the final high-pass (or bandpass or bandstop) filter. Clearly this process can be generalized to more complicated reactance functions; however, the symmetry conditions which result become very complicated

---

Figure 12-25 (a) and (b) The transformation of low-pass prototype elements into bandstop filter impedances; (c) the transformation of the frequency variable; (d) the effect of the transformation on the loss response.

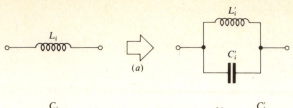

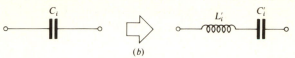

(a)

(b)

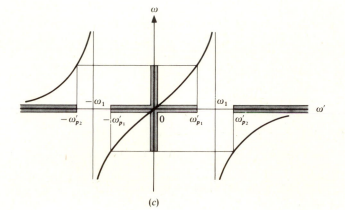

(c)

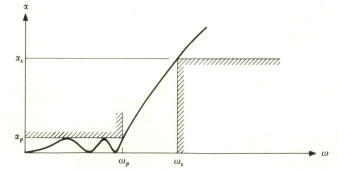

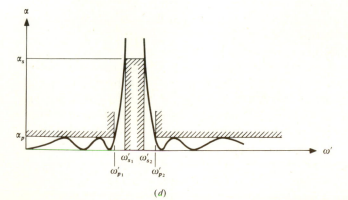

(d)

and restrict the applicability of the filters thus obtained. Since these transformations introduce reactance functions to replace ω, they are called *reactance transformations*.

In addition to reactance transformation, it is possible to use other simple mappings of the frequency variable. Consider the transformation

$$\omega'^2 = k \frac{\omega^2 - \omega_0^2}{\omega^2 - \omega_\infty^2} \tag{12-117}$$

This is one form of the *Moebius mapping*. The relation between ω' and ω is now irrational, and hence the reactances of a low-pass prototype cannot be replaced by physical reactances to obtain the final filter. However, assume that a realizable low-pass $|K|^2$ is available in terms of ω^2. Solving (12-117) for ω^2 gives

$$\omega^2 = \frac{\omega_\infty^2 \omega'^2 - k\omega_0^2}{\omega'^2 - k} \tag{12-118}$$

If we replace ω^2 in $|K|^2$ by the right-hand side of (12-118), a new realizable $|K|^2$ is obtained in terms of ω'^2.

The relation (12-117) contains three parameters: k, ω_0^2, and ω_∞^2. They have the following interpretations:

$$k = \text{ value of } \omega'^2 \text{ for } \omega^2 \to \infty$$

$$\omega_0^2 = \text{ value of } \omega^2 \text{ for } \omega'^2 = 0$$

$$\omega_\infty^2 = \text{ value of } \omega^2 \text{ for } \omega'^2 \to \infty$$

Since we have three free parameters, we can prescribe three pairs of ω, ω' values which map into each other.

Example 12-8 Figure 12-26a shows the $|K|^2$ response of a fourth-degree Chebyshev filter. Since $|K|^2 \neq 0$ for $\omega = 0$, $\alpha(0) \neq 0$ and hence, as the example solved in Sec. 12-3 showed, the terminating resistors of the two-port must be unequal. To allow both terminations to have the same value, $|K|^2$ must be changed so as to obtain $|K|^2 = 0$ for $\omega = 0$ but also so as to retain the low-pass filter character of the response. These aims can be achieved if the smallest reflection zero which, by (12-49), is located at $\omega_2^{(r)} = \omega_p \cos(3\pi/8)$, is shifted to $\omega' = 0$. To preserve the other features of the response, we can also prescribe that $\omega'_p = \omega_p$ and $\omega' \to \infty$ for $\omega \to \infty$. These three conditions are satisfied by the mapping

$$\omega'^2 = \omega_p^2 \frac{\omega^2 - \omega_2^{(r)2}}{\omega_p^2 - \omega_2^{(r)2}} \tag{12-119}$$

as simple substitutions show. The resulting response is shown in Fig. 12-26b. Note that the rise of $|K|^2$ in the stopband is slightly slower than before; i.e., the selectivity has been decreased somewhat by the transformation.

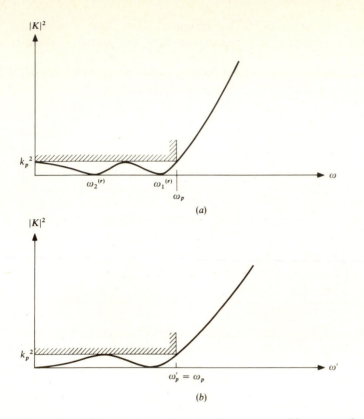

Figure 12-26 (*a*) Fourth-degree Chebyshev filter response; (*b*) response after Moebius transformation.

Example 12-9 Figure 12-27*a* shows the response of a fourth-degree elliptic filter. Now not only is $|K|^2 \neq 0$ for $\omega = 0$ (which prevents equal terminations) but also $0 < |K|^2 < \infty$ for $\omega \to \infty$, which prevents realization in the form of a ladder circuit without mutual inductance. To remedy both shortcomings, we require the smaller reflection zero $\omega^{(r)}$ to shift to zero and the larger loss pole $\omega^{(p)}$ to move to infinity on the ω' axis (Fig. 12-27*b*). Then, Eq. (12-117) indicates that $\omega_0 = \omega^{(r)}$, $\omega_\infty = \omega^{(p)}$, and if we require $\omega'_p = \omega_p$ again, from

$$\omega_p^2 = k \frac{\omega_p^2 - \omega^{(r)2}}{\omega_p^2 - \omega^{(p)2}} \tag{12-120}$$

we get

$$k = \omega_p^2 \frac{\omega_p^2 - \omega^{(p)2}}{\omega_p^2 - \omega^{(r)2}} \tag{12-121}$$

Therefore, the required mapping is

$$\omega'^2 = \left[\omega_p^2 \frac{\omega_p^2 - \omega^{(p)2}}{\omega_p^2 - \omega^{(r)2}} \right] \frac{\omega^2 - \omega^{(r)2}}{\omega^2 - \omega^{(p)2}} \tag{12-122}$$

which clearly has the required properties.

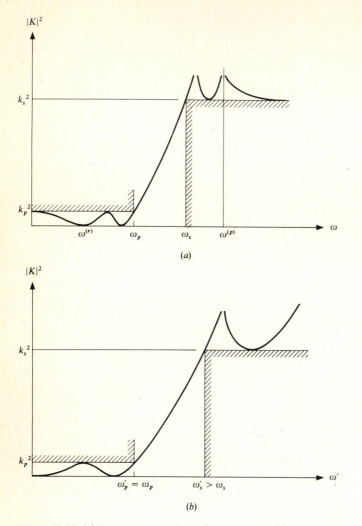

Figure 12-27 (*a*) Fourth-degree elliptic filter response; (*b*) response after Moebius transformation.

A different type of Moebius mapping is given by the formulas

$$\omega'^2 = k\frac{\omega - \omega_0}{\omega - \omega_\infty} \qquad \omega = \frac{\omega_\infty \omega'^2 - k\omega_0}{\omega'^2 - k} \qquad (12\text{-}123)$$

If now ω is replaced in $|K|^2$ by the expression on the right-hand side of (12-123), then a new realizable $|K|^2$ is obtained. The new $|K|^2$ in terms of ω' has a degree twice as high as that of the original one in terms of ω. The constants k, ω_0, and ω_∞ have an interpretation very similar to the parameters entering (12-117) and can be found similarly.

Example 12-10 Find the mapping transforming the third-degree elliptic filter response of Fig. 12-28a into the sixth-degree response shown in Fig. 12-28b.

To obtain the new response, we require $-\omega^{(r)}$ to shift to $\omega' = 0$; ω_p to shift to $\omega'_p = \omega_p$, and finally $-\omega^{(p)}$ to shift to $\omega' \to \infty$. Thus the three reflection zeros at $\omega' = 0$, $\omega_1^{(r)'}$, and $\omega_2^{(r)'}$ are the images of those at $\omega = -\omega^{(r)}$, 0, and $\omega^{(r)}$, respectively. The two loss poles at $\omega' = \omega_1^{(p)'}$ and $\omega_2^{(p)'}$ are the images of the loss poles at $\omega = \omega^{(p)}$ and ∞, respectively.

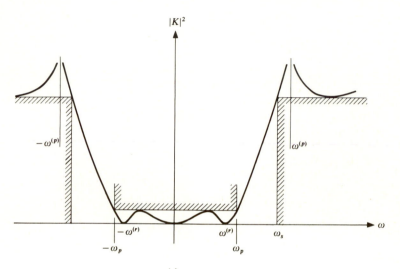

(a)

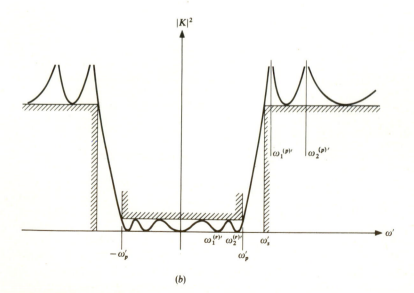

(b)

Figure 12-28 (a) Third-degree elliptic filter response; (b) response after Moebius transformation.

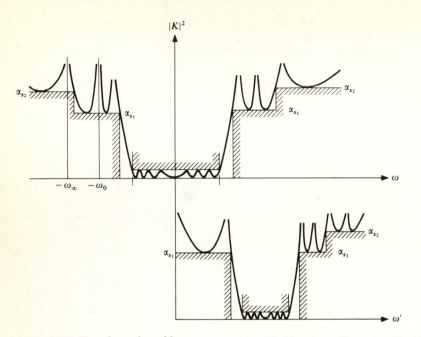

Figure 12-29 Transformation of low-pass to asymmetrical-bandpass filter using Moebius mapping.

The mapping accomplishing this transformation is obtainable in the same way as that of (12-122). The result is

$$\omega'^2 = \left[\omega_p^2 \frac{\omega_p + \omega^{(p)}}{\omega_p + \omega^{(r)}} \right] \frac{\omega + \omega^{(r)}}{\omega + \omega^{(p)}} \tag{12-124}$$

which clearly has the required properties.

The mapping of Eq. (12-123) can also be used, within limits, to obtain a frequency-asymmetrical bandpass filter response from a low-pass prototype response. An example is shown in Fig. 12-29, where the choice of ω_0 and ω_∞ is also illustrated. The details of the calculations are left as an exercise (Prob. 12-38).

12-6 LOW-PASS FILTERS WITH MAXIMALLY FLAT DELAY†

Often, a two-port is required to pass a time signal $v_1(t)$ from its input to its output without serious distortion. This requires that all sine-wave components of the signal be treated approximately the same way by the two-port. Thus, let $A \sin(\omega t + \varphi)$ be a signal component at the input, and let the corresponding output signal component be $kA \sin(\omega t - \beta + \varphi)$. Assume next that k is a

† The approach used in this section is somewhat unconventional. It is based on some useful discussions between one of the authors and H. J. Orchard. The usual exposition of this topic can be found in Ref. 9. Reference 12 gives a discussion similar to ours but applied to bandpass filters.

constant, independent of ω, and that the phase shift β equals ωT, where T is also a frequency-independent constant. Then the output component is $kA \sin [\omega(t - T) + \varphi]$: it is thus obtained by scaling the amplitude of the corresponding input component by k and replacing t by $t - T$. The latter operation corresponds to a delay of T seconds between output and input. Since all input components are thus scaled by the same value k and delayed by the same time interval T, obviously the output signal itself will be $kv_1(t - T)$, that is, a scaled and delayed (but undistorted) replica of the input.

As discussed in Sec. 6-2, the phase lag between the output and input voltages of a doubly terminated two-port is $\beta = \angle H(j\omega)$. The argument just completed thus indicates that for distortionless signal transmission the condition

$$\beta = \angle H(j\omega) = \omega T \tag{12-125}$$

must hold in the frequency range of the signal.† Thus, the *phase must be a linear function of* ω. Equivalently, the *phase delay*

$$T_p(\omega) \triangleq \frac{\beta(\omega)}{\omega} \tag{12-126}$$

and the *group delay* (also often called *envelope delay* or *differential delay*)

$$T_g(\omega) \triangleq \frac{d\beta(\omega)}{d\omega} \tag{12-127}$$

must be *constant* in the frequency range of the signal.

Note that $T_p(\omega)$ is the actual time displacement between the input and output signals for a steady-state sine wave of frequency ω; $T_g(\omega)$, on the other hand, does not have such direct physical meaning for low-pass circuits.‡ However, the variations of $T_g(\omega)$ provide a sensitive measure for the departures of $\beta(\omega)$ from the ideal linear characteristics around the frequency ω. Furthermore, if the input signal is a sine wave of frequency ω, modulated by a second sine wave of much lower frequency ω_L, then after demodulation the phase delay of the low-frequency signal is $[\beta(\omega + \omega_L) - \beta(\omega)]/\omega_L \approx T_g(\omega)$. Thus, $T_g(\omega)$ can readily be measured by measuring the phase delay of the demodulated signal. For these reasons, $T_g(\omega)$ is often used in design.

The calculation of $T_g(\omega)$ can proceed from Eq. (6-7)

$$\beta = \angle H(j\omega) = \mathrm{Im}\,[\ln H(j\omega)] \tag{12-128}$$

By (12-127)

$$T_g = \frac{d\beta}{d\omega} = \mathrm{Im}\left[\frac{d}{d\omega}\ln H(j\omega)\right] = \mathrm{Im}\left[j\frac{d}{d(j\omega)}\ln H(j\omega)\right] = \mathrm{Re}\left[\frac{d}{ds}\ln H(s)\right]_{s=j\omega} \tag{12-129}$$

---

† In addition, of course, α must be a constant in the same range to assure that the scale factor k is also constant.

‡ Although it is possible to find physical interpretation of $T_g(\omega)$ in narrow-band frequency-multiplex systems.

Since the even (odd) part of a rational function of s is real (imaginary) for $s = j\omega$, it follows that

$$T_g(\omega) = \text{Re} \left[\frac{dH(s)/ds}{H(s)} \right]_{s=j\omega} = \left\{ \text{Ev} \left[\frac{1}{H(s)} \frac{dH(s)}{ds} \right] \right\}_{s=j\omega} \qquad (12\text{-}130)$$

Here, as before, Ev $[f(s)]$ denotes the even part of $f(s)$.

Assume now a polynomial transducer factor $H(s) \equiv E(s)$. Then (12-130) gives

$$T_g(\omega) = \frac{1}{2} \left[\frac{E'(s)}{E(s)} + \frac{E'(-s)}{E(-s)} \right]_{s=j\omega} \qquad (12\text{-}131)$$

where $E'(s) \triangleq dE(s)/ds$, and $E'(-s)$ is obtained by substituting $-s$ for s in $E'(s)$. Evidently, we can define $F(s)$, an even rational function of s such that for $s = j\omega$, $F(j\omega) = T_g(\omega)$. From (12-131)

$$F(s) = \frac{1}{2} \left[\frac{E'(s)}{E(s)} + \frac{E'(-s)}{E(-s)} \right] = \frac{E'(s)E(-s) + E'(-s)E(s)}{2E(s)E(-s)} \qquad (12\text{-}132)$$

Assume now that the input signal contains most of its energy in the low-frequency region. Then, $T_g(\omega)$ should be constant around $\omega = 0$. Using maximally flat approximation, therefore, we require that the conditions

$$T_g(\omega) = T$$

$$\frac{dT_g(\omega)}{d(\omega^2)} = 0$$

$$\frac{d^2 T_g(\omega)}{d(\omega^2)^2} = 0 \qquad (12\text{-}133)$$

$$\cdots\cdots\cdots\cdots$$

$$\frac{d^{(n-1)}T_g(\omega)}{d(\omega^2)^{n-1}} = 0$$

hold for $\omega = 0$, where n is the degree of $E(s)$. These conditions are analogous to (12-11). Note that since $T_g(\omega)$ is an *even* rational function of ω, only the derivatives with respect to ω^2 need to be included in (12-133). In terms of s, (12-133) becomes

$$F(s) = T \qquad \frac{d^k F(s)}{d(s^2)^k} = 0 \qquad k = 1, 2, \ldots, n-1 \qquad (12\text{-}134)$$

for $s = 0$.

We shall now use Eqs. (12-132) and (12-134) to derive the maximally flat delay polynomial $E(s)$. As (12-132) shows, the denominator of $F(s)$ is $2E(s)E(-s)$. Hence, we shall assume a solution of (12-134) in the form

$$F(s) = T + \frac{P(s)}{2E(s)E(-s)} \qquad (12\text{-}135)$$

where $P(s)$ is some even polynomial. If the degree of $E(s)$ is n, then that of $P(s)$ is, by (12-133) and (12-135), at most $2n$; hence we can write

$$P(s) = p_0 + p_1 s^2 + \cdots + p_n s^{2n} \tag{12-136}$$

By (12-134), in close analogy to (12-12), we obtain then

$$p_0 = p_1 = p_2 = \cdots = p_{n-1} = 0 \tag{12-137}$$

Hence, from (12-135),

$$F(s) = T + \frac{p_n s^{2n}}{2E(s)E(-s)} = \frac{(2T)E(s)E(-s) + p_n s^{2n}}{2E(s)E(-s)} \tag{12-138}$$

But as (12-132) shows, the numerator polynomial of $F(s)$ is only of degree $2n - 1$. Hence, the coefficient of s^{2n} in the numerator on the right-hand side of (12-138) must vanish. Therefore, if $E(s)$ is in the form

$$E(s) = \sum_{i=0}^{n} a_i s^i \tag{12-139}$$

the cancellation of the highest-order coefficient in (12-138) requires

$$2Ta_n^2(-1)^n + p_n = 0 \qquad p_n = (-1)^{n+1} 2Ta_n^2 \tag{12-140}$$

Now any constant factor can be associated with $E(s)$ without affecting the group delay, as is evident, for example, from (12-131). Hence we choose at this stage $a_n = 1$. Furthermore, we normalize the time scale such that $T = 1$. This implies, by Eq. (1-24) of Sec. 1-4, that the ω and f values are scaled by $1/T$; the frequency and radian frequency units are thus both equal to $\omega_0 = 1/T$.

With these assumptions (12-140) gives $p_n = (-1)^{n+1} 2$ and hence, by Eqs. (12-132) and (12-138),

$$E'(s)E(-s) + E'(-s)E(s) = 2E(s)E(-s) + 2(-1)^{n+1} s^{2n}$$

$$s^{-2n}[E'(s)E(-s) + E'(-s)E(s) - 2E(s)E(-s)] = 2(-1)^{n+1} \tag{12-141}$$

$$s^{-2n} \operatorname{Ev} \{[E'(s) - E(s)]E(-s)\} = (-1)^{n+1}$$

Differentiating both sides of the last equation with respect to s and noting that

$$\frac{d}{ds} \operatorname{Ev} f(s) = \operatorname{Od} \frac{df(s)}{ds} \tag{12-142}$$

and

$$\frac{dE(-s)}{ds} = -\frac{dE(-s)}{d(-s)} = -E'(-s) \tag{12-143}$$

we obtain after a simple calculation

$$\operatorname{Ev} \{[sE''(s) - 2(n + s)E'(s) + 2nE(s)]E(-s)\} = 0 \tag{12-144}$$

Let us denote the polynomial in the square brackets by $D(s)$:

$$D(s) \triangleq sE''(s) - 2(n + s)E'(s) + 2nE(s). \tag{12-145}$$

Note that by (12-139) the degree of $D(s)$ appears to be n; this, however, is an illusion, since the coefficient of s^n in $D(s)$ is, by (12-145),

$$d_n = -2na_n + 2na_n \equiv 0 \tag{12-146}$$

Thus, $D(s)$ is only of degree $n - 1$.

By (12-144), Ev $[D(s)E(-s)] \equiv 0$, so that $D(s)E(-s)$ must be a *pure odd* real polynomial. Hence, its zeros must occur in quadrantal symmetry. Furthermore, as discussed in Sec. 6-2, realizability requires $E(s)$ to be a strictly Hurwitz polynomial; hence all n zeros of $E(-s)$ must lie in the RHP. Consequently, $D(s)$ must contain n LHP zeros to complete the symmetric pattern. However, as shown above, $D(s)$ is only of degree $n - 1$.

The only possible escape out of this contradiction is to assume either that $E(s) \equiv 0$ or that $D(s) \equiv 0$. The former leads to a useless result; hence, we must set

$$D(s) \triangleq sE''(s) - 2(n + s)E'(s) + 2nE(s) \equiv 0 \tag{12-147}$$

This second-order differential equation forms the basis of the solution of our problem. Substituting (12-139) into (12-147) gives the relation

$$s \sum_{i=2}^{n} i(i - 1)a_i s^{i-2} - 2(n + s) \sum_{i=1}^{n} ia_i s^{i-1} + 2n \sum_{i=0}^{n} a_i s^i \equiv 0 \tag{12-148}$$

Hence, the coefficient of s^k $(0 \le k \le n - 1)$ satisfies†

$$(k + 1)ka_{k+1} - 2n(k + 1)a_{k+1} - 2ka_k + 2na_k = 0$$

$$a_k = \frac{(k + 1)(2n - k)}{2(n - k)} a_{k+1} \tag{12-149}$$

Since we have already set $a_n = 1$, the remaining coefficients are given by

$$a_{n-1} = \frac{n(n + 1)}{(1)(2)} a_n = \frac{n(n + 1)}{(2^1)(1!)}$$

$$a_{n-2} = \frac{(n - 1)(n + 2)}{(2)(2)} a_{n-1} = \frac{(n - 1)n(n + 1)(n + 2)}{(2^2)(2!)} \tag{12-150}$$

$$a_{n-3} = \frac{(n - 2)(n + 3)}{(2)(3)} a_{n-2} = \frac{(n - 2)(n - 1)n(n + 1)(n + 2)(n + 3)}{(2^3)(3!)}$$

. . . .

The coefficient of s^{n-k} is thus clearly

$$a_{n-k} = \frac{(n - k + 1)(n - k + 2) \cdots n(n + 1) \cdots (n + k)}{2^k k!} \tag{12-151}$$

---

† Notice that (12-149) holds also for $k = n$ in a trivial way since $a_{n+1} \equiv 0$.

or, using $i = n - k$ and indicating the order of $E(s)$ as a superscript of a_i,

$$a_i^n = \frac{(i + 1)(i + 2) \cdots (2n - i)}{2^{n-i}(n - i)!} = \frac{(2n - i)!}{2^{n-i}i!\,(n - i)!} \tag{12-152}$$

The coefficients calculated from (12-152) are given, for $1 \le n \le 11$, in Table 12-9. The polynomials defined by these coefficients are closely related to the Bessel polynomials. Since they were first described by W. E. Thomson of the British Post Office Research Station, the filters derived from these transfer functions are called *Bessel filters*, *Thomson filters*, or (as a compromise) Bessel-Thomson filters.

By the construction of the formula for a_i^n, it is clear that under the stated assumptions [polynomial transfer function, strictly Hurwitz $E(s)$] the derived solution to the maximally flat delay approximation problem is *unique*.

Next, rewrite the formula (12-152) for a_i^n as follows

$$a_i^n = \frac{(2n - i - 2)!}{2^{n-i-1}(i - 2)!\,(n - i - 1)!} \; \frac{(2n - i)(2n - i - 1)}{2i(i - 1)(n - i)}$$

$$= \frac{(2n - i - 2)!}{2^{n-i-1}(i - 2)!\,(n - i - 1)!} \; \frac{2(n - i)(2n - 1) + i(i - 1)}{2i(i - 1)(n - i)}$$

$$= (2n - 1)\frac{(2n - i - 2)!}{2^{n-i-1}i!\,(n - i - 1)!} + \frac{(2n - i - 2)!}{2^{n-i}(i - 2)!\,(n - i)!} \tag{12-153}$$

Using (12-152), with n replaced first by $n - 1$ and then by $n - 2$, (12-153) gives

$$a_i^n = (2n - 1)a_i^{n-1} + a_{i-2}^{n-2} \tag{12-154}$$

It is easy to show (see Prob. 12-41) that (12-154) holds for all i between 0 and n, that is, for $0 \le i \le n$. Hence, the polynomials of degrees n, $n - 1$, and $n - 2$ satisfy

$$E^n(s) = \sum_{i=0}^{n} a_i^n s^i = (2n - 1)\sum_{i=0}^{n-1} a_i^{n-1} s^i + s^2 \sum_{i=0}^{n} a_{i-2}^{n-2} s^{i-2}$$

$$= (2n - 1)E^{n-1}(s) + s^2 E^{n-2}(s) \tag{12-155}$$

Here, we indicated the degree of $E(s)$ by its superscript.

Equations (12-154) and (12-155) provide a useful recurrence process for calculating the $E^n(s)$. For $n = 0$, $a_0^0 = a_n^0 = 1$; for $n = 1$, $a_1^1 = a_n^1 = 1$; and from (12-152), $a_0^1 = 2!/[(2)(0!\,1!)] = 1$ since $0! = 1! = 1$. Hence $E^0(s) = 1$, $E^1(s) = s + 1$ and therefore by (12-155)

$$E^2(s) = 3E^1(s) + s^2 E^0(s) = s^2 + 3s + 3$$

$$E^3(s) = 5E^2(s) + s^2 E^1(s) = s^3 + 6s^2 + 15s + 15$$

etc. Thus, all entries of Table 12-9 can be obtained.

Another useful relation obtainable from (12-152) concerns the *derivative* of $E^n(s)$:

$$\frac{dE^n(s)}{ds} = \sum_{i=1}^{n} a_i^n i s^{i-1} = \sum_{k=0}^{n-1} b_k^n s^k \tag{12-156}$$

Table 12-9 Coefficients of the maximally flat-delay polynomial $E^n(s)$†

Note that $a_n(s)$ is chosen as 1 for all n

n	a_0	a_1	a_2	a_3	a_4	a_5	a_6	a_7	a_8	a_9	a_{10}
1	1	1									
2	3	3									
3	15	15	6								
4	105	105	45	10							
5	945	945	420	105	15						
6	10,395	10,395	4,725	1,260	210	21					
7	135,135	135,135	62,370	17,325	3,150	378	28				
8	2,027,025	2,027,025	945,945	270,270	51,975	6,930	630	36			
9	34,459,425	34,459,425	16,216,200	4,729,752	945,945	135,135	13,860	990	45		
10	654,729,075	654,729,075	310,134,825	91,891,800	18,918,900	2,837,835	315,315	25,740	1,485	55	
11	13,749,310,575	13,749,310,575	6,547,290,750	1,964,187,225	413,513,100	64,324,260	7,567,560	675,675	45,045	2,145	66

† Reproduced by permission from L. Weinberg, "Network Analysis and Synthesis," p. 500, McGraw-Hill, New York, 1962; reprinted by Robert E. Krieger Publishing Co., Inc., Huntington, N.Y., 1975.

Table 12-10 Zeros (natural modes) for maximally flat-delay filters†

n	
1	-1.0000000
2	$-1.5000000 \pm j0.8660254$
3	$-2.3221854; \; -1.8389073 \pm j1.7543810$
4	$-2.8962106 \pm j0.8672341; \; -2.1037894 \pm j2.6574180$
5	$-3.6467386; \; -3.3519564 \pm j1.7426614; \; -2.3246743 \pm j3.5710229$
6	$-4.2483594 \pm j0.8675097; \; -3.7357084 \pm j2.6262723; \; -2.5159322 \pm j4.4926730$
7	$-4.9717869; \; -4.7582905 \pm j1.7392861; \; -4.0701392 \pm j3.5171740; \; -2.6856769 \pm j5.4206941$
8	$-5.5878860 \pm j0.8676144; \; -2.8389840 \pm j6.3539113; \; -4.3682892 \pm j4.4144425; \; -5.204408 \pm j2.6161751$
9	$-6.2970193; \; -6.1293679 \pm j1.7378484; \; -5.6044218 \pm j3.4981573; \; -4.6384399 \pm j5.3172717; \; -2.9792608 \pm j7.291637$
10	$-6.9220449 \pm j0.8676651; \; -3.1089162 \pm j8.2326995; \; -6.6152916 \pm j2.6115683; \; -5.9675282 \pm j4.3849471; \; -4.886195 \pm j6.2249855$
11	$-7.6223398; \; -6.3013375 \pm j5.2761917; \; -5.1156483 \pm j7.1370208; \; -7.4842299 \pm j1.737028; \; -7.058924 \pm j3.489145;$ $-3.297221 \pm j9.1771116$

† Reproduced by permission from L. Weinberg, "Network Analysis and Synthesis," p. 500, McGraw-Hill, New York, 1962; reprinted by Robert E. Krieger Publishing Co., Inc., Huntington, N.Y., 1975.

where $k \triangleq i - 1$. Clearly, from (12-152),

$$b_k^n = (k + 1)a_{k+1}^n = (k + 1)\frac{(2n - k - 1)!}{2^{n-k-1}(k + 1)!(n - k - 1)!}$$

$$= \frac{(2n - k)!}{2^{n-k}k!(n - k)!} - \frac{(2n - k - 1)!}{2^{n-k}(k - 1)!(n - k)!} \qquad (12\text{-}157)$$

Using (12-152) again to identify the two terms on the right-hand side, we find

$$b_k^n = a_k^n - a_{k-1}^{n-1} \qquad (12\text{-}158)$$

For $k = 0$, $a_{k-1}^{n-1} \equiv 0$, and hence (12-158) gives $b_0^n = a_0^n$, which is valid since $b_0^n = a_1^n = a_0^n$ for all n. Similarly, for $k = n$, (12-158) predicts $b_n^n = a_n^n - a_{n-1}^{n-1} = 1 - 1 \equiv 0$ for all n which is also true. Hence, (12-158) holds for $0 \leq k \leq n$ and thus

$$\frac{dE^n(s)}{ds} = \sum_{k=0}^{n} a_k^n s^k - s \sum_{k=0}^{n} a_{k-1}^{n-1} s^{k-1} = E^n(s) - sE^{n-1}(s) \qquad (12\text{-}159)$$

Equations (12-155) and (12-159) can be used (see Probs. 12-43 to 12-45) to show that $E^n(s)$ does indeed satisfy (12-141) and thus the maximally flat-delay property. Using these relations, it can also be readily shown† that all $E^n(s)$ are strictly Hurwitz, as required by physical realizability and as assumed earlier in the derivation. Dividing both sides of (12-155) by $sE^{n-1}(s)$ gives

$$\frac{E^n(s)}{sE^{n-1}(s)} = \frac{2n - 1}{s} + \frac{sE^{n-2}(s)}{E^{n-1}(s)} \qquad (12\text{-}160)$$

For $n = 2$, $E^{n-2}(s) \equiv 1$, and $E^{n-1}(s) = s + 1$, as shown earlier. Hence,

$$\frac{E^2(s)}{sE^1(s)} = \frac{3}{s} + \frac{s}{s + 1} \qquad (12\text{-}161)$$

is a PR function. So is its reciprocal, and hence so is

$$\frac{E^3(s)}{sE^2(s)} = \frac{5}{s} + \frac{sE^1(s)}{E^2(s)} \qquad (12\text{-}162)$$

Thus, by (12-160), $E^4(s)/sE^3(s)$ is also PR. Proceeding this way, we see that $E^n(s)/sE^{n-1}(s)$ is PR for all n. Hence, $E^n(s)$ and $E^{n-1}(s)$ both have at least the Hurwitz property for all n. To show that they are *strictly* Hurwitz we must prove that they do not have any $j\omega$-axis zeros, i.e., any factors of the forms $s^2 + \omega_0^2$ or s. Such factors (as shown in Chap. 4) are present in both the even and odd parts of, say, $E^{n-1}(s)$ and therefore of $sE^{n-1}(s)/E^n(s)$.

By (12-159),

$$\frac{dE^n(s)/ds}{E^n(s)} = 1 - \frac{sE^{n-1}(s)}{E^n(s)} \qquad (12\text{-}163)$$

† This derivation is due to H. J. Orchard.

and hence, from Eqs. (12-132) and (12-138) with $T = 1$,

$$F(s) = \text{Ev} \frac{dE^n(s)/ds}{E^n(s)} = 1 - \text{Ev} \frac{sE^{n-1}(s)}{E^n(s)} = 1 + \frac{p_n s^{2n}}{2E^n(s)E^n(-s)}$$

$$\text{Ev} \frac{sE^{n-1}(s)}{E^n(s)} = \frac{-p_n s^{2n}}{2E^n(s)E^n(-s)}$$

(12-164)

This shows that $j\omega$-axis zeros of $E^{n-1}(s)$ can only occur for $s = 0$. However, by (12-152), the constant term of $E^{n-1}(s)$ is

$$a_0^{n-1} = \frac{(2n-2)!}{2^{n-1}(n-1)!} \neq 0$$

(12-165)

and hence $E^{n-1}(s)$ is not zero for $s = 0$. Thus, $E^{n-1}(s)$ has *no* $j\omega$-axis zeros and is therefore strictly Hurwitz. Since our derivation is valid for all n, the strictly Hurwitz character of the maximally flat-delay polynomials has thus been proved.†

It is also reassuring to note that, by (12-152),

$$\frac{a_i^n}{a_0^n} = \frac{(2n-i)!}{2^{n-i}i!(n-i)!} \frac{2^n n!}{(2n)!} = \frac{n!/(n-i)!}{2^{-i}(2n)!/(2n-i)!} \frac{1}{i!}$$

$$= \frac{(n-i+1)(n-i+2)\cdots(n-1)n}{2^{-i}(2n-i+1)(2n-i+2)\cdots(2n-1)2n} \frac{1}{i!}$$

(12-166)

and hence for $n \to \infty$

$$\frac{a_i^n}{a_0^n} \to \frac{n^i}{2^{-i}(2n)^i} \frac{1}{i!} = \frac{1}{i!}$$

(12-167)

Hence

$$E^n(s) = a_0^n \sum_{i=0}^{n} \frac{a_i^n}{a_0^n} s^i \to a_0^n \sum_{i=0}^{\infty} \frac{s^i}{i!} \qquad E^n(s) \to a_0^n e^s$$

(12-168)

that is, $E^n(s)$ tends (apart from the scale factor a_0^n) to the unit delay operator e^s.

The asymptotic behavior of the loss $\alpha(\omega) = 10 \log [E(j\omega)E(-j\omega)]$ corresponding to $E^n(s)$ can also be found for large values of n. We have

$$E^n(s)E^n(-s) = \left(\sum_{i=0}^{n} a_i^n s^i \right) \left[\sum_{i=0}^{n} a_i^n(-1)^i s^i \right]$$

(12-169)

---

† The alert reader will have noted that the above proof of the strictly Hurwitz property of $E^n(s)$ does not exclude the possibility of a *common* zero of $E^n(s)$ and $E^{n-1}(s)$ in the RHP or on the $j\omega$ axis. It can, however, be easily shown that $E^n(s)$ and $E^{n-1}(s)$ cannot have any zeros in common. The proof is based on (12-155) and is hinted at in Prob. 12-49.

This gives a pure even polynomial of the form $b_0 + b_1 s^2 + b_2 s^4 + \cdots$. To find the first few coefficients, we obtain, by repeatedly using (12-149),

$$a_1^n = a_0^n$$

$$a_2^n = \frac{n-1}{2n-1} a_0^n$$

$$a_3^n = \frac{n-2}{3(2n-1)} a_0^n \qquad (12\text{-}170)$$

$$a_4^n = \frac{(n-2)(n-3)}{6(2n-1)(2n-3)} a_0^n$$

Hence, the lowest-order coefficients of $E^n(s)E^n(-s)$ are

$$b_0 = (a_0^n)^2$$

$$b_1 = 2a_0^n a_2^n - (a_1^n)^2 = -\frac{(a_0^n)^2}{2n-1} \qquad (12\text{-}171)$$

$$b_2 = 2a_0^n a_4^n - 2a_1^n a_3^n + (a_2^n)^2 = \frac{n-1}{2n-3} \frac{(a_0^n)^2}{(2n-1)^2}$$

Thus for $s = j\omega$, that is, for $s^2 = -\omega^2$,

$$E^n(j\omega)E^n(-j\omega) = (a_0^n)^2 \left[1 + \frac{\omega^2}{2n-1} + \frac{n-1}{2n-3}\left(\frac{\omega^2}{2n-1}\right)^2 + \cdots \right] \qquad (12\text{-}172)$$

It is of interest to compare this expression with the Taylor series of the *gaussian function*†

$$(a_0^n)^2 e^{\omega^2/(2n-1)} = (a_0^n)^2 \left[1 + \frac{\omega^2}{2n-1} + \frac{1}{2}\left(\frac{\omega^2}{2n-1}\right)^2 + \cdots \right] \qquad (12\text{-}173)$$

Clearly, the relative difference is

$$\frac{(a_0^n)^2 e^{\omega^2/(2n-1)} - E^n(j\omega)E^n(-j\omega)}{(a_0^n)^2 e^{\omega^2/(2n-1)}} = e^{-\omega^2/(2n-1)}\left[\frac{-\omega^4}{2(2n-3)(2n-1)^2} \right.$$

$$\left. + \text{ higher-order terms} \right] \qquad (12\text{-}174)$$

Hence, for $n \geq 3$, and for ω values not in excess of, say, 3,‡ the approximation

$$E^n(j\omega)E^n(-j\omega) \approx (a_0^n)^2 e^{\omega^2/(2n-1)} \qquad (12\text{-}175)$$

holds. The loss is then

$$\alpha = 10 \log \left[E^n(j\omega)E^n(-j\omega) \right] \approx 20 \log a_0^n + \frac{10 \log e}{2n-1} \omega^2 \qquad (12\text{-}176)$$

---

† A gaussian function $f(x)$ is of the form $f(x) = Ae^{-Bx^2}$.
‡ In terms of denormalized units, this condition is $\omega < 3/T$.

The approximation of (12-175) can also be made in (12-138). Using (12-140), we get then (with $T = 1$)

$$T_g(j\omega) = 1 - \frac{\omega^{2n}}{E(j\omega)E(-j\omega)} = 1 - (a_0^n)^{-2}\omega^{2n}e^{-\omega^2/(2n-1)} \qquad (12\text{-}177)$$

Often, we require the zero-frequency loss to equal zero.† If we choose $H(s) = E^n(s)/a_0^n$, this will be achieved. Then (12-176) will be replaced by

$$\alpha \approx \frac{10 \log e}{2n - 1}\omega^2 \qquad (12\text{-}178)$$

The group delay will, of course, remain unchanged.

† This permits $R_G = R_L$ in the resulting circuit.

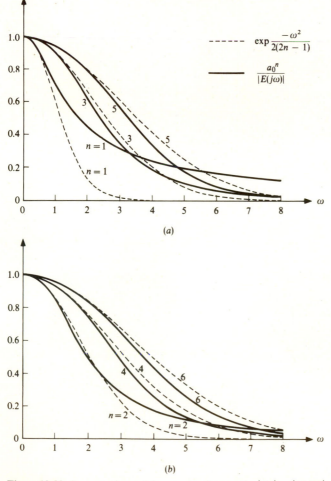

(a)

(b)

Figure 12-30 A comparison of the exact voltage ratio $|A_v| = |V_2/V_1|$ calculated from the maximally flat delay function $H(s) = E^n(s)/a_0^n$ with the approximate response found from Eq. (12-175); (a) for odd degrees n; (b) for even degrees.

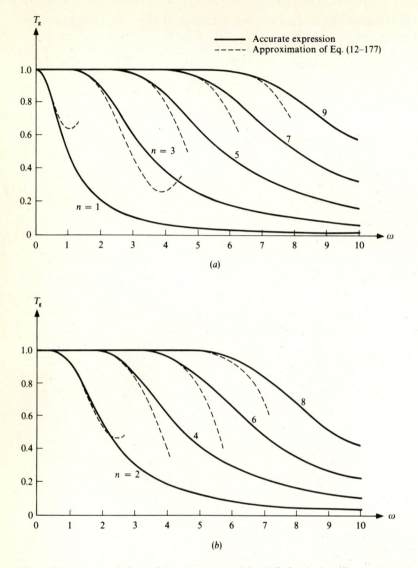

Figure 12-31 A comparison of the exact group delay $T_g(j\omega)$ calculated from the maximally flat delay transfer function $H(s) = E^n(s)/a_0^n$ with the response obtained from the approximation (12-177) (a) for odd degrees n; (b) for even degrees.

The output-input amplitude response $a_0^n/|E(j\omega)|$ and the group-delay response $T_g(j\omega)$ associated with $H(s) = E^n(s)/a_0^n$ are plotted for low values of n in Figs. 12-30 and 12-31, respectively. The maximally flat property is evident for the delay response. To illustrate the quality of the approximations given in Eqs. (12-175) and (12-177), these figures also show the responses computed from the approximate expressions.

Example 12-11 Find a transfer function such that the group delay is 10 μs at low frequencies, with a deviation less than 0.5 percent for $|f| \leq 30$ kHz. The loss should be zero at zero frequency and should remain less than 1 dB for $|f| \leq 30$ kHz.

To satisfy the zero-loss condition for $\omega = 0$, we choose the frequency-normalized transfer function in the form

$$H(s) = \frac{E^n(s)}{a_0^n}$$

The degree n can be found from the specifications on the loss and delay deviations. Since $\omega T = (2\pi 3 \times 10^4)(10^{-5}) \approx 1.885$, with (12-178) the loss requirement gives at the normalized limit frequency ωT

$$\frac{10 \log e}{2n-1} (\omega T)^2 \approx \frac{4.343}{2n-1} [(2\pi 3 \times 10^4)(10^{-5})]^2 < 1$$

This leads to

$$n > 8.2$$

Hence, $n = 9$ is the minimum usable degree. For $n = 9$, the relative delay deviation at the limit frequency is, by (12-177),

$$\frac{\Delta T}{T} \approx (a_0^9)^{-2}(\omega T)^{18} e^{-(\omega T)^2/17} \approx 6.17 \times 10^{-11}$$

This is much less than the prescribed 0.5 percent and hence meets the delay specification.

Using Table 12-9, we find frequency-normalized transfer function to be

$$H(s) = \frac{E^9(s)}{a_0^9} = \frac{1}{a_0^9} \sum_{i=0}^{9} a_i^9 s^i$$

$$\approx 2.90196 \times 10^{-8}s^9 + 1.30588 \times 10^{-6}s^8$$

$$+ 2.87294 \times 10^{-5}s^7 + 4.02212 \times 10^{-4}s^6$$

$$+ 3.92157 \times 10^{-3}s^5 + 2.74510 \times 10^{-2}s^4$$

$$+ 0.137256s^3 + 0.470588s^2 + s + 1$$

To denormalize $H(s)$, s must be replaced by sT. This changes a_i^n into $a_i^n T^i$; in our example the leading coefficient becomes 2.90196×10^{-53}. Hence normalized calculation is advisable to avoid over- or underflows.

It is interesting to note that the output-input transfer function $1/|H(j\omega)|$ and the impulse response of the maximally flat-delay filter are both gaussian functions of ω and t, respectively. The proof is outlined in Prob. 12-48.

As mentioned earlier, the Bessel-Thomson (maximally flat-delay) approximation is analogous to the Butterworth (maximally flat-loss) approximation discussed in Sec. 12-2. It can be anticipated that a more efficient function may be obtained if an equal-ripple, rather than a maximally flat, approximation to a constant delay is achieved. Such an approximation leads to a function analogous to the Chebyshev function described in Sec. 12-3. It is illustrated in Fig. 12-32. The derivation of such functions can be performed using the Z variable in-

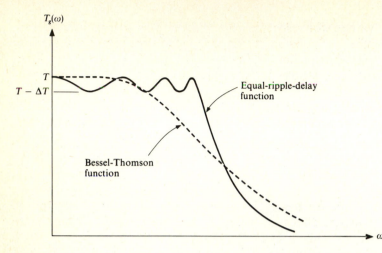

Figure 12-32 A comparison of the Bessel-Thomson and the equal-ripple delay functions.

troduced in Sec. 12-4; it is described in Ref. 11, which also contains an extensive tabulation of the corresponding natural modes. The derivation is straightforward but lengthy and hence is not included here; the reader should consult Ref. 3 for a discussion of the theory and Ref. 10 for the tables of natural modes. Reference 3 also contains a description of the rational function $H(s)$ which combines flat group delay in the passband with a Chebyshev loss in the stopband.

12-7 APPROXIMATION METHODS FOR BANDPASS FILTERS

Often the desired passband is not centered around the frequency origin but between two positive frequencies ω_{p_1} and ω_{p_2}, so that a bandpass rather than a low-pass filter is required. As discussed in Sec. 12-5, it is possible to design such a filter using a reactance transformation if the desired loss response has a geometric (logarithmic) symmetry around the band center frequency $\sqrt{\omega_{p_1}\omega_{p_2}}$. Also, a Moebius transformation can sometimes be used to obtain filter responses which are asymmetrical. Often, however, neither of these techniques is applicable, and the filter must be designed directly. In this section, therefore, a brief description will be given of direct approximation methods applicable for bandpass filters.

To obtain a bandpass loss response which is maximally flat at ω_0, we can use the function

$$|K(j\omega)|^2 = C_n \frac{(\omega^2 - \omega_0^2)^n}{\prod_{i=1}^{k}(\omega^2 - \omega_i^2)^2} \qquad K(s) = \pm C_n^{1/2} \frac{(s^2 + \omega_0^2)^{n/2}}{\prod_{i=1}^{k}(s^2 + \omega_i^2)} \qquad (12\text{-}179)$$

where n is even and $k \leq n/2$. The characteristic function of (12-179) is clearly a modified version of that given in Eq. (12-36). It is easy to verify that $|K(j\omega)|^2$ and its first $n - 1$ derivatives with respect to ω^2 are all zero at $\omega = \pm\omega_0$, whatever the values chosen for the loss poles ω_i. An example of the loss response obtainable with this type of characteristic function is shown in Fig. 12-33. An iterative design process for finding the loss poles ω_i for prescribed α_p and specified minimum stopband loss is given in Ref. 2.

Equal-ripple passband general stopband bandpass filters can also be obtained by modifying the design technique described in Sec. 12-4. Now the transformation

$$Z \triangleq \sqrt{\frac{S^2 + 1}{S^2 + \Omega_{p_1}^2}} = \sqrt{\frac{1 + S^{-2}}{1 + \Omega_{p_1}^2 S^{-2}}} \qquad \text{Re } Z \geq 0 \qquad (12\text{-}180)$$

is used to obtain the frequency-normalized response type illustrated in Fig. 12-34a. Following the discussion of Sec. 12-4, it can again be shown that the transformation of (12-180) maps the passband (which here extends from Ω_{p_1} to 1) to the jY axis. The upper stopband is mapped to $X_{s_2} \leq X \leq 1$, where X_{s_2} is the transformed value of Ω_{s_2}; the lower stopband is mapped to $1/\Omega_{p_1} \leq X \leq X_{s_1}$, where X_{s_1} is the transformed value of the lower stopband limit Ω_{s_1}. When $P(Z)$ is defined, as in Sec. 12-4, by (12-78), and $|K|^2$ by (12-79), the derivation proving the equal-ripple-passband character of the loss remains valid.

The only change in the design procedure given in Sec. 12-4 is that now there are loss poles at $\Omega = 0$, which contribute factors of the form $Z + 1/\Omega_{p_1}$ to the expression of $P(Z)$ given in (12-87). Hence, (12-87) is replaced by

$$P(Z) = \prod_{i=1}^{n} (Z + X_i) = \left(Z + \frac{1}{\Omega_{p_1}}\right)^{n_0} (Z + 1)^{n_\infty} \prod_{i=1}^{(n - n_0 - n_\infty)/2} (Z + X_i)^2$$

$$(12\text{-}181)$$

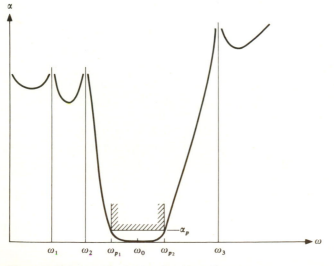

Figure 12-33 Maximally flat loss bandpass filter response.

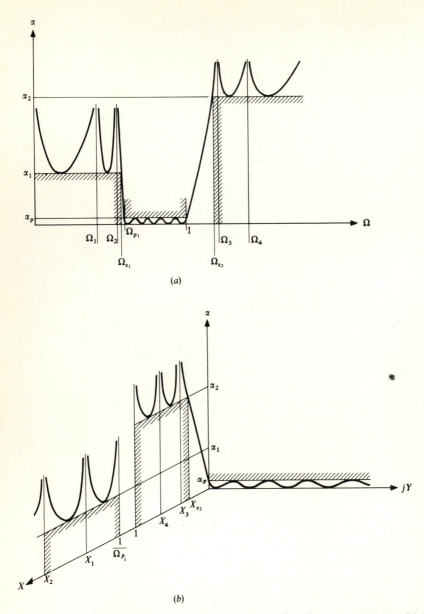

Figure 12-34 (a) Equal-ripple passband general stopband bandpass loss response; (b) loss response transformed into the Z plane.

where n_0 is the number of loss poles at $\Omega = 0$. Substituting into (12-79) gives

$$|K|^2 = k_p^2 \frac{[P_e(Z)]^2}{(-Z^2 + \Omega_{p_1}^{-2})^{n_0}(-Z^2 + 1)^{n_\infty} \displaystyle\prod_{i=1}^{(n-n_0-n_\infty)/2}(-Z^2 + X_i^2)^2} \tag{12-182}$$

Using (12-180), it is easily shown that

$$\Omega_{p_1}^{-2} - Z^2 = \frac{(1 - \Omega_{p_1}^2)S^2}{\Omega_{p_1}^2(S^2 + \Omega_{p_1}^2)}$$

$$1 - Z^2 = \frac{-(1 - \Omega_{p_1}^2)}{S^2 + \Omega_{p_1}^2} \tag{12-183}$$

$$X_i^2 - Z^2 = \frac{1 - \Omega_{p_1}^2}{\Omega_{p_1}^2 - \Omega_i^2} \frac{S^2 + \Omega_i^2}{S^2 + \Omega_{p_1}^2}$$

It follows that the denominator of $|K|^2$ in (12-182) equals

$$D(S) = C \frac{S^{2n_0} \displaystyle\prod_{i=1}^{(n-n_0-n_\infty)/2} (S^2 + \Omega_i^2)^2}{(S^2 + \Omega_{p_1}^2)^n} \tag{12-184}$$

where C is a constant. Unless n is *even*, $D(S)$ is not a full square and hence no real rational $K(s)$ can be found from (12-182). Since $n = n_0 + n_\infty + 2m$, where m is the number of finite nonzero loss poles for positive Ω, this indicates that $n_0 + n_\infty$ must be even if this approximation is used.†

It will be noted that for $\Omega_{p_1} = 0$ and $n_0 = 0$, all formulas and hence the whole design procedure simplify to the low-pass filter approximation described in Sec. 12-4.

Next, the design of maximally flat-delay bandpass filters will be briefly discussed. The "obvious" process of obtaining such a filter by applying the low-pass–to–bandpass transformation (12-97) to the maximally flat-delay function derived in Sec. 12-6 is, unfortunately, useless. This can be seen by noting that the desired bandpass phase response is linear; the low-pass response found in Sec. 12-6 is indeed a linear-phase one, but the transformation (12-97) distorts the frequency axis in a nonlinear manner, as Fig. 12-22c illustrates. Hence, a phase response linear in ω around $\omega = 0$ becomes nonlinear in ω' around ω_1 after the transformation.

·An exact solution to the flat-delay problem is possible, both for maximally flat and equal-ripple delay response.[3,11,12] However, the necessary derivations are lengthy and hence are omitted. Instead, a simple if inexact procedure will be described for obtaining an approximately flat-delay bandpass filter function from a flat-delay low-pass prototype. Let the natural modes of the low-pass filter be those shown in Fig. 12-35a. If the s variable is scaled so that the bandwidth of the low-pass response becomes equal to that of the desired bandpass bandwidth, the zero pattern of Fig. 12-35b is obtained. This scaling can be accomplished if we replace s by $[(\omega_{p_2}' - \omega_{p_1}')/2\omega_p]s$ in the linear-phase low-pass response $H(s)$. Next,

---

† The reader is referred for a detailed discussion of the design process to Ref. 2, which also describes a modified technique for synthesizing filters with odd degrees.

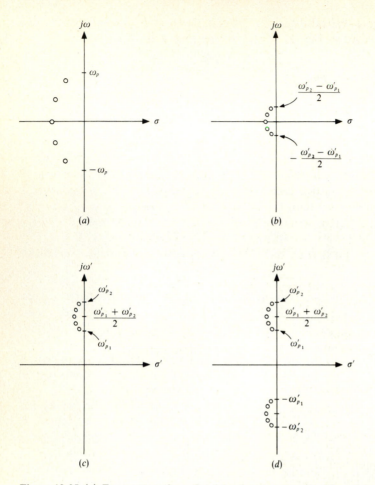

Figure 12-35 (*a*) Zero pattern for a flat-delay low-pass circuit; (*b*) scaled zero pattern; (*c*) scaled and shifted zero pattern; (*d*) realizable, approximately flat-delay bandpass zero pattern.

the zero pattern is *shifted* to the bandpass band center by adding $j(\omega'_{p_1} + \omega'_{p_2})/2$ to the scaled s variable (Fig. 12-35c). Thus, the new variable s' is given by

$$s' = \frac{\omega'_{p_2} - \omega'_{p_1}}{2\omega_p} s + j\frac{\omega'_{p_1} + \omega'_{p_2}}{2} \tag{12-185}$$

Solving for s and substituting into $H(s)$ gives the new response

$$H_1(s') = H(s) = H\left(\frac{2\omega_p}{\omega'_{p_2} - \omega'_{p_1}} s' - j\omega_p\frac{\omega'_{p_1} + \omega'_{p_2}}{\omega'_{p_2} - \omega'_{p_1}}\right) \tag{12-186}$$

Since all operations on s up to this point were linear, the new function $H_1(s')$ still possesses the linear-phase property. However, it is no longer realizable since it contains complex coefficients, as can be seen directly from (12-186) or from

Fig. 12-35c. To restore realizability, we must include the conjugates of all zeroes in the pattern, as indicated in Fig. 12-35d. The resulting transducer function

$$H_2(s') = H\left(\frac{2\omega_p}{\omega'_{p2} - \omega'_{p1}} s' - j\omega_p \frac{\omega'_{p1} + \omega'_{p2}}{\omega'_{p2} - \omega'_{p1}}\right)$$

$$\times H\left(\frac{2\omega_p}{\omega'_{p2} - \omega'_{p1}} s' + j\omega_p \frac{\omega'_{p1} + \omega'_{p2}}{\omega'_{p2} - \omega'_{p1}}\right) \qquad (12\text{-}187)$$

is now real and hence realizable. However, the arbitrary addition of the conjugate zero pattern in the lower half plane is going to disturb the flat-delay property of $H_1(s')$ somewhat. If the relative bandwidth is very narrow, i.e., if the bandwidth $(\omega'_{p2} - \omega'_{p1})$ is much smaller than the band-center frequency $(\omega'_{p1} + \omega'_{p2})/2$, then the upper-half-plane zeros will be much closer to the passband portion $\omega'_{p1} \leq \omega' \leq \omega'_{p2}$ of the $j\omega'$ axis than the lower-half-plane zeros. Hence, for $(\omega'_{p2} - \omega'_{p1})/(\omega'_{p1} + \omega'_{p2}) \ll \frac{1}{2}$, the flat-delay property will be, to a good approximation, retained.

It should be noted that the transformation (12-187) from $H(s)$ to $H_2(s')$ alters the minimum absolute value of the function along the $j\omega$ axis. Hence, $|H(j\omega)| \geq 1$ for all ω does not guarantee $|H_2(j\omega')| \geq 1$ for all ω. Therefore, in general, a constant scaling factor C must also be included in $H^2(s')$.

The design process will be illustrated by a simple example.

Example 12-12 Find a transducer function $H_2(s')$ such that the specifications

$$T_g(\omega') = 0.5 \pm 0.05 \ \mu s \qquad \alpha(\omega') \leq 1 \ dB$$

are met in the frequency range 9.8 MHz $\leq f' \leq$ 10.2 MHz.

Since here the narrow-band condition $(\omega'_{p2} - \omega'_{p1})/(\omega'_{p1} + \omega'_{p2}) = 2\pi 0.4/2\pi 20 = 0.02 \ll \frac{1}{2}$ holds, the approximation of Eqs. (12-185) to (12-187) can be used. We can scale the time variable such that the required T_g becomes equal to 1 for the low-pass prototype function. This will enable us to use the results of Sec. 12-6 in the calculation of the prototype function. By Sec. 1-4, scaling the time by $T_g = 0.5 \ \mu s$ is equivalent to scaling the radian frequency by $\omega_0 = 1/T_g = 2.0$ Mrad/s. By (12-185), $\omega_0 = (\omega'_{p2} - \omega'_{p1})/2\omega_p$, and hence the passband limit of the low-pass prototype is given by

$$\omega_p = \frac{\omega'_{p2} - \omega'_{p1}}{2\omega_0} = \frac{1}{2}(\omega'_{p2} - \omega'_{p1})T_g \approx 0.6283185$$

Since a 1-dB loss corresponds to a voltage ratio of about 0.89125, it is easy to establish from Figs. 12-30 and 12-31 that $n = 2$ barely meets the specifications. In order to allow for the distortion of the loss and delay responses inherent in this approximating process, we choose $n = 3$. Then, from Table 12-9,

$$H(s) = E^3(s) = s^3 + 6s^2 + 15s + 15$$

Next, we use (12-187) to obtain the bandpass function. After we define

$$\omega_1 = \omega_p \frac{\omega'_{p1} + \omega'_{p2}}{\omega'_{p2} - \omega'_{p1}}$$

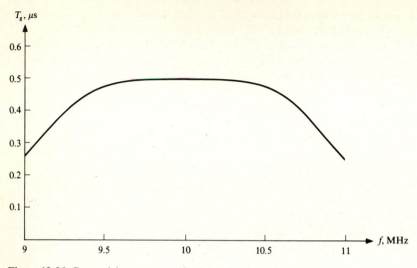

Figure 12-36 Group-delay response of a flat-delay bandpass filter.

(12-187) leads to

$$H_2(s') = H\left(\frac{s'}{\omega_0} - j\omega_1\right) H\left(\frac{s'}{\omega_0} + j\omega_1\right)$$
$$= C(s'^6 + a_5' s'^5 + a_4' s'^4 + a_3' s'^3 + a_2' s'^2 + a_1' s' + a_0')$$

Here, calculation gives

$$a_5' = 2.4 \times 10^7 \qquad a_4' \approx 1.210753 \times 10^{16}$$
$$a_3' \approx 1.911765 \times 10^{23} \qquad a_2' \approx 4.789984 \times 10^{31}$$
$$a_1' \approx 3.76908 \times 10^{38} \qquad a_0' \approx 6.1905886 \times 10^{46}$$

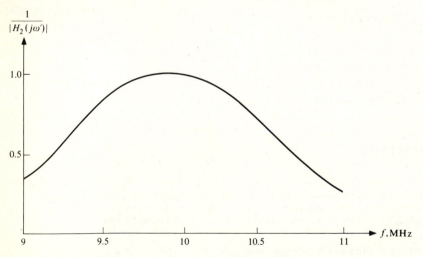

Figure 12-37 Amplitude response of a flat-delay bandpass filter.

The factor C can be found from the condition that the minimum value of $|H_2(j\omega')|$ must equal 1. This gives $C \approx 4.22011 \times 10^{-45}$.

The group-delay and amplitude responses corresponding to $H_2(j\omega')$ are shown in Figs. 12-36 and 12-37, respectively. All specifications are obviously met. In fact, the loss varies between 0 and 0.672 dB in the passband; the delay is between 0.50038 and 0.50081 μs.

Additional material pertaining to the solution of the approximation problem for bandpass filters can be found in Refs. 2, 3, and 11.

All discussions of this chapter have been carried out in terms of $H(s)$ and $K(s)$, that is, in terms of the parameters of a doubly terminated reactance two-port. These parameters satisfy the Feldtkeller equation $|H|^2 = 1 + |K|^2$ for $s = j\omega$, and hence $H(s)$ is subject to the restriction $|H(j\omega)| \geq 1$. For other types of circuits, e.g., active-RC filters, there may not exist any such limitation on the magnitude of the transfer function. Nevertheless, the derivations and results given in this chapter remain applicable, with a minor modification, which will be explained by example.

Example 12-13 To illustrate the process, assume that a third-degree active low-pass filter is to be designed with maximally flat-loss response. Let the desired voltage gain at zero frequency be $A_v(0) = 10$. Then we can introduce $H(s)$ through the relation

$$A_v(s) = \frac{A_v(0)}{H(s)} \tag{12-188}$$

Since $|A_v(j\omega)|$ has its maximum value at $\omega = 0$, the function $H(s)$ in (12-188) satisfies $|H(j\omega)| \geq 1$, just as before. Hence, we can use the Butterworth function obtained in Sec. 12-2 and tabulated in Table 12-2. This gives

$$A_v(s) = \frac{10}{s^3 + 2s^2 + 2s + 1}$$

which has all required properties.

It should also be pointed out that the whole of this chapter has been devoted to the approximation of prescribed *frequency* responses. Approximation of a prescribed *time* response is also possible. This subject is, however, beyond the scope of this book. The interested reader should consult Refs. 4 and 13.

12-8 SUMMARY

In this chapter, analytical methods were described for finding realizable network functions. They were applicable to lumped linear circuits, active or passive. The main emphasis was on filter transfer functions: low-pass, high-pass, and bandpass characteristics were considered in detail. The topics discussed included:

1. The definition of filters and their usual specifications (Sec. 12-1).
2. The choice of the transfer function to be approximated (Sec. 12-1).

3. Approximation criteria: maximally flat approximation, min-max approximation (Sec. 12-1).
4. The Butterworth approximation: the form of $|K|^2$, the location of the natural modes, the definition of the selectivity and discrimination parameters, the calculation of the necessary degree, asymptomatic loss behavior, inclusion of finite loss poles in the transfer function (Sec. 12-2).
5. Chebyshev approximation: the form of $|K|^2$, the locus of the natural modes, a comparison of Butterworth and Chebyshev loss responses, the calculation of the degree (Sec. 12-3).
6. The inverse Chebyshev filter: the derivation of $|K|^2$, the loss response (Sec. 12-3).
7. The solution of the approximation problem for equal-ripple passband and general stopband filters: the mapping of the frequency variable and the resulting transformation of the frequency axis, the appropriate choice for $|K|^2$, the relation between Chebyshev and general-parameter characteristic functions, the design process (Sec. 12-4).
8. Elliptic filters (Sec. 12-4).
9. Reactance transformations: low-pass to high-pass, low-pass to bandpass, and low-pass to band-stop, the resulting changes in the response and in the circuit elements, design procedures for high-pass, bandpass, and bandstop filters using a low-pass prototype (Sec. 12-5).
10. Moebius transformations: degree-preserving and degree-doubling transformations and their applications (Sec. 12-5).
11. Maximally flat-delay approximation: the relation between low distortion in the time domain and constant loss and delay in the frequency domain, the differential equation for the maximally flat-delay polynomial and its solution, recurrence relation, and relation between the polynomial and its derivative, asymptotic behavior of the solution (Sec. 12-6).
12. Approximation methods for bandpass filters: maximally flat-loss approximation, equal-ripple-passband–general-stopband approximation, the approximate design of maximally flat-delay bandpass filters, application to active filter transfer functions (Sec. 12-7).

Approximation theory is an important branch of both circuit theory and applied mathematics. The material presented in this chapter represents only a few results of special interest in circuit synthesis. The interested reader is referred to the literature (the references quoted at the end of the chapter provide a good starting point) for more detailed discussions on this fascinating subject.

PROBLEMS

12-1 (a) Show that no rational function

$$|K|^2 = \frac{a_0 + a_1\omega^2 + a_2\omega^4 + \cdots + a_n\omega^{2n}}{b_0 + b_1\omega^2 + b_2\omega^4 + \cdots + b_m\omega^{2m}}$$

can be zero at all points of a frequency range $0 \le \omega \le \omega_p$ unless $a_i = 0$ for all i and hence $|K|^2 \equiv 0$. *Hint:* How many zeros does an nth-order polynomial have? (b) What are the implications?

12-2 Use the reasoning of Prob. 12-1 to prove that $1/|K|^2$ cannot be zero at every point of a frequency range unless $1/|K|^2 \equiv 0$. What are the implications?

12-3 Prove that of the s_k given in Eq. (12-16), the first n ($k = 1, 2, \ldots, n$) lie in the LHP and the rest in the RHP.

12-4 Calculate the element values in Fig. 12-9 from the given z_{11} and y_{11} functions.

12-5 Calculate the natural modes S_k of the $H(S)$ given in (12-31). Compare your results with those given in Table 12-1 for $n = 5$.

12-6 Calculate the loss $\alpha(f_p)$ at the passband limit f_p for the circuits of Fig. 12-9. How much safety margin is there?

12-7 Recalculate the 3-dB radian frequency ω_0 for the circuits of Fig. 12-9 if the condition $\alpha(f_p) = \alpha_p$ is exactly met. How much is $\alpha(f_s)$?

12-8 Design a Butterworth filter from the following specifications:
    (a) Both terminating resistors equal 75 Ω.
    (b) $\alpha \leq 0.5$ dB for $f \leq 0.7$ MHz.
    (c) $\alpha \geq 40$ dB for $f \geq 6$ MHz.
Assume $\alpha(f_s) = \alpha_s$. How much are ω_0, $\alpha(f_p)$? Calculate the element values.

12-9 Find a general expression for the leading coefficients (that of S^n) in $H(S)$ and $K(S)$ of a Chebyshev filter. *Hint:* Use (12-41) and the relation

$$\binom{n}{0} + \binom{n}{2} + \binom{n}{4} + \cdots = 2^{n-1}$$

12-10 Design a Chebyshev filter from the specifications given in Prob. 12-8 for a Butterworth filter. Compare the degrees of the two circuits.

12-11 Show that for $\alpha_p \to 0$ the locus of the Chebyshev filter's natural modes tends to a circle. What conclusions can you draw?

12-12 Show that if $H(S) = C \prod_{k=1}^{n} (S - S_k)$ is the transducer function of a Chebyshev filter, then $C = \pm k_p 2^{n-1}$.

12-13 Prove that the formula (12-71) giving the minimum degree of a Chebyshev filter for prescribed k and k_1 remains valid for inverse Chebyshev filters.

12-14 Show the location of the natural modes of a Chebyshev filter in the $u = v + jw$ plane. *Hint: Use* Eq. (12-56).

12-15 Find the formulas giving the s-plane location of the natural modes of an inverse Chebyshev filter.

12-16 Show that the Chebyshev polynomials defined by $T_n(\Omega) = \cos (n \cos^{-1} \Omega)$ satisfy the recurrence relations

$$T_0 = 1$$

$$T_1 = \Omega$$

$$T_{n+1} = 2\Omega T_n - T_{n-1}$$

Hint: Use the formula $\cos (nu \pm u) = \cos nu \cos u \mp \sin nu \sin u$.

12-17 Prove that $|K|^2$ as given in (12-79) is nonnegative in the stopband where $Z = X, 0 < X \leq 1$. *Hint:* Use (12-78); note that each conjugate imaginary pole pair $\pm j\Omega_i$ contributes a factor $(Z + X_i)^2$.

12-18 Derive the second and third terms of Eq. (12-85). *Hint:* Use the trigonometric relation $j \tan \beta = \sqrt{1 - \cos^{-2} \beta}$ and Eq. (12-76).

12-19 Show the equal-ripple passband behavior of the general-parameter filter from the alternative definition (12-86).

12-20 Show that $|K|^2$, defined in Eq. (12-86), has poles at the loss-pole frequencies Ω_i.

12-21 Show that $|K|^2$, as defined in Eq. (12-86), is a real rational function in $S = j\Omega$. *Hint:* Follow the derivation of Eqs. (12-80) to (12-84) in reverse.

12-22 (a) Design the general-parameter filter obtainable in the example of Sec. 12-4 if the minus rather than the plus sign is chosen in the expression for $K(S)$.
    (b) Design a general-parameter filter with loss poles at $\Omega = 2$ and ∞ and $\alpha_p = 0.7$ dB.

12-23 Calculate the element values and the loss response of the filter of Fig. 12-19a from that of Fig. 12-17. Use $A = 2 \times 10^8$. Where is the new loss pole?

12-24 Show that ω_1 in (12-97) can be interpreted as the transform of the frequency origin of the low-pass response and A as the ratio of the bandwidths of the low-pass and bandpass filters.

12-25 Design a high-pass filter using a Butterworth low-pass filter prototype satisfying the following specifications: $\alpha \leq 0.5$ dB for $f \geq 5$ MHz, $\alpha \geq 45$ dB for $f \leq 0.4$ MHz, $R_1 = R_2 = 50$ Ω. Where is the 3-dB frequency f_0 of this filter?

12-26 Prove the equivalence of Fig. 12-24.

12-27 Show that it is possible to obtain the second network of Fig. 12-24 (with elements L'_1, C'_1, L'_2, and C'_2) directly by a low-pass–to–bandpass transformation from a parallel LC prototype impedance. Derive the design equations. *Hint:* The reactance of the low-pass impedance is $j\omega L/(1 - \omega^2 LC)$. Use (12-97) to obtain the corresponding bandpass impedance and expand into a Foster 1 circuit.

12-28 Design an elliptic bandpass filter from the following specifications:

$$\alpha \le 0.2 \text{ dB for } 15 \le f \le 17 \text{ kHz}$$

$$\alpha \ge 30 \text{ dB for } f \le 13 \text{ kHz and } f \ge 19 \text{ kHz}$$

Both terminations equal 75 Ω.

12-29 The low-pass filter shown in Fig. 12-38 has a selectivity of $k = 0.49$. The circuit is impedance-normalized (both terminations equal 1 Ω) and frequency-normalized such that $\omega_p \omega_s = 1$. Transform it into a bandpass filter satisfying $f'_{s_1} = 964.4$ Hz, $f'_{s_2} = 1089.4$ Hz, and $f'_{p_2} - f'_{p_1} \ge 60$ Hz. Both terminations should be 600 Ω.

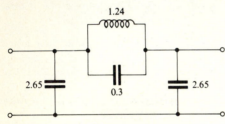

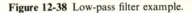

Figure 12-38 Low-pass filter example.

12-30 Design a Butterworth bandpass filter satisfying $\alpha_p = 3$ dB, $\alpha_s = 40$ dB, $\sqrt{f'_{p_1} f'_{p_2}} = \sqrt{f'_{s_1} f'_{s_2}} = 100$ kHz, and $f'_{p_2} - f'_{p_1} \ge 5$ kHz, $f'_{s_2} - f'_{s_1} = 30$ kHz. Both terminations are 100 Ω.

12-31 Calculate the transformed bandstop element values L'_i and C'_i in Fig. 12-25a and b. What are the resonant frequencies?

12-32 Calculate the bandstop limit frequencies ω'_{p_1}, ω'_{p_2}, ω'_{s_1}, and ω'_{s_2} in Fig. 12-25c and d. What symmetry condition must hold?

12-33 Construct a step-by-step design procedure for synthesizing a bandstop filter from given ω'_{p_1}, ω'_{p_2}, ω'_{s_1}, ω'_{s_2}, α_p, and α_s, using a low-pass prototype filter.

12-34 Prove or disprove the following statement: a low-pass–to–bandstop transformation is equivalent to a low-pass–to–high-pass transformation followed by a low-pass–to–bandpass one.

12-35 Transform the low-pass prototype filter shown in Fig. 12-23a using the reactance transformation

$$\omega = \omega' \frac{\omega'^2 - 3}{\omega'^2 - 2}$$

Draw the circuit diagram and plot schematically the anticipated loss response.

12-36 Calculate the explicit expressions for the $|K|^2$ of Fig. 12-26a and b, using $\alpha_p = 0.5$ dB and $\omega_p = 2$ Mrad/s. How much is α at $2\omega_p$ and $3\omega_p$ for the two responses? Where is the new reflection zero on the ω' scale?

12-37 Prove that in the transformed response of Fig. 12-27b, $\omega'_s > \omega_s$; that is, the selectivity has decreased. *Hint:* Use Eq. (12-122).

12-38 (a) Derive the mapping of (12-124) for the response of Fig. 12-28.

(b) Derive a mapping to achieve the low-pass–to–bandpass transformation shown in Fig. 12-29.

12-39 Prove that the phase $\beta(\omega) = \underline{/H(j\omega)}$ is an odd function of ω.

12-40 Derive Eq. (12-144) from (12-141). *Hint:* Differentiate (12-141) and use Eqs. (12-142) and (12-143).

12-41 Show that Eq. (12-154) remains valid for $i = 0$, 1 and n. *Hint:* For $i = 0$ and 1, $a_{-2} = a_{-1} \equiv 0$. For $i = n$, $a_n^{n-1} = a_n^{n-2} \equiv 0$. Use Eq. (12-152) for other values of i.

12-42 Use Eq. (12-154) or (12-155) to generate the coefficients a_i^n tabulated in Table 12-9.

12-43 From Eq. (12-155), prove the recurrence relation

$$E^n(-s)E^{n-1}(s) - E^n(s)E^{n-1}(-s) = (-s^2)[E^{n-1}(-s)E^{n-2}(s) - E^{n-1}(s)E^{n-2}(-s)]$$

12-44 From the result proved in Prob. 12-43, show that

$$E^n(-s)E^{n-1}(s) - E^n(s)E^{n-1}(-s) = 2(-1)^n s^{2n-1}$$

12-45 Show that $E^n(s)$, as defined by Eqs. (12-139) and (12-152), satisfies the maximally flat-delay condition (12-141). *Hint:* Use (12-159) to express $E'(s)$ and $E'(-s)$; then use the result of Prob. 12-44.

12-46 Calculate $T_g(\omega)$ from the maximally flat-delay polynomial $E^n(s)$ for $n = 1$, 2, and 3. Verify the maximally flat property.

12-47 Find an approximate formula for the 3-dB bandwidth $\omega_{3\,\text{dB}}$ of the Bessel filter.

12-48 Show that the asymptotic expression, giving the impulse response of a Bessel filter for $n \to \infty$ is a gaussian function. *Hint:* Combining Eqs. (12-177) and (12-178) gives $H(s) \to a_0^n \exp[\omega^2/2(2n-1) + j\omega]$.

Use also the Fourier-transform relations $f(t - T) \leftrightarrow F(\omega)e^{-j\omega T}$ and $\sqrt{\alpha/\pi}\,e^{-\alpha t^2} \leftrightarrow e^{\omega^2/4\alpha}$.

12-49 Show that the maximally flat polynomials $E^n(s)$ and $E^{n-1}(s)$ cannot have any zeros in common, for any value of n. *Hint:* The statement is obviously valid for $E^0(s) = 1$ and $E^1(s) = s + 1$. Use Eq. (12-155) to extend it to $E^2(s)$ and $E^1(s)$, then to $E^3(s)$ and $E^2(s)$, etc.

12-50 Design a maximally flat-delay filter with the following specifications. The group delay is 100 ± 15 ns in the $0 \le f \le 3$ MHz range. The loss is 0 dB at zero frequency and less than 6 dB in the $0 \le f \le 3$ MHz range. Plot the loss and delay response of your filter for $0 \le f \le 5$ MHz.

12-51 Prove that for any rational $H(S)$ the group delay $T_g(\omega) \to 0$ as const/ω^2 as $\omega \to \infty$.

12-52 Calculate the value of the constant factor C_n in Eq. (12-179) so that, for given ω_0 and ω_i, the loss at $\omega = \omega_{p_1}$ is equal to α_p.

12-53 Design an equal-ripple bandpass filter with the following specifications:

Passband: $f_{p_1} = 25$ kHz $f_{p_2} = 40$ kHz $\alpha_p = 0.2$ dB

Loss poles: $f_1 = 85$ kHz $f = 0$ (single pole) $f \to \infty$ (single pole)
Terminations: $R_1 = R_2 = 600\ \Omega$

12-54 Calculate the expression giving the constant C in Eq. (12-184).

12-55 (*a*) Compare the responses shown in Figs. 12-36 and 12-37 with those obtainable by appropriately scaling and shifting the responses of a third-degree maximally flat low-pass function [compare Eq. (12-186)]. How much distortion is introduced by the lower-half-plane zero pattern?

(*b*) Repeat all calculations for $f'_{p_1} = 8$ MHz and $f'_{p_2} = 12$ MHz. How much is the distortion now?

REFERENCES

1. Saal, R., and E. Ulbrich: On the Design of Filters by Synthesis, *IRE Trans. Circuit Theory*, December 1958, pp. 284–327.
2. Orchard, H. J., and G. C. Temes: Filter Design Using Transformed Variables, *IEEE Trans. Circuit Theory*, December 1968, pp. 385–408.
3. Temes, G. C.: Approximation Theory, in G. C. Temes and S. K. Mitra (eds.), "Modern Filter Theory and Design," Wiley, New York, 1973.
4. Vlach, J.: "Computerized Approximation and Synthesis of Linear Networks," Wiley, New York, 1969.

5. Grossman, A. J.: Synthesis of Tchebycheff Parameter Symmetrical Filters, *Proc. IRE*, April 1957, pp. 454–473.

6. Saal, R.: "Der Entwurf von Filtern mit Hilfe des Kataloges normierter Tiefpässe," Telefunken GmbH, Backnang, West Germany, 1961.

7. Zverev, A. I.: "Handbook of Filter Synthesis," Wiley, New York, 1967.

8. Christian, E., and E. Eisenmann: "Filter Design Tables and Graphs," Wiley, New York, 1966.

9. Weinberg, L.: "Network Analysis and Synthesis," McGraw-Hill, New York, 1962; reprinted by Robert E. Krieger Publishing Co., Inc., Huntington, N.Y., 1975.

10. Christian, E.: "Introduction to the Design of Transmission Networks," QuikCopy, Raleigh, N.C., 1975.

11. Ulbrich, E., and H. Piloty: "Über den Entwurf von Allpässen, Tiefpässen und Bandpässen mit einer im Tschebyscheffschen Sinne approximierten konstanten Gruppenlaufzeit," *Arch. Elektron. Uebertragungstech.*, October 1960, pp. 451–467.

12. Halpern, P. H.: Solution of Flat Time Delay at Finite Frequencies, *IEEE Trans. Circuit Theory*, March 1971, pp. 241–246.

13. Su, K.: "Time-Domain Synthesis of Linear Networks," Prentice-Hall, Englewood Cliffs, N.J., 1971.

14. Daniels, R. W.: "Approximation Methods for Electronic Filter Design," McGraw-Hill, New York, 1974.

INDEX